MgB_2 Superconducting Wires

Basics and Applications

World Scientific Series in Applications of Superconductivity and Related Phenomena

ISSN 2424-8533

Series Editor: Guy Deutscher *(Tel-Aviv University, Israel)*

Published

Vol. 1 Research, Fabrication and Applications of Bi-2223 HTS Wires
edited by Kenichi Sato

Vol. 2 MgB_2 Superconducting Wires: Basics and Applications
edited by René Flükiger

Editor-in-Chief
Guy Deutscher

World Scientific Series in Applications of Superconductivity and Related Phenomena
VOL **2**

MgB_2 Superconducting Wires

Basics and Applications

Edited by

René Flükiger
University of Geneva, Switzerland

World Scientific
NEW JERSEY • LONDON • SINGAPORE • BEIJING • SHANGHAI • HONG KONG • TAIPEI • CHENNAI • TOKYO

Published by

World Scientific Publishing Co. Pte. Ltd.
5 Toh Tuck Link, Singapore 596224
USA office: 27 Warren Street, Suite 401-402, Hackensack, NJ 07601
UK office: 57 Shelton Street, Covent Garden, London WC2H 9HE

Library of Congress Cataloging-in-Publication Data
Names: Flükiger, René, editor.
Title: MgB_2 superconducting wires : basics and applications / [compiled by]
René Flükiger (University of Geneva, Switzerland).
Description: [Hackensack] New Jersey : World Scientific, [2016] | Series: World Scientific
series in applications of superconductivity and related phenomena ; vol. 2 |
Includes bibliographical references.
Identifiers: LCCN 2015043818 | ISBN 9789814725583 (hc : alk. paper)
Subjects: LCSH: High temperature superconductors. | Magnesium diboride. |
Electric wire--Materials.
Classification: LCC TK7872.S8 M535 2016 | DDC 621.3/5--dc23
LC record available at http://lccn.loc.gov/2015043818

British Library Cataloguing-in-Publication Data
A catalogue record for this book is available from the British Library.

Desk Editor: Kalpana Bharanikumar

Typeset by Stallion Press
Email: enquiries@stallionpress.com

Printed in Singapore

Contents

Foreword

A big surprise opened the new millennium of superconductivity: the discovery of superconductivity in the simple intermetallic compound MgB_2 at the astounding temperature of 40 K, reported by the group of Prof. Jun Akimitsu in Tokyo (Japan).

A few years earlier, in 1994, a paper by Cava *et al.* (Princeton, USA), which documented the discovery of transition metal Boron Carbide, was entitled: "Good news from an abandoned gold mine: A new family of quaternary intermetallic superconductors." Quaternary systems of considerable structural complexity were reported which attained the critical temperature of 23 K, equal to the highest known transition temperature among the intermetallic compounds, for Nb_3Ge.

As history shows, that gold vein was not depleted yet, and still retained its most precious part, superconductivity at 40 K, in the simple binary system Magnesium Diboride (MgB_2). Paradoxically, the binary family to which this compound belongs has been well known, as well as its crystal structure, since the 1950's.

For the first time ever in our experience, we have witnessed the possibility to do leading edge research on a new superconducting material that could be directly bought on catalogue in bulk samples.

Unlike high temperature superconductivity, the mechanism of which has not been clarified yet, superconductivity in MgB_2 is of the traditional BCS-type, and the astonishingly high value of the critical temperature is due to the presence of two electronic bands, with very different characteristics, scarcely interacting, which produce two different superconducting gaps. This peculiarity had been predicted in the 1950's, but only now was it observed experimentally in such a spectacular way for the first time.

As for all the other intermetallic superconductors, its coherence length value is large compared to the crystal cell, and therefore grain boundaries don't impede the current flow.

This fact, together with the possibility of operating at temperatures in the range of around 20 K, has created many expectations about the applicability of this compound, especially for power applications. In a very short time lapse, the feasibility of metal-sheathed multifilamentary wires and tapes has been demonstrated.

Considerable potential applications of this compound were immediately apparent in a cryogen free temperature range or even using H as the cryogenic liquid. Unfortunately, the performances of MgB_2 wires in applied magnetic field turned out being not as spectacular as those reported for thin films, which defines the limit of applicability of MgB_2, but overcoming these limitations is still a goal of future research. A further task of the international research is to determine the variety of applications where MgB_2 can be used in the future. But already now there are some applications for which MgB_2 is competitive, very well identified and illustrated in this volume.

The book is edited by René Flukiger, an active and representative material scientist operating in this field, and gathers the contributions on the various aspects of the MgB_2 research, given by the very groups which were the protagonists of the exciting, sudden growth connected to the scientific, technological and application issues of this material. Hence, I believe that this volume is an indispensable tool for those who want to know the state-of-the-art for this material from the very voice of the protagonists of its development.

Carlo Ferdeghini
Genova, February 15, 2015

Prof. Guy Deutscher serves as scientific Editor-in-Chief of a new series of "Applications of Superconductivity". He wishes to acknowledge the help of the Executive Committee of the International Energy Agency (IEA) Agreement on Superconductivity in preparation of this Series.

Editor's Preface

The present book gives an overview of the work undertaken in the 15 years since the discovery of the superconducting compound MgB_2. With a transition temperature of 39 K, this compound occupies an intermediate position between low temperature superconductors (LTS), e.g., NbTi and Nb_3Sn, and the class of high temperature superconductors (HTS) materials with T_c values well above 77 K, the liquid nitrogen temperature.

Today, round MgB_2 wires can be produced in multifilamentary configurations, in lengths exceeding 10 km and working at temperatures up to 25 K. The main advantage of MgB_2 wires resides in the low production cost, resulting from a relatively simple, classical fabrication route and low material cost of the constituents. The costs for MgB_2 wires are comparable to those of NbTi wires and are markedly lower than those of Nb_3Sn wires. A considerably stronger difference is seen when comparing the costs with those of HTS coated conductors. Another advantage of MgB_2 wires is the absence of weak links at the grain boundaries.

However, several factors limit the range of MgB_2 based applications: the low value of the irreversibility field, B_{irr}, and the inherent anisotropy. For fields parallel to the surface, the pinning force of a thin, binary MgB_2 film is as high as that of Nb_3Sn, but is one order of magnitude lower for perpendicular applied fields. This anisotropy was found to be lowered by the addition of carbon, but has still a limiting effect. Indeed, since a distribution of grain orientations is always present in multifilamentary wire, the achievable critical current density resulting from percolation will always be limited to intermediate values.

Grain boundaries are the main pinning source in MgB_2 wires. However, the critical current density in MgB_2 wires depends on the connectivity between neighbor grains, the highest J_c values are obtained for MgB_2 filaments with the highest connectivity between neighbor grains. Due to the absence of weak links, a high connectivity reflects a higher mass density in the filaments.

At the present day, MgB_2 based MRI magnets and high current cables have already reached the market. As will be shown in the various chapters, a number of other applications are also envisaged. Intense efforts are undertaken for further enhancing J_c in multifilamentary wires and simultaneously reducing the AC losses, which will extend the range of possible applications.

The book is organized as follows:

The theoretical background of the two-band superconductor MgB_2 is treated in **Chapter 1**, showing how the interplay between the quasi-two-dimensional σ-band and the almost isotropic π-band results in anomalous and complex behavior of the characteristic lengths and the critical fields. The coherence length and penetration depth are magnetic-field dependent, have a different T dependence than for single-band materials and have different anisotropies.

Chapter 2 describes the preparation of MgB_2 single crystals and thin films and their physical properties. In (2a), various fundamental properties are measured on MgB_2 crystals grown under very high pressures (up to 3 GPa), both in the clean limit and with various substitutions. The deposition of high quality MgB_2 thin films with high critical current densities and critical fields is described in (2b). In carbon alloyed MgB_2 films prepared by hybrid physical-chemical vapor deposition, $B_{c2}(0)$ values well over 60 T have been obtained, in contrast to bulk or filamentary MgB_2, where the highest known B_{c2} value does not so far exceed 35 T. The mechanism behind these extremely high B_{c2} values is so far not well understood; in particular, it has not been possible to reproduce them in bulk or filamentary

MgB_2. This result shows that the potential of MgB_2 is not yet fully explored. However, a breakthrough is needed before MgB_2 wires can be envisaged for applications at higher fields.

Based on transmission electron microscopy (TEM) observations in MgB_2 thin films (2d), the correlations between growth conditions, the formation of nanoscale disorder, and the resulting variations of physical properties are studied. Microstructure parameters relevant for the critical current density are quantitatively measured by TEM in filamentary wires (2d), e.g., MgB_2 grain size, volume fraction of B rich secondary phases, and oxygen mole fraction.

In **Chapter 3**, the production methods of MgB_2 samples with a high mass density and their properties are discussed. The properties of bulk MgB_2 after reaction under high temperature/high pressure are discussed in (3a), where the effect of various additives, e.g., SiC, Ti, Ta, Zr on the critical current density is shown. The so-called Reactive Liquid Infiltration (RLI) method described in (3b) has a particular importance, being the basis for the production of wires with a mass density close to 100%, with the highest J_c values measured so far, discussed later in (4e) and (4f). The RLI method uses the peculiar diffusion properties of Mg in B and does not need a high pressure. The RLI method can be used for the fabrication of large, almost fully dense samples with a great variety of sizes. Their low cost may open the way for future applications.

Chapter 4 summarizes the wire fabrication processes known at the present day. Regardless of the fabrication technique, wires with the highest J_c values require the use of high quality boron: the fabrication of large quantities of nanosize, amorphous B with a very low oxygen content is described in (4a). Wire fabrication by the *ex situ* method (4a) starts with already formed MgB_2 powders, the *in situ* method with Mg + B mixtures (4b). In spite of lower J_c values, *ex situ* wires have the advantage of a higher homogeneity (lower J_c fluctuation), over lengths of several kilometers. In the case of *in situ* wires, J_c can be enhanced by a factor 4 at 20 K by applying pressures of the order of 1 GPa prior to the reaction heat treatment (4d). Two preparation methods based on the RLI process are the Internal Magnesium Diffusion (IMD) technique (4e) and the Advanced Internal Mg Infiltration technique (4f). In both cases, the mass density in the filaments approaches 100%, resulting in the highest J_c values measured so far in MgB_2 wires.

Chapter 5 focuses on selected aspects of the material properties and of the conductor design in view of various applications of MgB_2 wires. Physical measurements, e.g., specific heat, magnetic relaxation and thermal conductivity are used as operative tools for a better understanding, which will lead to further progress (5a). AC loss measurements in MgB_2 wires are described in (5b), while the effects of mechanical load on J_c are treated in (5c). Finally, the variation of the transport properties of MgB_2 wires under neutron irradiation is described in (5d).

Chapter 6 describes various applications, starting with the commercial fabrication of MRI magnets based *ex situ* MgB_2 wires. At present, the field in these systems is limited to values around 1 T, but an enhancement to considerably higher fields is expected in the next years. The fabrication of bulk, liquid He free quasi-permanent MgB_2 magnets producing fields exceeding 5 T for future cryomagnet applications is described in (6b). MgB_2 wires of 100 μm diameter and less have been developed as current leads for adiabatic demagnetization refrigerator (ADR) magnets, as valves in space-flight (satellites) or as LH_2 level sensors.

MgB_2 wires have been found to be competitive for high current cables working in the range of 10–20 K. The first 13 kA cable was developed for the High-Luminosity Large Hadron Collider (HL-LHC) accelerator at CERN (6d) involving a large quantity of MgB_2 wire (>1,000 km). Other applications requiring high power transmission are being envisaged. The first experiments with MgB_2 power cables cooled by LH_2 have been successfully carried out (6e). The simultaneous use of LH_2 as a cryogen and as a source of chemical energy could increase the total energy transfer volume.

The last MgB_2 application treated in this book concerns wind generator projects (6f), where the light mass gravity of MgB_2 could be a decisive advantage, in particular in view of future 10 MW generators.

A part of the work for this book was undertaken as Emeritus Professor of the University of Geneva, another part during my present activity as Associate Member at CERN, Geneva.

René Flükiger
15 September 2015

List of Contributors

Adriana Serquis
CAB-CNEA, Centro Atomico
Bariloche, Argentina
aserquis@cab.cnea.gov.ar

Akiyasu Yamamoto
Department of Applied Chemistry
University of Tokyo
7-3-1Hongo, Bunkyo, Tokyo 113-8656, Japan
yamamoto@appchem.t.u-tokyo.ac.jp

Amalia Ballarino
CERN, TE-MSC
1223 Meyrin, Switzerland
amalia.ballarino@cern.ch

Andrea Malagoli
CNR-SPIN
Corso Perrone 24
16152 Genova, Italy
andrea.malagoli@spin.cnr.it

Asger Bech Abrahamsen
Department of Wind Energy
Technical University of Denmark
Fredericksborgvej 399
4000 Roskilde, Denmark
asab@dtu.dk

Balaji Birajdar
Special Center For Nano Science
Jawaharlal Nehru University
New Delhi 110067, India
birajdar@mail.jnu.ac.in

Bernardo Bordini
CERN, TE-MSC
1223 Meyrin, Switzerland
bernardo.bordini@cern.ch

Carlo Ferdeghini
CNR-SPIN, Consiglio Nazionale della Ricerca
Corso Perrone 24
Genova, Italy
carlo.ferdeghini@spin.cnr.it

Carmine Senatore
Department of Quantum Material Physics (DQMP)
University of Geneva
24, Quai Ernest Ansermet
1211 Geneva, Switzerland
carmine.senatore@unige.ch

E. W. Collings
Center for Superconducting and Magnetic Materials
Ohio State University
2041 College Road North
Columbus, OH 43210, USA
collings.2@osu.edu

Giovanni Giunchi
Materials Science Consultant
Via Teodosio 8
20131 Milano, Italy
giovanni.giunchi@gmail.com

Guangze Li
Center for Superconducting and Magnetic Materials
Ohio State University
2041 College Road North
Columbus, OH 43210, USA
li.1423@osu.edu

Hiroaki Kumakura
National Institute for Material Science
Tsukuba, Ibaraki 305-0047, Japan
kumakura.hiroaki@nims.go.jp

Ilaria Pallecchi
Physics Department
University of Genova
Via Dodecaneso 33
Genova 16146, Italy
ilaria.pallecchi@spin.cnr.it

Ilkin Kokal
Pavezyum Chemicals
Tuzla, Istanbul, Turkey
info@pavezyum.tr

Jan Kovac
Institute of Electrical Engineering
Slovak Academy of Sciences
Dubravska cesta 9
Bratislava 84104, Slovakia
jan.kovac@savba.sk

Janusz Karpinski
Laboratory for Solid State Physics, ETH
8093 Zurich, Switzerland
karpinski@phys.ethz.ch

Leonardo Bertora
Paramed s.r.l., Magnet Technology
Corso F.M. Perrone 73r
Genova 16152, Italy
leonardo.bertora@paramed.it

Leonardo Civale
LANL
Los Alamos, NM 87545-0001, USA
lcivale@lanl.gov

Marco Bonura
Department of Quantum Material Physics (DQMP)
University of Geneva
24, Quai Ernest Ansermet
1211 Geneva, Switzerland
marco bonura@unige.ch

Marina Putti
Physics Department
University of Genova
Via Dodecaneso 33
Genova 16146, Italy
putti@fisica.unige.it

Matthaeus A. Wolak
Department of Physics, Temple University
1925 N. 12th Street
Philadelphia, PA 19122, USA
matthaeus.wolak@temple.edu

Mehmet S. Somer
Chemical Department
Koç University
Istanbul, Turkey
msomer@ku.edu.tr

Michael A. Susner
OakRidge National Laboratory
Oak Ridge, TN 35831-6056, USA
mike.susner@ornl.gov

Michael D. Sumption
Center for Superconducting and Magnetic Materials
Ohio State University
2041 College Road North
Columbus, OH 43210, USA
Sumption@matsceng.ohio-state.edu

Oliver Eibl
Institut für Angewandte Physik
Universität Tübingen
72076 Tübingen, Deutschland
oliver.eibl@uni-tuebingen.de

Paul M. Voyles
Materials Science and Engineering
University Wisconsin-Madison
Madison, WI 53706, USA
voyles@engr.wisc.edu

Pavol Kovac
Institute of Electrical Engineering
Slovak Academy of Sciences
Dubravska cesta 9
Bratislava 84104. Slovakia
pavol.kovac@savba.sk

René Flükiger
Department of Quantum Matter Physics
University of Geneva
1211 Geneva, Switzerland
rene.flukiger@unige.ch

Sebastiano Giannelli
CERN, TE-MSC
1223 Meyrin, Switzerland
sebastiano.giannelli@cern.ch

Selçuk Acar
Pavezyum Chemicals
Tuzla, Istanbul, Turkey
info@pavezyum.tr

Shariar M.A. Hossain
ISEM
University of Wollongong
Northfields Avenue
Wollongong, NSW 2522, Australia
shahriar@uow.edu.au

Sonja I. Schlachter
Institute of Technical Physics
Karlsruhe Institute of Technology
Hermann-von-Helmholtz-Pl. 1
Eggenstein-Leopoldshafen 76344, Germany
sonja.schlachter@kit.edu

Tatiana Prikhna
Institute for Superhard Materials
National Academy of Sciences
2, Avtozavodskaya Street
Kiev 04074, Ukraine
prikhna@mail.ru

Valeria Braccini
CNR-SPIN
Corso Perrone 24
16152 Genova, Italy
valeria.braccini@spin.cnr.it

Vitaly Vysotsky
Russian R and D Cable Institute
111024 Moscow, Russia
vysotsky@ieee.org

Wilfried Goldacker
Institute for Technical Physics
Karlsruhe Institute of Technology
Hermann-von-Helmholtz-Pl. 1
Eggenstein-Leopoldshafen 76344, Germany
wilfried.goldacker@kit.edu

Wolfgang Hässler
Material Research
Leibniz Institute for Solid State
P. O. Box 270116
01171 Dresden, Germany
w.haessler@ifw-dresden.de

XiaoXing Xi
Department of Physics, Temple University
1925 N. 12th Street
Philadelphia, PA 19122, USA
matthaeus.wolak@temple.edu
xiaoxing@temple.edu

Ye Zhu
Department of Materials Science and Engineering
South University of Science and Technology
Shenzhen, Guangdong 510855, China
zhuy@ymail.com

Chapter 1

Vortex Matter in the Two-band Superconductor MgB_2

Leonardo Civale[*,‡] and Adriana Serquis[†]

[*]*Los Alamos National Laboratory, Los Alamos, NM 87545, USA*
[†]*Centro Atómico Bariloche, Bariloche 8400, Argentina*
[‡]*lcivale@lanl.gov*

The physics of vortex matter in MgB_2, the first discovered technologically relevant multiband superconductor (SC), are determined by the parameters of the two gaps and by their coupling. The interplay of the quasi-two-dimensional (2D) σ-band and the almost isotropic π-band results in anomalous and complex behavior of the characteristic lengths and critical fields. The coherence length and penetration depth are magnetic-field dependent, have different T dependence from those in single-band materials, and have different anisotropies. At low T and H the critical currents (J_c) are high and the relaxation rates (S) are low, indicative of large pinning energy. In clean MgB_2, as H and T increases, the π-band contribution is strongly suppressed, so J_c decreases and S increases sharply, but C-doping allows maintaining strong pinning up to high fields as needed for applications.

1. Introduction

The subject of this book is the superconducting compound MgB_2. Certainly, only a fraction of the hundreds of identified superconductors (SCs) have whole books dedicated to them. What is special about MgB_2?

Since the discovery of superconductivity in 1911 until 1973, the highest known transition temperature (T_c) increased from $\sim$4K in Hg to $\sim$23 K in Nb_3Ge, at a modest average rate of $\sim$0.3 K/year.[1] By then the microscopic origin of superconductivity, based on electron–phonon interactions and described by the extremely successful BCS theory,[2] was well understood.[3] The general consensus in the early 1980s was that this pairing mechanism was unlikely to produce significantly higher T_c values,[4] so superconductivity research (at least from the fundamental science perspective) appeared close to a dead end. Everybody working in the condensed matter and material sciences communities at the time knows how this belief was dramatically shattered by Bednorz and Muller's discovery in 1986.[5] Not only did the highest T_c increase from ~30–164 K ($HgBa_2Ca_2Cu_3O_8$ under pressure) in 1993, at an average rate of ~20 K/year, but the proliferation of whole families of superconducting compounds opened up an unimagined richness of new physics, on top of the potential technological impact.[1]

By 2001, the field had calmed down significantly, so the discovery of superconductivity[6] in MgB_2, a simple and well-known binary compound, was a much welcomed surprise. It soon became clear that, in many regards, MgB_2 was more similar to conventional low T_c superconductors (LTS) than to the oxide SC (at that time simply called high temperature SCs, or High Temperature Superconductors (HTS)). For one thing, the observation of a significant boron (B) isotope effect[7] indicated that superconductivity in MgB_2 was phonon–mediated, and strong experimental evidence confirmed the s-wave symmetry of the order parameter.[8–14] However, with $T_c \sim 39$ K, this clearly contradicted the old concept that the T_c of Nb_3Ge was close to the upper limit for this mechanism,[4] indicating that MgB_2 was somehow unique and that it would force us to revise some previously accepted knowledge. The "uniqueness" of MgB_2 was reinforced by the fact that, in sharp contrast with the oxides, it was not just the first member of a large new family of SC. The initial hope that isostructural compounds with progressively higher T_c would soon be found did not materialize, only a few other borides and diborides turned out to be superconducting, but with much lower T_c.[15] Chemical doping also proved very hard.[15]

Vortex physics had become a major research field in condensed matter and statistical mechanics since the discovery of the oxide

HTS.[16] When MgB_2 arrived, the vortex matter community knew from the beginning some of the basic questions to ask and had well-developed experimental and theoretical tools to address them, which was certainly not the case in 1986. Do grain boundaries behave as weak links and strongly limit inter-grain critical currents? How large is the anisotropy? Are there vortex liquid phases and melting lines? Are the solid vortex phases crystalline or glassy? What are the dominant pinning centers and mechanisms? Is the vortex dynamics Anderson–Kim (A–K) like or glassy? The previous experience from HTS helped a lot to make fast initial progress. Although overall vortex physics in MgB_2 turned out to be closer to LTS than to HTS, there was still another surprise: adding to its uniqueness, it was the first technologically relevant multiband SC.[8–14,17,18] As will be discussed below, this resulted in novel vortex behavior.

More recently, iron-based SC were discovered.[19] Although they are multiband as MgB_2, important differences in the levels of hybridization soon became clear. In addition, they also constitute a huge new family, even larger than the oxides, and exhibit vortex dynamics as fast as the HTS,[20] again in sharp contrast with MgB_2.

2. Vortex Matter Concepts in Single-band SC

2.1 *Characteristic lengths and critical fields*

In the simplest case of single-band isotropic SC,[3] the equilibrium magnetic field (H) versus temperature (T) phase diagram is fully determined if we know T_c and the two temperature-dependent characteristic lengths, the penetration depth $\lambda(T)$ and the coherence length $\xi(T)$. SC can be classified in two groups depending on the value of the Ginzburg–Landau parameter $\kappa = \lambda/\xi$. Materials with $\kappa < 1/\sqrt{2}$ are *type I*; they have the simplest phase diagram with a single curve, the thermodynamic critical field $H_c(T) = \Phi_0/(2\sqrt{2}\pi\lambda\xi)$, where $\Phi_0 = 2.07 \times 10^{-7}\,\text{Gcm}^2$ is the flux quantum. In macroscopic samples and ignoring demagnetizing effects, below $H_c(T)$ the material is in the Meissner state, characterized by the absence of internal field, $B = 0$. SC with $\kappa > 1/\sqrt{2}$ are called *type II*. They have two critical fields, the lower critical field $H_{c1}(T) \approx [\Phi_0/(4\pi\lambda^2)]\ln(\kappa)$

below which the SC is in the Meissner state, and the upper critical field $H_{c2}(T) \approx \Phi_0/(2\pi\xi^2)$ above which the material becomes normal. In between the system is in the *mixed state*, where it minimizes the free energy by remaining superconducting but allowing the penetration of magnetic field in the form of vortices. A vortex is a quantized tubular object that carries a total flux Φ_0; it consists of a *core* of radius $\sim\xi$ where the superconducting order parameter is suppressed, surrounded by rotating supercurrents and associated axial magnetic fields that decay from the core on a radius $\sim\lambda$. Vortices repeal each other, so in the absence of other interactions they form a regular triangular lattice (the Abrikosov lattice) with parameter $a \sim \sqrt{\Phi_0/B}$, where $B \approx H$. All SC used for wires are type II. In fact they are typically *extreme type II*, with $\kappa \gg 1$, so $H_{c2}/H_{c1} \sim \kappa^2 \gg 1$ and the mixed state occupies most of the phase diagram. Although $H_c \sim \sqrt{H_{c1}H_{c2}}$ does not appear as an easily measurable line in the H–T plane, it still is a fundamental parameter that measures the condensation energy per unit volume of the superconducting state, $H_c^2/8\pi$.

Additional complexity occurs in non-cubic materials that exhibit electronic mass anisotropy.[16] The most common are the layered materials, where the effective electronic mass along the c-axis (m_c) is larger than in the ab plane (m_{ab}). This uniaxial anisotropy introduces one additional parameter $\gamma = \sqrt{m_c/m_{ab}} > 1$, which is *temperature independent* and describes the angular dependence of *both* characteristic lengths,[16,21,22] $\xi(\Theta) = \xi_{ab}\varepsilon(\Theta)$ and $\lambda(\Theta) = \lambda_{ab}/\varepsilon(\Theta)$, where Θ is the angle between $\mathbf{H}$ and the c-axis, and $\varepsilon(\Theta) = [\cos^2(\Theta) + \gamma^{-2}\sin^2(\Theta)]^{1/2}$. In particular, for the principal axes orientations $\xi_c = \xi_{ab}/\gamma$ and $\lambda_c = \lambda_{ab}\gamma$. The easiest way to confirm this behavior is to measure the angular-dependent $H_{c2}(\Theta) = H_{c2}^c/\varepsilon(\Theta)$, where H_{c2}^c is the value for $\mathbf{H}//c$ (H_{c1} is harder to measure, and its angular-dependence is more complicated due to the $\ln(\kappa)$ factor).[16]

In pure materials with few defects $\xi(T = 0) \sim \xi_0$, known as the Pippard coherence length (or the BCS coherence length; they are related by a numerical factor close to one, see Ref. 3 for details), which can be seen as the "intrinsic" parameter. Increasing disorder reduces the electronic mean free path l, resulting in a smaller coherence length: $\xi^{-1} = \xi_0^{-1} + l^{-1}$.[3] Adding defects to enhance H_{c2}, moving the material from the *clean limit* ($l \gg \xi_0$) to the *dirty limit* ($l \ll \xi_0$) is a well-established procedure in some LTS.[23] In contrast, the amount or type of disorder has little or no influence on H_{c2} in HTS such as $YBa_2Cu_3O_7$ (YBCO),[24] because the intrinsic ξ_0 is very short (~1.6 nm for $\mathbf{H}//c$).

2.2 *Vortex pinning*

Electric currents exert forces on vortices.[3,25] In the absence of disorder nothing opposes those forces, so an arbitrarily small current would make vortices move, producing dissipation and rendering the SC useless for power applications. Material disorder breaks the translational invariance, resulting in position-dependent vortex energy. This produces vortex pinning, i.e., vortices lay in low energy locations and it is necessary to apply a finite *critical current density* (J_c) to overcome the pinning force and initiate vortex motion.[25] In a simplified picture we can think of a vortex as a particle sitting inside a "defect" represented by a potential well whose depth and size are the pinning energy (U_p) and defect size respectively. However the reality is far more complex as vortices are elastic objects that interact with each other and with many defects simultaneously, so it is convenient to consider a global "pinning landscape" concept.[26]

All real SC samples have some level of disorder that depends on the growth method and produces a $J_c > 0$. However, J_c can be further increased (sometimes by orders of magnitude) by engineering the pinning landscape by chemical doping, introduction of second phases, irradiation, cold work, strain at the film/substrate interface, and other ways depending on the material. Ideally, vortex pinning could be strong enough to produce J_c as high as the depairing current density (J_d) at which Cooper pairs break apart,[16] however this theoretical limit has not been reached. It is common to use the fraction J_c/J_d as a measure of the strength of pinning. Achieving the highest possible J_c is of utmost importance for SC power applications, so enormous efforts have been devoted to this goal for decades, and a huge literature on the topic is available.[27] A systematic discussion of pinning is outside our scope.

The angular dependence of J_c is frequently complex as it depends on both γ and the geometry of the defects.[16,26] For instance, correlated disorder such as aligned columnar defects (linear) or twin boundaries (planar) tend to produce maxima in $J_c(\Theta)$ when the vortices are parallel to the defects, while randomly distributed localized disorder such as nanoparticles or point defects make more isotropic contributions. However, even in the latter case $J_c(\Theta)$ should not be used to determine γ.

2.3 *Flux creep*

Even for $J < J_c$ vortices can "jump" above the pinning barriers and escape from the pinning centers due to thermal fluctuations.[3,16,28,29] This produces a time decay of the "persistent current", a phenomenon known as thermal flux creep. There is also a quantum analog of this effect that allows vortices to depin by "tunneling" through the barrier even at $T = 0$.[16] Usually, quantum creep is significant only at low T (below $\sim$1 K).

The strength of the thermal fluctuations in SC is quantified through the dimensionless Ginzburg number (Gi) which derives from a comparison of the thermal energy with the condensation energy in the coherence volume, $Gi = \frac{1}{2}\left(\frac{\gamma T_c}{H_c^2(0)\xi^3(0)}\right)^2$. (We are using the common practice of omitting the Boltzmann constant factor in the thermal energy, so energies are expressed in units of K). LTS used in wires and magnets (such as NbTi), have $Gi \approx 10^{-8}$, which results in very slow vortex dynamics, whereas cuprates such as YBCO have $Gi \approx 10^{-2}$. The large Gi in cuprates manifests in high flux creep rates, phase diagrams that include crossover fields depending on the pinning landscape, and vortex-liquid phases at high temperatures.[16]

3. Effects of Two-band Superconductivity on Equilibrium Vortex Properties in MgB_2

Early *ab initio* calculations[17,18] and experimental evidence coming from scanning tunnel microscopy (STM),[8–10] Raman spectroscopy,[11] point contact,[12] heat capacity[13] and rf response[14] measurements indicated that MgB_2 has two s-wave superconducting gaps $\Delta_\sigma(0) \approx 7.2\,\mathrm{mV}$ and $\Delta_\pi(0) \approx 2.3\,\mathrm{mV}$. These gaps reside on four different disconnected Fermi surface sheets: two axial quasi-two-dimensional (2D) σ-band sheets and two contorted three-dimensional (3D) π-band sheets. The σ-band forms two concentric cylindrical sheets via in-plane sp^2 hybridization of the B valence electrons, and the π-band results from the strongly coupled covalent bonding and antibonding of the B P_z orbitals.[30] This Fermi surface allows for three different impurity scattering channels: intraband scattering within each band and interband scattering between them.[31]

The theory of two-band superconductivity was developed decades ago,[32,33] but became a research topic of high interest after the discovery of superconductivity in MgB_2.[17,18,22,31,34] As the characteristic lengths and critical fields are influenced not only by the parameters of the two gaps but also (and critically) by their coupling, the H, T and Θ dependences become quite complex. For instance, as we will discuss below, ξ and λ have unconventional temperature dependences and moreover they are also field dependent. The anisotropy is different for ξ and λ, it is *temperature and field dependent* in both cases, and the angular dependencies are not described by a simple $\varepsilon(\Theta)$ function.[22,31,34–45]

3.1 *Two-band effects on H_{c2}*

Let's first discuss the experimentally easier to access H_{c2}. Initial studies showed a broad spectrum of H_{c2} values and anisotropies (in the cases it could be measured) in a variety of bulk, wire, single crystal and thin film MgB_2 samples, as can be seen in the review by Buzea and Yamashita.[15] This is in sharp contrast to the YBCO case and immediately points to the influence of l, i.e., of the amount of disorder. The lowest H_{c2} should approximately represent the intrinsic properties of clean samples.

Early results in randomly oriented powders showed values as low as $H^c_{c2}(0) \sim 2.5T$ and $H^{ab}_{c2}(0) \sim 17T$ (Fig. 1(a)), and an anisotropy parameter, defined as $\gamma_\xi(T) = H^{ab}_{c2}(T)/H^c_{c2}(T)$, decreasing

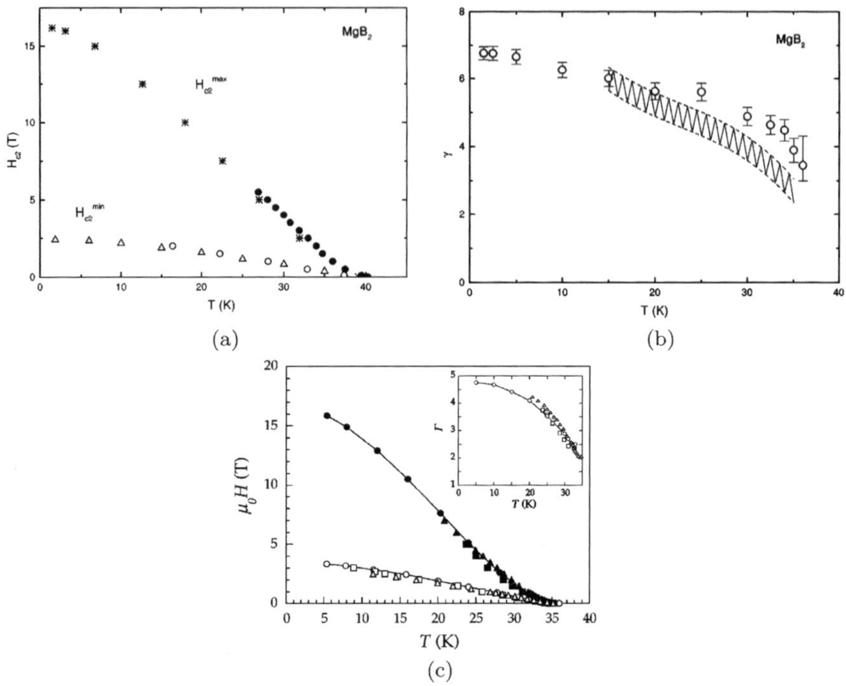

Fig. 1. (a) Anisotropic H_{c2} obtained from randomly oriented powders (Ref. 36), (b) T-dependent H_{c2} anisotropy for the data in (a), see text for details (Ref. 36). Reproduced with permission from American Physical Society. (c) H_{c2} for a MgB_2 single crystal obtained by transport (circles), ac-susceptibility (squares) and specific-heat (triangles) Inset: T-dependent anisotropy (Ref. 37).

monotonically from ~7 for $T \rightarrow 0$ to ~3 for $T \rightarrow T_c$ (Fig. 1(b)).[35,36] Very similar results were soon obtained for single crystals (Fig. 1(c)).[37] In a simple one-band picture, this corresponds to a rather large $\xi_0 \sim 11$ nm. When clean epitaxial thin films became available, they showed similar results.[38] An example from a recent study by J. Kim, N. Haberkorn, *et al.*[39] is shown in Fig. 2(a).

It must be realized that measurements of $H_{c2}^{c}(T)$ and $H_{c2}^{ab}(T)$ alone cannot capture the whole richness of the phase diagram, because the function $H_{c2}(\Theta)$ is not well described by a simple mass anisotropy parameter. The reason is that the response depends on the contribution of two bands, the approximately isotropic π-band

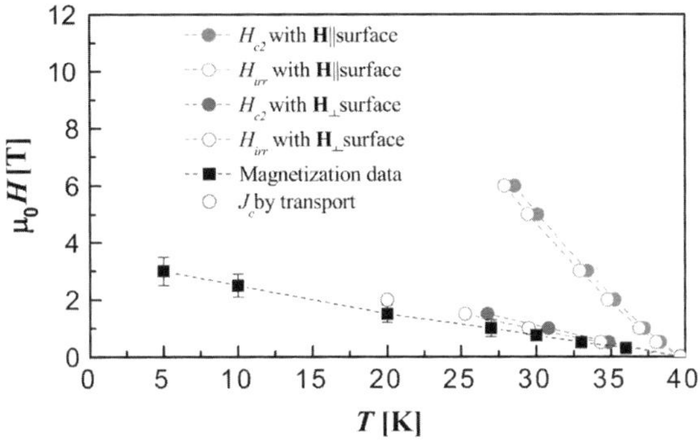

(a)

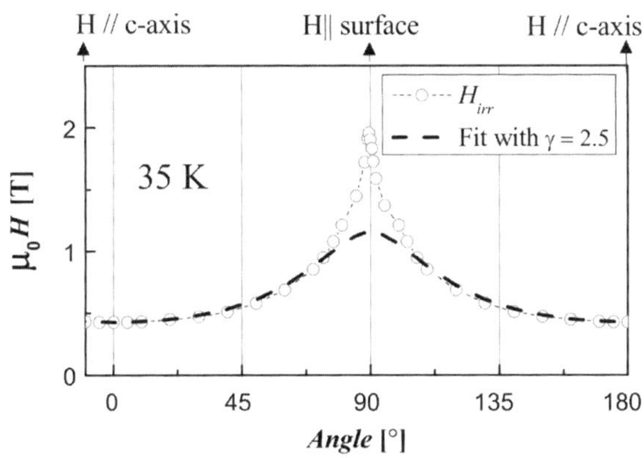

(b)

Fig. 2. (a) H_{c2} and H_{irr} versus T for an epitaxial MgB_2 thin film, for **H**//c and **H**⊥c, (b) angular dependence of H_{irr} at 35 K and fit to single-band anisotropic model, (Ref. 39). Reproduced with permission from Elsevier Ltd.

and the highly anisotropic σ-band. Similar effects have been found in pnictide SC.[40] The deviations from the single-band dependence $H_{c2}(\Theta) \propto 1/\varepsilon(\Theta)$ tend to become stronger as T decreases; near T_c it is frequently possible to fit $H_{c2}(\Theta)$ to a single-band anisotropic response, but the obtained γ_ξ is temperature-dependent. This is seen in Fig. 2(b), which shows $H_{c2}(\Theta)$ at $T = 35\,\mathrm{K}$ for the same film of Fig. 2(a).[39] The most conspicuous feature is a cusp centered at **H**//ab due to surface superconductivity, which appears for **H**//surface if the surface is smooth in the scale of ξ. Ideally the transition field for this geometry is $H_{c3} = 1.69 H_{c2}$,[3] so the ratio of the critical field in both principal axes would significantly overestimate γ_ξ (H_{c3} also occurs in single-band SC, so in general the determination of γ using only principal axes data should be taken with care). It is possible to obtain a good fit of the rest of the angular range (i.e., ignoring the cusp) to a single-band anisotropic response $H_{c2}(\Theta) \propto 1/\varepsilon(\Theta)$. This procedure gives a temperature-dependent γ_ξ increasing from $\sim$2.5 close to T_c to $\gamma_\xi \sim 4.5$ at 24 K,[39] consistent with the single crystal results.

In clean samples the π-band contribution decreases strongly as H and T increase, but the situation changes in dirtier samples. Figure 4 shows H_{c2} data for an epitaxial 500 nm thick film.[41] $H_{c2}^{c}(0) \sim 10T$ is larger than in previous examples (Fig. 3(a)) but γ_ξ is smaller (see inset of Fig. 3(a), so $H_{c2}^{ab}(T)$ values are similar. Again γ_ξ is temperature-dependent and decreases with T as in the previous cases. The $H_{c2}(\Theta)$ at two temperatures is shown in Fig. 3(b) (no evidence of H_{c3} was observed in this case). The dotted lines in Figs. 3(a) and 3(b) are fit the Gurevich model[31] of dirty two-band SC obtained using the same set of parameters. Based on the absence of T_c suppression as compared to clean single crystals the interband scattering was set equal to zero, and it was found that the ratio $\eta = D_\pi/D_\sigma$ of the intraband diffusivities in the π- and σ-bands was large, $\eta \sim 8$, consistent with the presence of non-magnetic impurities such as C and N.

3.2 *Two-band effects on* $\boldsymbol{\lambda}$

The experimental determination of H_{c1} is considerably more difficult than in the case of H_{c2}, as it cannot be measured by transport, but rather by magnetization techniques that are affected by vortex pinning as well as demagnetization and edge effects, turning them less reliable. The alternative is to measure λ. It was theoretically predicted[22] that $\gamma_\lambda = \lambda_c/\lambda_{ab}$ in clean MgB_2 should be much lower than γ_ξ and, in contrast with γ_ξ, it should increase with T, being almost isotropic for $T \to 0$. This was nicely confirmed when the

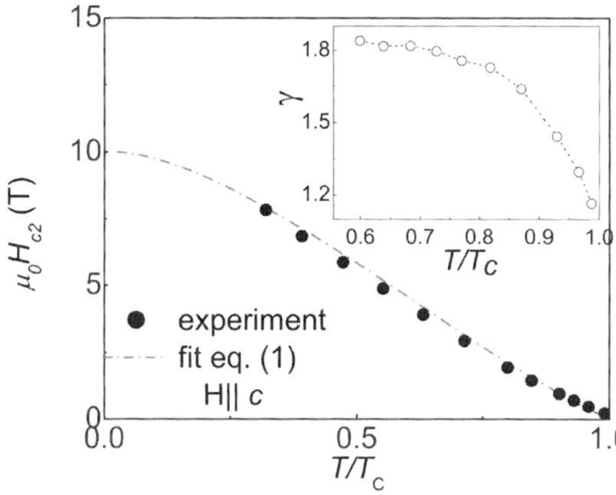

(a)

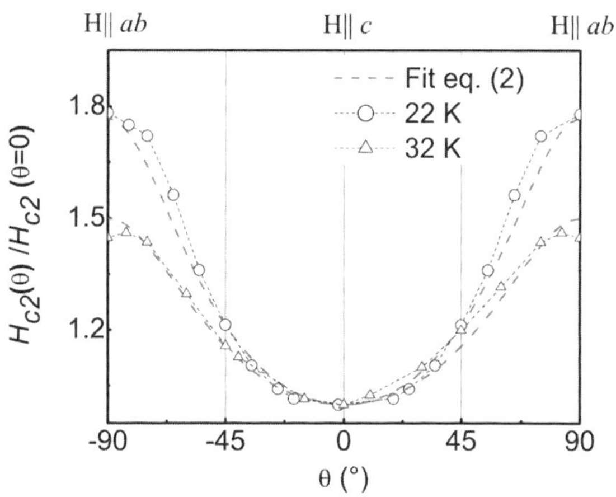

(b)

Fig. 3. (a) H_{c2} versus T for $\mathbf{H}//c$ for an epitaxial MgB_2 thin film. The dashed line is a fit to a two-band model. Inset: Temperature dependence of the H_{c2} anisotropy, (b) Angular dependence of H_{c2} at two different temperatures. The dashed lines are fits to the two-band model using the same parameters as in (a). (Ref. 41). Reproduced with permission from American Physical Society.

first study of neutron scattering from the flux line lattice in MgB_2 powder set an upper limit $\gamma_\lambda \sim 1.3$ at low T and H,[42] and a subsequent study by the same group on a single crystal showed that γ_λ increases both with T and H.[43] The almost isotropic λ at low T (with $\gamma_\lambda \sim 1.2$) was observed by direct imaging of the vortex

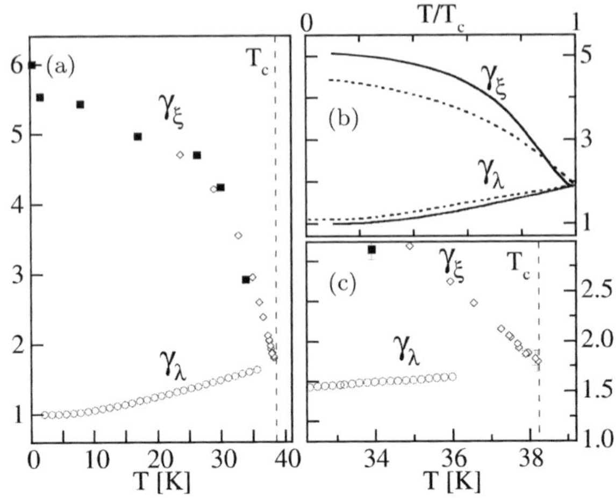

Fig. 4. (a) T-dependence of γ_λ and γ_ξ. For γ_ξ the open symbols are specific heat measurements and the closed symbols torque measurements, (b) Theoretical predictions for $\gamma_\xi(T)$ and $\gamma_\lambda(T)$ (see Ref. 45 for details), (c) Detail of the experimental data in (a) close to T_c (Ref. 45). Reproduced with permission from American Physical Society.

lattice in single crystals using scanning tunneling spectroscopy.[44] Fletcher *et al.*[45] also reported the increasing $\gamma_\lambda(T)$ in single crystals using a tunnel diode resonator, as shown in Fig. 4, as well as the *variations* of λ_{ab} and λ_c with T, which are inconsistent with a conventional BCS dependence but can be accounted for in a two-gap scenario.

The variations of λ with T provide a direct measure of the superfluid density $\rho(T) = \lambda(0)^2/\lambda(T)^2$. Kogan *et al.*[34] evaluated self-consistently $\rho(T)$ as well as the two gaps $\Delta_1(T)$ and $\Delta_2(T)$ for a two-band clean s-wave SC within the quasi-classical Eilenberger weak-coupling formalism, with two in-band and one interband pairing potentials. Their results for zero interband coupling is shown in Fig. 5(a), while Fig. 5(b) shows the experimental $\rho(T)$ measured for a MgB_2 single crystal and the fits obtained by adjusting the intra- and interband couplings.

The absolute values of λ are more difficult to measure than the temperature variations. D. Cunnane *et al.*[47] used two methods, based on Josephson junctions and Superconducting Quantum Interference

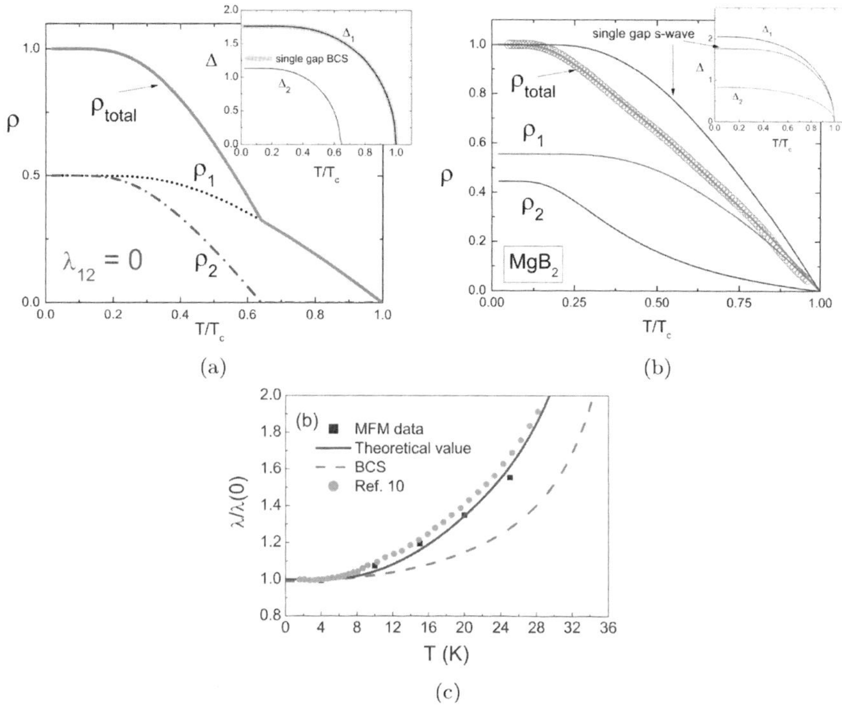

Fig. 5. (a) Calculated superfluid density and gaps versus reduced temperature (inset) for zero interband coupling, (b) Data and its of the superfluid density for a MgB_2 single crystal and corresponding temperature-dependent gaps (inset), (c) black squares: $\lambda(T)$ for the same film shown in Fig. 3, obtained by MFM. Solid curve: Calculated $\lambda(T)$ from the gap equations for two-band SC. Dashed curve: Conventional BCS model. Dotted curve: Tunnel diode resonator measurements from (Ref. 45). Figs. 5(a) and 5(b) from Ref. 34, Fig. 5(c) from (Ref. 41). Reproduced with permission from American Physical Society.

Device (SQUIDs), and obtained consistent results for clean thin films, $\lambda_{ab}(T \to 0) \sim 40\,\text{nm}$. Using magnetic force microscopy (MFM), we measured the absolute value of $\lambda_{ab}(T)$ in the same film shown in Fig. 3.[41] The result is shown in Fig. 5(c), which also includes the calculated temperature dependence based on the two-gap nature of the superconducting state.[34] This fit indicates larger intraband diffusivity in the 3D π-band compared to that in the 2D σ-band, leading to an increase of λ consistent with the $H_{c2}(T, \Theta)$ analysis. Direct comparison of the film results with those reported in Ref. 45

for single crystals (also included in Fig. 5(c)) show similar temperature dependence. The measurement of $\lambda_{ab}(T)$ and $H_{c2}(T,\Theta)$ in the same film provides a venue for thoroughly determining the superconducting properties and allows the calculation of the thermodynamic critical field H_c.

3.3 *Field dependences of ξ, λ and anisotropies*

Due to the interplay of the two bands, the characteristic lengths are not only T but also H dependent. In particular, the difference between γ_λ and γ_ξ, as well as the increase of γ_λ with H,[43,46] indicate the strong suppression of the π-band superconductivity with increasing field in clean systems.

To explore the $\xi(H)$ dependence in detail it is necessary to go beyond measuring H_{c2}, which is just a boundary in the H–T–Θ phase diagram, and look inside the mixed state. This is what Klein *et al.*[48] did on MgB_2 single crystals, by measuring the specific heat and extracting a field-dependent Sommerfeld coefficient. At $T = 2.5\,\mathrm{K}$ for $0.2T < \mu_0 H < 2.5T$ they found that $\xi(H)$ approximately decreases as $H^{-0.25}$. At the high end of their range they obtained $\xi(2.5T) \sim 11\,\mathrm{nm}$, consistent with H_{c2} results, but at their lowest accessible $H \sim 0.2T$ it increased to $\sim$23 nm. This has implications for vortex pinning and creep, as discussed below.

In the same study they measured the reversible magnetization (M_{rev}) in the mixed state.[48] In a single-band scenario, for $\kappa \gg 1$ materials in the London regime $H_{c1} \ll H \ll H_{c2}$, a dependence $M_{rev} \propto \lambda^{-2} \ln(H_{c2}/H)$ is approximately observed.[3] By extending this model to the two-band MgB_2, they extracted $\lambda(H)$, and found that it increased from $\lambda(H \to 0) \sim 45\,\mathrm{nm}$ to $\lambda(H \to H_{c2}) \sim 75\,\mathrm{nm}$.

3.4 *H_{c2} enhancement by controlled introduction of disorder*

Among the various advantages of MgB_2 for potential applications the low H_{c2} of early samples was a setback, as many applications, in particular magnets, require high H_{c2}. However, the fact that some samples had much higher values[15] indicated that there was a route to higher H_{c2} via the controlled introduction of disorder. Early studies already showed that modest H_{c2} increases could be produced by proton[49] and neutron[50] irradiations. However, chemical substitutions are clearly the method most amenable to industrial scale production. Very high H_{c2} values have been obtained by alloying MgB_2 samples with magnetic impurities.[51,52] Analysis in the context of Gurevich model[31] of two-gap superconductivity in the dirty limit indicate that MgB_2 offers special opportunities for H_{c2} increases by tuning of the impurity scattering by selective alloying on Mg and B sites.

Many researches have reported attempts to increase not only H_{c2} but also J_c by adding impurities in bulk samples. However, only doping with carbon (C), C compounds, Al, or oxygen alloying effectively improve H_{c2} and J_c for practical MgB_2 uses.[53–60] For C compound addition, it is worth mentioning that in most cases it is possible to simultaneously increase these two important parameters for high-field magnet applications. Although the effect of C substitution is one of the most studied in MgB_2, the results on C solubility and the effect of C-doping on T_c, J_c and H_{c2} reported so far vary significantly due to precursor materials, fabrication techniques and processing conditions used.[58–63] Doping MgB_2 with single and double-wall Carbon nano tubes (CNT) are capable of this simultaneous improvement.[64–66] The reason for this is the dual role of the CNT. They partially dilute into the MgB_2 matrix, acting as a source of C that increases H_{c2} (producing the highest H_{c2} values observed so far in bulk MgB_2). At the same time, the fraction of the CNT that retain their structural integrity are ideally suited to act as strong vortex pinning centers due to their tubular geometry and their diameter close to the superconducting coherence length of MgB_2, producing a large J_c enhancement. As an additional result, the measured H_{c2} versus T in all samples are successfully described using the theoretical model for a two-gap SC in the dirty limit proposed by Gurevich.[31] This has strong fundamental impact, as it provides clear evidence in support of the basic scenario currently used to describe the superconducting behavior of MgB_2, and opens a path for future research in H_{c2} enhancement.

The $H_{c2}(T)$ and $H_{irr}(T)$ dependence were determined from four-probe transport measurements in the mid-pulse magnet of NHMFL — LANL, capable to generate asymmetric field pulses up to 50T, performed at temperatures between 1.4 and 34 K. The proposed model for a two-gap SC in the dirty limit by Gurevich *et al.*[31,51] and Golubov *et al.*[67] considers that non-magnetic impurities affect the *intra*band electron diffusivities D_σ and D_π, and the *inter*band scattering rates $\Gamma_{\pi\sigma}$ and $\Gamma_{\sigma\pi}$. The H_{c2} dependence on temperature can be obtained from the Usadel equations taking into account the *intraband* and *interband* scattering effects. In this case we optimized the diffusivity ratio $\eta = D_\pi/D_\sigma$ and *inter*band scattering parameter $g = (\Gamma_{\sigma\pi} + \Gamma_{\pi\sigma})\nabla/2\pi k_B T_{c0}$, where $T_{c0} = T_c(g = 0)$ to fit the measurements using the Eq. (1) from Ref. 31 as described in Ref. 64. Figure 6 exhibits the dependence of H_{c2} defined as the onset (extrapolation of maximum slope up to normal state resistivity) of the ρ versus H data as a function of the reduced temperature $t = T/T_{c0}$ for samples with several dwCNT contents.

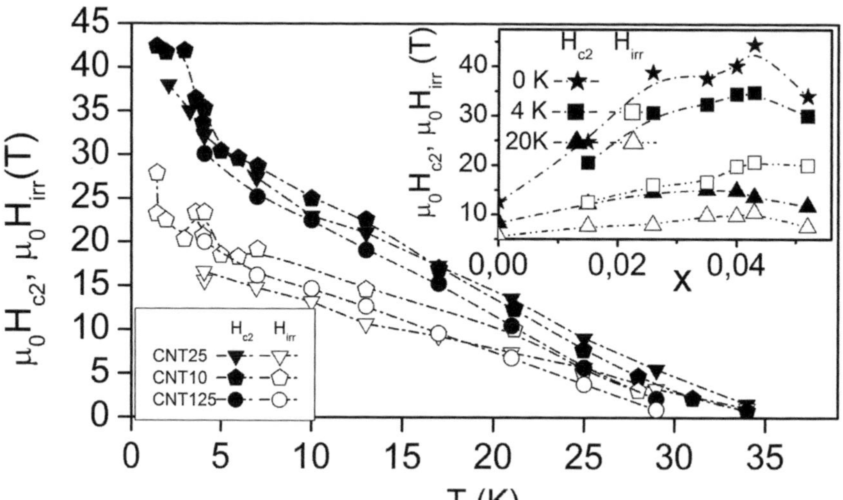

Fig. 6. Transport measurements of the upper critical field (H_{c2}) and the beginning of dissipation (H_{irr}) in the $R(H)$ curves for samples CNTdw01, CNTdw75 and CNTdw10. Inset: H_{c2} and H_{irr} versus x at 4 K (stars) and 20 K (circles) (Ref. 64).

It is apparent in these H_{c2}(T) data that the upward curvature signals as a characteristic of the presence of two gaps. The inset displays H_{c2} as function x for 4 K and 20 K for all samples. We observe that $H_{c2}(0)$ has a maximum near $x \approx 0.045$ for sample CNTdw10. This enhancement indicates that the incorporation of C in the lattice affects the scattering mechanism, increasing H_{c2} up to 44 T at $T = 0$, as predicted by the theoretical models. It is worth to note that, although η does not follow a clear tendency, we observe that $1/D_\pi$ follows the same dependence of $H_{c2}(0)$ as a function of x (see Table 1) signifying that C-doping is affecting the π-band.

It is also interesting to explore the effect of simultaneous substitution of Mg and B sites and we report here an example of MgB_2 samples prepared by solid-state with $CaCO_3$ additions. In this case, Ca substitution on Mg site produced an increase in the a lattice parameter[68] in contrast with the effect of C substitution on B site. Therefore, in this case is not possible to evaluate the nominal composition of Ca in the MgB_2 matrix (see inset in Fig. 7). However, it is also possible to fit the H_{c2} dependence on temperature using the Usadel equations. As can be observed in Fig. 7, H_{c2}(T) data of simultaneous substitution of Mg(Ca) and B(C,O) does not present

Table 1: Nominal composition, actual C content (x) calculated from a-axis values, critical temperature measured by magnetization (T_c^{mag}) and transport (T_c^{trans}), extrapolation of upper critical field (H_{c2}) at 0K and fit parameters $\eta = D_\pi/D_\sigma$ and $g = (\Gamma_{\sigma\pi} + \Gamma_{\pi\sigma})\nabla/2\pi k_B T_{c0}$ for the CNTdw doped samples.

Sample	$at.\%$	x	T_c^{mag}	T_c^{trans}	H_{c2} at $0K[T]$	$\eta = D_\pi/D_\sigma$	$g = (\Gamma_{\sigma\pi} + \Gamma_{\pi\sigma})\nabla/2\pi k_B T_{c0}$
CNT00	0	0	38.5	38.7	17.7	0.1786	0.00988
CNTdw01	1	0.015	37.5	38.4	23.8	0.1695	0.0018
CNTdw25	2.5	0.026	36	37.5	38.7	0.1288	0.0161
CNTdw05	5	0.035	34	35	37.5	0.1862	0.0291
CNTdw75	7.5	0.040	34	35.8	40.0	0.1409	0.0275
CNTdw10	10	0.043	33.5	—	44.4	0.1212	0.0355
CNTdw125	12.5	0.052	31	31.1	33.9	0.2075	0.0536

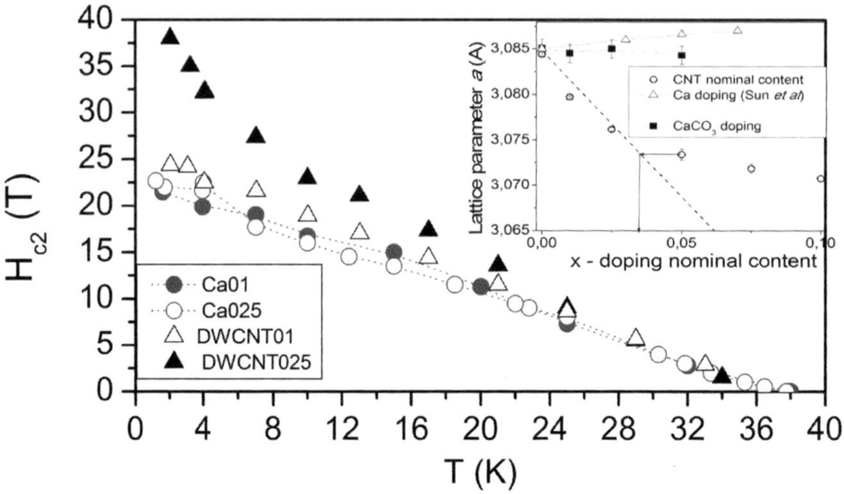

Fig. 7. Transport measurements of H_{c2} for Ca01 and Ca025 in comparison with CNTdw01 and CNTdw25 samples. Inset: The lattice parameter a as a function of nominal content x for $CaCO_3$ and CNT additions.

Table 2: Nominal composition, critical temperature, extrapolation of upper critical field (H_{c2}) at 0K and fit parameters $\eta = D_\pi/D_\sigma$ and $g = (\Gamma_{\sigma\pi} + \Gamma_{\pi\sigma})\nabla/2\pi k_B T_{c0}$ for the $CaCO_3$-doped samples.

Sample	*at.%*	T_c^{trans}	H_{c2} at $0K[T]$	$\eta = D_\pi/D_\sigma$	$g = (\Gamma_{\sigma\pi} + \Gamma_{\pi\sigma})\nabla/ 2\pi k_B T_{c0}$
Ca01	1	38.1	22.2	0.1695	0.001
Ca025	2.5	37.6	22.6	0.1492	0.002

the upward curvature but the fitting parameters shown in Table 2 follows the same trend as those of CNT samples.

4. Two-band Effects on Vortex Pinning

Due to the relevance for applications, many studies focused on pinning and methods to enhance J_c in MgB_2 have been published. In particular, the topic is addressed in other chapters of this book. In this short section we will only address a few issues that relate to the two-band nature and are relevant for the creep discussion below.

Let's start by estimating the depairing limit.[3,16] In single-band materials $J_c(T) = cH_c(T)/[3\sqrt{6}\pi\lambda(T)]$, but the complication in the case of MgB_2 is that ξ and λ are field dependent. For our purpose it is appropriate to use the values for $H \to 0$ that will reflect the full contribution from both bands. Based on the results presented above, taking $\xi(T \to 0, H \to 0) \sim 20$–$25\,\mathrm{nm}$ and $\lambda(T \to 0, H \to 0) \sim 40$–$50\,\mathrm{nm}$ we obtain $H_c(0) \sim 1800$–2500 Oe and $J_d(0) \sim 160$–$240\,\mathrm{MA/cm^2}$. Transport measurements in very clean epitaxial *narrow* films (width $<2\lambda$) that are expected to be vortex-free, indeed give $J_c \sim 160\,\mathrm{MA/cm^2}$, which can be taken as an estimate of J_d.[69]

Magnetization measurements in large films[39] (i.e., with vortices) from the same source as in Ref. 69 show that vortex pinning at low T and H can be very strong, with $J_c(T = 5\,\mathrm{K}, H = 0) \sim 26\,\mathrm{MA/cm^2}$, or ~16% of J_d. At low T there is a small H-independent regime that is the fingerprint of single vortex pinning, but may also involve self-field effects. As H increases J_c decays as H^{-1}; then drops dramatically at $H \sim 2\ T$. As T increases the $J_c \propto H^{-1}$ regime remains, but the H-independent regime disappears, and both the value of J_c and the field at which it suddenly drops decrease strongly. These observations point again to the fast suppression of π-band superconductivity with increasing H and T.

Clearly, wire and magnet applications require quite a different performance, with high J_c sustained up to higher fields (several Tesla) and temperatures ($>20\,\mathrm{K}$). Fortunately these conditions can be achieved in very dirty samples, which have much higher H_{c2}, lower anisotropy and stronger interband coupling as discussed in the previous section.

As will be also reviewed in other chapters, the J_c–H performance can be greatly improved by adding different C sources. In particular, the effect of CNT and SiC on the superconducting properties of MgB_2 bulk and wire samples, based on the known literature data and our own results, are within the best options.[70] As already mentioned CNT inclusions, with diameters close to the MgB_2 coherence length are good candidates for vortex pinning if they do not completely dissolve in the matrix but remain as tubes acting as columnar defects.

The distinct effect of C incorporation through different routes using various CNT types leads to a simultaneous improvement in J_c and H_{c2}, but their effectiveness change with temperature or applied field according to each type of addition.[54,59,60,64,65,71] This effect was reported in many studies, not only for bulk samples but also for wires and tapes prepared by the powder-in-tube (PIT) method. Other successful examples were the addition of C sources such as C-doped MgB_2 filaments,[61–63] nanocarbon,[72–74] amorphous C,[75] diamond,[76] hydrocarbons[77–79] and B_4C.[80–82]

5. Flux Creep

Flux creep was first observed and described by Kim and Anderson in 1962.[28,29] In the A–K scenario either a portion of a vortex or a bundle of vortices is modeled as a particle inside a pinning well. In this description J decays logarithmically with time and the relaxation speed, measured by the normalized relaxation rate $S = [\mathrm{d}J/\mathrm{d}\ln t]/J_c$, results $S(T) = T/U_p(T)$. For $T \ll T_c$ usually $U_p(T) \sim const$, so S increases linearly with T. This is observed in oxide and Fe-based HTS, except that $S(T)$ extrapolates to a finite positive value at $T = 0$ in contrast to the expectation, a feature that is considered as evidence of a quantum creep contribution.[16]

As T increases $U_p(T)$ decreases and, if the creep rate is large, J becomes significantly smaller than J_c. Now the elastic properties of the vortex matter start to play an important role and the A–K model fails, because the size of the "critical nucleus" that undergoes a thermal jump grows as J decreases and the "particle in a pinning well" description does not capture this physics. The dynamics becomes *glassy*, and the creep rate, redefined as $S = \mathrm{d}\ln J/\mathrm{d}\ln t$, becomes $S(T) = T/[U_p(T) + \mu T \ln(t/t_0)]$, where t_0 is a microscopic characteristic time and $\mu > 0$ is a *glassy exponent* that characterizes the dynamic regime, which to first approximation is independent of T and H within each regime.[16] Thus for high enough temperatures $U_p(T) \ll \mu T \ln(t/t_0)$, so $S \sim 1/[\mu \ln(t/t_0)] \sim const$ within each regime, resulting in the "plateaus" in $S(T)$ frequently observed in YBCO.[16]

Studies of the relaxation of the irreversible magnetization in randomly oriented MgB_2 powders were performed very soon after the discovery of superconductivity in this compound by J.R. Thompson *et al.*[83] At fields lower than ∼1 T they observed creep rates significantly below those in HTS, strongly improving the prospects for applications. A later study by the same group[84] on polycrystalline high J_c films showed even lower creep rates, as seen in Fig. 8(a). It is clear from the figure that $S(T)$ increases linearly with an A–K

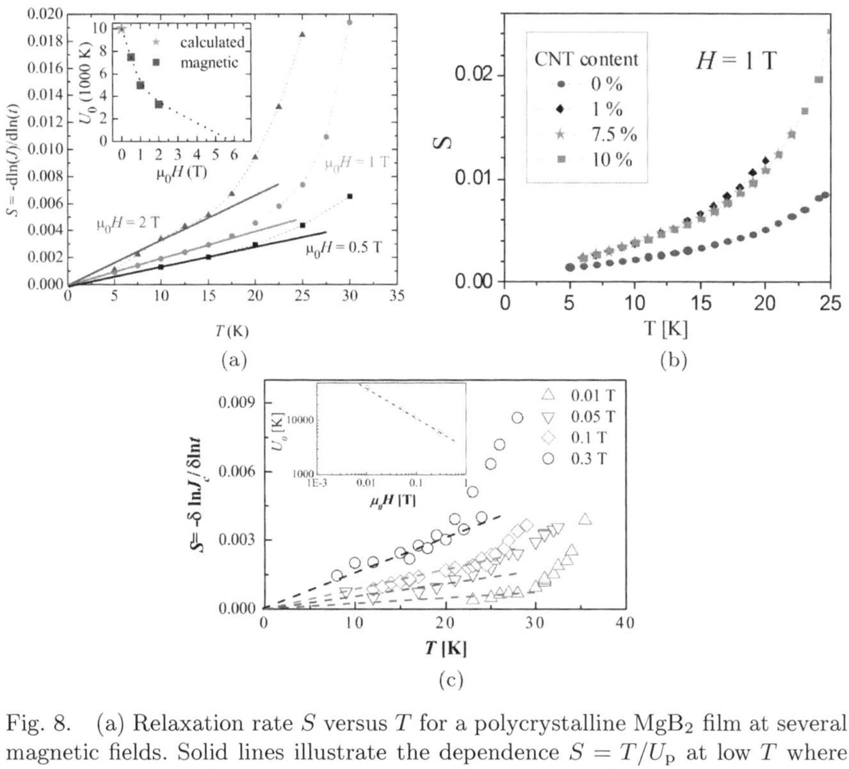

Fig. 8. (a) Relaxation rate S versus T for a polycrystalline MgB_2 film at several magnetic fields. Solid lines illustrate the dependence $S = T/U_p$ at low T where the pinning energy U_p is T-independent. Inset: Magnetic field dependence of U_p. (Ref. 84), (b) S versus T at $\mu_0 H = 1T$ for bulk polycrystalline samples with various contents of DWCNT for H_{c2} and J_c enhancement (Ref. 89). Reproduced with permission of AIP Publishing LLC, (c) S versus T for $\mathbf{H}//c$ at several low fields for a clean epitaxial film. Inset: Pinning energy obtained from $S = T/U_p$ (Ref. 39). Reproduced with permission from Elsevier Ltd.

description. The slope $S/T \sim 1/U_p(T\ 0)$ is small and increases monotonically with H, indicating that $U_p(T \to 0)$ is quite large at low H and decreases with H, as shown in the inset of the figure. Finally, the extrapolations to $S(T \to 0) \sim 0$ point to a very small quantum creep contribution.

Other early studies on bulk polycrystalline samples[85,86] and on partially oriented films[87] consistently found low S values at low T and H, but also pointed out that S increases drastically as H increases above a couple of Tesla. In particular one of those studies, by Y.Z. Zhang *et al.*,[88] focused on the magnetic relaxation in polycrystalline MgB_2 in the low T region, extending the measurements down to ~2 K. They confirmed that the $S(T \to 0)$ extrapolated values are indeed small but nonzero, and also showed that they increase with H.

In the work by Kim *et al.*,[39] we performed creep studies in the very clean film discussed above in the low H regime, as shown in Fig. 8(b). We observed very low creep rates, with $S(T)$ increasing linearly with T and extrapolating to $S(T \to 0) \sim 0$ consistent with the A–K scenario and almost no quantum creep contribution. The slope $S/T \sim 1/U_p(T \to 0)$ increases with H. For the lowest field $H = 100$ Oe we obtained the large value $U_p(T \to 0) \sim 60000\,\mathrm{K}$, decreasing approximately as $U_p(T \to 0) \propto H^{-1/2}$. This indicates the presence of a low density of very strong pinning centers.

Overall, the various studies on pinning and flux creep in not-too-dirty MgB_2 indicate that the suppression of the π-band strongly reduces U_p as H and T increase. A simple way to look at this is to consider that the scale of U_p is set by the condensation energy in a minimum superconducting volume, $U_p \propto U_p \propto (H_c^2/8\pi)\xi^3 \propto \xi/\lambda^2$, so both the increase in λ and the reduction in ξ as H increases and the π-band is suppressed contribute to the U_p reduction. This adds a richness to the vortex pinning and dynamics that is unique to two-band materials and clearly deserves further studies. For instance, for the analysis of vortex dynamics a careful reinterpretation of the role played by *Gi* (that becomes H-dependent) is needed. Another open question is whether creep in MgB_2 ever becomes glassy.

Finally, we will return to the polycrystalline bulk samples with CNT additions already presented above. In a study by Pasquini *et al.*,[89] we explored flux creep in samples with and without CNT. Some of the results are shown in Fig. 8(c), where $S(T)$ is compared for a sample without CNT and one with 10% of CNT. The slopes S/T increase with H and indicate $U_p(H)$ values similar to those in polycrystalline films. One conclusion of the study is that at low T the results are well described by a single vortex pinning regime where the correlation length (the length of the vortex that "jumps" in the creep process is limited by the grain size. An interesting consequence is that the incorporation of CNT increases both J_c and S, due to the combined effect of the reduction of both x and the grain size.

6. Conclusions

The two-band nature of superconductivity in MgB_2 has important effects on the behavior of vortex matter and results in a complex phase diagram and rich phenomenology. In clean samples the contribution of the almost isotropic π-band is strongly suppressed by magnetic fields, so the almost 2D σ-band dominates at high H. This results in-field dependences in λ and ξ (not present in single-band materials) as well as anomalous temperature dependences. The anisotropies of ξ and λ are different and both are also T and H dependent, again in contrast with a simple single-band description. The angular dependence of H_{c2} is not well described by the standard single-band $\varepsilon(\Theta)$ function. In single-band SC the increase of disorder may reduce the electronic mean free path, pushing the material towards the dirty limit, i.e., increasing H_{c2} and λ and decreasing H_{c1} and ξ. In MgB_2 the effect is more complex, as it depends on the influence of the particular disorder on several intra and interband scattering mechanisms. One consequence is that H_{c2} can be dramatically increased, which is very good for applications, and also exhibits anomalous T dependences.

In terms of pinning and dynamics, the main effect in clean samples is that the suppression of the π-band with field produces a fast decrease in $J_c(H)$ and increase in $S(H)$. The increase in disorder has the effect of increasing the coupling between the bands allowing the retention of a large J_c up to higher H and T, which is again very convenient for applications.

Acknowledgments

The work of L. Civale in the preparation of this manuscript was supported by the U.S. Department of Energy, Office of Basic Energy Sciences, Division of Materials Sciences and Engineering and the work of A. Serquis was supported by CNEA, CONICET and ANPCyT.

References

1. J. L. Sarrao and W. K. Kwok, chairs, *Basic Research Needs for Superconductivity*, report of the Basic Energy Sciences. Workshop on Superconductivity (Washington, DC, May 8–11, 2006), Available at: http://www.sc.doe.gov/bes/reports/abstracts.html#SC.
2. J. Bardeen, L. N. Cooper and J. R. Schrieffer, Theory of superconductivity, *Phys. Rev.* **108**, p. 1175 (1957).
3. For a general discussion of superconductivity, M. Tinkham, *Introduction to Superconductivity*, McGraw Hill, New York, (1975).
4. W. L. McMillan, Transition temperature of strong–coupled superconductors, *Phys. Rev.* **167**, p. 331 (1968).
5. G. Bednorz and K. A. Muller, Possible high T_c superconductivity in the Ba–La–Cu–O system, *Z. Phys. B* **64**, p. 189 (1986).
6. J. Nagamatsu *et al.*, Superconductivity at 39 K in magnesium diboride, *Nature (London)* **410**, p. 63 (2001).
7. S. L. Bud'ko *et al.*, Boron isotope effect in superconducting MgB_2, *Phys. Rev. Lett.* **86**, pp. 1877–1880 (2001).
8. G. Karapetrov, M. Iavarone, W. K. Kwock, G. W. Crabtree and D. G. Hinks, Scanning tunneling spectroscopy in MgB_2, *Phys. Rev. Lett.* **86**, p. 4374 (2001).
9. F. Giubileo, D. Roditchev, W. Sacks, R. Lamy, D. X. Thanh, J. Klein, S. Miraglia, D. Fruchart, J. Marcus and Ph. Monod, Two-gap state density in MgB_2: A true bulk property or a proximity effect?, *Phys. Rev. Lett.* **87**, p. 177008 (2001).
10. M. R. Esklidsen, M. Kugler, S. Tanaka, J. Jun, S. M. Kazakov, J. Karpinski and O. Fisher, Vortex Imaging in the π band of magnesium Diboride, *Phys. Rev. Lett.* **89**, p. 187003 (2002).

11. X. K. Chen, M. J. Kostantinovic, J. C. Irwin, D. D. Lawrie and J. P. Franck, Evidence for two superconducting gaps in MgB_2, *Phys. Rev. Lett.* **87**, p. 157002 (2001).
12. H. Schmidt, J. E. Zasadzinski, K. E. Gray and D. G. Hinks, Evidence for two-band superconductivity from break-junction tunneling on MgB_2, *Phys. Rev. Lett.* **88**, p. 127002 (2002).
13. F. Bouquet, R. A. Fisher, N. E. Phillips, D. G. Hinks and J. D. Jorgensen, Specific heat of $Mg^{11}B_2$: Evidence for a second energy gap, *Phys. Rev. Lett.* **87**, p. 047001 (2001).
14. J. J. Tu, G. L. Carr, V. Perebeinos, C. C. Homes, M. Strongin, P. B. Allen, W. N. Kang, E.-M. Choi, H.-J. Kim and S.-I. Lee, Optical properties of c-axis oriented superconducting MgB_2 films, *Phys. Rev. Lett.* **87**, p. 277001 (2001).
15. C. Buzea and T. Yamashita, Review of the superconducting properties of MgB_2, *Supercond. Sci. Technol.* **14**, pp. R115–R146 (2001).
16. G. Blatter, M. V. Feigel'man, V. B. Geshkenbein, A. I. Larkin and V. M. Vinokur, Vortices in high-temperature superconductors, *Rev. Mod. Phys.* **66**, p. 1125 (1994).
17. A. Y. Liu, I. I. Mazin, J. Kortus, Beyond Eliashberg superconductivity in MgB_2: Anharmonicity, two-phonon scattering, and multiple gaps, *Phys. Rev. Lett.* **87**, p. 087005 (2001).
18. H. J. Choi, D. Roundy, H. Sun, M. L. Cohen and S. G. Loule, The origin of the anomalous superconducting properties of MgB_2, *Nature (London)* **418**, p. 758 (2002).
19. Y. Kamihara, T. Watanabe, M. Hirano and H. Hosono, Iron-based layered superconductor $La[O_{1-x}F_x]FeAs$ ($x = 0.05$–0.12) with $T_c = 26\,K$, *J. Am. Chem. Soc.* **130**, p. 3296 (2008).
20. N. Haberkorn, J. Kim, K. Gofryk, F. Ronning, A. Sefat, U. Welp, L. Fang, W. K. Kwok and L. Civale, Enhancement of the critical current density by increasing the collective pinning energy in heavy ion irradiated Co-doped $BaFe_2As_2$ single crystals, *Supercond. Sci. Technol.* **28**, p. 055011 (2015) and references therein.
21. L. P. Gor'kov and T. K. Melik-Barkhudarov, *Sov. Phys. JETP* **18**, p. 1031 (1964).
22. V. G. Kogan, Macroscopic anisotropy in superconductors with anisotropic gaps, *Phys. Rev. B* **66**, p. 020509(R) (2002).
23. W. A. Fietz and W. W. Webb, Magnetic properties of some type-II alloy superconductors near the Upper Critical Field, *Phys. Rev.* **161**, p. 423 (1967); T. P. Orlando, E. J. McNiff, S. Foner and M. R. Beasley, Critical fields, Pauli paramagnetic limiting, and material parameters of Nb_3Sn and Vn_3Si, *Phys. Rev. B* **19**, p. 4545 (1979).

24. M. Miura *et al.*, Vortex liquid-glass transition up to 60 T in nanoengineered coated conductors grown by metal organic deposition, *Appl. Phys. Lett.* **96**, p. 072506 (2010).
25. A. M. Campbell and J. E. Evetts, Flux vortices and transport current in type-II superconductors, *Adv. Phys.* **21**, p. 194 (1972).
26. B. Maiorov *et al.*, Synergetic combination of different types of defects to optimize pinning landscape using $BaZrO_3$-doped $YBa_2Cu_3O_7$, *Nature Materials* **8**, p. 398 (2009).
27. S. R. Foltyn, L. Civale, J. L. MacManus-Driscoll, Q. X. Jia, B. Maiorov, H. Wang and M. Maley, Materials science challenges for high-temperature superconducting wire, *Nature Materials* **6**, p. 631 (2007).
28. Y. B. Kim, C. F. Hempstead and A. R. Strnad, Critical persistent currents in hard superconductors, *Phys. Rev. Lett.* **9**, p. 306 (1962).
29. P. W. Anderson, Theory of flux creep in hard superconductors, *Phys. Rev. Lett.* **9**, p. 309 (1962).
30. P. Samuely, P. Szabó, J. Kačmarčík, T. Klein and A. G. M. Jansen, Point-contact spectroscopy of MgB_2, *Physica C* **385**, p. 244 (2003).
31. A. Gurevich, Enhancement of the upper critical field by nonmagnetic impurities in dirty two-gap superconductors, *Phys. Rev. B* **67**, p. 184515 (2003).
32. H. Suhl, B. T. Matthias and L. R. Walker, Bardeen–Cooper–Schrieffer theory of superconductivity in the case of overlapping bands, *Phys. Rev. Lett.* **3**, p. 552 (1959).
33. A. A. Golubov and I. I. Mazin, Effect of magnetic and nonmagnetic impurities on highly anisotropic superconductivity, *Phys. Rev. B* **55**, p. 15146 (1997) and ref. therein.
34. V. G. Kogan, C. Martin and R. Prozorov, Superfluid density and specific heat within a self-consistent scheme for a two-band superconductor, *Phys. Rev. B* **80**, p. 014507 (2009).
35. F. Simon, A. Jánossy, T. Fehér, F. Murányi, S. Garaj, L. Forró, C. Petrovic, S. L. Bud'ko, G. Lapertot, V. G. Kogan and P. C. Canfield, Anisotropy of superconducting MgB_2 as seen in electron spin resonance and magnetization data, *Phys. Rev. Lett.* **87**, p. 047002 (2001).
36. S. L. Bud'ko and P. C. Canfield, Temperature-dependent Hc2 anisotropy in MgB2 as inferred from measurements on polycrystals, *Phys. Rev. B* **65**, p. 212501 (2002).
37. L. Lyard, P. Samuely, P. Szabo, T. Klein, C. Marcenat, L. Paulius, K. H. P. Kim, C. U. Jung, H.-S. Lee, B. Kang, S. Choi, S.-I. Lee, J. Marcus, S. Blanchard, A. G. M. Jansen, U. Welp, G. Karapetrov and W. K. Kwok, Anisotropy of the upper critical field and critical current in single crystal MgB_2, *Phys. Rev. B* **66**, p. 180502(R) (2002).

38. C. Zhuang, T. Tan, Y. Wang, S. Bai1, X. Ma, H. Yang, G. Zhang, Y. He, H. Wen, X. X. Xi, Q. Feng and Z. Gan, Clean MgB_2 thin films on different types of single-crystal substrate fabricated by hybrid physical–chemical vapor deposition, *Supercond. Sci. Technol.* **22**, p. 025002 (2009).
39. K. Jeehoon, N. Haberkorn, E. Nazaretski, R. dePaula, T. Tan f, X. X. Xi, T. Tajima, R. Movshovich and L. Civale, Strong magnetic field dependence of critical current densities and vortex activation energies in an anisotropic clean MgB_2 thin film, *Solid State Commun.* **204**, p. 56 (2015).
40. S. A. Baily *et al.*, Pseudoisotropic upper critical field in cobalt-doped $SrFe_2As_2$ epitaxial films, *Phys. Rev. Lett.* **102**, p. 117004 (2009); B. Maiorov *et al.*, Inversion of the upper critical field anisotropy in FeTeS films, *Supercond. Sci. Technol.* **27**, p. 044005 (2014).
41. K. Jeehoon, N. Haberkorn, Shi-Zeng Lin, L. Civale, E. Nazaretski, B. H. Moeckly, C. S. Yung, J. D. Thompson and R. Movshovich, Measurement of the magnetic penetration depth of a superconducting MgB_2 thin film with a large intraband diffusivity, *Phys. Rev. B* **86**, p. 024501 (2012).
42. R. Cubitt, S. Levett, S. L. Bud'ko, N. E. Anderson and P. C. Canfield, Experimental evidence for anisotropic double-gap behavior in MgB_2, *Phys. Rev. Lett.* **90**, p. 157002 (2003).
43. R. Cubitt, M. R. Eskildsen, C. D. Dewhurst, J. Jun, S. M. Kazakov and J. Karpinski, Effects of two-band superconductivity on the flux-line lattice in magnesium diboride, *Phys. Rev. Lett.* **91**, p. 047002 (2003).
44. M. R. Esklidsen, N. Jenkins, G. Levy, M. Kugler, Ø. Fischer, J. Jun, S. M. Kazakov and J. Karpinski, Vortex imaging in magnesium diboride with $H \perp c$, *Phys. Rev. B* **68**, p. 100508(R) (2003).
45. J. D. Fletcher, A. Carrington, O. J. Taylor, S. M. Kazakov and J. Karpinski, Temperature-dependent anisotropy of the penetration depth and coherence length of MgB_2, *Phys. Rev. Lett.* **95**, p. 097005 (2005).
46. F. Bouquet, Y. Wang, I. Sheikin, T. Plackowski, A. Junod, S. Lee and S. Tajima, Unusual effects of anisotropy on the specific heat of ceramic and single crystal MgB_2, *Phys. Rev. Lett.* **89**, p. 257001 (2002).
47. D. Cunnane, C. Zhuang, Ke Chen, X. X. Xi, J. Yong and T. R. Lemberger, Penetration depth of MgB_2 measured using Josephson junctions and SQUIDs, *Appl. Phys. Lett.* **102**, p. 072603 (2013).
48. T. Klein, L. Lyard, J. Marcus, Z. Holanova and C. Marcenat, Magnetic field dependence of the coherence length and penetration depth of MgB2 single crystals, *Phys. Rev. B* **73**, p. 184513 (2006).

49. Y. Boguslavsky, L. F. Cohen, G. K. Perkins, M. Polichetti, T. J. Tate, R. Gwilliam and A. D. Caplin, Enhancement of the high-magnetic-field critical current density of superconducting MgB_2 by proton irradiation, *Nature (London)* **411**, p. 561 (2001).
50. M. Eisterer, M. Zehetmayer, S. Tonies, H. W. Weber, M. Kambara, N. H. Babu, D. A. Cardwell and L. R. Greenwood, Neutron irradiation of MgB_2 bulk superconductors, *Supercond. Sci. Technol.* **15**, p. L9 (2002).
51. A. Gurevich, S. Patnaik, V. Braccini, K. H. Kim, C. Mielke, X. Song, L. D. Cooley, S. D. Bu, D. M. Kim, J. H. Choi, L. J. Belenky, J. Giencke, M. K. Lee, W. Tian, X. Q. Pan, A. Siri, E. E. Hellstrom, C. B. Eom and D. C. Larbalestier, Very high upper critical fields in MgB_2 produced by selective tuning of impurity scattering, *Supercond. Sci. Technol.* **17**, p. 278 (2004).
52. V. Braccini, A. Gurevich, J. E. Giencke, M. C. Jewell, C. B. Eom, D. C. Larbalestier, A. Pogrebnyakov, Y. Cui, B. T. Liu, Y. F. Hu, J. M. Redwing, Qi Li, X. X. Xi, R. K. Singh, R. Gandikota, J. Kim, B. Wilkens, N. Newman, J. Rowell, B. Moeckly, V. Ferrando, C. Tarantini, D. Marré, M. Putti, C. Ferdeghini, R. Vaglio and E. Haanappel, High-field superconductivity in alloyed MgB_2 thin films, *Phys. Rev. B* **71**, p. 012504 (2005).
53. X. Z. Liao, A. Serquis, Y. T. Zhu, J. Y. Huang, L. Civale, D. E. Peterson, H. F. Xu and F. M. Mueller, Optimization of critical current density in MgB_2, *J. Appl. Phys.* **93**, pp. 6208–6215 (2003).
54. S. X. Dou, S. Soltanian, J. Horvat, X. L. Wang, P. Munroe, S. H. Zhou, M. Ionescu, H. K. Liu and M. Tomsic, Enhancement of the critical current density and flux pinning of MgB_2 superconductor by nanoparticle SiC doping, *Appl. Phys. Lett.* **81**, p. 3419 (2002).
55. S. H. Zhou, A. V. Pan, M. J. Qin, H. K. Liu and S. X. Dou, Effect of the processing parameters of $MgB_{1.8}$ $(SiC)_{0.1}$/Fe tapes on the critical current density, *Physica C* **387**, p. 321 (2003).
56. A. Berenov, A. Serquis, X. Z. Liao, Y. T. Zhu, D. E. Peterson, Y. Bugoslavsky, K. A. Yates, M. G. Blamire, L. F. Cohen and J. L. MacManus-Driscoll, Enhancement of critical current density in low level Al-doped MgB_2, *Supercond. Sci. Technol.* **17**, p. 1093 (2004).
57. S. K. Chen, M. Wei and J. L. MacManus-Driscoll, Strong pinning enhancement in MgB_2 using very small Dy_2O_3 additions, *Appl. Phys. Lett.* **88(19)**, p. 192512 (2006).
58. S. X. Dou, W. K. Yeoh, J. Horvat and M. Ionescu, Effect of carbon nanotube doping on critical current density of MgB_2 superconductor, *Appl. Phys. Lett.* **83**, p. 4996 (2003).

59. W. K. Yeoh, J. Horvat, X. Xu, M. J. Qin, S. X. Dou, C. H. Jiang, T. Nakane, H. Kumakura and P. Munroe, Control of nano carbon substitution for enhancing the critical current density in MgB_2, *Supercond. Sci. Technol.* **19**, p. 596 (2006).
60. W. K. Yeoh, J. Horvat, S. X. Dou and P. Munroe, Effect of carbon nanotube size on superconductivity properties of MgB_2, *IEEE Applied Superconductivity* **15**, pp. 3284–3287 (2005).
61. R. H. T. Wilke, S. L. Bud'ko, P. C. Canfield, D. K. Finnemore, R. J. Suplinskas and S. T. Hannahs, Systematic effects of carbon doping on the superconducting properties of $Mg(B_{1-x}C_x)_2$, *Phys. Rev. Lett.* **92**, p. 217003 (2004).
62. R. H. T. Wilke, S. L. Bud'ko, P. C. Canfield, D. K. Finnemore, R. J. Suplinskas and S. T. Hannahs, Synthesis and optimization of $Mg(B_{1-x}C_x)_2$ wire segments, *Physica C* **424**, p. 1 (2005).
63. B. J. Senkowicz, J. E. Giencke, S. Patnaik, C. B. Eom, E. E. Hellstrom and D. Larbalestier, Improved upper critical field in bulk-form magnesium diboride by mechanical alloying with carbon, *Appl. Phys. Lett.* **86**, p. 202502 (2005).
64. A. Serquis, G. Serrano, M. S. Moreno, L. Civale, B. Maiorov, F. Balakirev and M. Jaime, Correlated enhancement of H_{c2} and J_c in carbon nanotube doped MgB_2, *Supercond. Sci. Technol.* **20**, p. L12 (2007).
65. G. Serrano, A. Serquis, S. X. Dou, S. Soltanian, L. Civale, B. Maiorov, F. Balakirev and M. Jaime, SiC and carbon nanotube distinctive effects on the superconducting properties of bulk MgB_2, *J. Appl. Phys.* **103**, p. 023907 (2008).
66. G. Serrano, A. Serquis, L. Civale, B. Maiorov, T. G. Holesinger, F. Balakirev and M. Jaim, Single-wall carbon nanotubes addition effects on the superconducting properties of MgB_2, *Int. J. Mod. Phys. B* **23**, pp. 3465–3469 (2009).
67. A. A. Golubov, J. Kortus, O. V. Dolgov, O. Jepsen, Y. Kong, O. K. Andersen, B. J. Gibson, K. Ahn and R. K. Kremer, Specific heat of MgB_2 in a one- and a two-band model from first-principles calculations, *J. Phys. Condensed Matter* **14**, p. 1353 (2002).
68. Y. Sun, D. Yu, Z. Liu, T. Wang, J. He and J. Xiang, Structural and superconducting properties of Ca-doped MgB_2 superconductors, *Supercond. Sci. Technol.* **20**, p. 261 (2007).
69. C. G. Zhuang, S. Meng, C. Y. Zhang, Q. R. Feng, Z. Z. Gan, H. Yang, Y. Jia, H. H. Wen and X. X. Xi, Ultrahigh current-carrying capability in clean MgB_2 films, *J. Appl. Phys.* **104**, p. 013924 (2008).
70. A. Serquis, G. Pasquini and L. Civale, *Carbon nanotubes addition effects on MgB_2 superconducting properties*. In J. M. Marulanda (Ed.) (2011), Electronic Properties of Carbon Nanotubes. InTech, Rijeka.

71. A. Vajpayee, V. P. S. Awana, S. Yuc, G. L. Bhalla and H. Kishan, Effect of single-walled carbon nano-tube (CNT) addition on superconducting properties of bulk MgB_2 superconductor, *Physica C* **470**, S653–S654 (2010).
72. S. Soltanian, J. Horvat, L. X. Wang, P. Munroe and S. X. Dou, Effect of nano-carbon particle doping on the flux pinning properties of MgB_2 superconductor, *Physica C* **390**, pp. 185–190 (2003).
73. Y. Ma, X. Zhang, G. Nishijima, K. Watanabe, S. Awaji and X. Bai, Significantly enhanced critical current densities in MgB_2 tapes made by a scaleable nanocarbon addition route, *Appl. Phys. Lett*, **88**, pp. 072502–072504 (2006).
74. W. K. Yeoh, J. H. Kim, J. Horvat, X. Xu, M. J. Qin and S. X. Dou, Control of nano carbon substitution for enhancing the critical current density in MgB_2, *Supercond. Sci. Technol.* **19**, p. 596 (2006).
75. B. J. Senkowicz, J. E. Giencke, S. Patnaik, C. B. Eom, E. E. Hellstrom and D. C. Larbalestier, Atmospheric conditions and their effect on ball-milled magnesium diboride, *Appl. Phys. Lett.* **86**, pp. 202502–202505 (2005).
76. C.H Cheng, H. Zhang, Y. Zhao, Y. Feng, X. F. Rui, P. Munroe, H. M. Zeng, N. Koshizuka and M. Murakami, *Supercond. Sci. Technol.* **16**, p. 1 (2003).
77. S. Zhou, A. V. Pan, D. Wexler and S. X. Dou, Sugar coating of Boron powder for efficient carbon doping of MgB_2 with enhanced current–carrying performance, *Advanced Materials* **19(10)**, pp. 1373–1376 (2007).
78. R. Flükiger, M. S. A. Hossain and C. Senatore, Strong enhancement of Jc and Birr in binary in situ MgB_2 wires after cold high pressure densification, *Supercond. Sci. Technol.* **22**, p. 085002 (2009).
79. H. Yamada, M. Hirakawa, H. Kumakura and H. Kitaguchi, Effect of aromatic hydrocarbon addition on in situ powder-in-tube processed MgB_2 tapes, *Supercond. Sci. Technol.* **19**, p. 175 (2006).
80. R. H. T. Wilke, S. L. Bud'ko, P. C. Canfield, D. K. Finnemore and S. T. Hannahs, Synthesis of $Mg(B_{1-x}C_x)_2$ powders, *Physica C* **432**, pp. 193–205 (2005).
81. S. Ueda, J. I. Shimoyama, A. Yamamoto, Y. Katsura, I. Iwayama, S. Horii and K. Kishio, Flux pinning properties of impurity doped MgB_2 bulks synthesized by diffusion method, *Physica C* **426–431**, pp. 1225–1230 (2005).
82. A. Yamamoto, J. I. Shimoyama, S. Ueda, I. Iwayama, S. Horii and K. Kishio, Effects of B_4C doping on critical current properties of MgB_2 superconductor, *Supercond. Sci. Technol.* **18**, pp. 1323–1328 (2005).

83. J. R. Thompson, M. Paranthaman, D. K. Christen, K. D. Sorge, H. J. Kim and J. G. Ossandon, High temporal stability of supercurrents in MgB_2 materials, *Supercond. Sci. Technol.* **14**, p. L17 (2001).
84. J. R. Thompson, K. D. Sorge, C. Cantoni, H. R. Kerchner, D. K. Christen and M. Paranthaman, Vortex pinning and slow creep in high-J_c MgB_2 thin films: a magnetic and transport study, *Supercond. Sci. Technol.* **18**, p. 970 (2005).
85. Y. Bugoslavsky, G. K. Perkins, X. Qi, L. F. Cohen and A. D. Caplin, Vortex dynamics in superconducting MgB_2 and prospects for applications, *Nature (London)* **410**, p. 563 (2001).
86. Hao Jin, H. H. Wen, H. P. Yang, Z. Y. Liu and Z. A. Ren, Rigid vortices in MgB_2, *Appl. Phys. Lett.* **83**, p. 2626 (2003).
87. H. H. Wen, S. L. Li, Z. W. Zhao, H. Jin, Y. M. Ni and W. N. Kang, Magnetic relaxation and critical current density of MgB_2 thin films, *Phys. Rev. B* **64**, p. 134505 (2001).
88. Y. Z. Zhang, R. Deltour, H. H. Wen and C. Q. Jin, Quantum tunneling of vortices in MgB_2 superconductor, *Appl. Phys. Lett.* **81**, p. 4802 (2002).
89. G. Pasquini, A. Serquis, A. J. Moreno, G. Serrano and L. Civale, Carbon nanotubes effects on the relaxation properties and critical current densities of MgB_2 superconductor, *J. Appl. Phys.* **114**, p. 023907 (2013).

Chapter 2a

Synthesis, Substitutions and Properties of MgB_2 Single Crystals

J. Karpinski

Laboratory for Solid State Physics, ETH, 8093 Zurich, Switzerland
Institute of Condensed Matter Physics, EPFL,
1015 Lausanne, Switzerland
karpinski@phys.ethz.ch

Conventional methods of crystal growth do not work for MgB_2. At ambient pressure, solubility of MgB_2 in Mg is too low for single crystal growth from solution. At pressures as high as of 3 GPa, the solubility increases rapidly and single crystals with a size up to $1.5 \times 1 \times 0.1\,\mathrm{mm}^3$ and a weight up to $230\,\mu$g have been grown from flux. Specific band structure of MgB_2 with two bands, π and σ, is strongly influenced by chemical substitutions. Substitutions of Al for Mg and C for B dope electrons, decrease T_c and lead to an increase of scattering within both π and σ bands, however, with different rates for both substituents. Therefore, different changes of the upper critical field, H_{c2}, and its anisotropy, γH_{c2}, for $Mg_{1-x}Al_xB_2$ and $MgB_{2-x}C_x$ are observed. In $MgB_{1.87}C_{0.13}$ single crystal, $H_{c2\|c}(0) \approx 8.5$ T is more than twice as large as that for binary MgB_2. Hole-doping with Li decreases T_c, in much slower rate, than electron-doping with C and Al. Substitution of magnetic Mn ions for Mg strongly suppresses T_c and H_{c2} due to magnetic pair breaking, in contrast to the substitution of Fe ions for Mg, at least for low Fe contents.

1. Introduction

This paper reviews our results previously published in several publications as well as some related results published by other authors. The structure of MgB_2 shown in Fig. 1 contains alternating layers of

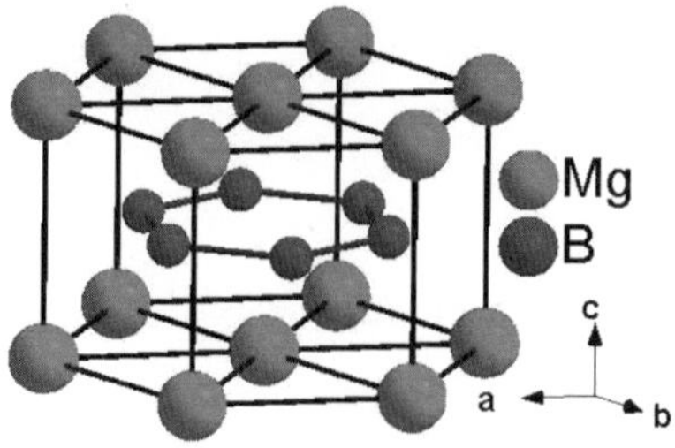

Fig. 1. Structure of MgB_2.

Mg and B. Weakly coupled electronic bands σ and π form quasi-two-dimensional σ sheets and quasi-three-dimensional π sheets of the Fermi surface. There are two energy gaps: $\Delta_\sigma \sim 7\,\text{meV}$ in σ bands and $\Delta_\pi \sim 2\,\text{meV}$ in π bands.[1–11] Strongly anisotropic structural and superconducting properties of MgB_2 require investigations of its intrinsic properties on single crystals rather than on polycrystalline samples with randomly oriented grains. However, it is very difficult to grow single crystals of MgB_2. Due to very low solubility of MgB_2 in metals (Mg, Al, Cu, etc.) or to the formation of other compounds, high temperature solution growth used for other borides is not possible at normal pressure. We have thus grown MgB_2 single crystals using the high pressure cubic anvil technique.[12–15] By substitutions in MgB_2, structure and electronic properties such as T_c, the upper critical fields and their anisotropy, defect structure, gaps width, etc., can be modified.[15–30] By the introduction of C, Al, Li, Mn and Fe in the melt, we were able to substitute partially, Mg or B in MgB_2 crystals and to dope the crystals with electrons or holes. In this paper, the results of crystal growth of pure and substituted MgB_2 crystals as well as investigations of the structure and superconducting properties will be reviewed.

2. Crystal Growth and Thermodynamics of the Mg–B–N System

Up to the boiling temperature of Mg (1107°C) at ambient pressure, the solubility of MgB_2 in Mg is extremely low (Fig. 2(a)). The

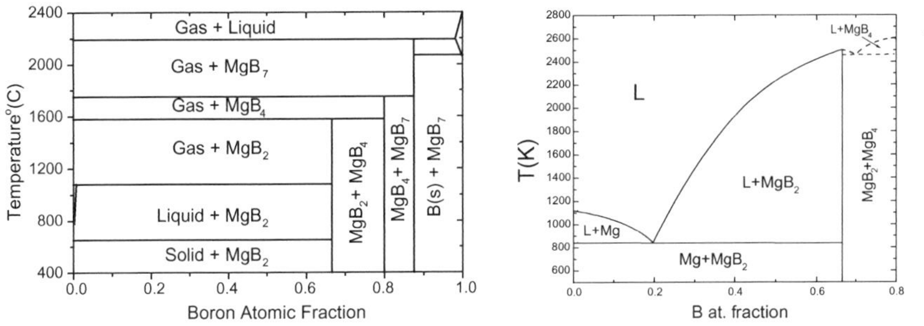

Fig. 2. T–x phase diagram of the Mg–B system at (a) P = 1 bar (from Ref. 31) and (b) p = 3 GPa plotted using our experimental data and some data of Ref. 32 (from Ref. 14). At p = 1 bar it is impossible to grow MgB_2 crystals from the solution in Mg due to low solubility. At P = 3 GPa the solubility increases and MgB_2 crystals can be grown from the solution.

solubility increases at higher temperature, but simultaneously the partial pressure of Mg vapor above molten Mg increases up to about 50 bar at the temperature of crystal growth (1800°C). At 1 bar pressure, MgB_2 decomposes above 1600°C. Therefore, to stabilize the structure high pressure technique should be applied. For the stability of MgB_2 compound at high temperature, both high partial pressure of Mg vapor and high hydrostatic pressure is necessary. The gas phase disappears at high pressure and in equilibrium, only solid and liquid phases remain. At the pressure of 3 GPa used for crystal growth, MgB_2 is in equilibrium with liquid and the solubility of MgB_2 in the Mg melt is very high. According to Turkevich,[32] the phase diagram contains Mg–MgB_2 eutectic with a melting temperature of 840 K. The phase diagram of the Mg–B system obtained from our data and from Ref. 32 is shown in Fig. 2(b). According to this phase diagram, MgB_2 single crystals can be grown from solution as well as from the stoichiometric melt. In the real experiments, samples are placed in BN container, which reacts with liquid Mg at the temperature of crystal growth. Therefore, at high temperatures, we have to consider reactions in the ternary Mg–B–N system, shown in Fig. 3.

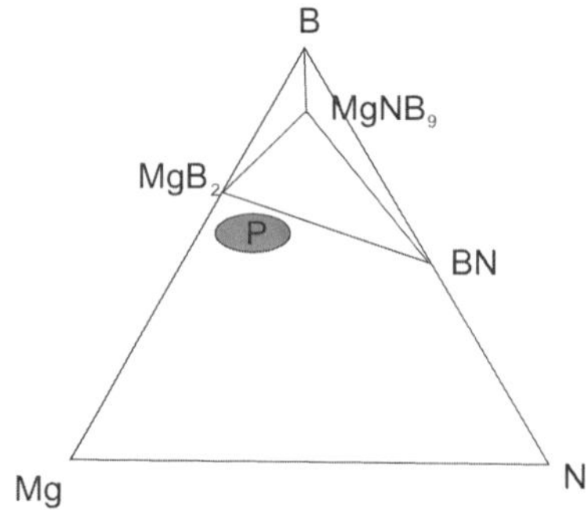

Fig. 3. Ternary Mg–B–N system with compounds involved in the crystal growth of MgB_2. Field P corresponds to the composition of precursor used for crystal growth.

The growth at high pressure from precursor containing Mg, B and BN turned out to be the most successful method to grow single crystals of MgB_2. Several groups have applied this method.[12–14,33–35] In some of our experiments, hexagonal black crystals of new $MgNB_9$ nitride phase[36] were obtained in the crystal growth process.

In our crystal growth experiments, mixture of Mg, B and BN was placed into a BN container of 6–8 mm internal diameter and 7–9 mm length. Pressures in the range of 2–3 GPa were applied using a pyrophyllite cube as a medium; the temperature was increased during one hour, up to the maximum of 1700–2200°C, kept for 0.5–3 h, and decreased during 1–2 h.

Figures 4(a) and 4(b) show the samples after typical cubic anvil growth experiment: a collection of MgB_2 and BN crystals, sticking together. Separate MgB_2 platelike crystals (Figs. 4(c) and 4(d)) were up to $1.5 \times 1 \times 0.2\,\text{mm}^3$ in size and up to $230\,\mu$g in weight. In our experiments, we used compositions with an excess of Mg with respect to stoichiometric MgB_2. Typically, we used a ratio Mg:B equal to 1:1.2.

The *P–T* phase diagram (Fig. 5) shows the phases obtained in our experiments for a composition with excess of Mg. At the conditions above the grey dashed line, MgB_2 plus BN and Mg are stable phases. Below this line, $MgNB_9$ and not MgB_2 is the stable phase for our composition of the samples.

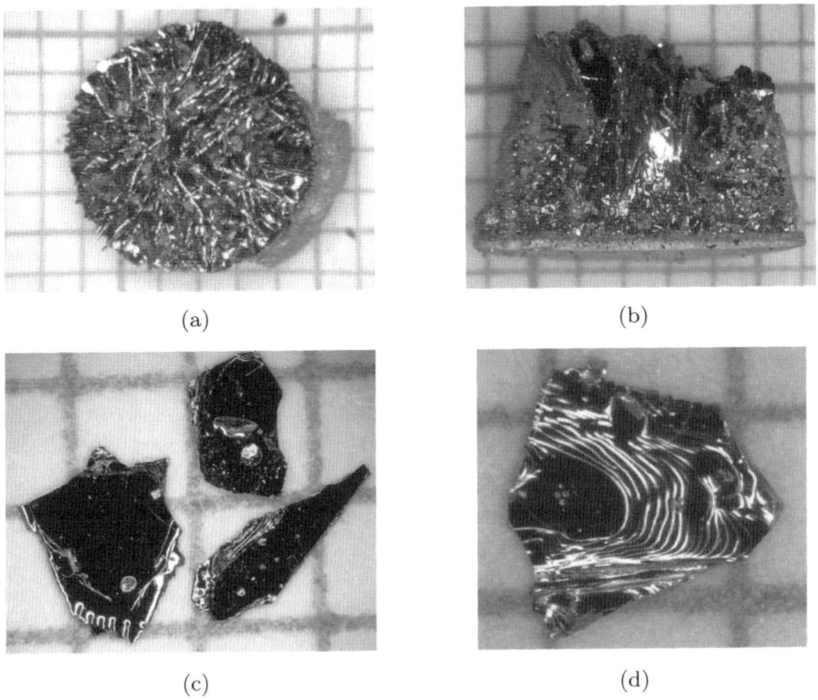

Fig. 4. MgB_2 crystals grown in BN crucible: (a) view from top, (b) view from side, (c) and (d) crystals after separation. Scale is 1 mm (from Ref. 14).

At $T > 1700°C$, sizeable MgB_2 crystals were grown. In the vicinity of the phase boundary (grey dashed line in Fig. 5), both $MgNB_9$ and MgB_2 phases can be found in the samples. We found MgB_2 crystals grown inside of $MgNB_9$ black crystals which can be seen in Figs. 6(a) and 6(b). Schematic drawing (Fig. 6(b)) explains the possible mechanism of crystal growth. MgB_2 crystals grown by decomposition of $MgNB_9$ may serve as seeds for the further growth of larger MgB_2 crystals from solution in Mg. After longer times at high temperature and pressure, $MgNB_9$ disappears and only MgB_2 and BN crystals, sticking together, remain in the crucible. The crystal growth process of MgB_2 is not a simple growth from a solution in molten Mg. MgB_2 crystals are the products of a reaction in the ternary system Mg–B–N.

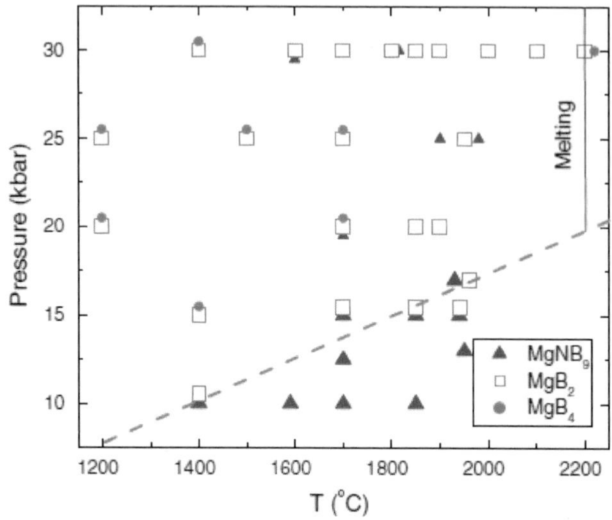

Fig. 5. *P–T* of Mg–B–N system for a composition with excess Mg (Mg:B = 1:1.2). Symbols show phases observed in the samples. MgB_4 appeared in some samples as a secondary phase. BN and Mg were also present in all samples (not shown on diagram). MgB_2 single crystals have been grown above 1700°C. Above the grey dashed line, the $MgNB_9$ compound appears only as a metastable secondary phase (from Ref. 14).

Fig. 6. (a) $MgNB_9$ crystal with MgB_2 crystals growing inside. The schematic drawing (b) explains the possible mechanism of MgB_2+BN crystal growth by decomposition of $MgNB_9$ (from Ref. 14).

3. Substitutions in MgB_2 Crystals

In general, one can expect two kinds of effects by substitutions:

(i) Modification of electronic structure and density of states (DOS) and

(ii) Introduction of defects in the structure, which increases scattering of quasi-particles and eventually causes superconducting pair breaking. Both these effects can lead to change of T_c. We have investigated different kind of substitutions in order to improve our understanding of these phenomena in multiband superconductor:

- Electron and hole-doping.
- Isovalent substitutions.
- Magnetic substitutions.

The results of our investigations about the influence of substitutions in MgB_2 single crystals on T_c are presented in Fig. 7. A short review of the results about the influence of substitutions on crystallographic structure, energy gap, T_c, as well as H_{c2} and its anisotropy is presented below.

3.1 *Carbon (C) substitution*

By the substitution of carbon (C) and aluminum (Al) for B and Mg, respectively, one can dope MgB_2 with electrons. One of the most interesting modifications is C substitution for B. Pure MgB_2

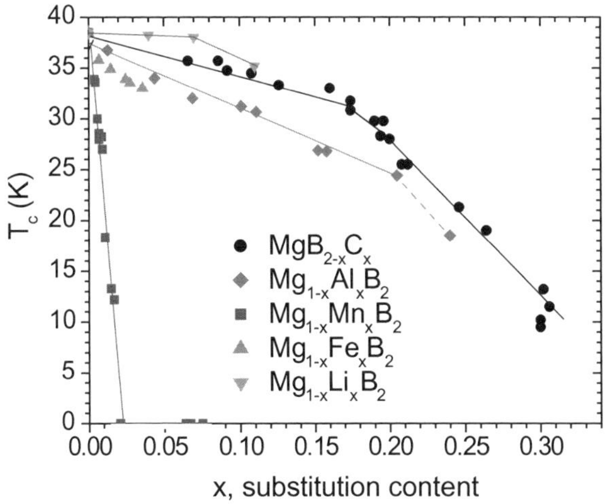

Fig. 7. Critical temperature dependence on substitutions in MgB_2 single crystals.

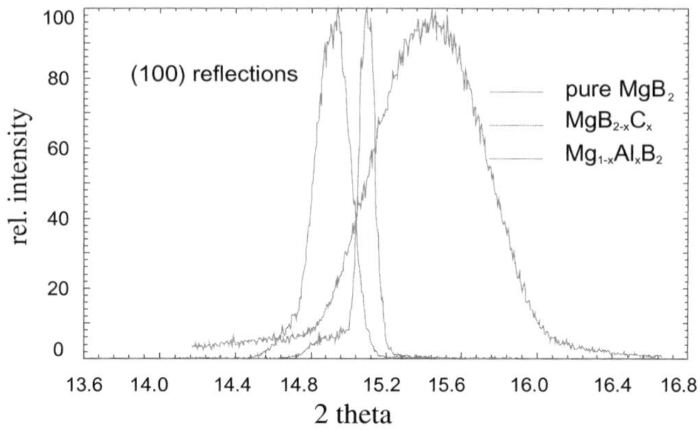

Fig. 8. (100) reflections shape for pure, Al and C substituted MgB_2. Al substitution does not cause broadening of reflections such as C substitution (from Ref. 14).

single crystals have golden color but after C substitution they became black and very brittle. Due to the introduction of defects in the MgB_2 structure, the X-ray reflections are much broader (Fig. 8). Magnetization curves of substituted crystals are rather sharp and do not show multistep transitions, thus suggesting that the length scale of the distribution of such defects caused by inhomogeneity in C distribution should be below the coherence length ξ. A macroscopic phase separation was not observed. Both upper critical fields $H_{c2\|c}$ and $H_{c2\|ab}$ are increased (Fig. 9) and the H_{c2} anisotropy γH_{c2} is decreased in C substituted MgB_2 single crystals.[16–18,21,25] A decrease of the mean free path due to enhanced intraband scattering in the σ band can be the reason of H_{c2} increase.[18,21,25] One may also expect a decrease of the coherence length ξ_{ab} as a result of C substitution in MgB_2. Indeed, ξ_{ab} changes from 10.7 nm for pure MgB_2 to 5.8 nm for 6.6% C substituted MgB_2.[21]

Gonnelli *et al.*,[27] performed point contact spectroscopy investigations on our crystals, showing that both gaps Δ_σ and Δ_π approach each other with increasing C content and merge at 13.5% C, only one gap being observed at higher C contents, thus indicating increased interband scattering.

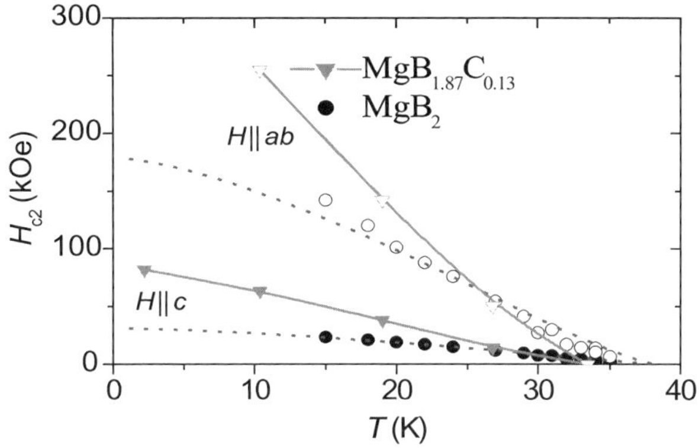

Fig. 9. Upper critical fields in MgB_2 and $MgB_{1.87}C_{0.13}$ crystals determined by torque magnetometry (from Ref. 14).

C substitution leads to electron-doping by additional electrons provided by the C atom. Electron-doping causes a decrease of T_c similarly for C and Al substitutions. The decrease of T_c with Al and C-doping has been attributed to σ hole band filling.[46] C and Al electron dopants reduce the number of holes at the top of the σ band together with the reduction of the electronic density of states, both leading to a decrease of T_c with a similar rate. However, H_{c2} increases with C substitution it decreases while with Al substitution, as will be shown later: this indicates that electron-doping is mainly responsible for T_c changes, whereas the behavior of H_{c2} can only be understood by taking into account the influence of impurities on the electron mean free path due to increased intraband scattering.

3.2 *Aluminum substitution*

Many groups studied substitution of Al for Mg in MgB_2 on poly-crystalline samples[37–41] and single crystals.[19,42,43] Al substitution chemistry is quite complicated because $Mg_{1-x}Al_xB_2$ crystals contain frequently a precipitation of the second phase. We observed a non-superconducting phase $MgAlB_4$, which precipitates as Al rich domains.[19] It was possible to obtain pure phase Al-doped crystals up to 30% Al content only.

As can be seen in Fig. 8, Al substitution does not cause the large broadening of the X-ray reflections observed after C substitution. Al atoms placed on Mg sites cause an out of plane distortion which can increase interband scattering. This may increase the Δ_π gap while decreasing the Δ_σ gap, which eventually leads to merging of both gaps. Slight increase of Δ_π for low Al-doping level has been observed by point contact spectroscopy and scanning tunneling spectroscopy.[44,45] Simultaneously, Δ_σ is significantly reduced, but up to the substitution level of 30% Al both gaps do not merge in our single crystals.

Investigation of upper critical fields gave different results than those for C substituted crystals. Indeed, the upper critical fields do not increase, with the exception of a small increase of $H_{c2\|c}$ for low Al-doping. The variation of H_{c2} as a function of temperature for 2.4% and 9.2% Al content is shown in Fig. 10: $H_{c2\|ab}$ decreases with increasing Al content, while $H_{c2\|c}$ does not change significantly. Therefore, H_{c2} anisotropy γH_{c2} decreases. In the case of C-doping, γH_{c2} decreases with increasing C content, while H_{c2} rises for both field directions. Al substitution for Mg increases scattering rather in

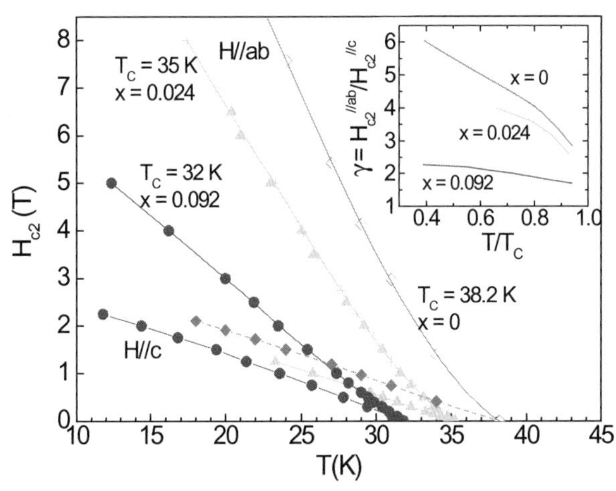

Fig. 10. Upper critical fields H_{c2} determined from magnetic and/or resistance (crosses) measurements for the field parallel to the *ab*-plane (open symbols) and the c (closed symbols) axis for two $Mg_{1-x}Al_xB_2$ crystals with different Al content. The inset shows the anisotropy of H_{c2} for the same crystals (from Ref. 19).

the π band and therefore does not enhance H_{c2}. The main reduction of the field is due to the decrease of T_c.

3.3 *Li hole-doping*

Hole doping in MgB_2 has been investigated very little and only a few published papers describe the system MgB_2:Li.[15,28,29] It was expected that for non-optimally doped hole-doping with Li^{+1}, T_c would increase by increasing the density of states, a decrease of T_c having been observed for electron doped samples with C^{4+} and Al^{3+}-doping. However, our experimental results do not support this prediction. Up to 8% substitution, T_c does not change significantly in Li-doped MgB_2.[15] Figure 7 shows a comparison of the T_c dependence on various substitutions in MgB_2, measured on our single crystals. Since T_c decreases with both electron and hole-doping, i.e., with Al, C and Li substitution, we may assume that pure MgB_2 is close to the optimally doped state. However, the decrease of T_c with hole-doping is much smaller than that one, almost identical, observed on the samples with C and Al substitution. This means that the T_c dependence is non-symmetric as a function of the hole and electron-doping. The strongly non-symmetric change of T_c due to the substitutions may indicate that MgB_2 is slightly under-doped.

3.4 *Magnetic element Mn and Fe substitution*

Manganese substitutes Mg in MgB_2 and decreases the c lattice parameter. The variation of T_c with Mn content and c lattice parameter is shown in Fig. 11.[20] Similarly to Al, T_c scales also well with the lattice parameter c, thus proving that Mn substitutes Mg. T_c is already decreased to zero with the substitution of 2% Mn. This is in contrast with the moderate decrease of T_c with non-magnetic elements such as C or Al shown in Fig. 7. To find out if this is caused by magnetic pair breaking or by electron-doping, one should know the valence state of Mn and the spin configuration of its d electrons. The magnetic state of Mn has been studied by the normal state magnetization and by X-ray absorption (XAS) involving 3D electrons, which unambiguously revealed that Mn in MgB_2 is in

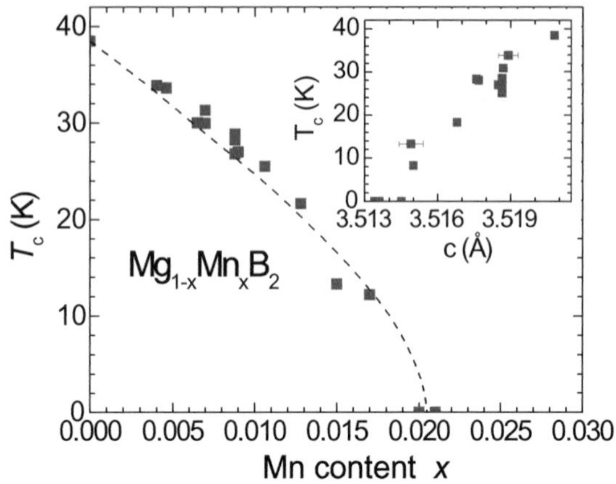

Fig. 11. Suppression of T_c with Mn substitution in $Mg_{1-x}Mn_xB_2$ single crystals. The inset shows T_c versus the lattice parameter c (from Ref. 20).

the divalent state, isovalent with Mg with spin S = 1/2.[20] The suppression of T_c seems to be a pure magnetic pair breaking effect, because essential changes of the electronic structure are not expected, Mn substituting as isovalent Mn^{2+}.

Iron has been substituted in MgB_2 crystals as a second possibly magnetic element. We succeeded to grow single phase, Fe-doped crystals only up to substitution level of 3–4%, Fe having a tendency to form iron rich precipitations in the MgB_2 crystals. Magnetic moment investigations and small changes of T_c for low concentrations of Fe (Fig. 7) suggest that Fe ions in MgB_2 are in the low spin state (S = 0) and that Fe is nonmagnetic in MgB_2, at least up to 3% substitution level.

4. Conclusions

The solubility of MgB_2 in Mg is extremely low at ambient pressure, but increases at high pressures and temperatures. Using the high pressure cubic anvil technique, sizeable single crystals of MgB_2 have been grown from flux in the ternary Mg–B–N system. Substitution effects of C, Al, Li, Mn and Fe on superconducting properties and structure have been studied on single crystals. Both C substitution for B and Al substitution for Mg dope electrons to MgB_2. The decrease of T_c with Al and C-doping has been attributed to σ hole band filling. Hole-doping by Li decreases T_c, but at a much slower rate than electron-doping by C and Al. The magnetic ions (Mn^{2+}) result in a strong suppression of the superconductivity ($T_c = 0$ for 2% of Mn), the main mechanism that controls T_c being magnetic pair breaking. The lack of significant changes in T_c for low Fe contents may suggest that Fe ions in MgB_2 are in the low spin state (S = 0).

Acknowledgments

I would like to thank my past and present collaborators at the Laboratory for Solid State Physics, ETH in Zurich, Switzerland and at the Institute of Condensed Matter Physics, EPFL in Lausanne, Switzerland, without whom it would have been impossible to perform this work.

References

1. J. Nagamatsu, N. Nakagawa, T. Muranaka, Y. Zenitani and J. Akimitsu, *Nature* **410**, p. 63 (2001).
2. J. Kortus, I. I. Mazin, K. D. Belashchenko, V. P. Antropov and L. L. Boyer, *Phys. Rev. Lett.* **86**, p. 4656 (2001).
3. A. Y. Liu, I. I. Mazin and J. Kortus, *Phys. Rev. Lett.* **87**, p. 087005 (2001).
4. H. J. Choi, D. Roundy, H. Sun, M. L. Cohen and S. G. Louie, *Nature* **418**, p. 758 (2002).
5. J. Kortus, O. V. Dolgov, R. K. Kremer and A. A. Golubov, *Phys. Rev. Lett.* **94**, p. 027002 (2005).
6. R. S. Gonnelli, D. Daghero, G. A. Ummarino, V. A. Stepanov, J. Jun, S. M. Kazakov and J. Karpinski, *Phys. Rev. Lett.* **89**, p. 247004 (2002).
7. P. Samuely, Z. Hol'anová, P. Szabó, J. Kačmarčík, R. A. Ribeiro, S. L. Bud'ko and P. C. Canfield, *Phys. Rev. B* **68**, p. 020505 (2003).
8. M. Iavarone, G. Karapetrov, A. E. Koshelev, W. K. Kwok, G. W. Crabtree, D. G. Hinks, W. N. Kang, Eun-Mi Choi, Hyun Jung Kim, Hyeong-Jin Kim and S. I. Lee, *Phys. Rev. Lett.* **89**, p. 187002 (2002).
9. M. R. Eskildsen, M. Kugler, S. Tanaka, J. Jun, S. M. Kazakov, J. Karpinski and Ø. Fischer, *Phys. Rev. Lett.* **89**, p. 187003 (2002).

10. M. R. Eskildsen, N. Jenkins, G. Levy, M. Kugler, Ø. Fischer, J. Jun, S. M. Kazakov and J. Karpinski, *Phys. Rev. B* **68**, p. 100508(R) (2003).
11. M. Angst, R. Puzniak, A. Wisniewski, J. Jun, S. M. Kazakov, J. Karpinski, J. Roos and H. Keller, *Phys. Rev. Lett.* **88**, p. 167004 (2002).
12. J. Karpinski, M. Angst, J. Jun, S. M. Kazakov, R. Puzniak, A. Wisniewski, J. Roos, H. Keller, A. Perucchi, L. Degiorgi, M. R. Eskildsen, P. Bordet, L. Vinnikov and A. Mironov, *Supercond. Sci. Technol.* **16**, p. 221 (2003).
13. J. Karpinski, S. M. Kazakov, J. Jun, M. Angst, R. Puzniak, A. Wisniewski and P. Bordet, *Physica C* **385**, p. 42 (2003).
14. J. Karpinski, N. D. Zhigadlo, S. Katrych, R. Puzniak, K. Rogacki and R. Gonnelli, *Physica C* **456**, p. 3 (2007).
15. J. Karpinski, N. D. Zhigadlo, S. Katrych, K. Rogacki, B. Batlogg, M. Tortello and R. Puzniak, *Phys. Rev. B* **77**, p. 214507 (2008).
16. S. M. Kazakov, R. Puzniak, K. Rogacki, A. V. Mironov, N. D. Zhigadlo, J. Jun, Ch. Soltmann, B. Batlogg and J. Karpinski, *Phys. Rev. B* **71**, p. 024533 (2005).
17. T. Masui, S. Lee and S. Tajima, *Phys. Rev. B* **70**, p. 024504 (2004).
18. S. Lee, T. Masui, A. Yamamoto, H. Uchiyama and S. Tajima, *Physica C* **397**, p. 7 (2003).
19. J. Karpinski, N. D. Zhigadlo, G. Schuck, S. M. Kazakov, B. Batlogg, K. Rogacki, R. Puzniak, J. Jun, E. Müller, P. Wägli, R. Gonnelli, D. Daghero, G. A. Ummarino and V. A. Stepanov, *Phys. Rev. B* **71**, p. 174506 (2005).
20. K. Rogacki, B. Batlogg, J. Karpinski, N. D. Zhigadlo, G. Schuck, S. M. Kazakov, P. Wägli, R. Puźniak, A. Wiśniewski, F. Carbone, A. Brinkman and D. van der Marel, *Phys. Rev. B* **73**, p. 174520 (2006).
21. C. Krutzler, M. Zehetmayer, M. Eisterer, H. W. Weber, N. D. Zhigadlo, J. Karpinski and A. Wisniewski, *Phys. Rev. B* **74**, p. 144511 (2006).
22. R. H. T. Wilke, S. L. Bud'ko, P. C. Canfield and D. K. Finnemore, *Phys. Rev. Lett.* **92**, p. 217003 (2004).
23. V. Braccini, A. Gurevich, J. E. Giencke, M. C. Jewell, C. B. Eom, D. C. Larbalestier, A. Pogrebnyakov, Y. Cui, B. T. Liu, Y. F. Hu, J. M. Redwing, Qi Li, X. X. Xi, R. K. Singh, R. Gandikota, J. Kim, B. Wilkens, N. Newman, J. Rowell, B. Moeckly, V. Ferrando, C. Tarantini, D. Marré, M. Putti, C. Ferdeghini, R. Vaglio and E. Haanappel, *Phys. Rev. B* **71**, p. 012504 (2005); *Phys. Rev. B* **71**, p. 179902(E) (2005).
24. M. Angst, S. L. Bud'ko, R. H. T. Wilke and P. C. Canfield, *Phys. Rev. B* **71**, p. 144512 (2005).

25. A. V. Sologubenko, N. D. Zhigadlo, S. M. Kazakov, J. Karpinski and H. R. Ott, *Phys. Rev. B* **71**, p. 020501(R) (2005).
26. D. Di Castro, M. Ortolani, E. Cappelluti, U. Schade, N. D. Zhigadlo and J. Karpinski, *Phys. Rev. B* **73**, p. 174509 (2006).
27. R. S. Gonnelli, D. Daghero, A. Calzolari, G. A. Ummarino, V. Dellarocca, V. A. Stepanov, S. M. Kazakov, N. Zhigadlo and J. Karpinski, *Phys. Rev. B* **71**, p. 060503 (2005).
28. Y. G. Zhao, X. P. Zhang, P. T. Qiao, H. T. Zhang, S. L. Jia, B. S. Cao, M. H. Zhu, Z. H. Han, X. L. Wang and B. L. Gu, *Physica C* **361**, p. 91 (2001).
29. S. Y. Li, Y. M. Xiong, W. Q. Mo, R. Fan, C. H. Wang, X. G. Luo, Z. Sun, H. T. Zhang, L. Li, L. Z. Cao and X. H. Chen, *Physica C* **363**, p. 219 (2001).
30. G. J. Xu, J.-C. Grivel, A. B. Abrahamsen, X. P. Chen and N. H. Andersen, *Physica C* **399**, p. 8 (2003).
31. Z.-K. Liu, D. G. Schlom, Q. Li and X. X. Xi, *Appl. Phys. Lett.* **78**, p. 3678 (2001).
32. V. Z. Turkevich, O. G. Kulik, P. P. Itsenko, A. N. Sokolov, A. N. Lutsenko and A. N. Vashchenko, *J. Superhard Materials* **25** pp. 6–10 (2003).
33. S. Lee, A. Yamamoto, H. Mori, Yu. Eltsev, T. Masui and S. Tajima, *Physica C* **378–381**, pp. 33–37 (2002).
34. S. Lee, *Physica C* **385** pp. 31–41 (2003).
35. K. H. P. Kim, C. U. Jung, B. W. Kang, K. H. Kim, H.-S. Lee, Sung-Ik Lee, N. Tamura, W. A. Caldwell and J. R. Patel, *Current Appl. Phys.* **4** pp. 272–275 (2004).
36. A. Mironov, S. M. Kazakov, J. Inn and J. Karpinski, *Acta Cryst. C* **58** p. i95 (2002).
37. J. Q. Li, L. Li, F. M. Liu, C. Dong, I. Y. Xiang and Z. X. Zhao, *Phys. Rev. B* **65** p. 132505 (2002).
38. R. J. Cava, H. W. Zandbergen and K. Inumaru, *Physica C* **385**, p. 8 (2003).
39. H. W. Zandbergen, M. Y. Wu, H. Jiang, M. A. Hayward, M. K. Haas and R. J. Cava, *Physica C* **366**, p. 221 (2002).
40. A. Cooley, J. Zambano, A. R. Moodenbaugh, R. F. Klie, J.-C. Zheng and Y. Zhu, *Phys. Rev. Lett.* **95**, p. 267002 (2005).
41. M. Putti, C. Ferdeghini, M. Monni, I. Pallecchini, C. Tarantini, P. Manfrinetti, A. Palenzona, D. Daghero, R. S. Gonnelli and V. A. Stepanov, *Phys. Rev. B* **71**, p. 144505 (2005).
42. T. Klein, L. Lyard, J. Marcus, C. Marcenat, P. Szabo, Z. Holanova, P. Samuely, B. W. Kang, H.-J. Kim, H.-S. Lee, H.-K. Lee and S.-I. Lee, *Phys. Rev. B* **73**, p. 224528 (2006).

43. A. V. Sologubenko, N. D. Zhigadlo, J. Karpinski and H. R. Ott, *Phys. Rev. B* **74**, p. 184523 (2006).
44. D. Daghero, D. Delaude, A. Calzolari, M. Tortello, G. A. Ummarino, R. S. Gonnelli, V. A. Stepanov, N. Zhigadlo, J. Karpinski and A. Palenzona, cond-mat/0608029.
45. C. Dubois, N. Jenkins, A. A. Manuel, O. Fischer, N. D. Zhigadlo and J. Karpinski, *Supercond. Sci. Technol.* **19**, pp. 695–698 (2006).
46. F. Bernardini and S. Massidda, *Phys. Rev. B* **74**, p. 014513 (2006).

Chapter 2b

Thin Film Deposition and Critical Fields

M. A. Wolak* and X. X. Xi

Department of Physics, Temple University
1925 N. 12th Street, Philadelphia PA 19122, USA
**matthaeus.wolak@temple.edu*

MgB_2 thin films remain an active area of research due to the potential of this material for numerous technological applications. In this chapter, we discuss the various deposition techniques used for MgB_2 in light of the key requirements for high quality MgB_2 thin films. While clean MgB_2 thin films have shown excellent superconducting properties characterized by high transition temperature and low residual resistivity, carbon-doping and particle irradiation are effective in increasing the upper critical field. In carbon (C) alloyed MgB_2 films prepared by hybrid physical–chemical vapor deposition (HPCVD), $B_{c2}(0)$ values well over 60 T have been obtained. It was also shown that the lower critical field B_{c1} of MgB_2 thin films could be enhanced through the reduction of the film thickness, which is important for the application of MgB_2 in superconducting radio frequency (SRF) cavities, in particular given its high thermodynamic critical field B_c.

1. Introduction

Soon after the discovery of superconductivity in MgB_2, Kang *et al.*, reported MgB_2 films with a high T_c of 39 K prepared by annealing an amorphous B precursor film in Mg vapor at 900°C.[1] The immediate attention to thin films was due to their importance for various technological applications as well as fundamental studies of MgB_2. MgB_2 thin films have played an important role in the investigation of two-band superconductivity[2] and have been explored in applications including superconducting devices and circuits,[3] coated conductors wires,[4,5] and superconducting radio frequency (SRF) cavity coatings.[6–8] Various deposition techniques have been adapted for the growth of MgB_2 thin films. Employing Mg in a vacuum deposition process is not an easy task as it is volatile and reacts strongly with oxygen (O_2).[9] The deposition of high purity MgB_2 films has to be performed either in an ultrahigh vacuum or in a reducing atmosphere, a high Mg vapor pressure being needed during the growth process. Very clean Mg deposition sources have to be used to prevent contamination. Extensive research has led to high quality MgB_2 films with characteristics matching or even exceeding those of bulk single crystals. The T_c value in clean epitaxial MgB_2 thin films prepared by hybrid physical–chemical vapor deposition (HPCVD) is almost 42 K, higher than that in the bulk.[10] This was achieved as a result of tensile strain formation due to the epitaxial growth process. This kind of strain engineering is an inherent advantage for MgB_2 thin films in comparison to their bulk counterparts.

It was also observed that MgB_2 films can exhibit considerably higher critical fields than bulk MgB_2. In C alloyed MgB_2 films prepared by HPCVD, $H_{c2}(0)$ values well over 60 T have been obtained, surpassing the bulk value, which continues to be of interest to researchers worldwide.[11,12] In this chapter, we review various deposition processes used for MgB_2 film growth and discuss the critical fields of MgB_2 thin films.

2. Deposition Methods

The volatile nature of Mg complicates the deposition of MgB_2, which decomposes easily at high temperatures. Deposition at low temperatures is feasible and is often employed, however, the crystal quality and the superconducting properties of these films are not optimal. Higher deposition temperatures are preferred for better film quality, however, these have to take place at a sufficiently high Mg vapor pressure in order to keep the stoichiometric MgB_2 phase thermodynamically stable.[9] The high temperatures employed for *in situ* epitaxial growth of MgB_2 necessitate a large Mg flux.

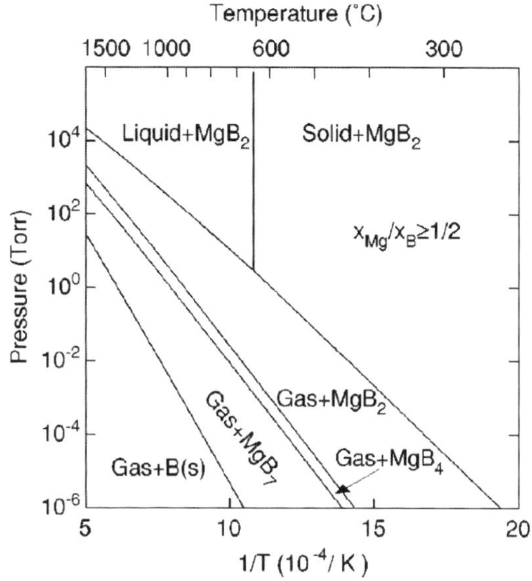

Fig. 1. Pressure–temperature phase diagram for the atomic ratio Mg: B $\geq$ 1:2. Thermodynamic stability of MgB_2 is only achieved in the region marked "Gas + MgB_2." (Reproduced with permission from Ref. 9.) © AIP Publishing.

Theoretical calculations show that depositing MgB_2 at temperatures of more than 750°C requires a Mg vapor pressure of more than 44 mTorr. As a result, high temperature growth is not possible for deposition processes where a large Mg flux is unfeasible.

The pressure–temperature phase diagram for an atomic ratio Mg:B of $\geq$ 1:2 in Fig. 1 shows that MgB_2 does not decompose and excess Mg does not condense on the MgB_2 surface only when the deposition is performed within the "Gas+MgB_2" window. In this case, the formation of single-phase MgB_2 is adsorption-controlled and automatic. An over-saturation with Mg is mostly preferred as it allows for complete reaction of B gas with Mg and any excessive Mg can be easily evacuated in its gas phase. It is highly important to avoid insufficient Mg supply, which will result in non-superconducting phases such as MgB_4 and MgB_7, or the solid B phase.

The prevention of O_2 contamination during film growth is extremely important, as the oxidation of Mg can diminish the actual Mg vapor pressure, leading to reduced film quality. Hence, deposition in either ultrahigh vacuum or in a reducing atmosphere is favorable. Contamination with C can significantly lower the critical temperature T_c.[13] Such contaminations are often encountered when using tainted source material. Nevertheless, both O_2 as well as C can considerably enhance the upper critical field of MgB_2 films and can be successfully used as dopants.[14,15]

2.1 *Two-step process*

The first approach developed for the deposition of MgB_2 was the two-step process, resulting in MgB_2 thin films showing the T_c value of 39 K.[1] In this process, the deposited precursor film consists either of B or an Mg–B mixture, which is then annealed at high temperatures in Mg vapor. While the annealing of B precursor films has to be performed in an enclosure in the presence of Mg vapor, Mg–B mixture precursor films can be optionally annealed without additional Mg. The deposition of the precursor film can be achieved by means of a wide range of commonly employed deposition processes.

2.1.1 *Annealing of B precursor films*

B precursor films can be grown by various deposition techniques. Pulsed laser deposition,[1,16] sputtering[17] thermal evaporation[18] and e-beam evaporation[19,20] have all been successfully used to achieve this task. The B precursor films have to be annealed in a high Mg vapor environment for MgB_2 formation. A common way is to seal the B films with high-purity Mg pieces in an enclosure made from Ta, Nb, Mo or quartz in an Ar atmosphere to prevent contamination of the films. The annealing takes place at elevated temperatures, often above 800°C. This step can take several hours to complete and higher temperatures allow for shorter annealing times. Films deposited in this manner can result in critical temperatures T_c of up to 39 K.

The purity of the B source is important for the quality of the MgB_2 films. Further, it is preferred that the deposition takes place in ultrahigh vacuum or in an inert atmosphere to reduce

O_2 contamination. A chemical vapor deposition (CVD) process can combine both requirements by using hydrogen (H) as a carrier gas and diborane (B_2H_6) as the B source. With this method, B films of high quality have been obtained.[21] Diborane decomposes readily at elevated temperatures, providing a very clean source of boron. MgB_2 films formed by annealing CVD deposited B films displayed a sharp superconducting transition at 40 K with a low residual resistivity of 1.25 $\mu\Omega$cm.

A novel method of depositing boron films for fabricating flexible MgB_2–CNT yarns has been recently demonstrated by.[22] In the described method, a CNT sheet was continuously exposed to BBr_3 in H atmosphere, while it was also locally heated with an external laser. The laser induced heating leads to temperatures up to 1670°C, providing crystalline B and hydrobromic acid vapor through a reaction of the BBr_3 and H on the surface of the CNT sheet. A subsequent annealing of the boron–coated CNT sheets Mg vapor led to MgB_2–CNT networks with critical temperatures of up to 37.8 K, providing flexible, porous yarns with better characteristics than bulk pure-MgB_2 superconducting wires.

The B precursor film can also be deposited by non-vacuum techniques. Kuhberger *et al.*,[22] screen-painted a boron–terpineol paste on the substrates in the ambient atmosphere, which were then dried at 300°C for 3 h, resulting in a uniform B film. The film was subsequently annealed in an enclosure with Mg pieces, leading to MgB_2 thin films with T_c of 35 K. This approach can be used for large-scale B film deposition in ambient atmosphere without needing a vacuum equipment, although the annealing of large films in Mg vapor is challenging.

2.1.2 *Annealing of Mg–B mixture precursor films*

MgB_2 films can also be obtained through the annealing of an Mg–B mixture precursor film, which can be deposited either from a single source containing Mg and B or from separate Mg and B sources. The precursor films can either consist of a mixture of both materials or multilayers of Mg and B which have been subsequently deposited.

Due to the presence of Mg in the precursor film, the annealing does not need to be conducted in an Mg vapor environment. It can be performed *in situ* in the vacuum chamber following the deposition of the mixture precursor film.

The precursor films are typically deposited at substrate temperatures up to 300°C employing various deposition methods including pulsed laser deposition,[14,16,23–28] sputtering,[29–31] thermal evaporation[18,32] and e-beam evaporation.[33,34] Numerous annealing processes have been attempted. While the *in situ* post-annealing in Ar atmosphere is fairly simple, the effective Mg vapor pressure changes with time, making the control of the annealing process difficult. Annealing the mixture precursor films in an enclosure with high-purity Mg pieces provides MgB_2 films of higher quality, with T_c up to 39 K.

Similar to the screen-printing of B films, a sol–gel technique has been developed for the deposition of Mg–B precursor film mixtures in ambient atmosphere.[35] These precursor film mixtures are spin-coated from a $Mg(BH_4)_2 \cdot Et_2O$ solution, dried for 10 min at 140°C and subsequently annealed in a quartz tube with Mg pieces for 15 min at 700°C. The resulting MgB_2 films showed T_c of above 34 K.

2.2 *In situ deposition: reactive co-evaporation (RCE), molecular beam epitaxy (MBE), sputtering, HPCVD*

Several deposition techniques allow *in situ* growth of MgB_2 films in a single step, without the need for further annealing. While most of these techniques are based on standard deposition processes, some were established specifically for the deposition of MgB_2.

2.2.1 *RCE*

RCE reported by Moeckly and Ruby[36] is a very successful physical vapor deposition process for the growth of MgB_2. It utilizes a 'pocket heater' with a rotating substrate holder placed inside the pocket, where a high Mg vapor pressure is provided by a thermally heated Mg source. An opening of the pocket is exposed to an e-beam evaporator from which B is being evaporated and deposited on the substrate. During the deposition, the substrate is being rotated, alternatively exposing the substrate to the B deposition source and the high Mg vapor pressure. The deposition of the boron layer is followed by the reaction with Mg to form MgB_2 when the substrate is rotated to the inside of the pocket heater, and the cycle repeats itself. The substrate holder rotates at several hundred RPMs during deposition, leading to an alternate B deposition/Mg annealing cycle at a very small thickness scale. Very high quality MgB_2 films have been reported using this technique, and films with a T_c of over 39 K have been obtained at substrate temperatures of 550°C.

Several variations of the RCE technique have been developed by other groups, with the main difference being the types of boron sources. Instead of employing e-beam evaporation, Wang *et al.* used a CVD process with B_2H_6 precursor in a H_2 carrier gas.[37] Very clean MgB_2 films were grown at higher substrate temperatures than regular RCE. Homogenous films on substrates of two inches in diameter were achieved with T_c of about 40 K. The same process has been successfully employed by Nishiyuki *et al.* using $B_{10}H_{14}$ as the B source [see Fig. 2(a)]. $B_{10}H_{14}$ is solid at ambient temperatures and stable in air. When heated, it decomposes between approximately 350°C and 415°C. The products of the decomposition are B and H, providing a clean B source. During the deposition, $B_{10}H_{14}$ is vaporized outside the vacuum chamber at 150–160°C and fed into the deposition system through stainless steel tubing. As can be seen in Fig. 2(b), the MgB_2 films deposited with this technique show T_c up to 39 K.[38] It is safer to use $B_{10}H_{14}$ as the B source than the toxic and highly explosive B_2H_6.

2.2.2 *MBE*

MgB_2 films have been deposited *in situ* using MBE. The MgB_2 thin films are fabricated at low substrate temperatures to avoid the need for a high Mg vapor pressure. This is possible because the UHV background pressure of better than 10^{-8} mmHg in the MBE systems

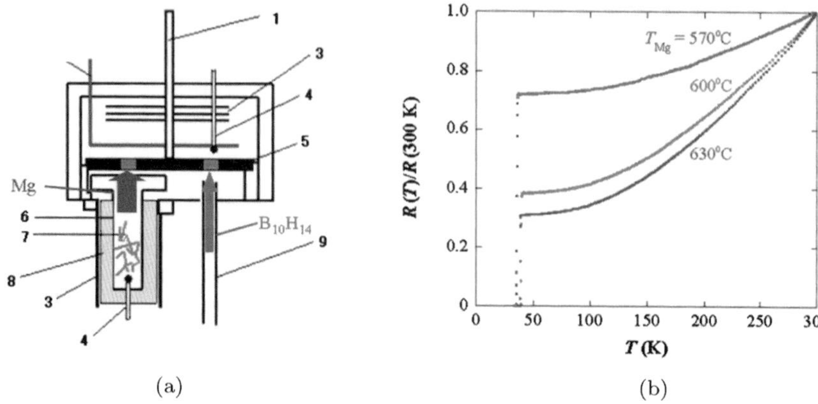

Fig. 2. (a) Schematic of the reactive coating technique employed by Nishiyuki *et al.*[38]; 1: axis of rotation, 2: the substrate heater, 3: radiation shields, 4: thermocouple, 5: substrate holder, 6: Mg cell, 7: Mg source, 8: Mg source heater and 9: $B_{10}H_{14}$ gas supply through stainless steel tubing. Panel (b) shows the resulting R–T curves for samples grown at various substrate temperatures. (Reproduced from Ref. 38). © Elsevier, 2011.

prevents O_2 contamination of the films. Pure Mg metal sources are employed to avoid C contamination. Superconducting MgB_2 films deposited with MBE at temperatures between 150°C and 320°C exhibit critical temperatures of up to 36 K.[39,40] In order to achieve high quality MgB_2 films with MBE, the composition of the film has to be precisely controlled to ensure a perfect stoichiometry. The stoichiometry is further dependent on the deposition temperature as the sticking coefficient of Mg changes with temperature. One of the most prominent characteristics of MBE fabricated MgB_2 thin films is their surface morphology. The low deposition temperature and the slow growth rate of MBE result in extremely smooth MgB_2 films, which is beneficial for device fabrications.

2.2.3 *Sputtering*

Sputter deposition process has also been effectively used to deposit MgB_2 films *in situ*. Even though sputtering had been already used in two-step processes, its use for *in situ* MgB_2 growth was challenging.

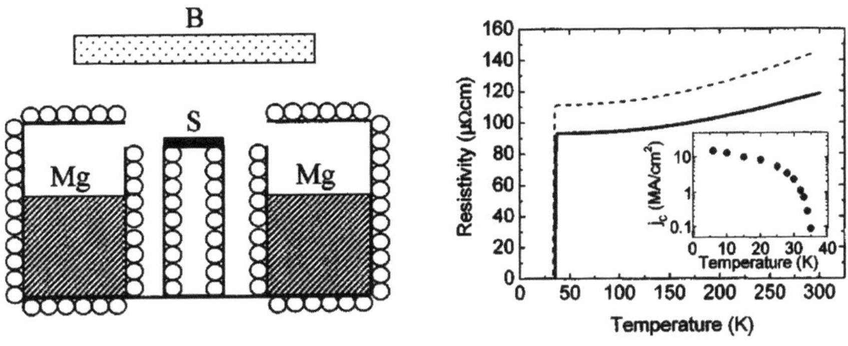

(a) (b)

Fig. 3. (a) Cross-section schematic of a sputter system with a Mg vapor source, a substrate heater (S), and a B sputter target (Ref. 42). © from Elsevier, 2005. (b) The temperature-dependent resistivity curve of a 37 nm thick as-grown (dashed line) in comparison to a 63 nm thick *in situ* post annealed (solid line) MgB_2 film on r-plane sapphire.

While a co-sputtering process using a multitarget carousel holder has been employed[41] for substrate temperatures below 300°C, films with good properties require a higher deposition temperature and high Mg vapor pressure.

Schneider *et al.*[42] provided a novel approach combining a Mg oven for thermally evaporating pure Mg pieces and sputtering from a B target (as shown in Fig. 3). With this combination, substrate temperatures of up to 440°C were reached, resulting in a substantial improvement of the film quality. Substrate temperatures above 440°C were difficult to realize because the necessary Mg vapor pressure is too high for the setup to produce.

2.2.4 *HPCVD*

Amongst all the MgB_2 deposition techniques, HPCVD remains the most effective.[43,44] In this technique, Mg is thermally evaporated from bulk Mg pieces while B is provided by the decomposition of the precursor gas B_2H_6. The deposition takes place in a high H_2 pressure of up to 100 Torr, making it possible to have a high Mg vapor pressure of the same magnitude.

In the original setup, a susceptor is used to heat both the substrate and the Mg pieces placed next to the substrate. The technique provides a high Mg vapor pressure, creating a thermodynamically stable environment for MgB_2 at the deposition temperature. Oxidation of Mg is suppressed through a reducing H atmosphere during deposition. The growth is regulated by the flow rate of the B_2H_6 precursor gas and is terminated once the gas is turned off.

The reducing H_2 environment and the use of high purity Mg and B sources allow the deposition of very clean epitaxial MgB_2 films.[45] The films can reach a T_c of over 40 K and exhibit a low residual resistivity ρ_0 of less than 0.1 $\mu\Omega$cm.[43] HPCVD films display a superior purity in comparison to single crystals[47] and bulk samples.[48] The higher T_c value is a result of strain induced E_{2g} phonon softening through an epitaxial growth induced strain which enhances the electron–phonon coupling.[10] The highest T_c is almost 42 K recorded for films deposited on (0001) SiC substrates.

The original HPCVD setup uses an inductively heated susceptor (see Fig. 4(a)) and can produce MgB_2 films no larger than 1 cm^2 in size. In order to achieve uniform coating on larger samples, several adjustments were made in the modified versions. In an inductively heated susceptor, the temperature is higher at the edge

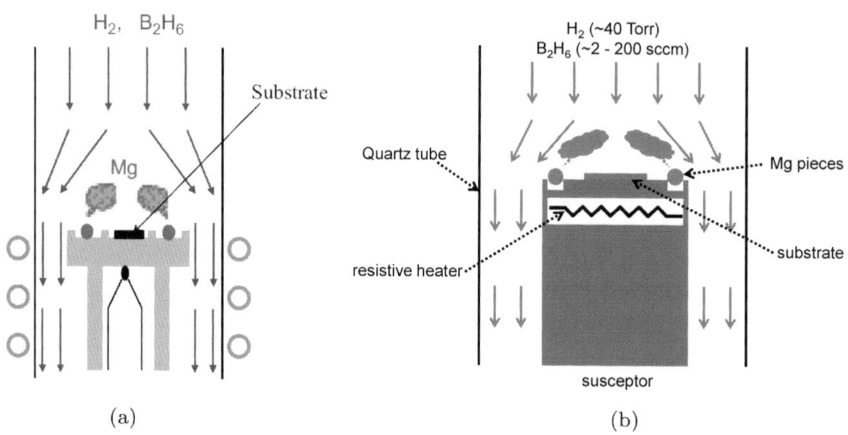

Fig. 4. Schematic of (a) a inductively heated HPCVD system (from Ref. 46). © from Elsevier, 2007, and (b) a resistively heated HPCVD system.

than in the center of the susceptor. For a larger scale setup, resistive heating is preferred, which can provide a more even heating of the susceptor. Achieving a sufficient Mg vapor pressure is one of the key requirements of the HPCVD process and in order to realize that at a larger scale, it is necessary to keep the center of the susceptor slightly cooler than its edge and drive Mg vapor from the edge towards the center. This task can be accomplished with a resistive heater by adjusting the radial density of the heating element in the heater coil. Figure 4(b) shows a typical HPCVD setup with a resistively heated susceptor. High quality MgB_2 films on substrates of up to 2" in diameter have been fabricated using this setup.[49] The susceptors described above act both as the substrate heater and the Mg source. As a result, the Mg partial pressure is ultimately linked to the substrate temperature. To widen the parameter space for MgB_2 thin film depositions, a separation of the Mg source temperature and substrate temperature is necessary.

Several modified HPCVD configurations have been developed. In one approach,[50] an inductively heated ring-shaped Mg reservoir surrounds the water-cooled B_2H_6 gas injection tube. The substrate was heated resistively and mounted facing down, above the B_2H_6 injection tube and Mg oven. This method resulted in the MgB_2 films growth at the very high rates of up to 2 μm/min. In another setup, a resistively heated Mo tube was used as the Mg oven, which was placed facing upwards. B_2H_6 gas was supplied through a water-cooled stainless steel tubing running parallel to the tube oven. The resistively heated substrate was mounted facing down above the oven. With this setup, MgB_2 films were grown at various substrate temperatures ranging from 350°C to 750°C. Films deposited at 620°C displayed the highest T_c of 40.8 K.

Another variation of the HPCVD process has been employed to coat the interior of cylindrical objects with the prospects of depositing a MgB_2 layer on the inside of a RF cavity.[8] The HPCVD setup was modified as shown in Fig. 5(a). Mg is being evaporated from a tube shaped oven located at the bottom of the cavity, while B is supplied through a water cooled line from the top. The cavity is heated to the appropriate deposition temperature by clam shell

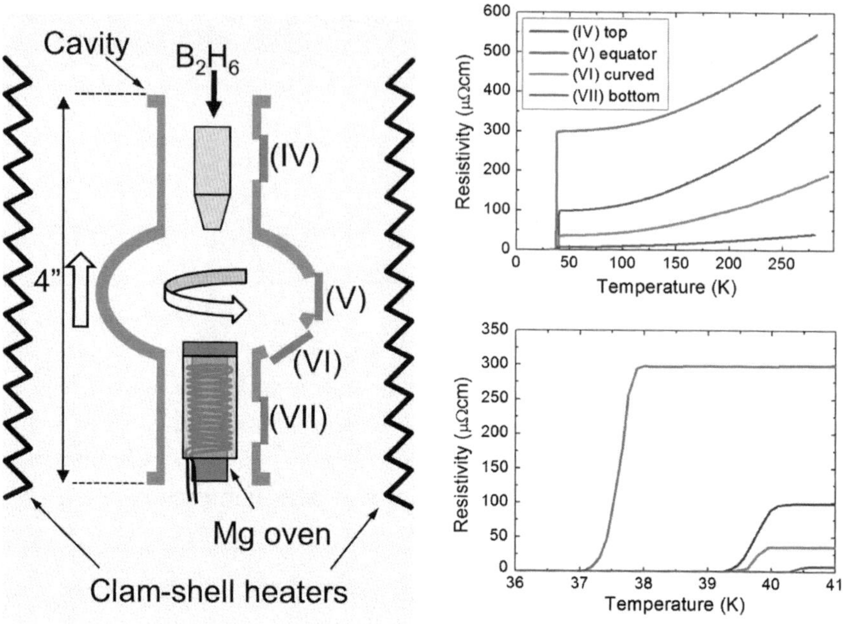

(a) (b)

Fig. 5. (a) Schematic representation of the HPCVD system used for coating of the RF cavity (b) resistivity versus temperature curves. (Reproduced from Ref. 8). © from *Phys. Rev. Spec Top-Accel. Beams.*

heaters, which enclose it completely. During the process, the cavity is moved upwards, leading to a deposition of MgB_2 on the interior of the cavity through the reaction of Mg vapor and decomposing B_2H_6 gas. To test the coating of a stainless steel mock cavity, several sapphire samples were mounted at different locations as shown in the figure, facing the inside of the mock cavity. MgB_2 films were successfully coated with these substrates and displayed relatively high T_c of above 37 K, as shown in Fig. 5(b).

3. Upper Critical Fields

The upper critical field B_{c2} of a type II superconductor is a critical material parameter for its applications in magnetic fields. At H_{c2}, the number of vortices in the superconductor is such that their normal cores overlap and no superconductivity can be found anymore.

A decrease in the electron mean free path reduces the size of the vortices, thus causing the number of vortices for their normal cores to overlap and therefore increasing B_{c2}. Doping the superconductor with impurities is an effective way to reduce the electron mean free path, thus to enhance the B_{c2} value.[52]

The B_{c2} value of MgB_2 single crystals is low, around 5 T for $H//c$ and 20 T for $H//ab$ at low temperature.[2] The temperature dependence of B_{c2} for a clean MgB_2 film deposited by HPCVD is shown in Fig. 6. Its values in both field directions are slightly higher than in bulk MgB_2 samples.

The inclusion of impurities in the two-band superconductor MgB_2 leads to increases in B_{c2} at a much larger extent than in single-band superconductors due to the modifications of both interband and intraband scattering.[53] Doping or alloying with carbon has been the most effective in increasing B_{c2}. In thin films fabricated by HPCVD, alloying with C has led to the largest enhancement of B_{c2} observed in MgB_2, to above 60 T at low temperature in parallel fields.[51] In the HPCVD process, doping or alloying is achieved by mixing C-containing precursor gas with the carrier gas during the deposition, and the C content in the films can be easily controlled.

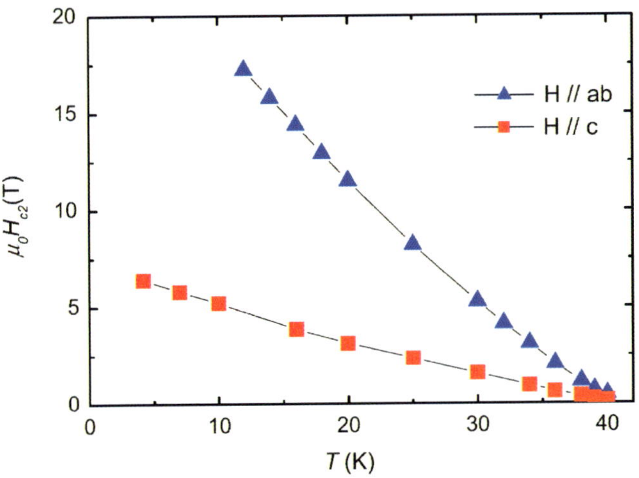

Fig. 6. H_{c2} plotted against temperature for a clean HPCVD MgB_2 film (from Ref. 53).

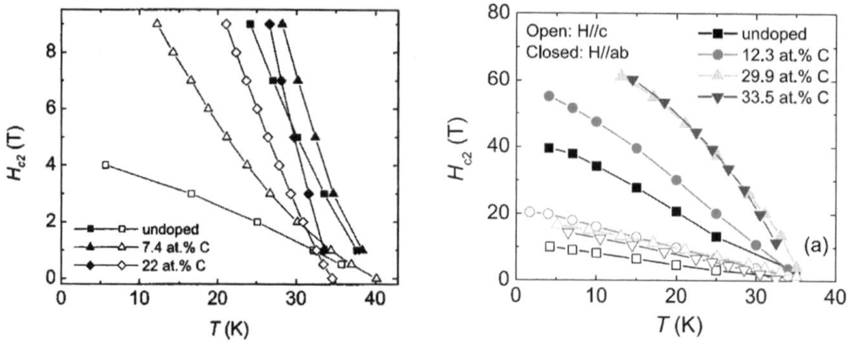

(a) (b)

Fig. 7. HPCVD films doped by C using (a) $(MeCp)_2Mg$ (from Ref. 56) and (b) TMB (from Ref. 11) precursors. Open symbols are for $H//c$ and closed symbols are for $H//ab$.

Several C-containing precursor gases have been successfully used.[54,55] Figure 7 shows the variation of B_{c2} versus T of HPCVD films doped with different concentrations of C using (a) $(MeCp)_2Mg$[56] and (b) trimethylboron (TMB)[55] as the C sources. The films doped using TMB show low temperature B_{c2} in parallel field well over 60 T. This high B_{c2} in C-doped MgB_2 films is considerably higher than those achieved in bulk C-doped MgB_2 samples.

Understanding the mechanism of the large B_{c2} enhancement in C-doped HPCVD films has been a focus of research as it may offer clues to similar effects in practical MgB_2 high-field conductors. X-ray diffraction measurements of the films showed a significantly different dependency of the lattice constants on the C concentration from bulk samples. In bulk samples, the a-axis lattice constant usually decreases while the c-axis lattice constant remains nearly unchanged with increasing C concentration.[15,57] For C-alloyed HPCVD films, however, the c-axis and a-axis constants both expand until about a nominal C concentration of 30 at.% is reached, at which point the film structure changes significantly.[10] Transmission electron microscopy study showed rotations about the c-axis and small tilts of the c-axis with a grain size of 5–20 nm in such films, pointing to extensive structural disorder. The very high B_{c2} in C-alloyed HPCVD films could therefore be a result of strong electron scattering from the nanoscale disorder.[58] Zhuang *et al.*[54] have shown that there is a correlation between the grain size and the improved high-field properties.

O_2 has also been suggested as the impurity that causes the enhancement of B_{c2} in MgB_2 thin films.[52] The high B_{c2} of 49 T in fiber-textured MgB_2 films with high O_2 content produced using the two-step process[14] was attributed to the O impurity or MgO precipitates in the films. It has been suggested that doping with O_2 leads to a noticeable buckling of the Mg planes along the c-axis. This results most likely in a strong out-of-plane π-band scattering, causing the increase in B_{c2}.[52] A post-growth annealing in O_2 process has been used by Singh *et al.* in order to systematically incorporate O_2 in MgB_2 films.[59] With this approach, an enhancement of B_{c2} with O_2 doping was achieved.

Particle irradiation has also been employed to improve the high-field properties of MgB_2 films. Irradiation with heavy ions,[60] α particles[61,62] and neutrons[63,64] can lead to a disorder in the film. The disorder strongly depends on the type, energy and fluence of the particles. It has been shown that this approach can significantly enhance B_{c2}[62,64] of MgB_2 films. In Fig. 7, temperature dependence of B_{c2} of HPCVD films irradiated by O_2 ions are plotted, showing an enhancement of the temperature slope with O_2 ion fluence (from Ref. 65).

The enhancement of B_{c2} in C-doped MgB_2 films has also been observed after coatings of fibers. Figure 9(a) shows the result of SiC fibers coated with polycrystalline MgB_2 by HPCVD.[4] The clean coating showed a low B_{c2} of 20 T at low temperatures, which was increased to 55 T in C-alloyed coating. Figure 9(b) shows the result for the undoped HPCVD coating on Hastelloy tapes.[66] Their low B_{c2} is consistent with the clean MgB_2 films. It can be expected that C-doping will enhance B_{c2} in these coated conductor wires and tapes, which can be explored for various high-field applications.[67] Superconducting MgB_2–CNT yarns recently showed a low H_{c2} similar to that of SiC fibers coated with MgB_2.[22] However, this novel approach provides extremely flexible superconductors which could be employed in a broad range of applications. Further more, the porous

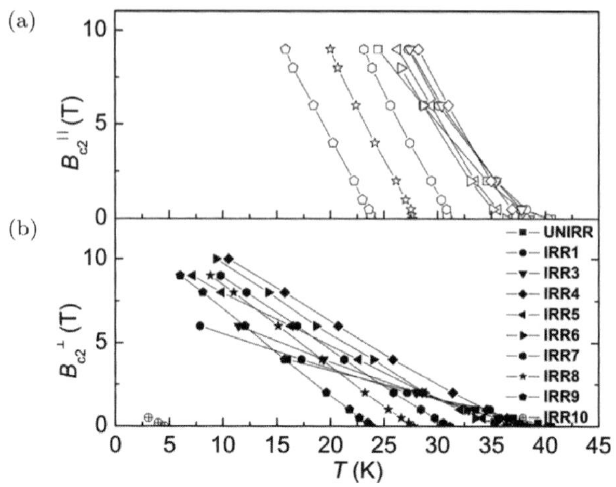

Fig. 8. Temperature dependence of B_{c2} of HPCVD films in (a) parallel field and (b) perpendicular field after irradiation by O_2 ions (from Ref. 65). © from AIP Publishing LLC.

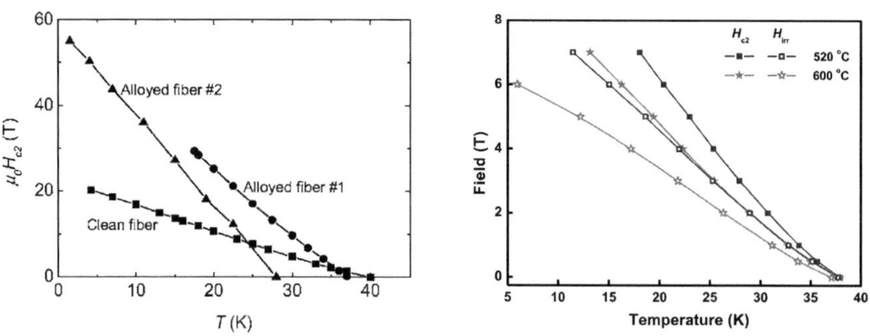

Fig. 9. (a) Upper critical field of MgB_2 coated fibers (from Ref. 4). © from AIP Publishing and (b) of coated conductors (from Ref. 66). © from Elsevier Ltd.

nature of the yarns provided means to incorporate materials into the MgB_2–CNT network, potentially enhancing the upper critical field.

4. Lower Critical Field and Thermodynamic Critical Field

The lower critical field H_{c1} of a type II superconductor is also of fundamental and technological interest. At the field H_{c1} vortices enter in the superconductor, causing dissipation in electromagnetic fields. This is an important aspect for SRF cavities. Another critical parameter for SRF cavities is the thermodynamic critical field H_c, which measures the difference in free energy between the normal and superconducting states.

4.1 *SRF cavity and the acceleration gradient*

SRF cavities are a key component in accelerators used for high energy physics, nuclear physics, storage–ring-based synchrotron radiation sources and free electron lasers.[68,69] While Nb is currently the most common material used for SRF cavities, its maximum accelerating field E_{acc}^{MAX}, which is ultimately determined by B_c where the Cooper pairs are broken by the magnetic field, has been improved to values near to its theoretical limit.[69,70] The employment of MgB_2 thin films in SRF cavities could lead to maximum accelerating fields exceeding that of Nb.[6] because of its higher thermodynamic critical field H_c (200 mT for Nb versus over 350 mT for MgB_2[71]). According to the BCS theory, the surface resistance R_s depends on the residual resistivity ρ_0 and the energy gap Δ. MgB_2 has larger energy gaps than Nb, and MgB_2 films deposited by HPCVD exhibit a very low residual resistivity ρ_0 as shown in Fig. 10, which is higher but close to the optimal ρ_0 for low surface resistance. Therefore, MgB_2 could have a lower R_s than Nb, thus higher quality factor Q of the SRF cavity. The RF measurement with a sapphire resonator technique at 18 GHz by Moeckly *et al.*[72] has demonstrated that the R_s of a clean HPCVD film at 15 K was only three times that of Nb at 4.2 K when scaled to 0.5 GHz. By further reducing the temperature of the MgB_2 films to 4.2 K, an improvement of more than an order of magnitude is anticipated. The R_s values of MgB_2 films deposited by reactive evaporation have been shown to be lower than that of sputtered niobium films.[73]

Using a superconducting thin film coating for SRF cavities also brings other advantages. As the RF field decays exponentially from the surface of a superconductor into its bulk, the RF properties of the cavity are only determined by a thin layer of the superconductor on the cavity surface. The bulk of the cavity body provides mostly

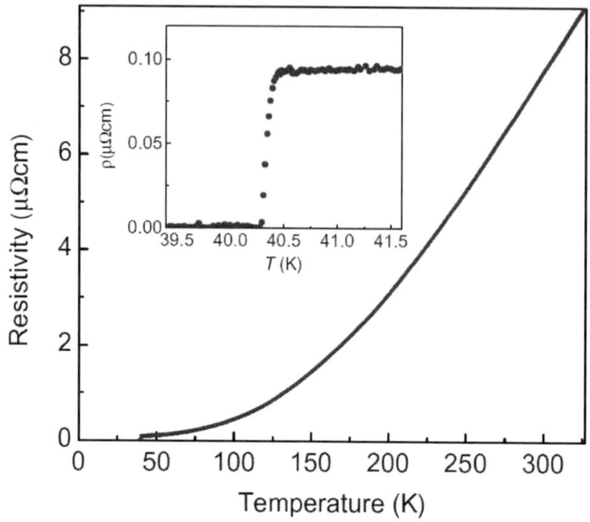

Fig. 10. The resistivity versus temperature curve for a 770 nm thick MgB_2 film deposited with HPCVD on a sapphire substrate (from Ref. 43). © from Elsevier Ltd.

mechanical support and is not of importance to its RF performance. Employing high purity bulk niobium for RF cavities results in high fabrication costs, making thin film coated cavities more attractive. In the case of MgB_2, the body of the cavity could be manufactured from a material with a high thermal conductivity, leading to better thermal stability of the cavity. Pulsed laser deposition,[74–76] RCE,[72,75,77] and HPCVD have all been investigated for SRF applications of MgB_2 thin films.

4.2 H_{c1} and its thickness dependence

It has been suggested that the vortex dissipation loss above H_{c1} is the limiting factor of the RF breakdown field in SRF cavities, not H_c. Research is being conducted to enhance H_{c1}, in particular using multilayered structures consisting of alternating insulator and high H_c superconductor layers with a thickness smaller than its London penetration depth l. The small superconducting film thickness results in increase in H_{c1}. An enhancement of H_{c1} has been demonstrated in MgO/NbN multilayers,[78] although the benefit of the multilayer structure is still being actively debated.[79–81]

For MgB_2, $H_{c1}(0)$ values from 300–1100 Oe have been reported for bulk samples,[82–84] lower than that of Nb (1700 Oe). This casts some doubt on the potential of MgB_2 as a replacement of Nb. Using a DC magnetization measurement, Tajima *et al.* have shown that the vortex penetration field (B_{vp}) of epitaxial MgB_2 films 300 nm and thinner, prepared by RCE and HPCVD, exceeds the B_{vp} of cavity-quality bulk Nb [Ref. 7, see Fig. 11(a)]. Figure 11(b) shows the experimental result of $H_{c1}(0)$ for epitaxial MgB_2 films deposited by HPCVD with thickness down to 6 nm along with the theoretically predicted thickness dependence.[85] The $H_{c1}(0)$ value is as high as 6500 Oe for the 60 nm-thick epitaxial thin films. The result agrees well with the theoretical prediction. These results suggest that H_{c1} is not an insurmountable barrier to the SRF applications of MgB_2.

In order to investigate the prospects of MgB_2 thin films for SRF applications, the thickness dependence of H_{c1} has been recently studied in both epitaxial and polycrystalline MgB_2 films deposited by HPCVD.[86] The study of polycrystalline films is important because MgB_2 coatings on cavity walls and the top MgB_2 layers in the

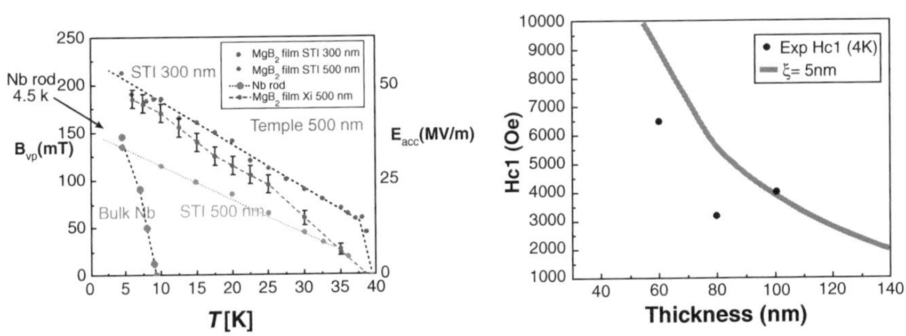

(a) (b)

Fig. 11. (a) B_{vp} versus T for a 500 nm film as prepared by HPCVD compared to two STI films of different thicknesses and bulk Nb (from Ref. 7), (b) H_{c1} as measured at 4 K for MgB_2 films of various thicknesses. The solid line displays the expected theoretical curve for the lower critical field enhancement according to a coherence length of 5 nm (from Ref. 85).

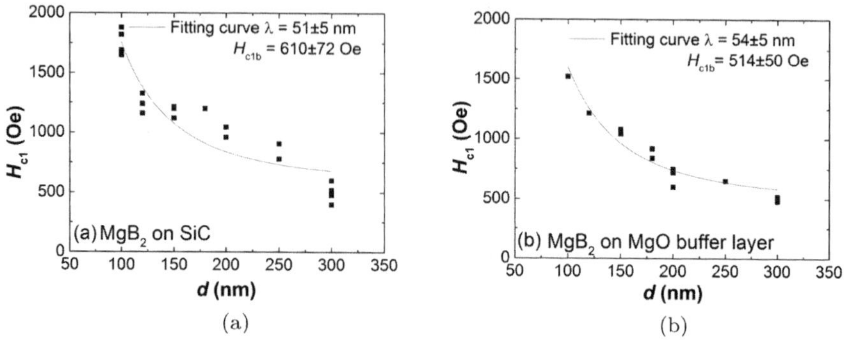

Fig. 12. Thickness dependence of H_{c1} (5 K) of (a) epitaxial MgB_2 films grown on SiC substrates and (b) polycrystalline MgB_2 films grown on MgO buffer layers (from Ref. 87).

multilayer structures are inevitably polycrystalline. In this study, the epitaxial MgB_2 films were deposited by HPCVD on (0001) SiC substrates and the polycrystalline MgB_2 films were deposited on MgO buffer layers on SiC prepared by sputtering.[87] In Fig. 12, H_{c1} (5 K) is plotted as a function of MgB_2 film thickness from 100 nm–300 nm for (a) epitaxial and (b) polycrystalline MgB_2 films. In both cases, H_{c1} increases with decreasing film thickness to values similar to that of bulk Nb. The solid lines in Fig. 5 are fitting to the theoretical thickness dependence of H_{c1} applicable to films thicker than λ, based on the free energy of a single vortex in a thin film.[88] The enhancement of H_{c1} depends exclusively on the ratio of the film thickness to the penetration depth d/λ. Good fittings are obtained for both the epitaxial and polycrystalline films with $\lambda = 51$ nm for the epitaxial films and $\lambda = 54$ nm for the polycrystalline films. The bulk H_{c1} values obtained are within the range of values reported in the literature. The λ values are slightly higher than the previous measurements on our clean film samples.[89]

It is difficult to measure H_{c1} of thin superconducting films accurately. Aligning the film surface parallel to the magnetic field is critical and the diamagnetic signal is small in thin superconducting films. The results from the magnetization method used in this study contain uncertainties in the absolute values of H_{c1}. However, the behavior of the film thickness dependence of H_{c1} was accurately revealed, which is of critical importance for SRF applications.

Conclusions

In summary, MgB_2 thin films have been deposited using many deposition techniques, including the two-step processes of annealing B or Mg–B precursor films in Mg vapor, or the *in situ* processes including RCE, MBE, sputter deposition and HPCVD. The key requirements for the MgB_2 deposition are a sufficiently high Mg vapor pressure in order to achieve thermodynamic phase stability of MgB_2, an O_2-free environment and pure sources of Mg and B. Amongst all the deposition techniques for MgB_2 thin films, HPCVD remains the most successful as it leads to the highest quality MgB_2 films.

Furthermore, C and O_2 doping, as well as particle irradiation can lead to an increase of the upper critical field H_{c2}. The highest H_{c2} value observed has been in carbon-alloyed MgB_2 thin films fabricated using HPCVD. Extensive structural disorder and small grain size have been suggested as the causes of the high B_{c2}. In addition, the lower critical field H_{c1} of MgB_2 thin films has been actively studied, in particular, the enhancement of H_{c1} by lowering the film thickness. The H_{c1} of thin MgB_2 films can be comparable to or even higher than the commonly used conventional superconductor Nb for RF cavities. Combined with higher thermodynamic critical field H_c, MgB_2 coating is promising as a higher T_c alternative to Nb for SRF applications.

Acknowledgments

We acknowledge support by the U.S. Department of Energy Office of Science, the Office of High Energy Physics under Grant No. DE-SC0004410 during the preparation of this chapter.

References

1. W. N. Kang, H. J. Kim, E. M. Choi, C. U. Jung and S. L. Lee, MgB_2 superconducting thin films with a transition temperature of 39 Kelvin, *Science* **25**, p. 1521 (2001).

2. X. X. Xi, Two-band superconductor magnesium diboride, *Reports on Progress in Physics* **71**, p. 116501 (2008).
3. A. Brinkman and J. M. Rowell, MgB_2 tunnel junctions and SQUIDs, *Physica C* **456**, pp. 188–195 (2007).
4. V. Ferrando, P. Orgiani, A. V. Pogrebnyakov, J. Chen, Q. Li and J. M. Redwing, High upper critical field and irreversibility field in MgB_2 coated conductor fibers, *Appl. Phys. Lett.* **87**, p. 252509 (2005).
5. M. Ranot and W. N. Kang, MgB_2 coated superconducting tapes with high critical current densities fabricated by hybrid physical–chemical vapor deposition, *Curr. Appl. Phys.* [*Review*] **12**, pp. 353–363 (2012).
6. E. W. Collings, M. D. Sumption and T. Tajima, Magnesium diboride superconducting RF resonant cavities for high energy particle acceleration. *Supercond. Sci. Technol.* **17**, pp. S595–S601 (2004).
7. T. Tajima, L. Civale, D. J. Devlin, R. K. Schulze, I. O. Usov and G. C. Martinez, Status of MgB_2 coating studies for SRF applications, *Proc. SRF2013*, Paris, France, pp. 777–781 (2013).
8. M. A. Wolak, T. Tan, A. Krick, E. Johnson, M. Hambe and K. Chen, Superconducting magnesium diboride coatings for radio frequency cavities fabricated by hybrid physical–chemical vapor deposition, *Phys. Rev. Spec. Top-Accel. Beams* **17**, p. 012001 (2014).
9. Z. K. Liu, D. G. Schlom, Q. Li and X. X. Xi, Thermodynamics of the Mg-B system: implications for the deposition of MgB_2 thin films, *Appl. Phys. Lett.* **78**, pp. 3678–80 (2001).
10. A. V. Pogrebnyakov, J. M. Redwing, S. Raghavan, V. Vaithyanathan, D. G. Schlom and S. Y. Xu, Enhancement of the superconducting transition temperature of MgB_2 by a strain-induced bond-stretching mode softening, *Phys. Rev. Lett.* **93**, p. 47006 (2004).
11. V. Braccini, A. Gurevich, J. E. Giencke, M. C. Jewell, C. B. Eom, D. C. Larbalestier, A. Pogrebnyakov, Y. Cui, B. T. Liu, Y. Hu, F. J. M. Redwing *et al.*, High-field superconductivity in alloyed MgB_2 thin films, *Phys. Rev. B* **71**, p. 012504 (2005).
12. C. Ferdeghini, V. Ferrando, C. Tarantini, E. Bellingeri, G. Grasso, A. Malagoli *et al.*, Upper critical fields up to 60 T in dirty magnesium diboride thin films, *IEEE Trans. Appl. Supercond.* **15**, pp. 3234–3237 (2005).
13. J. Kortus, O. V. Dolgov, R. K. Kremer and A. A. Golubov, Band filling and interband scattering effects in MgB_2: carbon versus aluminum doping, *Phys. Rev. Lett.* **94**, p. 027002 (2005).
14. C. B. Eom, M. K. Lee, J. H. Choi, L. J. Belenky, X. Song, L. D. Cooley *et al.*, High critical current density and enhanced irreversibility field in superconducting MgB_2 thin films, *Nature* **411**, pp. 558–560 (2001).

15. R. H. T. Wilke, S. L. Bud'ko, P. C. Canfield, D. K. Finnemore, R. J. Suplinskas and S. T. Hannahs, Systematic effects of carbon doping on the superconducting properties of $Mg(B_{1-x}C_x)_2$, *Phys. Rev. Lett.* **92**, p. 217003 (2004).
16. S. R. Shinde, S. B. Ogale, R. L. Greene, T. Venkatesan, P. C. Canfield, S. L. Bud'ko *et al.*, Superconducting MgB_2 thin films by pulsed laser deposition, *Appl. Phys. Lett.* **79**, pp. 227–229 (2001).
17. S. D. Bu, D. M. Kim, J. H. Choi, J. Giencke, E. E. Hellstrom, D. C. Larbalestier *et al.*, Synthesis and properties of c-axis oriented epitaxial MgB_2 thin films, *Appl. Phys. Lett.* **81**, pp. 1851–1853 (2002).
18. A. Plecenik, L. Satrapinsky, P. Kus, S. Gazi, S. Benacka, I. Vavra *et al.*, MgB_2 superconducting thin films on Si and Al_2O_3 substrates, *Physica C* **363**, pp. 224–230 (2001).
19. S. H. Moon, J. H. Yun, H. N. Lee, J. I. Kye, H. G. Kim, W. Chung *et al.*, High critical current densities in superconducting MgB_2 thin films, *Appl. Phys. Lett.* **79**, pp. 2429–2431 (2001).
20. M. Paranthaman, C. Cantoni, H. Y. Zhai, H. M. Christen, T. Aytug, S. Sathyamurthy *et al.*, Superconducting MgB_2 films via precursor postprocessing approach, *Appl. Phys. Lett.* **78**, pp. 3669–3671 (2001).
21. M. Hanna, S. Wang, A. D. Eck, R. H. T. Wilke, K. Chen, A. Soukiassian *et al.*, Clean epitaxial MgB_2 films fabricated by the ex situ annealing of chemical vapour deposition-grown B films in Mg vapour, *Supercond. Sci. Technol.* **21**, p. 045005 (2008).
22. M. Kuhberger, G. Gritzner, K. R. Schoppl, H. W. Weber, A. A. F. Olsen and T. H. Johansen, Preparation and characterization of superconducting MgB_2 films on alumina, *Supercond. Sci. Technol.* **17**, pp. 764–769 (2004).
23. A. Berenov, Z. Lockman, X. Qi, J. L. MacManus-Driscoll, Y. Bugoslavsky, L. F. Cohen *et al.*, Growth of strongly biaxially aligned MgB_2 thin films on sapphire by postannealing of amorphous precursors, *Appl. Phys. Lett.* **79**, pp. 4001–4003 (2001).
24. D. H. A. Blank, H. Hilgenkamp, A. Brinkman, D. Mijatovic, G. Rijnders and H. Rogalla, Superconducting Mg-B films by pulsed-laser deposition in an in situ two-step process using multicomponent targets, *Appl. Phys. Lett.* **79**, pp. 394–396 (2001).
25. C. Ferdeghini, V. Ferrando, G. Grassano, W. Ramadan, E. Bellingeri, V. Braccini *et al.*, Growth of c-oriented MgB_2 thin films by pulsed laser deposition: structural characterization and electronic anisotropy, *Supercond. Sci. Technol.* **14**, pp. 952–957 (2001).
26. H. J. Kim, W. N. Kang, E. M. Choi, M. S. Kim, K. H. P. Kim and S. Lee, High current-carrying capability in c-axis-oriented superconducting MgB_2 thin films, *Phys. Rev. Lett.* **87**, p. 087002 (2001).

27. X. H. Zeng, A. Sukiasyan, X. X. Xi, Y. F. Hu, E. Wertz, Q. Li *et al.*, Superconducting properties of nanocrystalline MgB_2 thin films made by an in situ annealing process, *Appl. Phys. Lett.* **79**, pp. 1840–1842 (2001).
28. H. Y. Zhai, H. M. Christen, L. Zhang, C. Cantoni, M. Paranthaman, B. C. Sales *et al.*, Superconducting magnesium diboride films on Si with T_{c0} similar to 24 K grown via vacuum annealing from stoichiometric precursors, *Appl. Phys. Lett.* **79**, pp. 2603–2605 (2001).
29. J. R. Ahn, S. G. Lee, Y. S. Hwang, G. Y. Sung and D. K. Kim, Fabrication of MgB_2 thin film by rf magnetron sputtering, *Physica C* **388**, pp. 127–128 (2003).
30. S. N. Ermolov, M. V. Indenbom, A. N. Rossolenko, I. K. Bdikin, L. S. Uspenskaya, N. S. Stepakov *et al.*, Superconducting MgB_2 films obtained by magnetron sputtering, *JETP Letters* **73**, pp. 557–561 (2001).
31. R. Vaglio, M. G. Maglione and R. Di Capua, High-quality MgB_2 thin films in situ grown by dc magnetron sputtering, *Supercond. Sci. Technol.* **15**, pp. 1236–1239 (2002).
32. S. Schiestel, C. A. Carosella, J. S. Horwitz, M. Osofsky, C. Kendziora, S. B. Qadri *et al.*, Ion beam assisted deposition of superconducting MgB_2 thin films, *Surface Coatings Technol.* **158**, pp. 203–207 (2002).
33. P. Kus, A. Plecenik, L. Satrapinsky, Y. Xu and R. Sobolewski, Superconducting properties of MgB_2 thin films prepared on flexible plastic substrates, *Appl. Phys. Lett.* **81**, pp. 2199–2201 (2002).
34. R. Rogai, V. Galluzzi, A. Mancini, G. Celentano, T. Petrisor, A. Rufoloni *et al.*, Growth of MgB_2 thin films by means of in situ deposition techniques, *Intl. J. Modern Physics B* **17**, pp. 703–708 (2003).
35. L. P. Chen, C. Zhang, Y. B. Wang, Y. Wang, Q. R. Feng, Z. Z. Gan *et al.*, A sol-gel method for growing superconducting MgB_2 films, *Supercond. Sci. Technol.* **24**, p. 015002 (2011).
36. B. H. Moeckly and W. S. Ruby, Growth of high-quality large-area MgB_2 thin films by reactive evaporation, *Supercond. Sci. Technol.* **19**, pp. L21–L24 (2006).
37. S. F. Wang, K. Chen, C. H. Lee, A. Soukiassian, D. R. Lamborn, R. DeFrain *et al.*, High quality MgB_2 thick films and large-area films fabricated by hybrid physical–chemical vapor deposition with a pocket heater, *Supercond. Sci. Technol.* **21**, p. 085019 (2008).
38. K. Nishiyuki, H. Sawaki, Y. Iwata and M. Naito, Simple route to grow high-quality MgB_2 thin films using decaborane as a boron source, *Physica C* **471**, pp. 1189–1192 (2011).

39. K. Ueda and M. Naito, As-grown superconducting MgB_2 thin films prepared by molecular beam epitaxy, *Appl. Phys. Lett.* **79**, pp. 2046–2048 (2001).
40. M. Hurata, T. Fujiyoshi, S. Kihara, T. Sueyoshi, K. Miyahara, Y. Harada *et al.*, High critical current density under magnetic fields in as-grown MgB_2 thin films deposited by molecular-beam epitaxy, *Supercond. Sci. Technol.* **20**, pp. L1–L4 (2007).
41. A. Saito, A. Kawakami, H. Shimakage and Z. Wang, As-grown deposition of superconducting MgB_2 thin films by multiple-target sputtering system, *Japanese Journal of Applied Physics, Part 2-Letters* **41**, pp. L127–L129 (2002).
42. R. Schneider, J. Geerk, G. Linker, F. Ratzel, A. G. Zaitsev and B. Obst, In situ deposition of MgB_2 thin films by magnetron cosputtering and sputtering combined with thermal evaporation, *Physica C* **423**, pp. 89–95 (2005).
43. X. X. Xi, A. V. Pogrebnyakov, S. Y. Xu, K. Chen, Y. Cui, E. C. Maertz *et al.*, MgB_2 thin films by hybrid physical-chemical vapor deposition, *Physica C* **456**, pp. 22–37 (2007).
44. X. H. Zeng, A. V. Pogrebnyakov, A. Kotcharov, J. E. Jones, X. X. Xi, E. M. Lysczek *et al.*, In situ epitaxial MgB_2 thin films for superconducting electronics, *Nat. Mater.* **1**, pp. 35–38 (2002).
45. A. V. Pogrebnyakov, J. M. Redwing, J. E. Jones, X. X. Xi, S. Y. Xu, Q. Li *et al.*, Thickness dependence of the properties of epitaxial MgB_2 thin films grown by hybrid physical-chemical vapor deposition, *Appl. Phys. Lett.* **82**, pp. 4319–4321 (2003).
46. D. R. Lamborn, D. W. Snyder, X. X. Xi and J. M. Redwing, Modeling studies of the chemical vapor deposition of boron films from B_2H_6, *J. Crystal Growth* **299**, pp. 358–364 (2007).
47. S. Lee, Crystal growth of MgB_2, *Physica C* **385**, pp. 31–41 (2001).
48. P. C. Canfield, D. K. Finnemore, S. L. Bud'ko, J. E. Ostenson, G. Lapertot, C. E. Cunningham *et al.*, Superconductivity in dense MgB_2 wire, *Phys. Rev. Lett.* **86**, pp. 2423–2426 (2001).
49. T. Tan, C. G. Zhuang, A. Krick, K. Chen and X. X. Xi, Large-area MgB_2 films fabricated by scaled-up hybrid physical-chemical vapor deposition, *IEEE Trans. Appl. Supercond.* **23**, pp. 7500304–7500304 (2013).
50. D. R. Lamborn, R. H. T. Wilke, Q. Li, X. X. Xi, D. W. Snyder and J. M. Redwing, Growth of thick MgB_2 films by impinging-jet hybrid physical-chemical vapor deposition, *Adv. Mater.* **20**, pp. 319–323 (2008).
51. X. X. Xi, MgB_2 thin films, *Supercond. Sci. Technol.* **22**, 043001 (2009).
52. A. Gurevich, S. Patnaik, V. Braccini, K. H. Kim, C. Mielke, X. Song *et al.*, Very high upper critical fields in MgB_2 produced by selective

tuning of impurity scattering, *Supercond. Sci. Technol.* **17**, pp. 278–286 (2004).
53. A. Gurevich and V. M. Vinokur, Interband phase modes and nonequilibrium liton structures in two-gap superconductors, *Phys. Rev. Lett.* **90**, p. 047004 (2003).
54. C. G. Zhuang, S. Meng, H. Yang, Y. Jia, H. H. Wen, X. X. Xi *et al.*, Significant improvements of the high-field properties of carbon-doped MgB_2 films by hot-filament-assisted hybrid physical-chemical vapor deposition using methane as the doping source, *Supercond. Sci. Technol.* **21**, p. 082002 (2008).
55. W. Q. Dai, V. Ferrando, A. V. Pogrebnyakov, R. H. T. Wilke, K. Chen, X. J. Weng *et al.*, High-field properties of carbon-doped MgB_2 thin films by hybrid physical-chemical vapor deposition using different carbon sources, *Supercond. Sci. Technol.* **24**, p. 125014 (2011).
56. A. V. Pogrebnyakov, X. X. Xi, J. M. Redwing, V. Vaithyanathan, D. G. Schlom, A. Soukiassian *et al.*, Properties of MgB_2 thin films with carbon doping, *Appl. Phys. Lett.* **85**, pp. 2017–2019 (2004).
57. S. Lee, T. Masui, A. Yamamoto, H. Uchiyama and S. Tajima, Carbon-substituted MgB_2 single crystals, *Physica C* **397**, pp. 7–13 (2003).
58. Y. Zhu, D. C. Larbalestier, P. M. Voyles, A. V. Pogrebnyakov, X. X. Xi and J. M. Redwing, Nanoscale disorder in high critical field, carbon-doped MgB_2 hybrid physical-chemical vapor deposition thin films, *Appl. Phys. Lett.* **91**, p. 082513 (2007).
59. R. K. Singh, Y. Shen, R. Gandikota, C. Carvalho, J. M. Rowell and N. Newman, Effect of oxygen incorporation on normal and superconducting properties of MgB_2 films, *Appl. Phys. Lett.* **93**, p. 242504 (2008).
60. S. R. Shinde, S. B. Ogale, J. Higgins, R. J. Choudhary, V. N. Kulkarni, T. Venkatesan *et al.*, Modification of critical current density of MgB_2 films irradiated with 200 MeV Ag ions, *Appl. Phys. Lett.* **84**, pp. 2352–2354 (2004).
61. R. Gandikota, R. K. Singh, J. Kim, B. Wilkens, N. Newman, J. M. Rowell *et al.*, Effect of damage by 2 MeV He ions on the normal and superconducting properties of magnesium diboride, *Appl. Phys. Lett.* **86**, p. 012508 (2005).
62. R. Gandikota, R. K. Singh, J. Kim, B. Wilkens, N. Newman, J. M. Rowell *et al.*, Effect of damage by 2 MeV He ions and annealing on H_{c2} in MgB_2 thin films, *Appl. Phys. Lett.* **87**, p. 072507 (2005).
63. R. Di Capua, H. U. Aebersold, C. Ferdeghini, V. Ferrando, P. Orgiani, M. Putti *et al.*, Role of interband scattering in neutron irradiated MgB_2 thin films by scanning tunneling spectroscopy measurements, *Phys. Rev. B* **75**, p. 014515 (2007).

64. V. Ferrando, I. Pallecchi, C. Tarantini, D. Marre, M. Putti, C. Ferdeghini *et al.*, Systematic study of disorder induced by neutron irradiation in MgB_2 thin films, *J. Appl. Phys.* **101**, p. 043903 (2007).
65. J. Wang, C. G. Zhuang, J. Li, Z. Wang, Q. R. Feng and D. N. Zheng, Upper critical field and Raman spectra of MgB_2 thin films irradiated with low energy oxygen ion, *J. Appl. Phys.* **106**, p. 093915 (2009).
66. M. Ranot, S. Oh, K. C. Chung and W. N. Kang, MgB_2 coated conductors directly grown on flexible metallic Hastelloy tapes by hybrid physical–chemical vapor deposition, *Curr. Appl. Phys.* **13**, pp. 1808–1812 (2013).
67. Y. Iwasa, D. C. Larbalestier, M. Okada, R. Penco, M. D. Sumption and X. X. Xi, A round table discussion on MgB_2 toward a wide market or a niche production? A summary, *IEEE Trans. Appl. Supercond.* **16**, pp. 1457–1464 (2006).
68. H. Padamsee, The science and technology of superconducting cavities for accelerators, *Supercond. Sci. Technol.* **14**, pp. R28–R51 (2001).
69. H. Padamsee, Accelerating applications of RF superconductivity — Success stories, *IEEE Trans. Appl. Supercond.* **15**, pp. 2432–2439 (2005).
70. K. Saito, ed., Gradient yield improvement efforts for single and multi-cells and progress for very high gradient cavities, *Proc. SRF2007*, Peking Univ., Beijing, China, pp. 82–93 (2007).
71. F. Bouquet, R. A. Fisher, N. E. Phillips, D. G. Hinks and J. D. Jorgensen, Specific heat of $Mg^{11}B_2$: evidence for a second energy gap, *Phys. Rev. Lett.* **87**, p. 047001 (2001).
72. B. H. Moeckly, K. E. Kihlstrom, A. T. Findikoglu and D. E. Oates, Microwave properties of MgB_2 thin films grown by reactive evaporation, *IEEE Trans. Appl. Supercond.* **15**, pp. 3308–3312 (2005).
73. D. E. Oates, Y. D. Agassi and B. H. Moeckly, Microwave measurements of MgB_2: implications for applications and order-parameter symmetry, *Supercond. Sci. Technol.* **23**, p. 034011 (2010).
74. S. Mitsunobu, S. Inagaki, H. Nakanishi, K. Saito, M. Wake and M. Yoshida, (Eds.), MgB_2 thin films on Nb cavity by pulse laser deposition, *Proceedings SRF2007*, Beijing, China, pp. 413–417 (2007).
75. T. Tajima, A. Canabal, Y. Zhao, A. Romanenko, B. H. Moeckly, C. D. Nantista *et al.*, MgB_2 for application to RF cavities for accelerators, *IEEE Trans. Appl. Supercond.* **17**, pp. 1330–1333 (2007).
76. Y. Zhao, M. Ionescu, M. Roussel, A. V. Pan, J. Horvat and S. X. Dou, Superconducting and microstructural properties of two types of MgB_2 films prepared by pulsed laser deposition, *IEEE Trans. Appl. Supercond.* **15**, pp. 3261–3264 (2005).

77. D. E. Oates, Y. D. Agassi and B. H. Moeckly, Intermodulation distortion and nonlinearity in MgB_2: experiment and theory, *IEEE Trans. Appl. Supercond.* **17**, pp. 2871–2874 (2007).
78. C. Z. Antoine, S. Berry, S. Bouat, J. F. Jacquot, J. C. Villegier, G. Lamura *et al.*, Characterization of superconducting nanometric multilayer samples for superconducting rf applications: first evidence of magnetic screening effect, *Phys. Rev. Special Topics — Accelerators and Beams* **13**, p. 121001 (2010).
79. A. Gurevich, On the theoretical RF field limits of multilayer coating structures of superconducting resonator cavities, *arXiv*:13095626 (2013).
80. S. Posen, G. Catelani, M. Liepe, J. Sethna and M. Transtrum, Theoretical field limits for multi-layer superconductors, *arXiv*:13093239 (2013).
81. S. Posen, G. Catelani, M. Liepe, J. Sethna and M. Transtrum, Response to comment on theoretical RF field limits of multilayer coating structures of superconducting resonator cavities, *arXiv*:13104479 (2013).
82. Y. Feng, G. Yan, Y. Zhao, A. K. Pradhan, C. F. Liu, P. X. Zhang *et al.*, Magnetic relaxation and lower critical field in MgB_2 wires, *J. Phys.-Condens. Mat.* **15**, p. 6395 (2003).
83. S. L. Li, H. H. Wen, Z. W. Zhao, Y. M. Ni, Z. A. Ren, G. C. Che *et al.*, Linear temperature dependence of lower critical field in MgB_2, *Phys. Rev. B.* **64**, p. 094522 (2001).
84. L. Lyard, T. Klein, J. Marcus, R. Brusetti, C. Marcenat, M. Konczykowski *et al.*, Geometrical barriers and lower critical field in MgB_2 single crystals, *Phys. Rev. B* **70**, p. 180504 (2004).
85. D. B. Beringer, C. Clavero, T. Tan, X. X. Xi, W. M. Roach and R. A. Lukaszew, Thickness dependence of H_{c1} in epitaxial MgB_2 thin films, *Personal communication* (2013).
86. D. B. Beringer, C. Clavero, T. Tan, X. X. Xi, W. M. Roach and R. A. Lukaszew, Thickness dependence and enhancement of H_{C1} in epitaxial MgB_2 thin films, *IEEE Trans. Appl. Supercond.* **23**, p. 750060 (2013).
87. T. Tan, M. A. Wolak, N. Acharya, K. Chen, A. Krick, S. J. May, K. Chen and A. Krick, Enhancement of lower critical field in epitaxial and polycrystalline MgB_2 thin films, *Appl. Phys. Lett.*, submitted (2014).
88. A. Gurevich, ed., Multilayer coating of superconducting cavities: challenges and opportunities, *Proc. 15th SRF Workshop*, Chicago (USA) (2011).
89. D. Cunnane, C. Zhuang, K. Chen, X. X. Xi, J. Yong and T. R. Lemberger, Penetration depth of MgB_2 measured using Josephson junctions and SQUIDs, *Appl. Phys. Lett.* **102**, p. 072603 (2013).

Chapter 2c

Nanoscale Disorder in MgB_2 Thin Films

Ye Zhu*,† and Paul M. Voyles†

**Department of Applied Physics*
The Hong Kong Polytechnic University
Hung Hom, Kowloon, Hong Kong
zhuy@ymail.com

†Department of Material Science and Engineering
South University of Science and Technology of China
Shenzhen, Guangdong, P.R. China 518055

MgB_2 thin films exhibit superior superconducting properties compared to bulk MgB_2. Resolving the structural origins of such property enhancements holds the promise for improving the performance of MgB_2 wires and tapes. Here we review electron microscopy studies on the microstructure of pure and doped MgB_2 thin films grown by different deposition methods. A side-by-side comparison of these results shed light on the correlations between growth conditions, the formation of nanoscale disorder and the resulting property enhancement in MgB_2 thin films. The effective electron scattering and vortex pinning centers in MgB_2 are also revealed in the cleaner and simpler thin film system.

1. Introduction

MgB_2 thin films are important for both fundamental research and magnetic and electronic applications. MgB_2 thin films can have much better superconducting properties, such as higher T_c, J_c and B_{c2} values than bulk MgB_2,[1] demonstrating the potential for further improving the performance of MgB_2 wires and tapes. Unravelling the structural origins of such property enhancements requires nanoscale or even atomic scale characterization, for which transmission electron microscopy (TEM) is a valuable tool. Moreover, compared to bulk MgB_2 which usually has a complicated small-grain, multiphase polycrystalline structure, MgB_2 epitaxial thin films provide cleaner systems for microscopic studies of the nature of the electron scattering and flux pinning disorder. In this chapter, we review the results of TEM structural characterization of MgB_2 thin films grown by a variety of deposition methods, including pulsed laser deposition (PLD), molecular beam epitaxy (MBE) and hybrid physical–chemical vapor deposition (HPCVD). Comparing the microstructure of these films provides insights into the formation mechanisms of various types of disorder in MgB_2, and the resulting structure-property correlations reveal effective electron scattering and vortex pinning centers in MgB_2, and point towards further possible enhancement in MgB_2 performance.

First, a note on preparing MgB_2 samples for TEM: the typical method to prepare TEM specimens from thin films is mechanical polishing followed by Ar-ion milling. However, MgB_2 will degrade when exposed to water and moisture.[2–6] Degradation can be prevented by depositing ∼10 nm of MgO or amorphous Si as a capping layer on top of the film before TEM sample preparation. Polishing oil is required as a lubricant for both sample cutting and polishing, instead of water- or alcohol-based solutions.[7–9] Ion milling should be limited in duration and at low energy to reduce the damage to the TEM sample surfaces. Following these guidelines, both cross-section and plan-view TEM specimens can be prepared, as shown in Fig. 1, providing multiple projections of microstructure of MgB_2 thin films. Plan-view samples need to be polished and ion-milled from the substrate side only, leaving free-standing MgB_2 from the top portion of the film.[2]

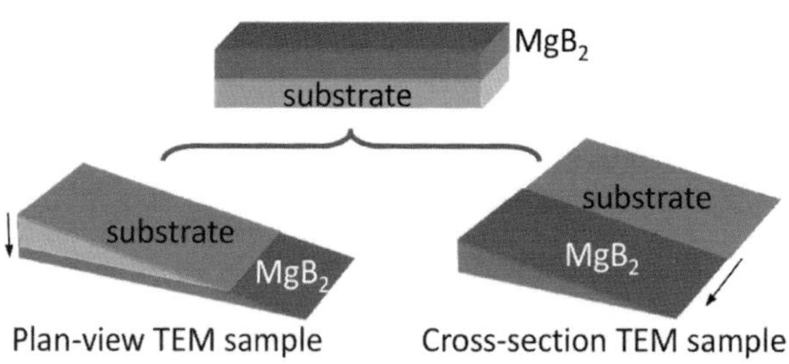

Fig. 1. Geometry of plan-view and cross-section TEM samples prepared from MgB_2 thin films. The arrows indicate the direction of the interface normal.

2. Morphology of MgB_2 Thin Films

This section reviews the basic morphology of various MgB_2 thin films grown with different methods and conditions, including nanocrystalline films grown at low temperatures, larger-grain polycrystalline films grown/annealed at higher temperatures and epitaxial single-crystal films with and without C-doping.

2.1 *Nanocrystalline MgB_2 thin films grown at low temperatures*

As described in the previous chapter, the Mg–B phase diagram indicates that MgB_2 films can be grown only under very high Mg vapor pressure or at fairly low growth temperatures. Low-temperature growth (room temperature to 300°C) often leads to nanocrystalline, non-epitaxial MgB_2 films, independent of the substrates and even after post-growth annealing[6,10–12] (with exceptions as seen in following sections). As illustrated by dark-field (DF) TEM imaging in Fig. 2 for a PLD film in plan-view[10] and Fig. 3 for an MBE film in cross-section,[6] room-temperature grown films contain $\leq$10 nm dispersed MgB_2 grains with (001) fiber texture as indicated by selected area diffraction (SAD) patterns. The extended diffraction arcs in Fig. 3(c) also indicate a large grain-to-grain variation of MgB_2 *c*-axis direction, which will be referred to here as *c*-axis tilt disorder. Such small MgB_2 grains with large orientation variation can be attributed to both the room-temperature growth and the intergrain MgO phase. The presence of MgO second phase has been explicitly identified using SAD in Fig. 2(b) and high-resolution TEM (HRTEM) imaging in Fig. 3(d). The MgO grains in Fig. 3(d) are $\sim$5 nm in diameter, comparable to the intrinsic MgB_2 coherence length ($\sim$6 nm in the *ab* plane). As stated in the previous chapter, oxygen (O_2) can be incorporated into MgB_2 by poorer vacuum during growth ($\geq 3 \times 10^{-8}$ Torr) or even just slowing the film growth rate. The formation of MgO during thin film deposition can further limit the growth of MgB_2 grains.

As expected from their similar microstructure, both films exhibit similar superconducting properties, i.e., higher-than-bulk $\rho(T_c)$, B_{c2} and J_c, but reduced T_c. These properties indicate that MgB_2

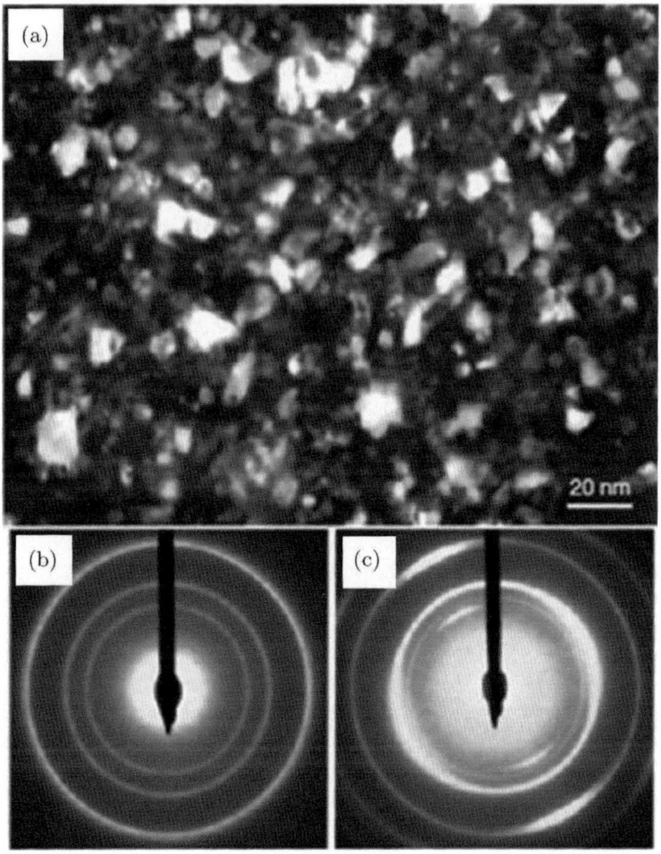

Fig. 2. (a) DF plan-view image of an MgB_2 film grown by PLD, showing ~10 nm grain size. (b, c) SAD patterns with incident electron beam (b) parallel and (c) inclined to the film normal. The diffraction rings correspond to MgO and (001)-textured MgB_2 phases. Reprinted by permission from Macmillan Publishers Ltd: From Ref. 10. © 2001 Nature Publishing Group.

grain boundaries and MgO nanoparticles can effectively enhance electron scattering and flux pinning. Even though the J_c enhancement through increased flux pinning must be compromised by the current-blocking effect from non-superconducting MgO grains, such O-incorporated MgB_2 films still exhibit one of the highest J_c among all MgB_2 materials, making them technologically appealing.[10–12] In addition to forming an intergrain MgO second phase, O_2 can also be

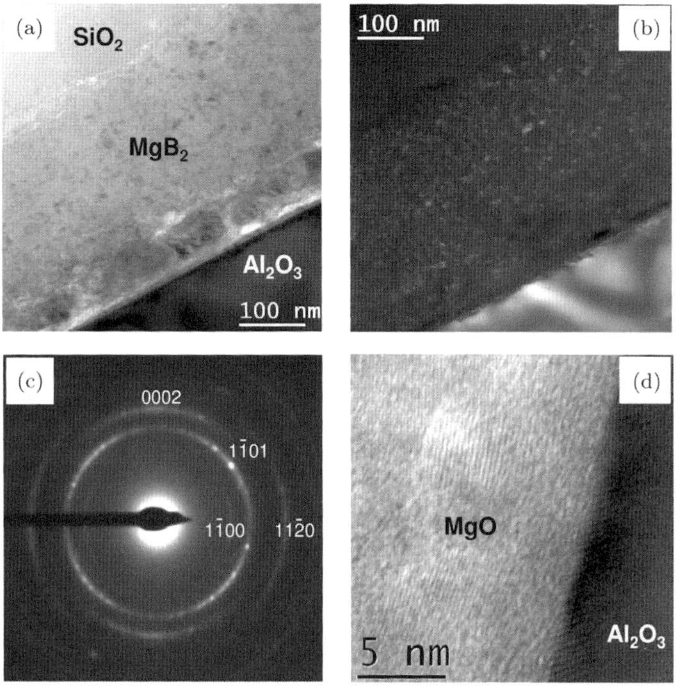

Fig. 3. (a) BF and (b) DF images taken from a cross-section MgB_2 film grown by MBE, (c) SAD pattern taken from the same sample, showing both (001) in-plane rotation and *c*-axis tilt disorder, (d) HRTEM image showing MgO particles around the interface. The fringes are MgO (111) planes in the upper grain and MgO (110) planes in the lower grain. From Ref. 6. Reproduced with permission from IOP Publishing. All rights reserved.

incorporated inside the MgB_2 lattice, as will be better demonstrated in Sec. 3.2.

2.2 *Large-grain MgB_2 thin films from growth/annealing at elevated temperatures*

In contrast to the nanocrystalline MgB_2 thin films grown at low temperatures, high-quality MgB_2 films can be made by thin film growth/annealing at elevated temperatures and sufficiently high Mg vapor pressure. The earliest high-quality MgB_2 films were obtained through post-growth annealing of boron precursor films in Mg vapor at high temperatures (600–900°C).[13–16] Such two-step grown films

Fig. 4. SEM images of (a) the surface and (b) the cross-section for a 3 μm thick MgB_2 film. From Ref. 17.

contain large MgB_2 grains with good crystallinity and the preferred orientation, in accordance with their high T_c, zero-field J_c and low $\rho(T_c)$. Clean MgB_2 films with grain sizes >100 nm have been fabricated by annealing CVD-deposited B films in Mg vapor,[17] as seen in Fig. 4. The high purity of the MgB_2 phase is due to the low-O_2 growth environment with H_2 as the carrier gas and B_2H_6 as the B source, and is evidenced by a sharp superconducting transition at 40 K and a low $\rho(T_c)$ of $1.25\mu\Omega \cdot$ cm measured from such films.

High-quality large-grain MgB_2 films can also be grown with single-step *in situ* deposition without annealing. To avoid the requirement of very high Mg vapor pressure, intermediate temperatures (300–600°C) are often used for such growth.[18–20] MgB_2 thin films with grain size ~40 nm and $T_c \sim 36$ K have been produced using MBE, an example of which is shown by the bright-field (BF) TEM image in Fig. 5.[18] The presence of 30°-rotated MgB_2 grains, will be discussed in more detail in Sec. 3.1. Despite the lower quality compared to two observed 12-fold symmetry reflects the preferred orientation alignment and the step-grown films, the single-step *in situ* deposition is of technological interest for fabricating multilayer superconducting devices such as Josephson junctions.[21]

Fig. 5. Plan-view TEM picture of the MgB_2 thin film on a *c*-plane sapphire substrate. Inset shows the SAD pattern of the sample. 12-fold symmetric bright segments are clearly shown, but there is significant mosaic spread. Reprinted with permission from Ref. 18. © 2002 AIP Publishing LLC.

2.3 *High quality epitaxial MgB_2 thin films*

High quality epitaxial MgB_2 films have been grown by both two-step growth and single-step *in situ* deposition.[7,15,16] Sapphire and (0001)-oriented SiC are the most used substrates to grow epitaxial MgB_2 oriented along the *c*-axis. SiC (a = 3.073 Å for 4H–SiC and 3.081 Å for 6H–SiC) offers the closest in-plane lattice match for the MgB_2 lattice (a = 3.086 Å), so that an epitaxial orientation relationship $[1\bar{1}00]MgB_2 \| [1\bar{1}00]SiC$ can be attained. To accommodate the large lattice mismatch (~23%) between MgB_2 and *c*-cut sapphire, the MgB_2 lattice needs to be rotated around the *c*-axis by 30°, so that a small lattice mismatch (~11%) can be achieved with $[1\bar{1}00]MgB_2 \| [11\bar{2}0]Al_2O_3$.[15,16] This orientation relationship is confirmed by SAD patterns in Figs. 6(b) and 6(c), which also demonstrate the high quality of the epitaxial two-step grown film from the sharp single-crystal-like MgB_2 diffraction.[16] The HRTEM image in Fig. 6(d) reveals epitaxial MgO and $MgAl_2O_4$ transition layers at the MgB_2/Al_2O_3 interface, with $(0001)[1\bar{1}00]MgB_2 \| (111)[1\bar{2}1]MgO \| (111)[1\bar{2}1]MgAl_2O_4 \| (0001)[11\bar{2}0]Al_2O_3$. The formation of MgO interfacial layer has been widely observed in MgB_2 films and will be discussed further in Sec. 3.3.

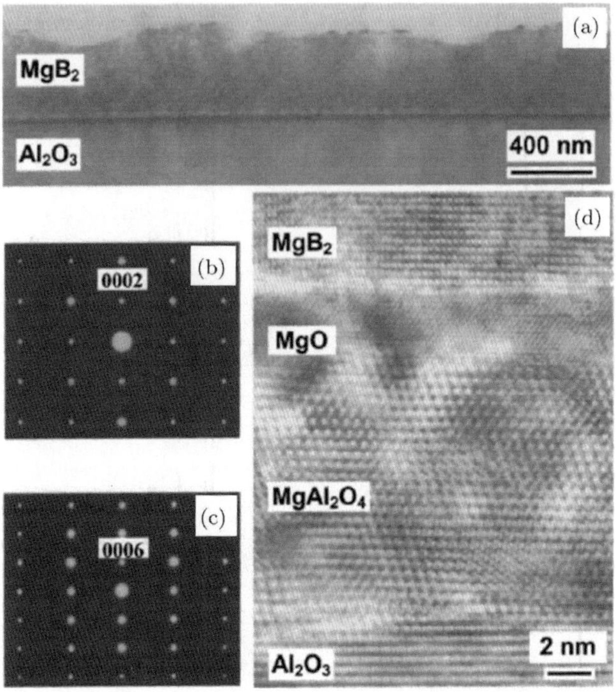

Fig. 6. (a) BF cross-sectional TEM image of a MgB_2 thin film grown on a (0001) Al_2O_3 substrate, (b) SAD of MgB_2 along the $[11\bar{2}0]$ zone axis, (c) SAD of Al_2O_3 substrate along the $[10\bar{1}0]$ zone axis and (d) cross-sectional HRTEM micrograph of anepitaxial MgB_2 film near the interface. Reprinted with permission from Ref. 16. © 2002, AIP Publishing LLC.

On the other hand, epitaxial MgB_2 thin films have also been obtained using other substrates including MgO and polycrystalline alumina. Figure 7 presents a microstructure comparison of epitaxial MgB_2 films grown on various substrates using the single-step *in situ* reactive evaporation deposition.[7,20] All films show columnar morphology, consistent with Vollmer–Weber island growth and coalescence.[22] For substrates without the in-plane six-fold symmetry, such as r-plane sapphire and polycrystalline alumina, the MgB_2 columns are generally inclined, indicating that the growth direction is tilted away from the substrate normal. MgO interfacial layers are identified on the sapphire and alumina substrates, but not on the SiC substrate.

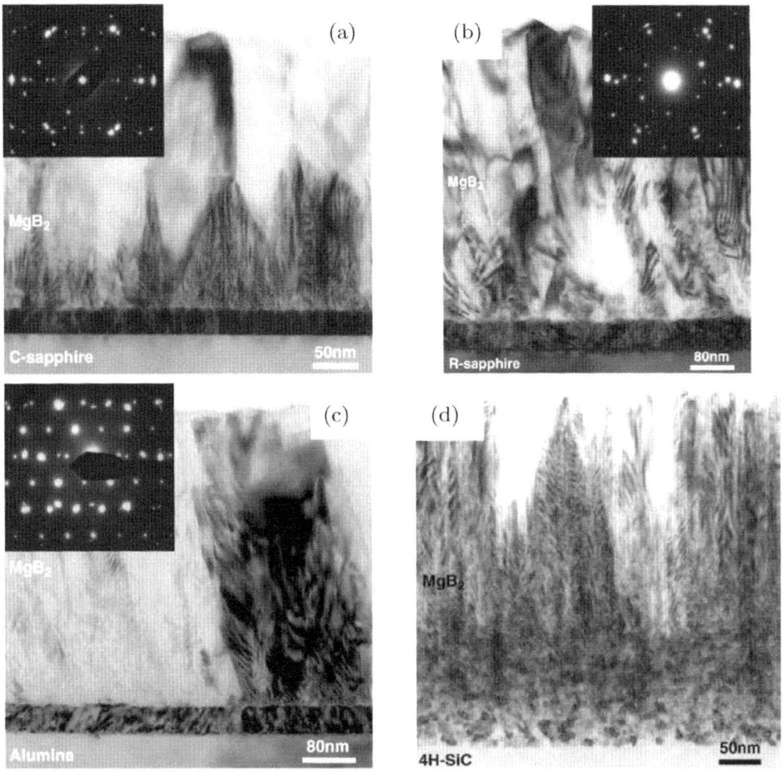

Fig. 7. Cross-sectional BFTEM images showing morphologies of epitaxial MgB_2 thin films grown by reactive evaporation deposition on (a) *c*-plane sapphire, (b) *r*-plane sapphire, (c) polycrystalline alumina and (d) 4H-SiC substrates. The insets show corresponding SAD patterns. Reprinted with permission from Ref. 7. © 2005 Elsevier.

All of these films exhibit high T_c(38–40 K) and low $\rho(T_c)$[7], showing that high-quality epitaxial MgB_2 can be achieved on a variety of substrates.

Among all the deposition techniques to grow epitaxial MgB_2 thin films, HPCVD has produced the highest-quality, cleanest films, due to its high T and Mg vapor pressure, high purity B and Mg sources, and the highly-reducing growth environment.[1] The cleanest known MgB_2 material has been the pure HPCVD film deposited around

Fig. 8. (a) BF image taken from a plan-view pure HPCVD film at the [0001] zone axis. The inset is a [0001] zone-axis SAD pattern taken from the same region, (b) HRTEM image taken along [11$\bar{2}$0] from a cross-section sample, (c) [11$\bar{2}$0] zone-axis SAD pattern taken from the same sample as in (b). Weak diffraction spots in the left part of the figure are from MgB_2 grains with their c-axis 90° rotated with respect to matrix MgB_2. From Ref. 6.

700°C in a highly reducing H_2 atmosphere on SiC and sapphire substrates. As shown in Fig. 8(a), these films contain single-crystal MgB_2 grains over 200 nm in size with no grain-boundary second phase, consistent with their very low $\rho(T_c)$ ($>1\mu\Omega \cdot$ cm). High density dislocation arrays at grain boundaries may form where MgB_2 columns coalesce during growth, putting neighboring MgB_2 grains in biaxial tension and giving rise to exceptionally high T_c (~41.8 K).[22,23] The SAD pattern taken from the same region (inset of Fig. 8(a)) also shows single-crystal-like MgB_2 diffraction with several weaker spots in-between. Some are due to the presence of a few MgB_2 grains rotated by 30° about the c-axis. Other weak diffraction spots can be attributed to double diffraction, which commonly occurs when there exist two different materials through the sample thickness, such as MgB_2 and the SiC substrate, as we have here.[6]

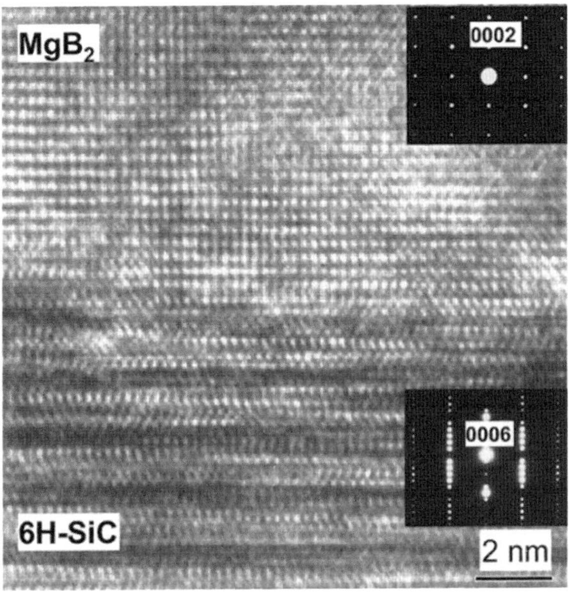

Fig. 9. HRTEM image of the film/substrate interface in a pure MgB_2 HPCVD film on (0001) 6H–SiC substrate. Top inset: SAD pattern from the film. Bottom inset: SAD pattern from the substrate. Reprinted with permission from Ref. 24. © 2007 Elsevier.

Cross-section samples prepared from the same pure HPCVD film show that most of the film has a perfect MgB_2 lattice as illustrated in the HRTEM image in Fig. 8(b). The $[11\bar{2}0]$ SAD pattern in Fig. 8(c) is also single-crystal-like, with some weak diffraction spots which seem to be generated from a small amount of MgB_2 grains with their c-axis parallel to the interface. A clean MgB_2/SiC interface is shown in the HRTEM image in Fig. 9, which also demonstrates the a-to-a orientation alignment between MgB_2 and SiC.[24]

2.4 *C-doped MgB_2 thin films grown by HPCVD*

C-doping in HPCVD MgB_2 films has been achieved, by introducing a secondary H_2 flow carrying a C-containing precursor molecule during HPCVD deposition.[24] Both the type of C precursors and the deposition conditions (such as H_2 flow rate) affect the morphology of MgB_2 grains in C-doped HPCVD films. To better illustrate the effect

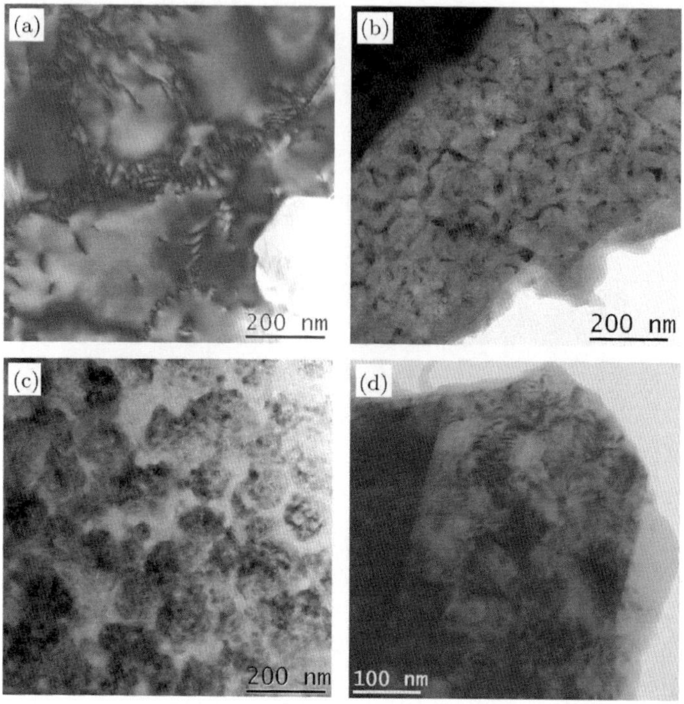

Fig. 10. BF-TEM images taken at the [0001] zone-axis from plan-view (a) pure, (b) $(MeCp)_2Mg$-sourced, (c) CH_4-sourced, (d) $B(CH_3)_3$ sourced MgB_2 HPCVD films. (a–c) Reprinted from Ref. 25. (d) From Ref. 6.

of different C precursors, we here present side-by-side comparisons of the microstructure of pure and doped HPCVD films with various C sources, including $(MeCp)_2Mg$, CH_4 and $B(CH_3)_3$. Figure 10 shows BF images taken from four HPCVD films in plan-view. Again, all films show columnar morphology, indicating the Volmer–Weber growth mechanism.[22] Despite similar growth temperatures (650–810°C), adding C into MgB_2 thin films clearly reduces the grain size and increases structural disorder. MgB_2 grain sizes follow the order: the $(MeCp)_2Mg$-sourced film (5–40 nm) $<$ the CH_4-sourced film (50–100 nm) $<$ the $B(CH_3)_3$-sourced film (over 100 nm) $<$ the pure film (over 200 nm). There is also a correlation between the MgB_2 grain size and the epitaxial quality of the film: SAD patterns taken from

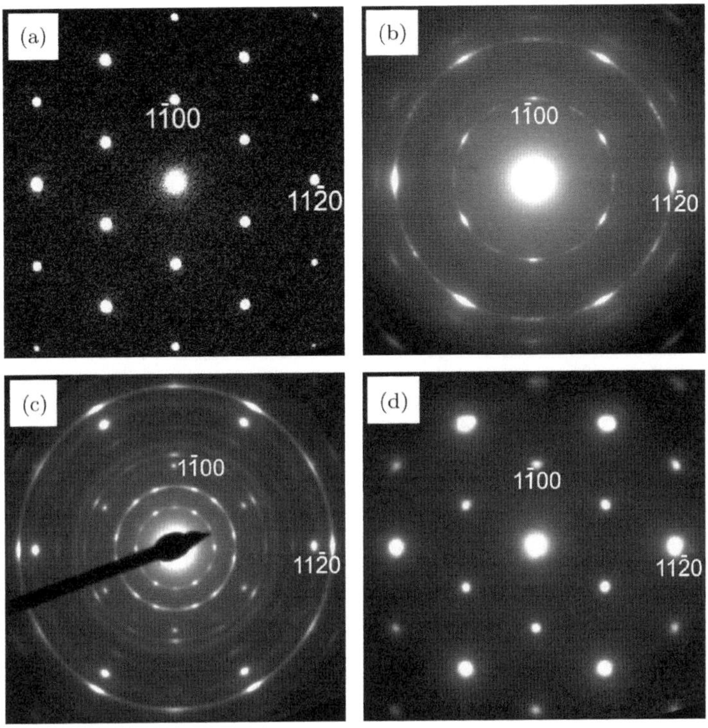

Fig. 11. [0001] zone-axis SAD patterns from plan-view (a) pure (reprinted from Ref. 25), (b) $(MeCp)_2Mg$-sourced (reprinted with permission from Ref. 9. AIP Publishing LLC), (c) 10 sccm CH_4-sourced (from Ref. 26. Reproduced with permission from IOP Publishing), (d) 10 sccm $B(CH_3)_3$-sourced MgB_2 HPCVD films (from Ref. 6. Reproduced with permission from IOP Publishing. The arrows indicate diffraction from second phases such as MgO).

these samples show sharp diffraction spots from epitaxial MgB_2 in the large-grain films, such as the pure (Fig. 11(a)) and $B(CH_3)_3$-sourced films (Fig. 11(d)); while diffraction arcs/streaks caused by small-angle rotations about the *c*-axis are observed in the small-grain films, such as the $(MeCp)_2Mg$-sourced (Fig. 11(b)) and CH_4-sourced films (Fig. 11(c)). Such in-plane orientation variation is up to $\pm 10°$, so it is much less than the orientation variation in the low-temperature grown MgB_2 films shown in Figs. 2 and 3. Diffraction from second phases such as MgO has been observed in small-grain

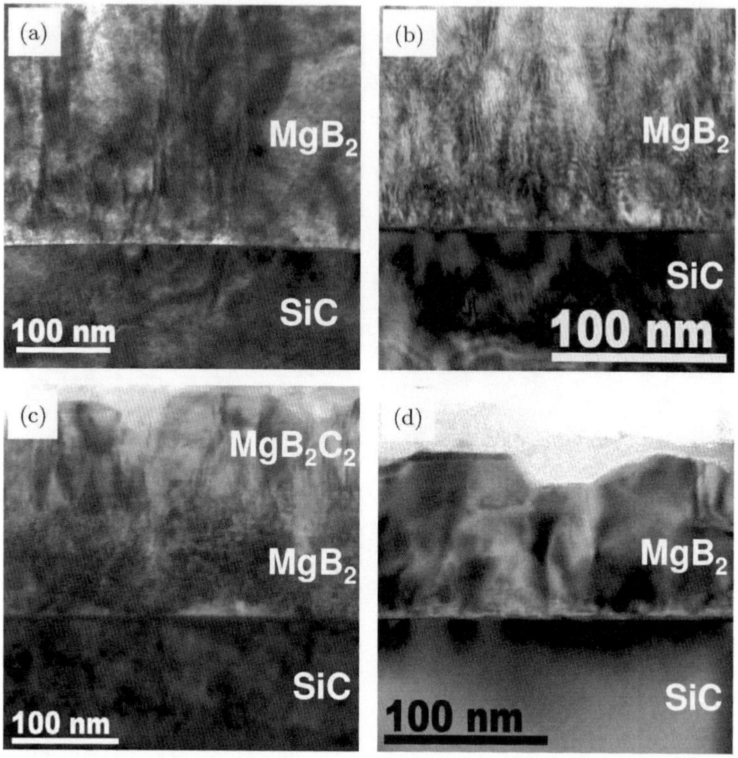

Fig. 12. BFTEM images taken at the [11$\bar{2}$0] zone-axis from cross-section (a) pure (from Ref. 6, Reproduced with permission from IOP Publishing), (b) $(MeCp)_2Mg$-sourced (reprinted with permission from Ref. 9. © 2007 AIP Publishing LLC), (c) 10 sccm CH_4-sourced (reprinted from Ref. 25), (d) 10 sccm $B(CH_3)_3$-sourced MgB_2 HPCVD films (from Ref. 6. © Reproduced with permission from IOP Publishing).

films as indicated by arrows in Figs. 11(b–c), but is not detectable in large-grain films.

TEM results from cross-section film samples, shown in Figs. 12 and 13, demonstrate the same trend. The $(MeCp)_2Mg$- and CH_4-sourced films have the highest density of defects, as indicated by the Moiré fringes and other contrast complexities present in Figs. 12(b) and 12(c). The CH_4-sourced film (Fig. 12(c)) has two layers: the top MgB_2C_2 layer is formed when too much carbon is introduced during growth, and the bottom MgB_2 layer shows a high degree

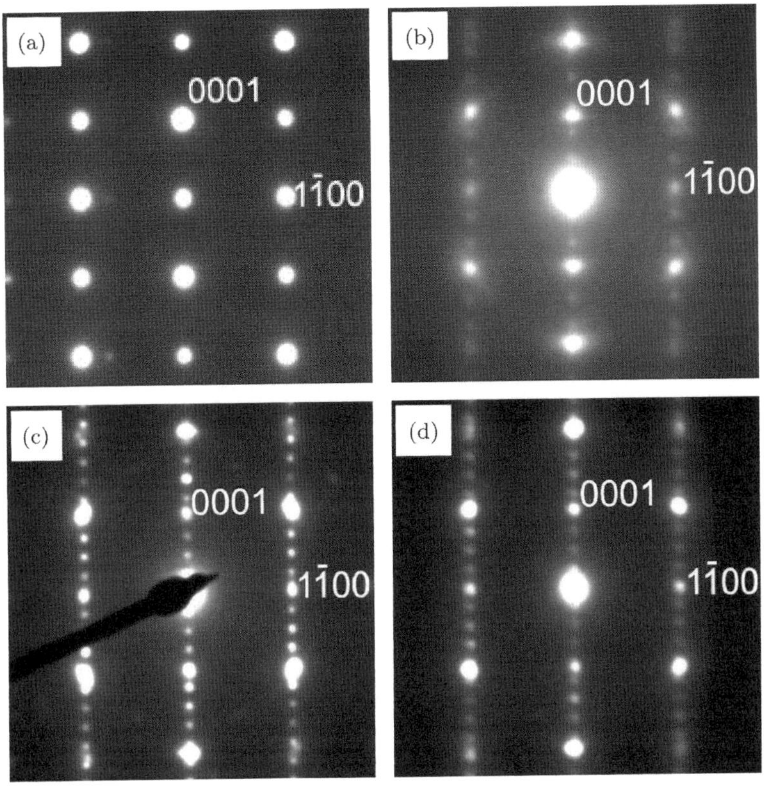

Fig. 13. [11$\bar{2}$0] zone-axis SAD patterns taken from cross-section (a) pure (from Ref. 6. Reproduced with permission from IOP Publishing. All rights reserved), (b) $(MeCp)_2Mg$-sourced (reprinted with permission from Ref. 9. © 2007 AIP Publishing LLC), (c) CH_4-sourced (reprinted from Ref. 25).

of disorder.[26] The $B(CH_3)_3$-sourced film (Fig. 12(d)) is 50–90 nm thick, and the large thickness variance is due to the higher deposition temperature (810°C versus ~700°C for others), which results in less control over the deposition rate.[6] On the other hand, the $B(CH_3)_3$-sourced film does not exhibit the complex contrast as observed in the $(MeCp)_2Mg$- and CH_4-sourced films, and is more like the pure HPCVD film shown in Fig. 12(a), indicating high epitaxial quality with few defects. The SAD patterns further show that the pure and $B(CH_3)_3$-sourced films are epitaxial with all MgB_2 grains having almost uniform orientation, while the $(MeCp)_2Mg$- and CH_4-sourced films contain c-axis tilt disorder as evidenced by the diffraction steaks associated with the MgB_2 (0001) diffraction spots in Figs. 13(b) and 13(c). The smaller intensity spread of the (0001) diffraction streaks in Fig. 13(c) indicates less c-axis tilt disorder in the CH_4-sourced film compared to the $(MeCp)_2Mg$-sourced film.

Overall, HPCVD films increase in crystal quality in the order: $(MeCp)_2Mg$-sourced films, CH_4-sourced films, $B(CH_3)_3$-sourced films, pure films. The film doped using $(MeCp)_2Mg$ has the smallest MgB_2 grains, extensive in-plane rotations and c-axis tilt disorder, while the pure film is an almost perfect single crystal with few defects. The quality of the films doped using CH_4 and $B(CH_3)_3$ is in between. These general observations are consistent with their superconducting properties, with $(MeCp)_2Mg$-sourced films showing the highest $\rho(40\,\mathrm{K})$ and $H_{c2}^{||}(0\,\mathrm{K})$, and the lowest T_c out of the group.[6,27] The $B(CH_3)_3$-sourced films have low $\rho(40\,\mathrm{K})$ and high grain connectivity and J_c values. The $\sim 10^6\,\mathrm{Acm}^{-2}$ $J_c(5\,\mathrm{K},\ 8\,\mathrm{T}\ \mathrm{H}\|ab)$ of $B(CH_3)_3$-sourced films is one of the highest in-field J_c values achieved in MgB_2. On the other hand, $B(CH_3)_3$-sourced films have the low H_{c2}, probably due to the lack of electron-scattering defects.[6]

2.5 *Tilted c-axis MgB_2 thin films*

Except for the inclined MgB_2 columns observed in Fig. 7, most of the high-quality large-grain or epitaxial MgB_2 films are grown with the c-axis perpendicular to the film plane, taking advantage of the anisotropic layered structure of MgB_2. However, changing the c-axis orientation of MgB_2 thin films is desirable for investigations of the anisotropic transport properties and two-gap properties of MgB_2.[28] Tilted c-axis growth of MgB_2 thin films has been achieved on (110)- and (211)-oriented MgO and $Y_{0.18}Zr_{0.82}O_2$ (YSZ) substrates using HPCVD.[28,29] On a (211)-oriented MgO substrate, MgB_2 film still tends to grow epitaxially with $(0001)MgB_2 \parallel (111)MgO$, such that the c-axis of MgB_2 is tilted by 19.5° (the angle between MgO(111) and (211)) away from the film normal, as seen in Fig. 14. The tilt angle can be increased to 41° by depositing MgB_2 on

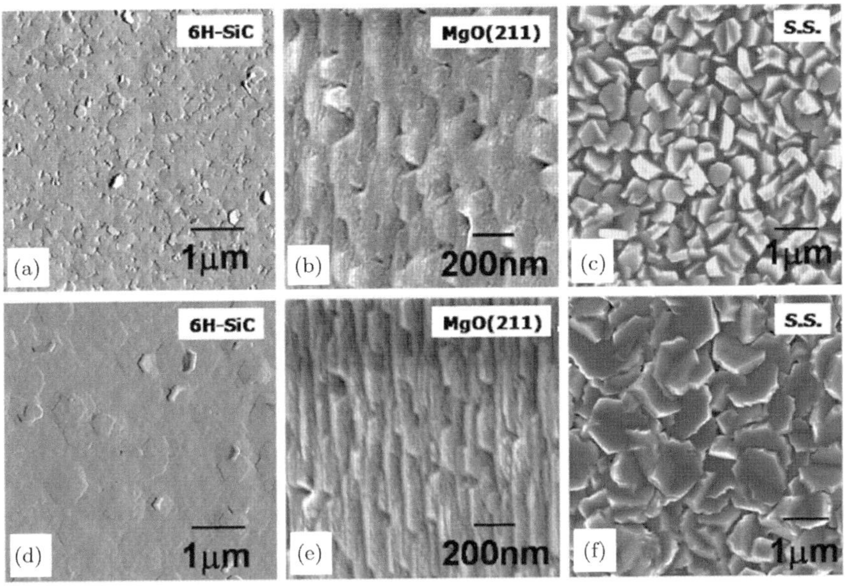

Fig. 14. Surface morphologies of MgB_2 films grown by HPCVD on three different substrates: 6H–SiC (0001) in (a) and (d), MgO (211) in (b) and (e) and stainless steel in (c) and (f). For each substrate, two films grown with different deposition flow rates are presented and the upper images refer to the lower deposition rate. From Ref. 29. © 2010 IOP Publishing. Reproduced with permission from IOP Publishing. All rights reserved.

(110)-oriented MgO. Similarly, a (001)-textured MgB_2 film can form on (110)-oriented YSZ with the *c*-axis tilted $\pm 32°$ from the film normal axis perpendicular to the film plane, taking advantage of the anisotropic layered structure of MgB_2. However, changing the *c*-axis orientation of MgB_2 thin films is desirable for investigations of the anisotropic transport properties and two-gap properties of MgB_2.[28] Tilted *c*-axis growth of MgB_2 thin films has been achieved on (110)- and (211)-oriented MgO and $Y_{0.18}Zr_{0.82}O_2$ (YSZ) substrates using HPCVD.[28,29] On a (211)-oriented MgO substrate, MgB_2 film still tends to grow epitaxially with (0001)MgB_2 $\parallel$ (111)MgO, such that the *c*-axis of MgB_2 is tilted by 19.5° (the angle between MgO(111) and (211)) away from the film normal, as seen in Fig. 14. The tilt angle can be increased to 41° by depositing MgB_2 on (110)-oriented

MgO. Similarly, a (001)-textured MgB_2 film can form on (110)-oriented YSZ with the *c*-axis tilted ±32° from the film normal.[28]

Figure 14 also shows the effect of deposition rate on MgB_2 grain size and surface morphology. For HPCVD MgB_2 films, the deposition rate is strongly affected by the B_2H_6 and H_2 flow rate (due to the adsorption-controlled growth) as well as the growth temperature.[29] For all three types of HPCVD films shown in Fig. 14 ((0001)-oriented epitaxial MgB_2 on SiC, *c*-axis tilted epitaxial MgB_2 on (211) MgO and randomly-oriented polycrystalline MgB_2 on stainless steel), higher deposition rates give rise to larger MgB_2 grains and higher surface roughness. The effect is clearer for the tilted *c*-axis MgB_2 film on (211) MgO, in which thinner grains in the *c*-axis direction and larger grains in the *ab* plane form with the higher deposition rate (Fig. 14(e)). On the other hand, both the crystallinity and superconducting properties of HPCVD MgB_2 films are not sensitive to such morphology change. Tuning the deposition rate therefore provides an effective way of fabricating smooth MgB_2 films without degrading the properties, useful for making multilayer superconducting devices.[29]

3. Disorder in MgB_2 Thin Films

This section presents the detailed microstructure of defects in MgB_2 thin films, and discusses their effects on superconducting properties.

3.1 *Grain boundaries between in-plane rotated MgB_2 grains*

Even in the very clean pure MgB_2 HPCVD films, there is structural disorder, such as high-density dislocation arrays at grain boundaries and MgB_2 grains rotated by 30° about the *c*-axis (Fig. 8). A 30° rotated MgB_2 grain in plan-view, identified by convergent-beam electron diffraction (CBED), is shown at the center of Fig. 15. The CBED pattern taken from the central grain shows two sets of diffraction disks rotated by 30° with respect to each other: one from matrix MgB_2 (the same as the upper-right CBED pattern) and the other from the central 30° rotated MgB_2 grain. Both patterns

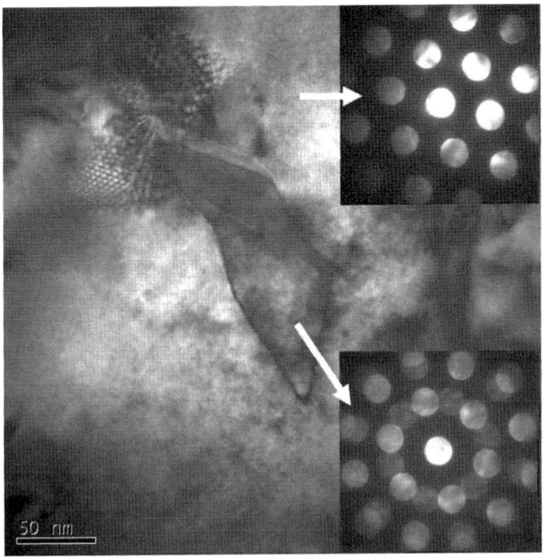

Fig. 15. BFTEM image taken from a plan-view pure MgB_2 HPCVD film at the [0001] zone axis. A rotated MgB_2 grain sits at the center, as indicated by the CBED patterns shown in the insets. From Ref. 6. © 2010 IOP Publishing. Reproduced with permission from IOP Publishing.

are present because the CBED interaction volume is larger than the central, 30° rotated grain.[6] A 30° rotated MgB_2 grain has also been imaged in a cross-section film sample, as shown in the HRTEM image taken around the MgB_2/SiC interface in Fig. 16(a). Besides the *ab* plane lattice fringes of MgB_2 and SiC, there are also Moiré fringes at the center of the image, which are generated from the overlap of two MgB_2 grains rotated by 30° around the *c*-axis with respect to each other (as confirmed by the power spectrum in Fig. 16(b)).[30] A sharp (001) grain boundary between the rotated and matrix MgB_2 grains is also visible in Fig. 16(a).

There are many more rotated MgB_2 grains in C-doped HPCVD films than in pure films. In the SAD pattern taken from a $(MeCp)_2Mg$-sourced film (Fig. 17(a)), the brightest arcs correspond to [0001] MgB_2 as expected for an epitaxial thin film with some orientation spread about the *c*-axis. The other strong diffraction arcs form another pattern rotated 30° with respect to the main

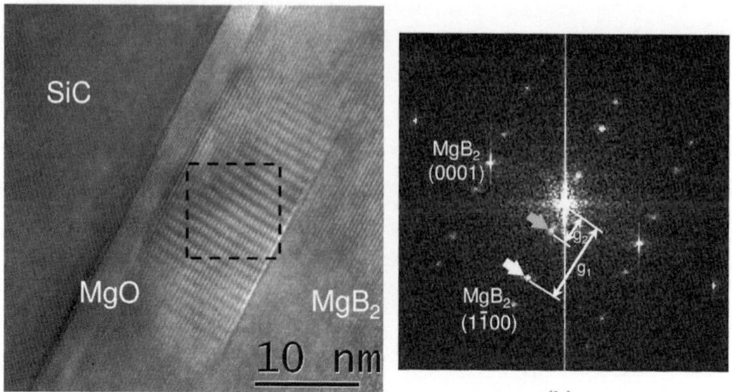

(a) (b)

Fig. 16. (a) HRTERM image taken along $[11\bar{2}0]$ in a neutron-irradiated MgB_2 HPCVD film, showing Moiré fringes generated from the overlap of two MgB_2 grains with a 30° rotation in between. (b) Power spectrum generated from the dashed square in (a). From Ref. 30.

MgB_2 diffraction, as illustrated by gray spots in Fig. 17(b). The schematic in Fig. 17(b) can also be used to describe the CBED pattern in the lower-right inset of Fig. 15. DF TEM images in Figs. 17(c)–17(d) further show that the 30° rotated grains comprise 10–20% of the doped film. Similar 30° rotated grains have also been reported in other hexagonal thin films, their origin not being yet fully understood.[7,18,31–33]

The atomic structure of rotated MgB_2 grains has been explicitly imaged using HRTEM, as shown in Fig. 18. There are two MgB_2 grains viewed along the [0001] direction, rotated by 30° with respect to each other, and with a ∼5 nm wide amorphous region in between. Such intergrain amorphous regions are common and give rise to dark regions in Figs. 17(c) and 17(d) (bright regions in Fig. 10(b)). EELS on this sample shows that the amorphous boundary phase is significantly C rich. At the middle of the EELS scan line in Fig. 19(d), which corresponds to an MgB_2 grain, the C content is ∼5%, while in the surrounding regions the C content is ∼40%. It is therefore evident that $(MeCp)_2Mg$ is not an effective C-doping source

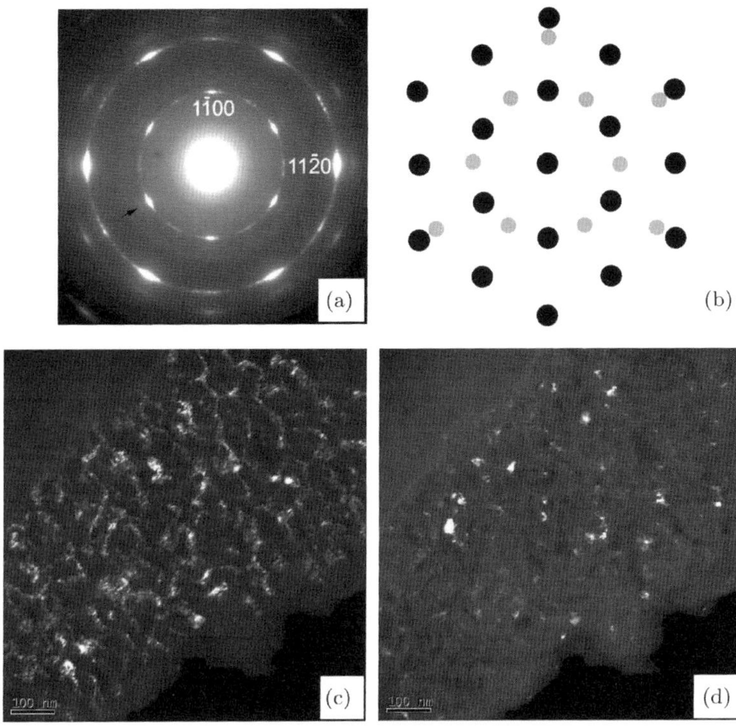

Fig. 17. (a) [0001] zone-axis SAD pattern taken from a plan-view C-doped film using $(MeCp)_2Mg$, (b) The model to explain the SAD pattern in (a). (a–b) Reprinted with permission from Ref. 9. © 2007 AIP Publishing LLC, (c) DF TEM image showing unrotated MgB_2 grains (bright regions), (d) DF image showing 30° rotated MgB_2 grains (bright regions). (c, d) Reprinted from Ref. 25.

and most of the C forms the amorphous boundary phase between MgB_2 grains. Such an amorphous boundary phase limits the current-carrying cross-section, significantly increasing $\rho(40\,K)$ and decreasing the measured J_c, which explains why $(MeCp)_2Mg$-sourced films have lower zero-field J_c than pure MgB_2 films.[33] It probably also limits the MgB_2 grain size in the $(MeCp)_2Mg$-sourced film, as evidenced by the behavior in CH_4- and $B(CH_3)_3$-sourced films, in which the MgB_2 grain size increases with a reduction of such a boundary phase.

Clean grain boundaries between rotated MgB_2 grains are also present. An example of 30°-rotated MgB_2 grains with a clean boundary is shown in the upper-right inset of Fig. 18. The lower-left

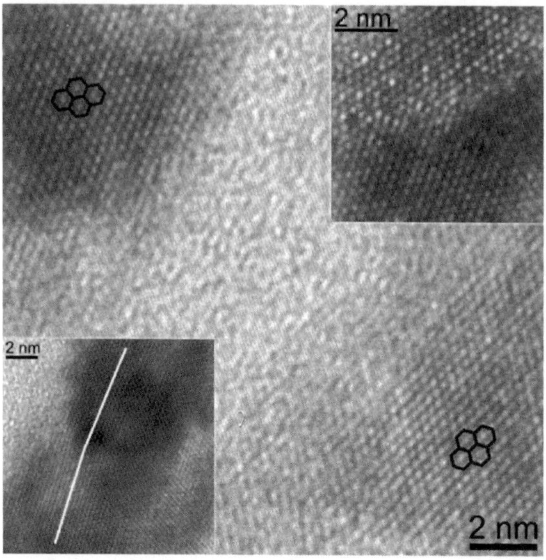

Fig. 18. HRTEM image taken along [0001] showing an amorphous region between two 30° rotated MgB_2 grains in a plan-view, $(MeCp)_2Mg$-sourced film. The upper-right inset shows a clean boundary between two MgB_2 grains, and the lower-left inset shows two MgB_2 grains with a 4.5° rotation angle with respect to each other. Reprinted with permission from Ref. 9. © 2007 AIP Publishing LLC.

inset shows two MgB_2 grains misoriented by 4.5° with respect to each other, which is consistent with the angle subtended by the extended diffraction streaks in Fig. 17(a). At the boundary of these rotated grains, interface dislocations are present. Such clean boundaries are expected to act as flux pinning centers without reducing the grain connectivity and the current-carrying cross-section.

3.2 *Out-of-plane disorder in MgB_2 thin films*

As seen in SAD patterns from cross-section samples in Fig. 13, another form of structural disorder is *c*-axis tilt, which is especially prominent in $(MeCp)_2Mg$-sourced HPCVD films. Figure 20(a) is a HRTEM image taken from a thin cross-section sample. The lattice on the left side is on the $[11\bar{2}0]$ zone-axis, while on the right, the lattice fringes of the MgB_2 *ab* plane are tilted by $\sim$3.6° with respect to the left. The right-hand grain is also rotated around the *c*-axis, which

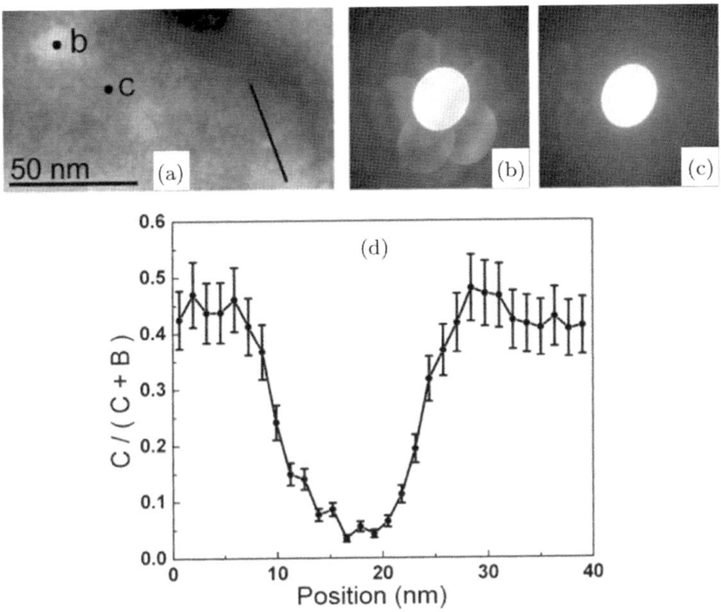

Fig. 19. (a) Z-contrast image taken from a plan-view, $(MeCp)_2Mg$-sourced film sample, (b) and (c) CBED patterns taken from spot b and c in (a), respectively, (d) C distribution derived from EELS line-scan along the black line in (a). Reprinted with permission from Ref. 9. © 2007 AIP Publishing LLC.

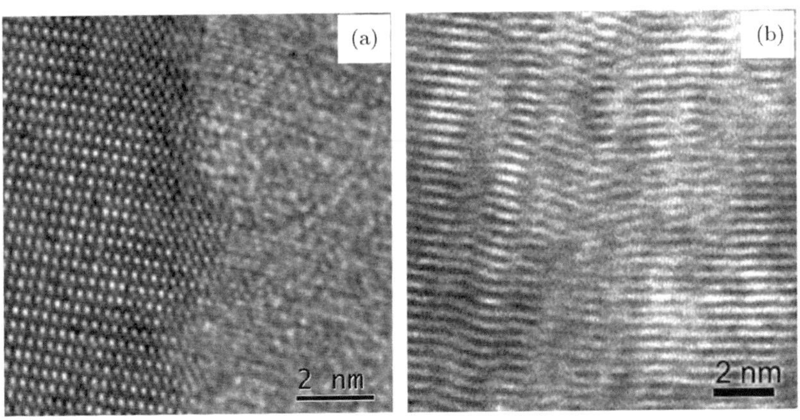

Fig. 20. HRTEM images taken along $[11\bar{2}0]$ from a cross-section film sample doped using $(MeCp)_2Mg$, showing tilted MgB_2 grains in (a) thin (from Ref. 6. Reproduced with permission from IOP Publishing), and (b) thick regions (reprinted with permission from Ref. 9. AIP Publishing LLC).

makes the viewing direction off the zone-axis, so it shows only weak lattice fringes. Dislocations with Burger's vector along the c-axis have been observed between tilted grains. In thicker cross-section samples, such tilted MgB_2 grains overlap through the thickness and give rise to a "buckled" appearance of the ab-plane lattice fringes shown in Fig. 20(b). This imaging artefact has been confirmed using HRTEM image simulations.[25] The SAD patterns in Fig. 13 shows many more tilted MgB_2 grains in $(MeCp)_2Mg$-sourced HPCVD films, so they are presumably facilitated by the amorphous boundary phase caused by $(MeCp)_2Mg$. Such disorder is of particular interest, as it disturbs the MgB_2 p_z π orbitals and may preferentially cause strong π-band scattering, which could lead to $H_{c2}^{\parallel}(0\,K)$ being greatly enhanced ($>60\,T$) in comparison to the effect of bulk alloying with C ($<45\,T$).[34]

Another observed out-of-plane disorder is semi-coherent MgO nanoplatelets in CH_4-sourced HPCVD films. Figure 21(a) is the

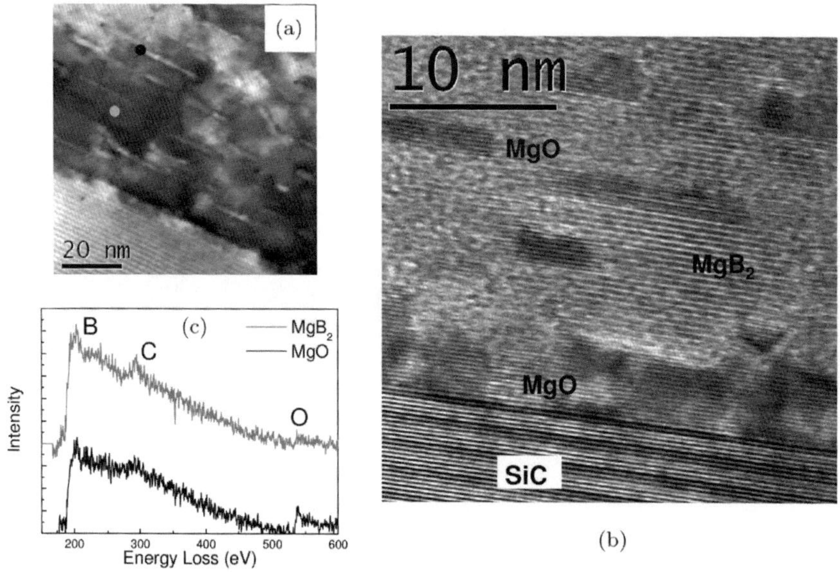

Fig. 21. (a) Z-contrast STEM image of a cross-section CH_4-sourced film sample, (b) HRTEM image taken along $[11\bar{2}0]$ from the same sample as (a), showing MgO platelets which are semi-coherent with the MgB_2 lattice, (c) EEL spectra from MgB_2 (gray) and MgO (black) regions as indicated in (a). From Ref. 26. Reproduced with permission from IOP Publishing.

amorphous boundary phase caused by $(MeCp)_2Mg$. Such disorder is of particular interest, as it disturbs the MgB_2 p_z π orbitals and may preferentially cause strong π-band scattering, which could lead to $H_{c2}^{\|}(0\,K)$ being greatly enhanced (>60 T) in comparison to the effect of bulk alloying with C (<45 T).[34]

Another observed out-of-plane disorder is semi-coherent MgO nanoplatelets in CH_4-sourced HPCVD films. Figure 21(a) is a scanning-TEM (STEM) Z-contrast image taken near the interface of a cross-section sample, showing many defects in the form of bright lines inside the MgB_2 film. The HRTEM image in Fig. 21(b) illustrates that these bright-line defects are several-nanometer-thick MgO platelets that are largely coherent with the surrounding MgB_2 lattice, as identified from the lattice fringes. A more detailed atomic structure of epitaxial MgB_2 on MgO is illustrated in Fig. 22 demonstrating the orientation relationship between the two phases

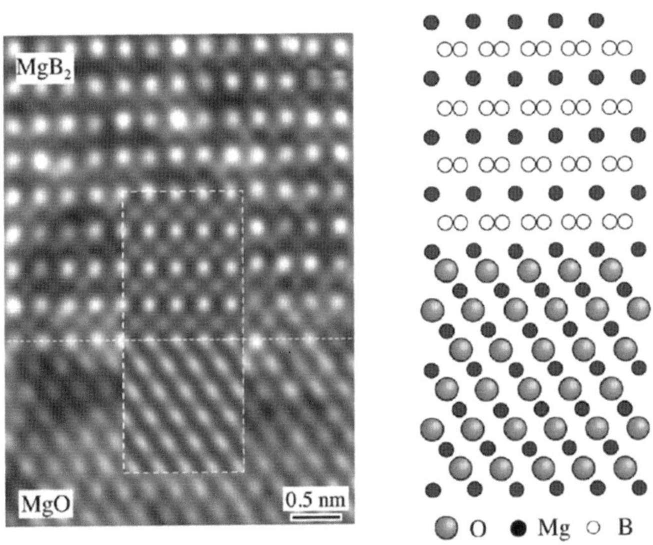

(a) (b)

Fig. 22. (a) HRTEM image of the MgB_2–MgO interface with the imaging electron beam along the $[11\bar{2}0]$ direction of MgB_2. The inset in the middle is a simulated image, (b) an atomic structure of the interface deduced from (a). Reprinted with permission from Ref. 15. © 2002 AIP Publishing LLC.

((0001)[1$\bar{1}$00]MgB_2 ∥ (111)[1$\bar{2}$1]MgO). As MgO has a higher average atomic number than MgB_2, the MgO platelets appear as bright lines in the Z-contrast image. EEL spectra taken from the bright line region and surrounding MgB_2 region (Fig. 21(c)) verify the excess O_2 associated with the defects. (The small O_2 signal from the MgB_2 region arises from surface oxidation of the TEM sample).

As MgO platelets have not been observed in other HPCVD films, either the CH_4 gas flow or the tungsten filament which was used to decompose CH_4 must introduce O_2 into the film. The size of the MgO nanoplatelets is close to the MgB_2 coherence length (~6 nm), so they should be effective pinning centers which enhance J_c. Although the in-plane mismatch between MgO (111) and MgB_2 (0001) is small (<3.5%), the spacing between MgO (111) planes is 2.44 Å, quite different from MgB_2 c lattice parameter (3.52 Å), so that the lattice mismatch at platelet boundaries causes c-axis dislocations and strain in MgB_2 lattice. The strain along the c-axis can be as large as 10% around MgO platelets and is still significant up to 10 nm away.[26] As the spacing between MgO platelets is less than 20 nm (Figs. 21(a) and 21(b)), a large fraction of the sample suffers this c-axis strain, which disturbs the out-of-plane p_z π orbitals, preferentially causing strong π scattering and potentially leading to exceptionally high $H_{c2}^{\parallel}(0\,\mathrm{K})$ (~60 T).[26] Such MgO nanoplatelets may also be present in low-temperature-grown O_2-incorporated MgB_2 films and contribute to their enhanced J_c and H_{c2}, but they are much easier to characterize in nearly epitaxial thin films. As chemical doping of MgB_2 often introduces intergrain second phases which reduce the grain connectivity and therefore J_c, this result suggests that introducing MgO nanoplatelets as in CH_4-sourced HPCVD films may be a more effective approach to enhance J_c and H_{c2} simultaneously in MgB_2.

3.3 *Other second phases in MgB_2 thin films*

MgO can form epitaxial layers at the film/substrate interface, especially for films grown on O_2-containing substrates regardless of deposition techniques.[7] For reactive evaporation deposition, Fig. 7

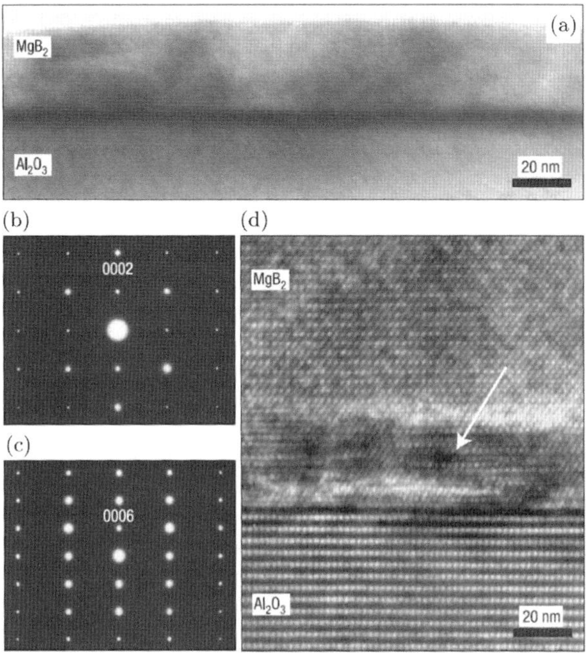

Fig. 23. (a) BF TEM image from a cross-section pure HPCVD film, (b) $[11\bar{2}0]$ zone-axis SAD pattern from the film, (c) an SAD pattern taken from the substrate along the same direction as in (b), showing a $[10\bar{1}0]$ zone-axis Al_2O_3 SAD pattern, (d) HRTEM image of the interface showing an intermediate layer, resulting from a superposition of MgO with MgB_2 lattice. Reprinted by permission from Macmillan Publishers Ltd: Nature Materials [Ref. 35]. Nature Publishing Group.

reveals continuous dark MgO interfacial layers on sapphire and alumina substrates (Figs. 7(a) and 7(c)), but not at the MgB_2/SiC interface (Fig. 7(d)). The same trend is observed for pure MgB_2 films grown by HPCVD: a thin MgO layer is present at the MgB_2/sapphire interface, as seen in Fig. 23(d), but not at the MgB_2/SiC interface (Fig. 9). Given that HPCVD films are grown in a highly reducing H_2 atmosphere, O_2 diffusion out of the oxide substrates is believed to be the origin of the MgO layer at interfaces. Furthermore, it has been reported that pre-depositing B onto the oxide substrate surface (before Mg deposition) prevents the formation of MgO interfacial layers, which confirms the O_2 out-diffusion mechanism.[7] Due to the similar Mg sublattices in MgO (111) and MgB_2 (0001), the

MgO interfacial layer does not change the orientation relationship between MgB_2 and sapphire ((0001)[1$\bar{1}$00]MgB_2 ∥ (111)[1$\bar{2}$1]MgO ∥ (0001)[11$\bar{2}$0]Al_2O_3). Misfit dislocations with Burgers vector $\frac{1}{2}[1\bar{1}00]$ can also form at the MgB_2/MgO interfaces.[15] As the product of reaction between MgO and Al_2O_3, $MgAl_2O_4$ epitaxial transition layers have also been observed at the MgB_2/sapphire interface (Fig. 6(d)). The formation of $Mg_{1-x}Al_xB_2$ associated with O_2 out-diffusion near the MgB_2/sapphire interface has also been reported.[36]

In C-doped MgB_2 HPCVD films grown on SiC substrates, however, the MgO phase is also present at interfaces. One example of a continuous MgO interfacial layer is shown in Fig. 21(b) for a CH_4-sourced film. Similar interfacial layers have been observed in $B(CH_3)_3$-sourced films (but not the MgO nanoplatelets). As in MgB_2/sapphire thin films, continuous MgO layers at interfaces do not affect the orientation relationship between MgB_2 and SiC ((0001)[1$\bar{1}$00]MgB_2 ∥ (111)[1$\bar{2}$1]MgO ∥ (0001)[1$\bar{1}$00]SiC), so the films are still able to grow epitaxially. On the other hand, MgO in $(MeCp)_2Mg$-sourced films forms separated interfacial islands instead of continuous layers, as shown in the HRTEM image in Fig. 24(a). As in the MgO nanoplatelets, the lattice spacing mismatch between

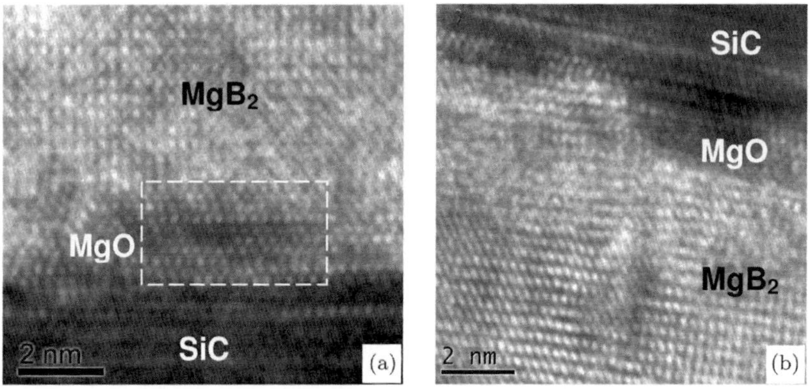

Fig. 24. HRTEM images taken from a [11$\bar{2}$0] cross-section $(MeCp)_2Mg$-sourced film sample, showing (a) a coherent MgO island at the interface (Ref. 6 © 2010 IOP Publishing. Reproduced with permission from IOP Publishing. All rights reserved), (b) c-axis tilt disorder originating from MgO interface islands (reprinted from Ref. 25).

MgO (111) planes (2.44 Å) and MgB_2 (0001) planes (3.52 Å) causes discontinuity of MgB_2 *ab* planes by forming *c*-axis dislocations and tilt boundaries as shown in Fig. 24(b). The high density of *c*-axis tilt grain boundaries observed in $(MeCp)_2Mg$-sourced films might be partially due to these MgO interfacial islands, in addition to the amorphous C-rich boundary phase. Since doped HPCVD films use the same SiC substrates as the pure HPCVD films, the observed MgO interfacial phase is likely to be caused by the additional gas flow carrying the C-doping source. On the other hand, the presence of coherent interfaces between MgO (111) and MgB_2 (0001) lattice has inspired the work of fabricating MgB_2/MgO/MgB_2 sandwich-type Josephson junctions using HPCVD (Fig. 25), which exhibit high J_c comparable to conventional Nb junctions.[37]

C-doping in MgB_2 also leads to various second phases, such as the C-rich amorphous boundary phase in $(MeCp)_2Mg$-sourced films (Figs. 18 and 19). Changing the C source to CH_4 effectively reduces the boundary phase, but gives rise to epitaxial MgB_2C_2 layers on top of MgB_2 films. The epitaxial MgB_2C_2 phase has been identified using the SAD pattern taken from the top part of the film: the outer diffraction spots indicated by arrows in Fig. 26(b) match the MgB_2C_2

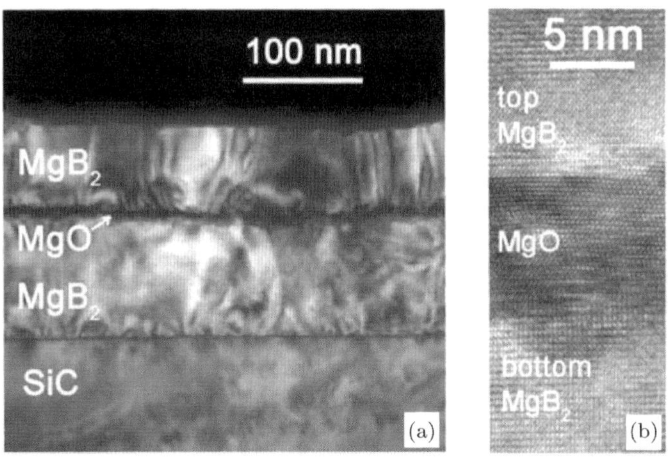

Fig. 25. (a) Low magnification DF-TEM and (b) HRTEM [11$\bar{2}$0] cross-sectional view of the MgB_2/MgO/MgB_2 trilayer structure grown by HPCVD. Reprinted with permission from Ref. 37. © 2010 AIP Publishing LLC.

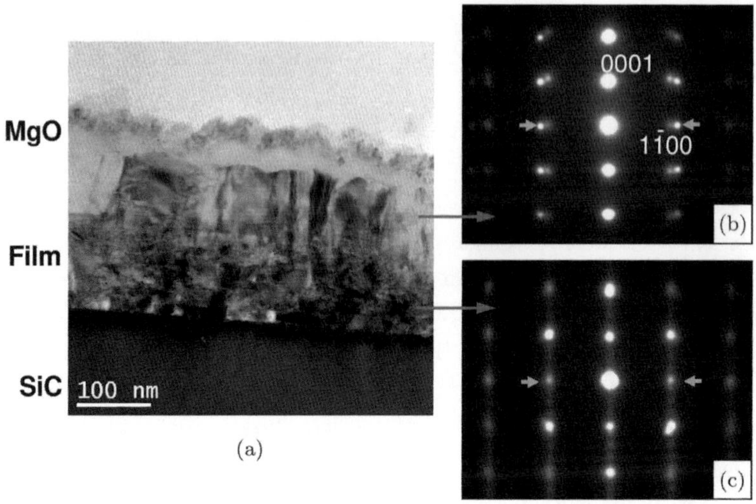

Fig. 26. (a) BF TEM image of a cross-sectional CH_4-sourced film sample (from Ref. 6, Publishing. Reproduced by permission of IOP Publishing. All rights reserved), (b) and (c) [11$\bar{2}$0] zone-axis SAD patterns taken from the upper layer (b) and lower layer (c), respectively (from Ref. 26, IOP Publishing. Reproduced with permission from IOP Publishing.). The upper layer shows the presence of MgB_2C_2 diffraction while the lower layer does not, as indicated by arrows.

structure, while the inner spots are from [11$\bar{2}$0] MgB_2. In the SAD pattern taken from the interface region (Fig. 26(c)), all the MgB_2C_2 diffraction spots disappear, leaving only MgB_2 diffraction. (Weak spots: come from the SiC substrate.) The MgB_2C_2 phase in the top layer is consistent with the lower density as revealed by Z-contrast imaging (Fig. 27(a)) and the ~50% C concentration measured by EELS (Fig. 27(b)). Figure 27(b) also shows that there is about 2–13 at.% C concentration in the lower MgB_2 layer, with a C-depletion region at the MgB_2/MgB_2C_2 interface. Another HPCVD film sample grown at lower CH_4 flow rate shows a much thinner MgB_2C_2 top layer (~20 nm for 7 sccm flow rate versus ~100 nm for 10 sccm flow rate). It indicates that MgB_2 films were overdoped even at 7 sccm CH_4 flow rate and that excess C is rejected from the $Mg(B_{1-x}C_x)_2$ to form the epitaxial MgB_2C_2 second phase.[6] The formation of MgB_2C_2 with the presence of excess C is predicted by the calculated Mg–B–C phase diagram.[38] As J_c was derived using the total film thickness, which

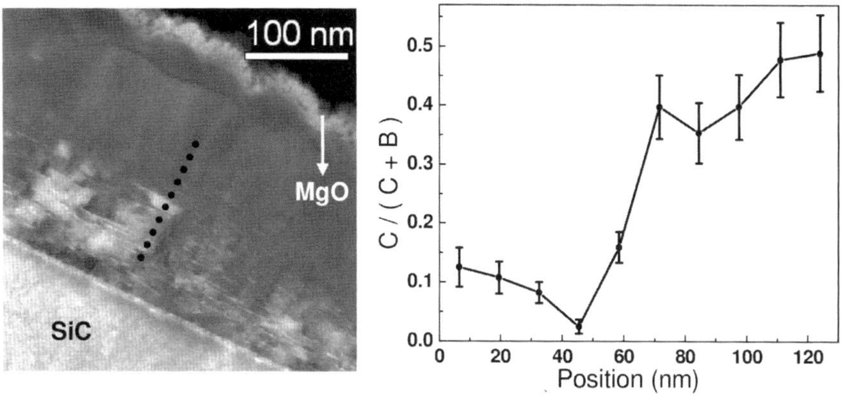

(a) (b)

Fig. 27. (a) Z-contrast STEM image of a cross-sectional CH_4-sourced film sample. (b) C distribution derived from EELS line scan in (a). From Ref. 26. Reproduced with permission from IOP Publishing. All rights reserved.

could be twice as high as the thickness of the MgB_2 superconducting layer, the actual J_c of these samples is even higher than the published values.[39] Coherent intergrowth of MgB_2C_2 nanostripes have also been observed in heavily-doped HPCVD MgB_2 films using $B(CH_3)_3$ as the C source.[40]

4. Conclusions

In summary, MgB_2 thin films exhibit various forms of structural disorder, including both defects (dislocation networks, boundaries between in-plane rotated and c-axis tilted MgB_2 grains) and second phases (MgO nano-platelets, C-rich amorphous boundary phases and degraded amorphous phase). Correlating the disorder with the superconducting properties allows to draw the following conclusions:

Both nanoscale second phase segregation at grain boundaries and the grain boundary itself are pinning defects in MgB_2. Even though it is unclear which form of defects provides pinning force more effectively, the current-blocking effect from the second phases suggests that small-grain MgB_2 without any boundary second-phase is desirable to achieve higher J_c. C-doped HPCVD thin films using $B(CH_3)_3$ provide such a microstructure. The low high-field J_c in pure HPCVD films suggests that dislocations provide little pinning force at high fields.

Effective electron scattering defects in MgB_2 include carbon dopants, grain boundaries and amorphous grain boundary phases. In particular, grain boundaries with small tilt angles between the *ab* planes of the grains (as seen in $(MeCp)_2Mg$-sourced C-doped HPCVD films) and semi-coherent MgO nanoplatelets (as seen in CH_4-sourced C-doped HPCVD films) may cause significant π scattering and give rise to exceptionally high $H_{c2}^{\parallel}(0\,K)$ ($\geq 60\,T$). Dislocations in pure MgB_2 HPCVD films are not very effective in causing electron scattering and increasing H_{c2}.

Results from C-doped MgB_2 films suggest that an epitaxial MgB_2 film with high-density *c*-axis disorder or large lattice strain and without second phases at grain boundaries may yield both high J_c and H_{c2}. HPCVD growth using $B(CH_3)_3$ as the C source provides a promising way to achieve such an optimal microstructure. Intentionally doping O_2 to form semi-coherent MgO nanoplatelets as in CH_4-sourced HPCVD films may also be an effective way to introduce lattice strain. Overall, MgB_2 thin films have provided a simpler and cleaner system than bulk MgB_2 for microstructural observation using TEM-based techniques.

References

1. X. X. Xi, *Supercond. Sci. Technol.* **22**, p. 043001 (2009).
2. H. Y. Zhai, H. M. Christen, L. Zhang, M. Paranthaman, P. H. Fleming and D. H. Lowndes, *Supercond. Sci. Technol.* **14**, p. 425 (2001).
3. A. Serquis, Y. T. Zhu, D. E. Peterson, F. M. Mueller, R. K. Schulze, V. F. Nesterenko and S. S. Indrakanti, *Appl. Phys. Lett.* **80**, p. 4401 (2002).
4. B. J. Senkowicz, R. Pérez Moyet, R. J. Mungall, J. Hedstrom, O. N. C. Uwakweh, E. E. Hellstrom and D. C. Larbalestier, *Supercond. Sci. Technol.* p. **19**, pp. 1173–1177 (2006).
5. Y. Cui, J. E. Jones, A. Beckley, R. Donovan, D. Lishego, E. Maertz, A. V. Pogrebnyakov, P. Orgiani, J. M. Redwing and X. X. Xi, *IEEE T. Appl. Supercon.* **15**, p. 224 (2005).
6. Y. Zhu *et al.*, *Supercond. Sci. Technol.* **23**, p. 095008 (2010).

7. L. Gu and B. H. Moeckly, D. J. Smith, *J. of Crystal Growth* **280**, p. 602 (2005).
8. J. S. Wu, N. Jiang, B. Jiang, J. C. H. Spence, A. V. Pogrebnyakov, J. M. Redwing and X. X. Xi, *Appl. Phys. Lett.* **85**, p. 1155 (2004).
9. Y. Zhu, D. C. Larbalestier, P. M. Voyles, A. V. Pogrebnyakov, X. X. Xi and J. M. Redwing, *Appl. Phys. Lett.* **91**, p. 082513 (2007).
10. C. B. Eom *et al.*, *Nature* **411**, pp. 558–560 (2001).
11. X. H. Zeng *et al.*, *Appl. Phys. Lett.* **79**, p. 1840 (2001).
12. S. Hata, H. Sosiati, N. Kuwano, Y. Tomokiyo, A. Matsumoto, M. Fukutomi, H. Kitaguchi, K. Komori and H. Kumakura, *IEEE Trans. Appl. Supercond.* **15**, p. 3238 (2005).
13. W. N. Kang, H.-J. Kim, E.-M. Choi, C. U. Jung and S.-I. Lee, *Science* **292**, p. 1521 (2001).
14. S. H. Moon, J. H. Yun, H. N. Lee, J. I. Kye, H. G. Kim, W. Chung and B. Oh, *Appl. Phys. Lett.* **79**, p. 2429 (2001).
15. W. Tian, X. Q. Pan, S. D. Bu, D. M. Kim, J. H. Choi, S. Patnaik and C. B. Eom, *Appl. Phys. Lett.* **81**, p. 685 (2002).
16. S. D. Bu *et al.*, *Appl. Phys. Lett.* **81**, p. 1851 (2002).
17. M. Hanna *et al.*, *Supercond. Sci. Technol.* **21**, p. 045005 (2008).
18. W. Jo, J.-U. Huh, T. Ohnishi, A. F. Marshall, M. R. Beasley and R. H. Hammond, *Appl. Phys. Lett.* **80**, p. 3563 (2002).
19. J. Kim, R. K. Singh, J. M. Rowell, N. Newman, L. Gu and D. J. Smith, *J. Crystal growth* **270**, pp. 107–112 (2004).
20. B. H. Moeckly and W. S. Ruby, *Supercond. Sci. Technol.* **19**, L21–L24 (2006).
21. J. Rowell, *Nat. Mater.* **1**, p. 5–6 (2002).
22. A. V. Pogrebnyakov *et al.*, *Phys. Rev. Lett.* **93**, p. 147006 (2004).
23. R. Koch, *J. Phys. Condens. Matter* **6**, p. 9519 (1994).
24. X. X. Xi *et al.*, *Physica C* **456**, pp. 22–37 (2007).
25. Y. Zhu, *TEM Characterization of Microstructure and Chemistry in Magnesium Diboride Superconductor.* PhD Thesis (2008).
26. Y. Zhu, F. Hunte, C. G. Zhuang, Q. R. Feng, Z. Z. Gan, X. X. Xi, D. C. Larbalestier and P. M. Voyles, *Supercond. Sci. Technol.* **22**, p. 125001 (2009).
27. V. Braccini *et al.*, *Phys. Rev. B* **71**, p. 012504 (2005).
28. P. Orgiani *et al.*, *Supercond. Sci. Technol.* **23**, p. 025012 (2010).
29. C. G. Zhuang, K. Cheng, J. M. Redwing, Q. Li and X. X. Xi, *Supercond. Sci. Technol.* **23**, p. 055004 (2010).
30. I. Pallecchi *et al.*, *Supercond. Sci. Technol.* **22**, p. 015023 (2009).
31. A. Posadas, J.-B. Yau, C. H. Ahn, J. Han, S. Gariglio, K. Johnston, K. M. Rabe and J. B. Neaton, *Appl. Phys. Lett.* **87**, p. 171915 (2005).

32. R. C. Powell, N.-E. Lee, Y.-W. Kim and J. E. Greene, *J. Appl. Phys.* **73**, p. 189 (1993).
33. A. V. Pogrebnyakov *et al.*, *Appl. Phys. Lett.* **85**, p. 2017 (2004).
34. A. Gurevich, *Phys. Rev. B* **67**, p. 184515 (2003).
35. X. H. Zeng *et al.*, *Nature Materials* **1**, p. 1 (2002).
36. K. Yamazaki and M. Naito, *Physica C* **495**, pp. 84–87 (2013).
37. K. Chen, C. G. Zhuang, Q. Li, Y. Zhu *et al.*, *Appl. Phys. Lett.* **96**, p. 042506 (2010).
38. A. Saengdeejing, J. E. Saal, Y. Wang and Z. K. Liu, *Appl. Phys. Lett.* **90**, p. 151920 (2007).
39. C. G. Zhuang, S. Meng, H. Yang, Y. Jia, H. H. Wen, X. X. Xi, Q. R. Feng and Z. Z. Gan, *Supercond. Sci. Technol.* **21**, p. 082002 (2008).
40. W. Dai *et al.*, *Supercond. Sci. Technol.* **24**, p. 125014 (2011).

Chapter 2d

Structure-Property Correlation of MgB_2 Wires and Tapes

Balaji Birajdar* and Oliver Eibl†

*Special Centre for Nanoscience
Jawaharlal Nehru University
New Delhi-110067, India

†Institut für Angewandte Physik, Eberhard Karls Universität
Tübingen, Auf der Morgenstelle 10, 72076
Tübingen, Germany
*birajdar@mail.jnu.ac.in

The critical current density of MgB_2 wires and tapes has been continuously enhanced by increasing B_{c2} and decreasing the B_{c2} anisotropy, mainly by introducing structural disorder. Applications of MgB_2 for superconducting magnets and long distance power transport are now in sight. Thanks to their unique ability to reveal the nanostructure, electron microscopy techniques have played an important role in the technological development of MgB_2. In this chapter the microstructure of MgB_2 wires and tapes, prepared by different variants of the powder-in-tube (PIT) technology, was investigated using quantitative electron microscopy and spectroscopy methods: combined scanning electron microscopy (SEM), electron probe microanalysis and transmission electron microscopy (TEM) analysis with artefact-free sample preparation, elemental mapping and advanced chemical quantification. The purity of the B precursor powder, the SiC doping and the metallic sheath affect the MgB_2 microstructure. The following microstructure parameters are relevant for critical current density and were quantitatively measured: (i) MgB_2 grain size, (ii) MgB_2 colony size, (iii) volume fraction of B-rich secondary phases and (iv) oxygen (O_2) mole fraction. These quantities are then correlated with the critical current density at zero field and the decay field, which are extracted from the linear behavior of logarithm of critical current density as a function of the external magnetic field.

1. Introduction

After the discovery of superconductivity at 39 K in MgB_2 by Nagamatsu *et al.*[1] this compound yielded new physics of superconductivity in terms of two-band superconductivity, anisotropy of the electron–phonon interactions and anhormonicity of the phonon system.[2–4]

The powder-in-tube (PIT) technology was used for fabricating superconducting wires and tapes and it was quickly realized that large critical current densities can be reached.[5,6] Innovative applications of MgB_2 wires and tapes in the magnet and energy technology were successfully realized in a short period of time some of these good results have been reported in Refs. 7 and 8. More recently, in the framework of the upgrade of the Large Hadron Collider (LHC machine), a world-record current of 20 kA was reached at 24 K in a transmission line consisting of two 20 m long MgB_2 long cables. This result demonstrated the potential of the MgB_2 technology for long distance power transport and its direct impact for the LHC at CERN[9] (see also Chapter 6d).

As a consequence, applications of MgB_2 for superconducting AC cables are in sight and measuring AC losses in MgB_2 conductors has become a research field of increasing importance,[10] requiring a detailed knowledge of the microstructure of the individual wires. Critical current densities, technologically the most relevant parameter for applications, are limited by B_{c2} and its strong anisotropy in this compound with hexagonal symmetry. The lower B_{c2} value is found along the c-axis in the range of about 5 T, the higher B_{c2} value is found in the basal plane going up to 25 T.[11] Because of this large anisotropy, percolative current paths exist when field levels exceed the lower B_{c2} values. In MgB_2, the critical current density, J_c, has been continuously enhanced by increasing B_{c2} and decreasing the B_{c2} anisotropy (mainly by introducing structural disorder, e.g., carbon-doping, low sintering temperatures), increasing the number of pinning-relevant defects (grain boundaries, nanometer sized secondary phases) and decreasing the microstructural granularity.

Electron microscopy techniques have played an important role in this effort, revealing the nanostructure and the local chemical composition of B-based compounds. Measuring the chemical composition of B-based compounds turned out to be difficult but is a key technique for understanding microstructures. Quantitative B analysis by WDX spectroscopy[12] and EDX spectroscopy[13] was successfully applied. In comparison, X-ray diffraction because of the comparatively weaker interaction between X-rays and matter has only limited capability to detect B-rich secondary phases in MgB_2. In summary, electron microscopy is able to reveal structural and chemical information with sub-μm lateral spatial resolution. A combination of scanning electron microscopy (SEM) and transmission electron microscopy (TEM) based techniques was applied for covering the relevant length scales from 100 μm down to a few nanometres.[13]

In this chapter, Secs. 2 and 3 briefly describe the PIT processing and superconducting properties of six different MgB_2 wires and tapes. Section 4 gives the methodology and application of advanced electron microscopy technique for the analysis of MgB_2 wires and tapes. Section 5 describes a correlation model between critical current density and microstructure of MgB_2 wires and tapes.

2. PIT Processing of MgB_2 Wires and Tapes

The PIT technology, a well-known technology for fabricating multi-filamentary superconducting wires, has been used for MgB_2. For a review and context to other superconducting materials relevant for wire production see Refs. 14–17. Recently, a strong enhancement of the high-field critical current densities was reported for wires prepared by the PIT technology with C coated B powder.[18]

Six different MgB_2 wires and tapes with different variants of the PIT technique or different precursor powders and processing parameters are listed in Table 1. Samples 1 and 2 are monofilament tapes in which a mechanically alloyed mixture of precursor powders in the form of B and Mg particles is used. Overview scanning electron images of cross-sections of these tapes are shown in Fig. 1.

Samples 3 and 4 are *in situ* 4-filament wires using a rectangular wire-in-tube technique. The precursor powders are in the form of

Table 1(a): Synthesis parameters of MgB_2 wires and tapes: synthesis technique, precursor powder, annealing temperature/time (θ), for samples 1–6.

S.no.	Synthesis technique	Powder	θ(°C/h)
		Mechanical alloying (MA); without SiC additives	
1[a]	MA	Mg (Goodfellow)+2B (Fluka)	600/3
2[a]	,,	Mg (Goodfellow)+2B (Strem)	650/3
		In situ; with SiC additives	
3[b]	RWIT	Mg (Alpha Aesar)+2B (Geneva)+11.3 mol.% SiC (Alpha Aesar)	650/0.5
4[b]	,,	Mg+2B+11.3 mol.% SiC (Oxford)	650/0.5
		Ex situ; without and with SiC additives	
5[c]	14-filament	MgB_2 (Lamia)	980
6[d]	HE+TAR	MgB_2 (Alpha Aesar) + 1.5 mol.% SiC	950/0.5

Note: (Rectangular wire-in-tube) RWIT; (Hydrostatic extrusion and two axial rolling) HE+TAR; Ref. 19[a], Ref. 21[b], Ref. 20[c], Ref. 22[d]. Reproduced with permission from Ref. 23. © 2009, AIP Publishing LLC.

Table 1(b): Superconducting and microstructure parameters: T_c, extrapolated critical current density at zero field J_{c0}, magnitude of the reciprocal of the slope of the log J_c(B) plot (B_p), MgB_2 grain size (g), colony size (S), volume fraction of B-rich phases (f) and O_2 mole fraction of samples 1–6.

S.no.	T_c(K)	J_{c0} Acm^{-2}	B_p(T)	g(nm)	S(μm)	f(%)	O(at.%)
			Mechanical alloying; without SiC additives				
1[a]	36.5	1.25×10^6	1.99	25	—	1.4	3.75
2[a]	32.5	3.62×10^6	1.66	20	15	0.5	9.34
			In-situ; with SiC additives				
3[b]	33.7	3.3×10^5	2.5	35	—	2.7	10.4
4[b]	35.5	2.6×10^5	1.95	75	—	4.6	15.46
			Ex-situ; without and with SiC additives				
5[c]	36.5	2.9×10^6	0.79	500	6	∼0	14.57
6[d]	38.7	3.1×10^5	1.32	150	3	1.1	8.89

Note: RWIT; HE+TAR; Ref. 19[a], Ref. 21[b], Ref. 20[c], Ref. 22[d]. Reproduced with permission from Ref. 23. © 2009, AIP Publishing LLC.

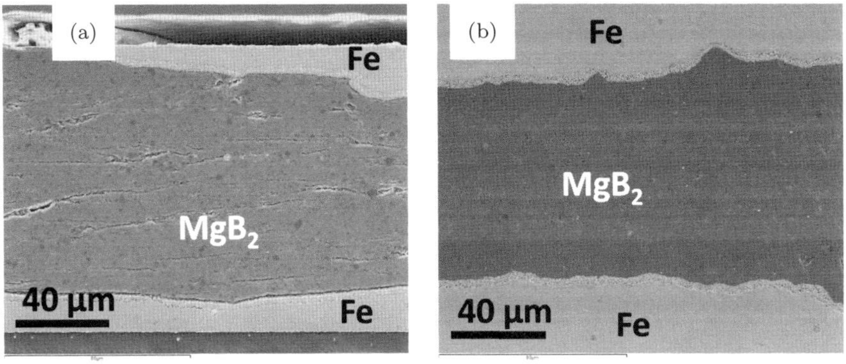

Fig. 1. SEM-Overview images of cross-sections of MgB_2 tapes. (a) sample 1, (b) sample 2. © IOP Publishing. Reproduced with permission from Ref. 19.

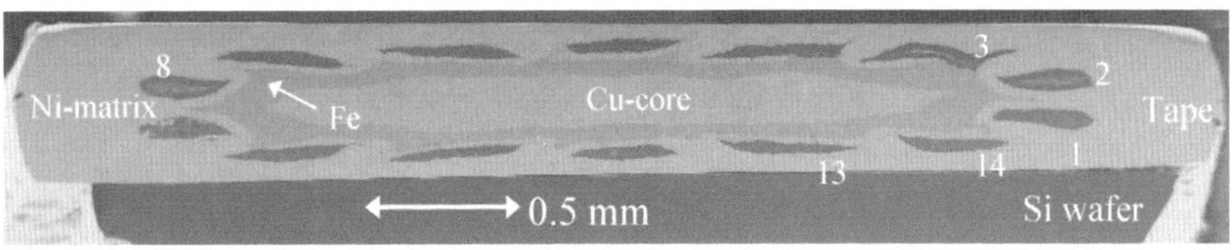

Fig. 2. SEM-overview image of the 14-filament tape (sample 5) in cross-section. The Cu core, the Fe diffusion barrier, 14 MgB_2 filaments and the Ni matrix are shown. A Si wafer is glued on one side of the tape. © IOP Publishing. Reproduced with permission from Ref. 20.

elemental B and Mg, and SiC powders. Samples 5 and 6 are prepared using prereacted MgB_2 powder (*ex situ* samples). Sample 5 is a 14-filament tape. As shown in the SEM image in Fig. 2, each tape consists of a Cu core for thermal stabilization, surrounded by 14 Ni-sheathed superconducting filaments with an Fe barrier to prevent the diffusion of Cu into Ni. Sample 6 is a monofilament wire prepared using a mixture of prereacted MgB_2 and SiC. The *ex situ* samples 5, 6 were annealed as usual at temperatures >900°C for obtaining grain growth and compacting of the prereacted MgB_2 grains.

More details about synthesis and processing parameters of these MgB_2 wires and tapes can be found in the references included in Table 1.

3. Superconducting Properties of MgB_2 Wires and Tapes

The superconducting transition temperature, T_c, of samples 1–6 was determined by resistivity measurements.[23] J_c was measured by transport measurements at 4.2 K using a 1 μV·cm^{-1} criterium at different applied magnetic fields. It was shown by Dhallé *et al.*[24] and van Weeren[25] that the J_c(B) data can be fitted as an exponential function $J_c = J_{c0} \exp(-B/B_p)$, where J_{c0} and B_p correspond to the extrapolated J_c at B = 0, and the reciprocal of the slope of the ln $J_c(B)$ plot. The J_c(B) plots are given in Fig. 3, while the T_c, J_{c0}, B_p are listed in Table 1(b).

4. Electron Microscopy Analytics for the Wire Technology

4.1 *Methodology of electron microscopy*

Methodology for the quantitative electron microscopy and spectroscopy of MgB_2 wires and tapes has been described by Birajdar

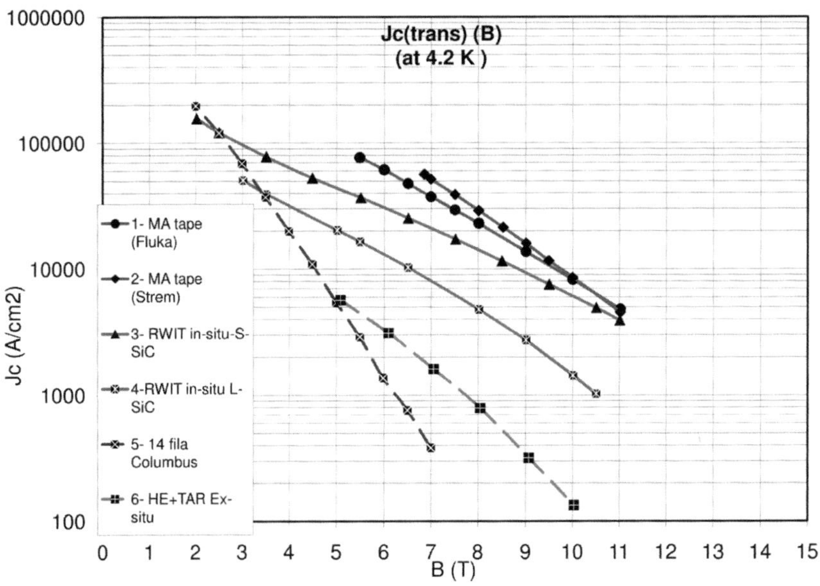

Fig. 3. The critical current density (J_c (B)) of samples 1–6 at 4.2 K. Reproduced with permission from Ref. 23. © 2009, AIP Publishing LLC.

et al.[13]: (i) carbon and O_2 contamination artifacts during sample preparation and under the electron beam in microscope, (ii) low fluorescence yield of B–K_α X-rays, (iii) B–K_α and C–K_α X-ray peak overlap in EDX, (iv) preferential absorption of B–K_α x-rays in the sample, (v) formation of MgB_2 colonies and secondary phases (MgB_{4+x}, MgB_{7+x}, $MgSi_xO_y$ and Mg_2Si) yielding structural granularity from μm to nm length scale which needs to be quantitatively measured.

The methodology of the quantitative electron microscopy and spectroscopy analysis of MgB_2 is established by meeting the above challenges. This methodology involves quantitative electron microscopy and spectroscopy using a judicious combination of SEM and TEM techniques with contamination free sample preparation, chemical mapping with good counting statistics and advanced chemical quantification using X-ray microanalysis (Fig. 4). Application of this methodology to samples 1–6 is summarized in Fig. 5.

C contamination during sample preparation was minimized by avoiding the use of organic resin as embedding material. In the last

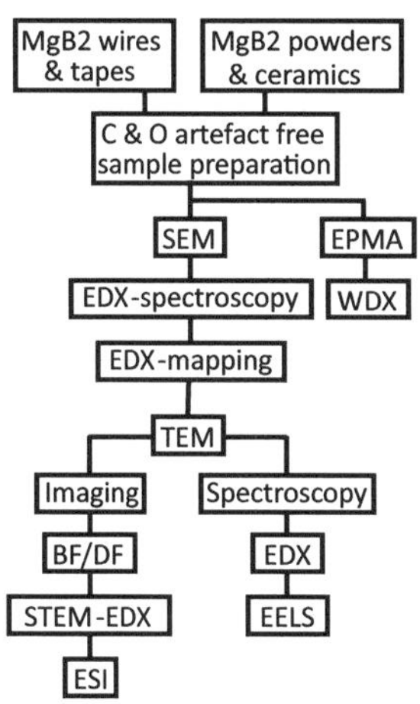

(a)

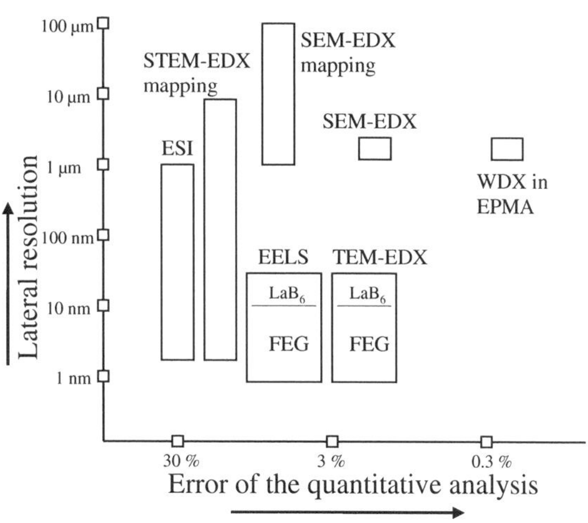

(b)

Fig. 4. (a) Schematic diagram of strategy for the microstructure analysis of MgB_2, (b) Lateral resolution and accuracy (in terms of error of the quantitative analysis) for different electron microscopy methods.

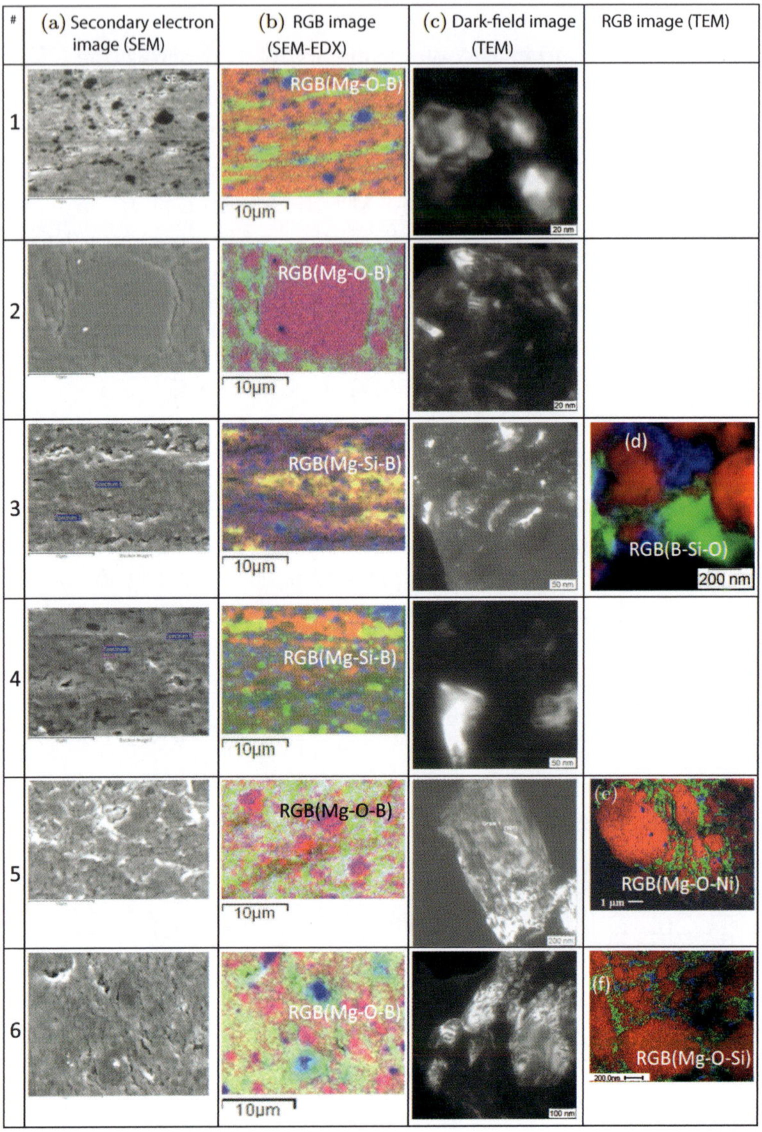

Fig. 5. Microstructure of samples 1–6 analyzed by, (a) secondary electron images, (b) the corresponding SEM–EDX RGB(Mg–O–B) or RGB(Mg–Si–B) images, (c) TEM dark-field images showing MgB_2 grain size. (d) TEM–ESI RGB(B–Si–O) image of sample 3. (e, f) STEM–EDX RGB(Mg–O–Ni) and RGB(Mg–O–Si) images of samples 5 and 6, respectively. (note: dark-field image of sample 2 was acquired from a similarly prepared bulk sample). Reproduced with permission from Ref. 23. © 2009, AIP Publishing LLC.

step of electron microscopy sample preparation the sample surface was cleaned By ion milling at 4 kV Ar. Minimizing C contamination is especially essential for quantitative EDX analysis of B due to overlap of B–K_α and C–K_α X-rays. In order to avoid artefacts due to oxidation, the sample after preparation was immediately transferred to the SEM or TEM.

Secondary electron imaging and backscattered electron imaging in SEM is used to analyze surface topography (cracks, voids) and secondary phases, on the length scale of 100 nm to few hundred micrometers (column a of Fig. 5). Diffraction contrast imaging in TEM is used to determine bulk micro and nanostructure (grain size, precipitate size and density, planar defects, dislocations, etc.) of a thin sample. (column c) of Fig. 5).

The spatial distribution of chemical phases like oxides, B-rich secondary phases, Mg_2Si etc., is revealed using EDX elemental mapping on the 100 nm to few tens of micrometers length scale in the SEM (column b of Fig. 5) and on the nm scale using TEM (column d of Fig. 5). In order to obtain EDX elemental maps and point spectra with good counting statistics, it is essential to overcome the limitation of low fluorescence yield of light elements by optimizing the acquisition conditions. Long measurement time, large probe current and size, aperture size, spectral processing time need to be adjusted by considering the effects of specimen drift, carbon contamination and irradiation damage under the electron beam, spatial resolution etc. The precise chemical quantification of these phases is carried out using point EDX spectra. It is extremely important to take into account the absorption of low energy X-rays like B–K_α X-rays during the quantitative evaluation of EDX spectra. Effect of sample tilt, and inhomogeneity in the sample composition, need to be taken into account.[13]

4.2 *Effect of B precursor powder on the MgB2 microstructure*

It has been widely reported in the literature[26–29] that the properties of MgB_2 depend on the quality of the precursor powder. Jiang *et al.*[26] presented a method of removing or avoiding oxide impurities in B

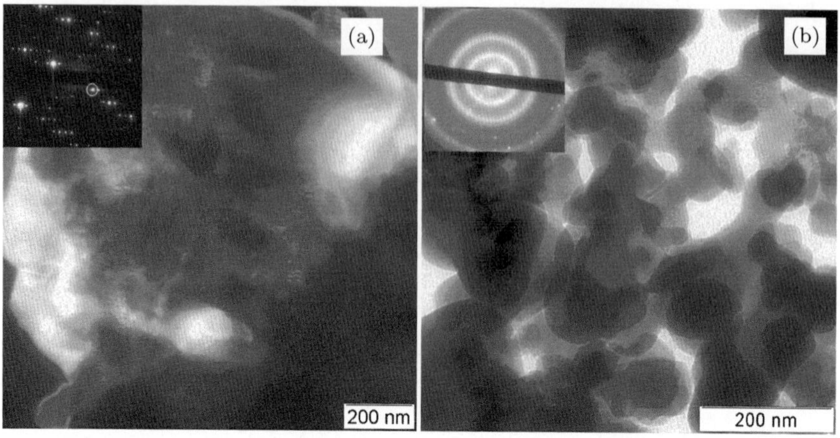

Fig. 6. B powder analysis. (a) Dark-field (DF) TEM image of grains of B powder used in sample 1, (b) bright-field (BF) TEM image of grains of B powder used in sample 2. The corresponding diffraction patterns are shown in the inset.

and Mg powder. Influence of the quality of B precursor powder on the MgB_2 tapes was studied by Hässler *et al.*[19] using electron microscopy and other techniques. DF TEM image of boron (B) powder used in sample 1 and BF TEM image of B powder used in sample 2 are shown in Figs. 6(a) and 6(b), respectively. They clearly reveal that the size of the (B) powder used in sample 2 (~100 nm) is smaller than in sample 1 (~500 nm). Although many of the B precursor powders are nominally referred to as amorphous, the diffraction patterns shown as inset in Fig. 6 clearly reveal that the B powder are nanocrystalline (sample 2) or crystalline (sample 1). Using TEM–EDX as well as other analytical techniques the boron powder in sample 2 was found to contain less metallic impurities in comparison to B powder in sample 1.[19] In comparison to sample 1, sample 2 was found to require a lower annealing temperature, contain smaller density of cracks and voids (see also Fig. 1), and well formed MgB_2 phase with smaller volume fraction and size of B-rich secondary phases, and smaller MgB_2 grains (row 1 and 2 of Fig. 5).

4.3 *Reactions at MgB_2 metal sheath interface*

Majority of the MgB_2 superconductors are prepared by the PIT technology due to mechanical and thermal considerations. Many of these sheath metals however react with the MgB_2 core, especially for *ex situ* powders for which invariably high annealing temperature (~900°C) have to be used. The reactions zones of iron boride, chromium boride, Nb–B solid solution have been inferred at the interface of MgB_2 with iron, steel and Nb respectively using Electron Probe Microanlaysis (EPMA) by Grovenor *et al.*[30] In the following we summarize an EPMA study of MgB_2–Ni interface in sample 5.[20] A backscattered electron image acquired in the EPMA in Fig. 7(a) shows an inner and an outer reaction zone. The inner reaction zone shows voids which might have formed due to the Kirkendall effect.[31] To determine the precise chemical composition of these intermetallic reactions zone, the B–K_α peak position in the wavelength dispersive X-ray spectrum needed to be calibrated using MgB_2 ceramics as a standard. The correlation diagram of B mole fraction and, Mg, O and Ni mole fractions is shown in Fig. 7(b). The composition of the inner reaction layer was found to be $MgB_2Ni_{2.9}$ in the middle and $MgB_{0.6}Ni_2$ at the edges. Outer reaction layer had a homogenous composition of $MgB_{3.3}Ni_{10}$.

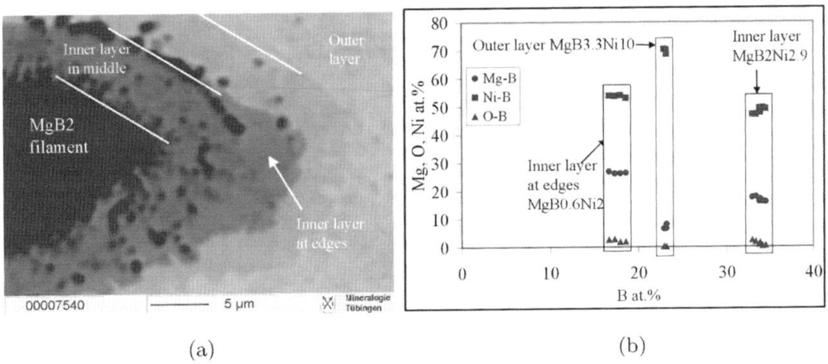

Fig. 7. Reaction zone at the MgB_2–Ni interface of a MgB_2 filament. (a) Back scattered electron image, (b) Correlation diagrams of Mg–B (•), Ni–B(■) and O–B(▲) mole fractions determined using EPMA.

4.4 *Typical features of MgB_2 filament microstructure*

A consistent picture regarding the microstructure of *ex situ* and *in situ* MgB_2 PIT superconductors has emerged.[13,19–21,23,32] In *ex situ* samples, 3–10 micrometer large MgB_2 colonies are embedded in a granular, porous matrix of small MgB_2 grains (or small MgB_2 colonies) as shown in Fig. 8. MgB_2 colonies are dense arrangement of MgB_2 grains (Fig. 9(a)), have perfect MgB_2 stoichiometry and are almost free of O_2. Few extremely large colonies may contain B-rich secondary phases but the volume fraction of such B-rich secondary phases is small. Colony formation might be attributed to sintering of MgB_2 particles to a ceramic. The MgB_2 grains in the porous matrix are surrounded by oxide layers. This might be attributed to various oxidation mechanism like surface oxidation,[33,34] internal oxidation due to gaseous O_2[35] and due to O_2 dissolved in MgB_2 structure.[36] Different MgB_2 grains in a 5 micrometer large colony are clearly revealed due to the diffraction contrast in Fig. 9(a). Long straight dislocations are seen in one of the grains. DF image (Fig. 9(b)) of another MgB_2 grain reveals 15–70 nm precipitates which were confirmed to be MgO using TEM–EDX spectroscopy.

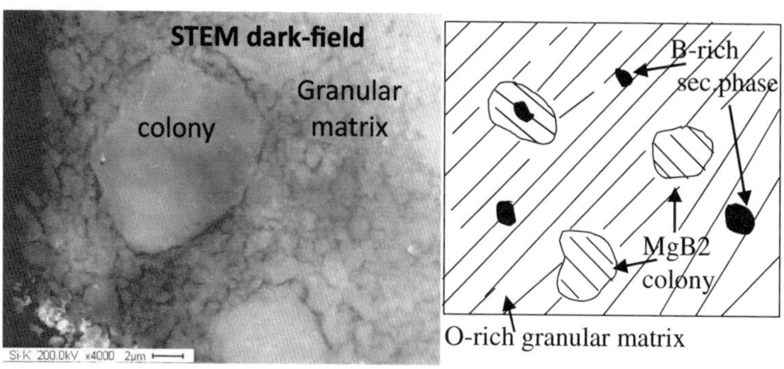

(a) (b)

Fig. 8. Microstructure of sample 5. (a) STEM dark-field image. (b) Schematic of the microstructure of *ex situ* MgB_2. Reproduced with permission from Ref. 23. © 2009, AIP Publishing LLC.

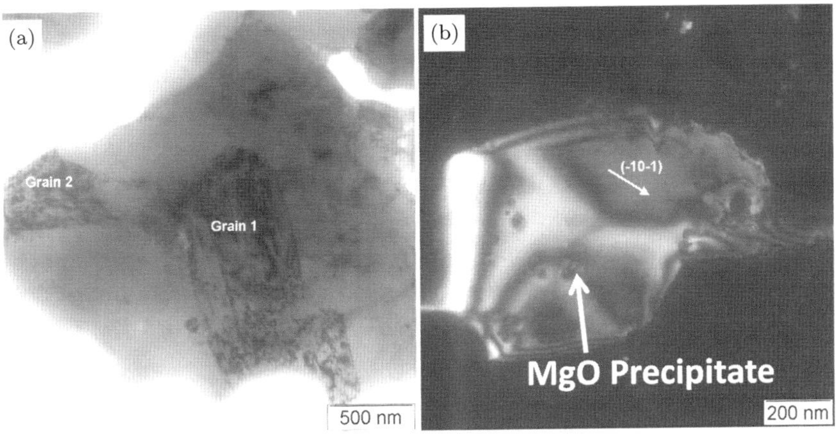

Fig. 9. Microstructure of sample 5. (a) BF TEM image of a MgB_2 colony showing grains with different diffraction contrast (b) DF TEM image showing MgO precipitate. © IOP Publishing. Reproduced with permission from Ref. 20. All rights reserved.

Unlike the *ex situ* samples in which prereacted MgB_2 powder is used, precursor B, Mg elemental powders are heat treated at lower temperatures of about 650 C in *in situ* samples. The MgB_2 colonies embedded in a matrix of porous MgB_2 matrix do not occur in such *in situ* samples. Instead (see row 1 of Fig. 5) (i) cracks and voids are observed due to the shrinkage of the material in less dense regions of the sample during *in situ* phase formation reaction. (ii) B-rich secondary phases occupy a significant volume fraction of the *in situ* sample. (iii) Oxides are primarily concentrated at cracks and voids which can be attributed to the oxidation of unreacted Mg. (iv) The size (30 nm) of MgB_2 grains is about an order of magnitude smaller than the *ex situ* samples. It can thus be inferred that the microstructure in *in situ* sample is primarily determined by inhibited phase formation due to the low annealing temperatures used in *in situ* samples. More detailed investigation of the microstructure of MgB_2 samples with well controlled synthesis, for example bulk samples, have been reported in the literature by Refs. 35–41.

4.5 *SiC-doping*

The enhancement in the B_{c2} and J_c values of MgB_2 wires and tapes by SiC-doping reported by Dou *et al.*[42] is a technology of great practical relevance. Two reactions, (i) release of C through reaction of Mg and SiC to form Mg_2Si and (ii) formation of C-doped MgB_2 by the reaction of Mg, B and C, both of which take place simultaneously at a temperature of about 600°C, have been proposed to explain the C-doping of MgB_2 by SiC.[43] In this section we report the electron microscopy findings supporting the above mentioned dual reaction model in the *in situ* samples 3 and 4.

Mg–K_α, Si–K_α, C–K_α and B–K_α SEM–EDX elemental maps acquired from sample 4, at a magnification of 1000× are shown in Fig. 10. It can be seen that the rectangular marked region is rich in Si and Mg and deficient in C and B. There is no correlation between

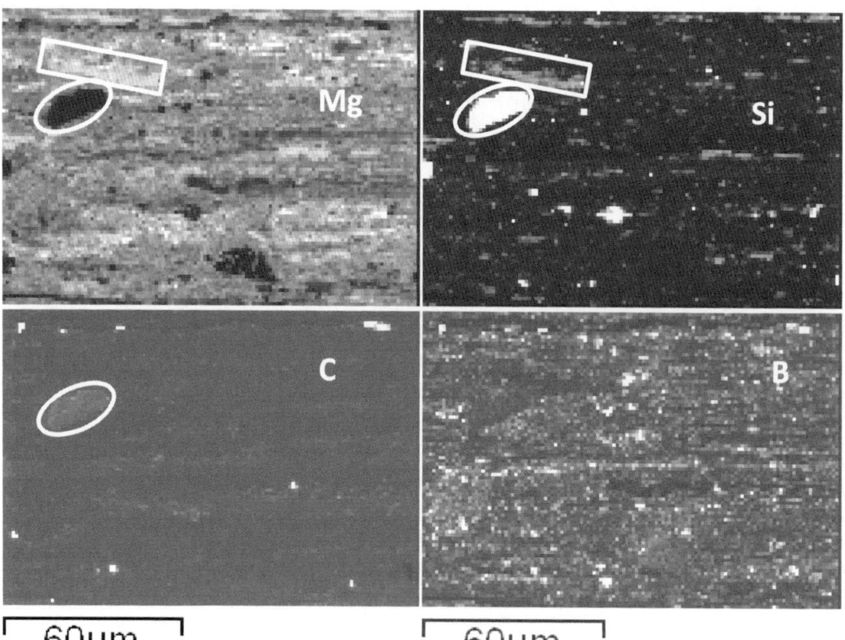

Fig. 10. Microstructure of sample 4. Mg–K_α, Si–K_α, C–K_α and B–K_α SEM-EDX elemental maps acquired at a magnification of 1000×.

Si and C in this region. Such regions were observed in Sample 3 also. EDX point spectra acquired from such regions indicated them to be Mg_2Si. This was supported also by RGB EFTEM overlay image in 3rd row of Fig. 5 and confirmed by selected area electron diffraction.[13] This therefore confirms the reactions of Mg with Si to form Mg_2Si while releasing C into the surrounding MgB_2 matrix. On the other hand, the elliptically marked region in Fig. 10 shows a region rich in Si and C. EDX spectra indicated these regions to be unreacted SiC. Such SiC particles were observed only in Sample 4, because bigger agglomerates of SiC were used in Sample 4 than in Sample 3.[21] Due to their large size, these SiC particles could not be completely converted into Mg_2Si by their reaction with Mg.

SiC nanoparticles added to *ex situ* MgB_2 superconductors did not form Mg_2Si but instead $MgSi_xO_y$. This documents the advantages of *in situ* processing conditions.

5. Microstructure-critical Current Density Model for MgB_2 Wires and Tapes

Two important observations were made from the J_c measurements of wires and tapes at 4.2 K (see Fig. 3):

- J_c varies over orders of magnitude for the different tapes at a certain field, however,
- The logarithm of the critical current density as a function of external magnetic field of an individual tape follows to a good approximation a straight line for all investigated MgB_2 wires and tapes.

Therefore, a parametrization of J_c data according to the following equation was carried out

$$J_c = J_{c0} \exp(-b_2), \quad \text{with } b_2 = \frac{B}{B_p}. \tag{1}$$

The two parameters, J_{c0} and B_p, are tape specific and were correlated to microstructural data obtained from these tapes.[23]

The goals of the micro and nanostructure analysis were first to measure quantitatively the relevant microstructure parameters for

J_c, namely (i) MgB_2 grain size (g), (ii) colony size (S), (iii) volume fraction of B-rich secondary phases (f) and (iv) O_2 mole fraction (O). Second, to correlate these parameters with the parameters of the critical current density described in section 3 namely, J_{c0} and B_p which were extracted from the $J_c(B)$ plots. These parameters are listed in Table 1.

The size of MgB_2 colonies is determined from the RGB (Mg–O–B) images or RGB (Mg–Si–B) images (column b of Fig. 3) and the largest observed colony size of these samples is reported in Table 1. The volume fraction of the B-rich secondary phases is determined by quantitative analysis of Mg and B SEM–EDX elemental maps by creating Mg–B concentration histograms.[44] The O_2 mole fraction was determined from spectra acquired from the entire area of an image (about 600 μm^2) during the SEM–EDX elemental mapping. The size of the MgB_2 grains was determined from a few DF TEM images (column c of Fig. 3) acquired at different regions of the sample and the average of the largest and smallest MgB_2 grains is taken as the representative grain size which is listed in Table 1.

The parameter B_p is related to flux-line pinning. The pinning can occur due to MgB_2 grain boundaries but also due to secondary phase boundaries or about ten nm sized secondary phases. Samples 1, 4, 5 and 6 clearly show an increase in B_p with a decrease in grain size. The anomalously high and low B_p of sample 1 and 3, respectively, could be attributed to the presence and absence of additional pinning due to the B-rich secondary phases.

The parameter J_{c0} is related to the connectivity which is expected to increase with (i) increase in colony size (S) (ii) decrease in volume fraction of B-rich secondary phases (f) and, (iii) decrease in O mole fraction (O). Indeed J_{c0} was found to increases with increasing colony size in *ex situ* like samples 2, 5 and 6, (sample 2 is considered *ex situ* like because of significant formation of MgB_2 phase formation during mechanical alloying itself). Also J_{c0} was found to decrease with increase in volume fraction of B-rich secondary phases. This indicates that poor connectivity rather than enhanced pinning due to secondary phases is the dominant factor influencing J_{c0}. The role of the different microstructural parameters discussed above is

Table 2: Microstructure parameters and their effect on J_c. Reproduced with permission from Ref. 23.

Variation in microstructure parameter	Effect on J_{c0} or B_p
Decrease in MgB_2 grain size (increase in pinning)	B_p Increases
Increase in superconducting-normal interfaces (increase in pinning)	B_p Increases
Decrease in O_2 mole fraction (increase in connectivity)	J_{c0} Increases
Increase in colony size (increase in connectivity)	J_{c0} Increases
Decrease in volume fraction of B-rich secondary phases (increase in connectivity)	J_{c0} Increases

summarized in Table 2. The detailed correlation diagrams are shown in Birajdar and Eibl.[23]

This parametrization was closely linked to a parametrization introduced by Eisterer *et al.*[45–47] and Eisterer,[48] which considers parameters of a flux line pinning model. The following parameters were introduced: (i) B_{c2}, (ii) connectivity (A_{con}), (iii) pinning strength (F_{p0}), (iv) B_{c2} anisotropy, (v) percolation threshold (p_c). Grain boundary pinning was assumed to be the most relevant pinning mechanism. It was pointed out that even a microstructurally homogenous polycrystalline MgB_2 conductor will show granularity at higher magnetic fields,[45,46] due to the anisotropy of B_{c2}. In addition, the microstructure of MgB_2 shows a significant granularity and is not homogenous. Further, the methods used to increase B_{c2}, like processing temperature, C-doping[49–51] and neutron irradiation, themselves introduce structural granularity due to the formation of secondary phases. Indeed, the effectively conducting cross-section of the MgB_2 filaments was calculated to be only one third of the geometrical cross-section of filaments in pure MgB_2 wires and one sixth of the geometrical cross-section in SiC added MgB_2 wires, however, the microstructure of these wires was not analyzed in detail.[47]

Birajdar and Eibl[23] have shown that the parameter B_p is linearly related to $B_{\rho=0}$ (the field up to which a continuous superconducting path exists through the sample),[47] while J_{c0} is proportional to the pinning strength (F_{p0}) and connectivity (A_{con}), while inversely proportional to $B_{\rho=0}$. In this way it could be shown that microstructural parameters when acquired quantitatively can be correlated quantitatively with $J_c(B)$ data and can be related quantitatively to more fundamental parameters of the flux line pinning theory of anisotropic superconductors.

It is noteworthy that new J_c data of tapes prepared with C coated B powder that extend the field range for J_c still show to a good approximation, a linear behavior of $\ln(J_c)$ as a function of the magnetic field.[18] This strengthens the relevance of Eq. (1) and the parametrization carried out in this chapter.

Acknowledgment

Most of the work reported in this chapter was supported by the EU-FP6 Specific Targeted Research Project HIPERMAG (STRP 505724-1). We thank HIPERMAG project organizers Andries den Ouden and Marc Dhalle for very helpful discussions and excellent project organization. Authors are thankful to the following partners of the HIPERMAG consortium for providing samples and for the measurement of their superconducting properties (T_c and J_c): W. Haessler of IFW Dresden, P. Kovach of IEE Bratislava, V. Braccini of INFM Genova, W. Pachla and A. Morawaski of IHPP Warsaw.

References

1. J. Nagamatsu *et al.*, *Nature* **410**, pp. 63–64 (2001).
2. T. Dahm and N. Schopohl, *Phys. Rev. Lett.* **91**, p. 017001 (1–4) (2003).
3. H. J. Choi *et al.*, *Phys. Rev. B* **66**, p. 020513 (1–4) (2002).
4. P. C. Canfield and G. W. Crabtree, *Physics Today* **34**, pp. 34–40 (2003).
5. W. Goldacker *et al.*, *Supercond. Sci. Technol.* **14**, pp. 787–793 (2001).
6. G. Grasso *et al.*, *Appl. Phys. Lett.* **79**, pp. 230–232 (2001).
7. V. Braccini *et al.*, *Physica C* **456**, p. 209 (2007) and Columbus Superconductors SpA. Webpage: http://www.columbussuperconductors.com/.
8. M. Tomsic *et al.*, *Int. J. Appl. Ceram. Technol.* **4**, p. 250 (2007) and Hyper Tech Research, Inc. Webpage: http://www.hypertechresearch.com/mgb2wire.htm

9. A. Ballarino, *Supercond. Sci. Technol.* **2**, p. 044024 (1–7) (2014).
10. F. Grilli *et al.*, *Physica C* **504**, pp. 167–171 (2014).
11. C. Buzea and T. Yamashita, *Supercond. Sci. Technol.* **14**, pp. R115–R146 (2001).
12. T. Wenzel *et al.*, *Phys. Stat. Sol.* **198**, pp. 374–386 (2003).
13. B. Birajdar *et al.*, *Supercond. Sci. Technol.* **21**, p. 073001 (1–20) (2008).
14. D. Larbalestier, The long road to high current density superconducting conductors, *Proc. of the centennial conference "100 years of superconductivity 2011, The Hague"*, CRC press, ISBN 978-1-4398-4946-0 (2012).
15. R. Flükiger and H. Kumakura, The MgB_2 conductor story, *Proc. of the centennial conference "100 years of superconductivity 2011, The Hague"*, CRC press ISBN 978-1-4398-4946-0 (2012).
16. H. Kumakura, *J. J. Appl. Phys.* **51**, p. 010003 (2012).
17. K. Tachikawa, Development of metallic superconductors in Japan, IEEE/CSC & ESAS European Superconductivity News Forum (ESNF) (2011).
18. J. Y. Shu *et al.*, *Supercond. Sci Technol.* **27**, p. 085012 (2014).
19. W. Hässler *et al.*, *Supercond. Sci. Technol.* **19**, pp. 512–520 (2006).
20. B. Birajdar *et al.*, *Supercond. Sci. Technol.* **19**, pp. 916–923 (2006).
21. P. Kováč *et al.*, *Supercond. Sci. Technol.* **21**, p. 045011 (1–8) (2008).
22. W. Pachla *et al.*, *Supercond. Sci. Technol.* **18**, pp. 552–556 (2005).
23. B. Birajdar and O. Eibl, *J. Appl. Phys.* **105**, p. 033903 (1–12) (2009).
24. M. Dhallé *et al.*, *Physica C* **363**, pp. 155–165 (2001).
25. H. van Weeren, (2007). Magnesium Diboride Superconductors for Magnet Applications, *Ph.D. thesis, University of Twente*, ISBN 978-90-365-2519-0.
26. J. Jiang *et al.*, *Supercond. Sci. Technol.* **19**, pp. L33–L36 (2006).
27. S. K. Chen *et al.*, *Supercond. Sci. Technol.* **18**, pp. 1473–1477 (2005).
28. R. A. Ribeiro *et al.*, *Physica C* **385**, pp. 16–23 (2003).
29. S. Zhou *et al.*, *Supercond. Sci. Technol.* **17**, pp. S528–S532 (2004).
30. C. R. M. Grovenor *et al.*, *Supercond. Sci. Technol.* **17**, pp. 479–484 (2004).
31. R. E. Smallman, Modern Physical Metallurgy (London: Butterworths) p. 147 (1985).
32. D. Eyidi *et al.*, *Supercond. Sci. Technol.* **16**, pp. 778–788 (2003).
33. R. P. Vasquez, *Phys. Rev. B* **64**, p. 052510 (1–4) (2001).
34. A. Serquis *et al.*, *Appl. Phys. Lett.* **80**, pp. 4401–4403 (2002).
35. D. Eyidi *et al.*, *Micron* **34**, pp. 85–96 (2003).
36. X. Z. Liao *et al.*, *J. Appl. Phys.* **93**, pp. 6208–6215 (2003).
37. Y. Zhu *et al.*, *Physica C* **356**, pp. 239–253 (2001).
38. J. C. Idrobo *et al.*, *Phys. Rev. B* **70**, p. 172503 (1–4) (2004).

39. R. F. Klie *et al.*, *Journal of Electron Spectroscopy and Related Phenomena* **143**, pp. 105–115 (2005).
40. S. Li *et al.*, *Appl. Phys. Lett.* **83**, pp. 314–316 (2003).
41. X. Song *et al.*, *J. Mater. Res.* **19**(8), pp. 2245–2255 (2004).
42. S. X. Dou *et al.*, *Appl. Phys. Lett.* **81**, pp. 3419–3421 (2002).
43. S. X. Dou *et al.*, *Phys. Rev. Lett.* **98**, p. 097002 (1–4) (2007).
44. A. Kenda *et al.*, *Micron* **30**, pp. 85–97 (1999).
45. M. Eisterer *et al.*, *Phys. Rev. Lett.* **90**, p. 247002 (1–4) (2003).
46. M. Eisterer *et al.*, *Journal of Applied Physics* **98**, p. 033906 (1–5) (2005).
47. M. Eisterer *et al.*, *Supercond. Sci. Technol.* **20**, pp. 117–122 (2007).
48. M. Eisterer, *Supercond. Sci. Technol.* **20**, pp. R47–R73 (2007).
49. A. Serquis *et al.*, *Supercond. Sci. Technol.* **20**, pp. L12–L15 (2007).
50. B. J. Senkowicz *et al.*, *Supercond. Sci. Technol.* **20**, pp. 650–657 (2007).
51. A. Matsumoto *et al.*, *Appl. Phys. Lett.* **89**, p. 132508 (1–3) (2006).

Chapter 3a

Structure and Properties of Bulk MgB_2

Tatiana Prikhna

Institute for Superhard Materials
National Academy of Sciences of Ukraine
2, Avtozavodskaya Str., Kiev 04074, Ukraine
prikhna@mail.ru

Practical application of MgB_2 requires mechanically stable, high quality materials for high superconducting (SC) performance. Such materials can easily be manufactured under high pressure conditions. The present chapter considers the formation of highly dense bulk MgB_2 materials with high connectivity between SC grains. Correlations are established between manufacturing conditions and materials properties and perspectives are discussed in view of their possible practical application, e.g., in electrical motors or inductive fault current limiters (FCL). SC and mechanical characteristics are compared between MgB_2 bulk materials prepared under a wide range of pressures and temperature conditions: by high quasi-hydrostatic pressure/high temperature, hot pressing, spark plasma sintering and "pressureless" synthesis and sintering. The effect of additions such as SiC, Ti, Ta and Zr on critical current density, connectivity, distribution of admixture oxygen (O_2) and appearance of higher magnesium borides in MgB_2 structure is discussed. The methods for estimating the MgB_2 properties are considered as well.

1. Introduction

Devices based on the superconductor MgB_2 ($T_c = 39\,K$) can be operated at the temperatures of liquid helium (He), hydrogen (H) or even neon (Ne). This compound can thus be envisaged for many applications, as MRI, cables, electromotors, magnetic bearings and fault current limiters (FCL).[1–6] In addition to optimizing superconducting

(SC) properties, the materials chosen for the mentioned applications should also fulfill the conditions for thermally and mechanically stable operations. They should also be stable against degradation owing to their high reactivity to moisture, as to stresses caused by mechanical load or by temperature and magnetic field gradients. There are many factors which can influence the functional properties of bulk MgB_2, the most important ones being: SC phase content, mass density of the conductor and connectivity between SC grains.

The application of pressure during the manufacturing process is quite effective in solving most of these problems. Here we discuss the preparation procedures, structure and properties of bulk MgB_2 produced by heating under high quasi-hydrostatic pressure (1–2 GPa), hot pressing (30 MPa), spark plasma sintering (16–96 MPa) and "pressureless" conditions (using cold densified precursor mixtures under usual or high pressure). Great attention will be paid to the estimation of the critical current density, J_c, by magnetic and inductive methods. Other properties of interest are: trapped magnetic field, upper critical magnetic field B_{c2}, irreversibility field, B_{irr}, ac losses, rate of SC–normal state (S–N) transition, hardness and fracture toughness. The results of tests of small reluctance motors (1,3–1,6 kW) with SC inserts into rotors based on bulk MgB_2 or MT–YBCO at 15–20 K are presented, as well as the results of tests of MgB_2 rings in the model of inductive fault current limiter.

2. Formation of MgB_2 SC Structure Under a Wide Range of Pressure–Temperature Conditions

In spite of the comparatively low SC transition temperature (39 K) of MgB_2 with respect to High Temperature Superconductors (HTS) materials, the interest in this material in view of practical applications is based on the transparency of clean grain boundaries to the SC current, as a consequence of comparatively high coherence lengths ($\xi_{ab}(0) = 3.7 \div 12$ nm and $\xi_c(0) = 1.6 \div 3.6$ nm) for MgB_2.[7] The possibility to achieve high values of J_c and moderately high B_{c2} and B_{irr} in polycrystalline materials allows economical preparation techniques: MgB_2 thus appears as an alternative to HTS. Besides, this interest can be explained by the development of liquid H technologies, either as a coolant for long power transmission lines or as an alternative fuel for motors.

The functional characteristics of SC MgB_2 depend on the pinning of vortices by nanostructural defects commensurable with the coherence length. The critical current density, J_c, of polycrystalline MgB_2 based materials is strongly influenced by the connectivity between the grains as well as by the concentration and distribution of the SC phase. Additional factors are the distribution of pinning centers, originating from structural inhomogeneities, e.g., grain boundaries, artificially added nanosize inclusions or distortions in the MgB_2 lattice due to the presence C or O_2 atoms. Finally, J_c is also influenced by extrinsic parameters, such as porosity and cracks. A positive effect in enhancing J_c in MgB_2 materials is obtained by dopants, in particular by C, C nanotubes, nanodiamonds or by C-based additives, e.g., SiC, ethyltoluene, carbohydrates, malic acid and many others.[8–13] In all cases, the enhancement of J_c is connected to an increase of the normal state resistivity. A considerably smaller enhancement of J_c is caused by metallic additives, e.g., Ti, Ta, Zr and others[14–18] (listed by Collings *et al.*[8]), but the mechanism of their effect is still under the discussion.

Based on many papers treating the manufacture and properties of MgB_2, it has been recognized that the main reasons why the critical current densities of wires or bulk MgB_2 materials are still lower than expected are the low connectivity and the low degree of compaction (low total mass density) of the superconductor. The application of pressure (30–50 MPa or 2 GPa) during material preparation can suppress the evaporation of magnesium (Mg), increase mass density up to theoretical limits and the connectivity between grains.[19,20] The high affinity of Mg to O_2 makes it almost impossible to manufacture MgB_2 based products without the presence of a certain amount of O_2, even taking B and Mg powders of the highest available purity and using a protective atmosphere for handling and reacting. It may be of importance how the O_2 is distributed in MgB_2, as follows from various works reporting a positive role of O_2 on J_c by forming nanosize inclusions of various structures, e.g., MgO, $Mg(B,O)_2$,

$MgB_{0.6-0.8}O_{0.8-0.9}$ or $MgB_{1.2-2.7}O_{1.8-2.5}$ nanolayers, which may act as pining centers in MgB_2.[20–24]

No correlation could be found in our laboratories[25] between the average grain size (0.8–9 μm) and the O_2 content (0.66–1.9 wt.%) of the initial powders of B or MgB_2 on one side and the amount of O_2 in consolidated MgB_2 (7–24 wt.% for *in situ* and 4–12 wt.% for *ex situ*) the sizes of crystallites (estimated from X-ray line broadening) or the critical current density, J_c (estimated base on hysteresis magnetization loops), of *in situ* or *ex situ* materials manufactured up to pressures of 2 GPa. The average crystallite size of MgB_2 samples increases with increasing reaction temperature. For reaction temperatures within 700–1000°C, sizes between 19 and 25 nm were observed in *ex situ* and between 15 and 37 nm for *in situ* synthesized MgB_2 samples. However, as mentioned above, the variation of J_c did not in all cases correlate with the grain size.[25] Usually J_c is somewhat higher for *in situ* than that of *ex situ* manufactured MgB_2: a possible explanation is the larger grain size in the *ex situ* prepared materials, sintering occurring at higher temperatures than for the *in situ* reaction (around 900°C and 650°C, respectively). Another reason seems to be the less homogeneous distribution of admixtures in MgB_2 matrix in the case of *ex situ* preparation (see Ref. 25).

2.1 *Effect of manufacturing temperature and pressure on the SC properties of MgB_2*

According to Bohnenstiel *et al.*[26] the reaction of B and Mg to MgB_2 starts around ~600°C. The reaction temperature of SC MgB_2 can vary between 600°C and 1,150°C, also depending on the applied pressure. The manufacturing process of wires at atmospheric pressure is conducted at lower temperatures, but the application of high pressure during their manufacturing process of bulk MgB_2 leads to an increase in critical current density.[27] The higher manufacturing pressure prevents the evaporation of Mg, thus allowing the phase formation at higher temperatures. The higher synthesis temperatures lead to higher J_c values, especially at low and medium external

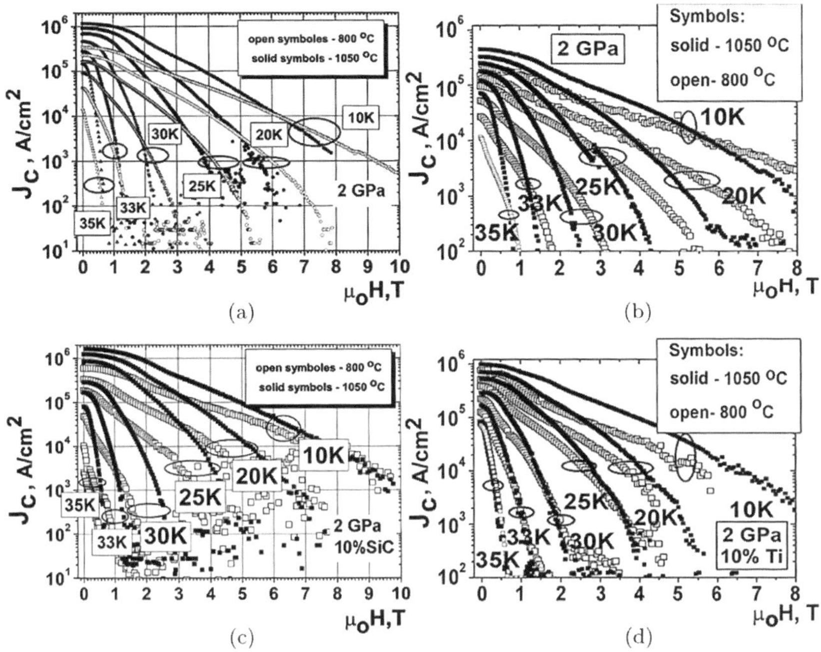

Fig. 1. Critical current density, J_c, versus applied magnetic field of bulk MgB_2 prepared under 2 GPa for 1 h: (a) from Mg:2B(I), (b) Mg:2B(III) at 800 and 1050°C, (c) from Mg:2B(I) with 10% SiC (200–800 nm) and (d) Mg:2B(III) with 10% Ti (1–3 μm) (see Refs. 19, 20 and 25).

magnetic fields, but also to lower J_c at high fields (Figs. 1 and 2(a)). At low manufacturing temperatures and high pressures, higher critical magnetic fields B_{c2} and B_{irr} can be achieved (Fig. 2(b)). This tendency has been observed for different types of B powders (Table 1) and for various pressures and preparation techniques (Table 2). It is accompanied by a shift of grain boundary pinning toward point pinning with increasing synthesis temperature,[20] reflected by the shift of the peak position of the volume pinning force.[28]

Characteristical properties of SC materials as J_c, B_{c2}, B_{irr}, connectivity (A_F) and the shielding fraction (S) can be estimated using commonly accepted methods which are briefly described in the following. In bulk samples, J_c is determined based on the

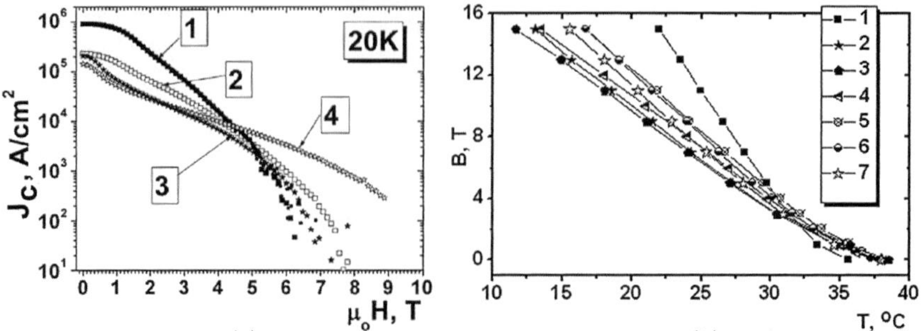

(a) (b)

Fig. 2. (a) J_c at 20 K in bulk MgB_2 prepared under various T at 2 GPa: 1, 2: samples made of B(I) at 1050°C and 800°C; 3, 4 of B(II) at 800°C and 600°C, respectively.[20] (b) B_{c2} of bulk $MgB_2^{6,20}$ (Mg:2B) using 1: B(II) at 2 GPa, 600°C/1 h, (HP); 2: B(III) at 30 MPa, 800°C/2h, (HotP); 3: B(III) at 50 MPa, 600–1050°C/1 h, (SPS); 4: B(III) at 2 GPa, 900°C/1 h (HP); 5: B(V)+10 wt.% Zr at 2 GPa, 800°C/1 h (HP); 6: B(V)+10 wt.% Ti at 2 GPa, 800°C/1 h (HP); 7: B(I)+10wt.% SiC at 2 GPa, 1050°C/1 h (HP).

Table 1: (1) The amounts of C, H and N in the initial B marked by the stars (*) were obtained by using the Universal Micro Analyzer "vario MICRO cube"; (2) The amount of O, C and N (which are not marked by stars) and the grain size were provided by the manufacturer. (3) The higher amount of C and N with respect to the producer's estimation may be explained by chemical reactions during storage.

Parameter			Amorphous B			MgB_2
Type	I	II	III	IV	V	VI
Grain size, μm	<5	<1	4	1.4	0.8	4.2
Purity, %	—	—	96.4	—	97	98
B, wt.%	—	—	—	—	—	44.1
Mg, wt.%	0.49	—	0.5	—	—	51.6
O, wt.%	0.66	—	1.5	1.9	1.7	1.9
			0.3		—	0.9
C, wt.%	0.31*	3.5*	0.47*	0.27*	—	—
	—	—	0.1	—	—	—
N, wt.%	0.48*	1.02*	0.40*	0.43*	—	—
H, wt.%	0.32*	0.87*	0.37*	0.11*	—	—
H_2O, wt.%	—	—	0.2	—	—	—
B–H_2O, wt.%	—	—	0.1	—	—	—
B–H_2O_2, wt.%	—	—	0.5	—	—	—

Table 2: Characteristics (J_c, amount of MgB_2, MgO and MgB_4, mass density ρ) of MgB_2-based materials prepared under different p–T conditions from Mg:2B mixtures (*in situ*) or from MgB_2 powder (*ex situ*). Preparation methods: HP — high quasi-hydrostatic pressure, HotP — hot pressing, SPS — spark plasma sintering, PL — pressureless sintering under 0.1 MPa Ar.[20] Numbering in Tables 2 and 3 is the same.

No	Preparation	Type of B or MgB_2	P, MPa	T, °C	J_c, MA/cm^2, at 0–1 T, at 20 K	MgB_2/MgO/ MgB_4, wt.%	Density, ρ, %
1.	*in situ*, HP	I	2000	1050	0.9–0.7	94/6/0	99
2.	*in situ*, HP	I	2000	800	0.2–0.15	91/5.5/0	98
3.	*in situ*, HP	III	2000	1050	0.4–0.3	87/13/0	99
4.	*in situ*, HP	III	2000	800	0.12–0.07	73/12/0	97
5.	*in situ*, HP	II	2000	600	0.14–0.05	64/30/0	83
6.	*in situ*, SPS	III	50	1050	0.5–0.4	83/4.5/12.5	94
7.	*in situ*, SPS	III	50	800	0.4–0.36	—	74
8.	*ex situ*, SPS	VI	50	1050	0.4–0.3	83/6.5/10.5	96
9.	*in situ*, HotP	III	30	1050	0.08–0.016	46/8.5/45.5	99
10.	*in situ*, HotP	III	30	800	0.3–0.2	—	72
11.	*in situ*, PL	IV	0.1	800	0.08–0.03	90/10/0	55
12.	*in situ*, PL*	III	0.1	600	0.26–0.13	94.5/5.5/0	65

Preliminarily cold densified at 2 GPa.

magnetization hysteresis loops obtained with a vibrating sample magnetometer using the Bean model.[30]

For VSM measurements on samples with typical sizes of a few mm, the value of J_c is calculated by Eq. (1):

$$J_c = \frac{2\Delta m}{V \cdot b \cdot \left(1 - \frac{b}{3a}\right)}, \tag{1}$$

where Δm is the hysteresis of the magnetic moment, V is the sample volume and a and b are the sample dimensions perpendicular to the applied field, with $a > b$.

The connectivity, A_F, is estimated from the difference in resistivity at 40 K and 300 K, ρ_{300}–ρ_{40}, measured by the four probe technique (2):

$$A_F = \frac{9\mu\Omega \cdot \text{cm}}{\rho 300 - \rho 40}, \tag{2}$$

where $9\,\mu\Omega \cdot$ cm is assumed to be the electrical resistivity of MgB_2 of a polycrystalline sample (see Ref. 31 and references therein).

The values of B_{c2} and B_{irr} are estimated from resistivity measurements (90% and 10% of the resistivity just above the beginning of the transition) using the four probe technique at fields from 0 to 15 T. The shielding fraction, S, is calculated from the ac susceptibility at 5 K with a numerical correction for the demagnetization of the actual sample geometry. In the case of *in situ* preparation, Mg chips and B powder in the ratio 1:2 were mixed and milled for 3–5 min in a high speed planetary activator under Ar before being compacted and the high J_c in low magnetic fields can be attained after synthesis (at 2 GPa) under high temperatures, but high J_c in high magnetic fields can be demonstrated by the materials prepared under low temperatures. In the case of *ex situ* preparation (from MgB_2 powder) the low manufacturing temperatures (below 900°C) under high pressures (2 GPa) generally lead to very low J_c, so for attaining high J_c the sintering process should be conducted under higher temperatures.

The critical current density of pressureless (PL) synthesized materials at comparatively low temperature (800°C) can be increased by increasing the pressure of the preliminary cold densification: for example, from 30 MPa up to 2 GPa. However, even after cold densification at 2 GPa, it is impossible to synthesize the MgB_2 phase from a Mg:2B mixture at 1050°C without applying a high pressure: after 15 min holding time at an Ar pressure of 0.1 MPa the MgB_2 phase disappeared (as shown by X-ray and SEM) due to Mg evaporation, resulting in the non-SC MgB_4 phase. An increase of the holding time up to 2 h resulted in the formation of the MgB_7 phase which is non-SC as well.

Materials prepared under medium pressures by SPS (spark plasma sintering) (50 MPa) or by hot pressing (30 MPa) exhibited high values of J_c (Fig. 3, Table 2) and B_{c2} (Fig. 2(b)), but pressures of 16–96 MPa were not sufficient to fully suppress the evaporation of Mg; thus, a rather high amount of MgB_4 phase was still present in the sample, as shown by X-ray (see Table 2, Nos. 6, 8, 9) and SEM (see Fig. 4 showing rather big areas with MgB_{4-6} stoichiometry in the SPS sample). But connectivity of the SPS sample No. 6 was near to the theoretical limit: $A_F = 98\%$, which could be the reason of the relatively high J_c (Fig. 3(a)).

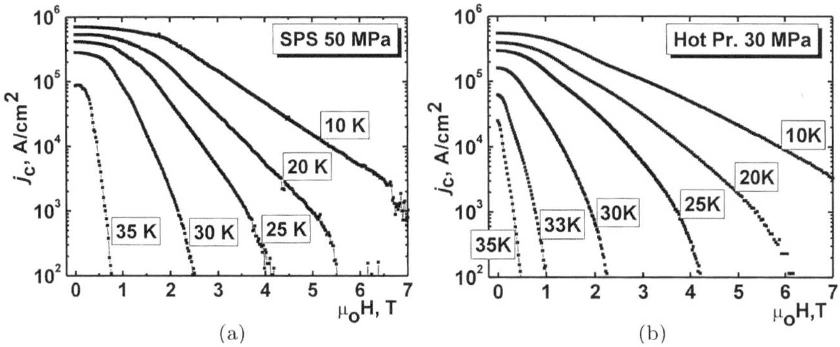

Fig. 3. Critical current density J_c versus magnetic field B of MgB_2 obtained (a) from B(III) by SPS at 50 MPa, 600°C for 0.3 h and 1050°C for 0.5 h (No. 6, Tables 2,3) and (b) from B(III)+10 wt.% Ti by HotP at 30 MPa, 1000°C, for 15 min.[6]

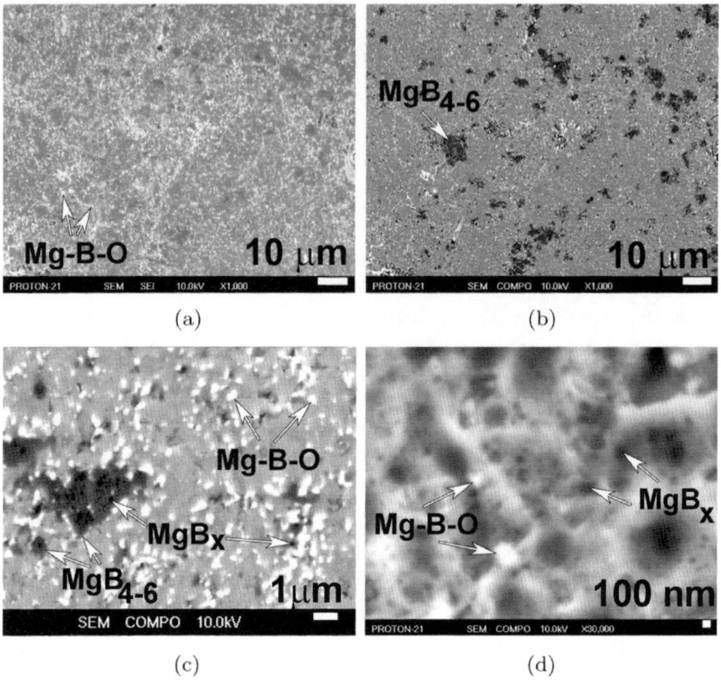

Fig. 4. SEM images of bulk MgB_2 (prepared by SPS) from B(III) at 50 MPa and 1050°C (No. 6, Tables 2, 3); connectivity $A_F = 98\%$) taken after etching in Ar under different magnifications and microscopic regimes: (a, b): same area under the same magnification as in SEI and BEI, (c,d): BEI images under higher magnification. The most bright inclusions Mg–B–O contain a high amount of O_2, the darkest inclusions are MgB_x ($x > 7$) and rather large medium grey inclusions or areas have the MgB_{4-6} composition.

At temperatures of 600–800°C, SC MgB_2 can be prepared in a wide range of pressures from 0.1 MPa up to 2 GPa, with different densities and rather high critical currents. In spite of low connectivity values of about 18% (No. 5, Table 3), samples prepared at 600°C, 2 GPa from Mg:2B(II) or using boron with a higher carbon content (3.5 wt.% C) exhibited high J_c at high magnetic fields (No. 4 in Fig. 2(a)) and high upper critical magnetic fields $B_{c2}(22\text{K}) = 15\,\text{T}$ (No. 1 in Fig. 2(b)), the extrapolation giving $B_{c2}(0\text{K}) = 42.1\,\text{T}$ and irreversibility fields: $B_{irr}(18.5\,\text{K}) = 15\,\text{T}$ and $B_{irr}(0\,\text{K}) = 32.5$ T.[32] According to X-ray diffraction, data materials prepared under high

Table 3: Connectivity, A_F, amount of shielding fraction, S, of bulk MgB_2 prepared under various conditions.[20] The numbering in Tables 2 and 3 is the same.

No	Preparation	Type of B	P, MPa	T, °C	A_F, %	S, %
3.	*in situ*, HP	III	2000	1050	79	94
4.	*in situ*, HP	III	2000	800	57	91–100
5.	*in situ*, HP	II	2000	600	18	90
6.	*in situ*, SPS	III	50	1050	98	91
8.	*ex situ*, SPS	VI	30	1050	80	100
9.	*in situ*, HotP	III	30	1050	32	—
10.	*in situ*, HotP	III	30	800	73	—
12.	*in situ*, cold dens. 2 GPa, PL	III	0.1	600	—	75

pressure (2 GPa) contain only MgB_2 and MgO phases. After synthesizing at low temperatures (800°C and below), some Mg (unreacted) and MgH_2 can be present as well.

Additions of SiC, Ti, Ta, Zr can increase J_c of MgB_2 (see Fig. 1, compare 1(a) with 1(c), 1(b) with 1(d), but the effect depends to a high extent on the quality of initial B or MgB_2 powders and on the preparation method. The highest B_{c2} were attained in the samples prepared at 2 GPa with C doped boron (No. 1, Fig. 2(b)), while somewhat lower B_{c2} values were observed in samples with Zr, Ti additions (Nos. 5, 6, respectively, Fig. 2(b)) or SiC (No. 7, Fig. 2(b)).

2.2 *Correlations between distribution of B- and O-enriched inclusions in the MgB_2 matrix and J_c*

The increase of J_c at low and medium magnetic fields with increasing reaction temperature (Figs. 1(a) and 1(b)) correlates with the structural transformations shown in Figs. 5 and 6. Materials prepared at 800°C contain O_2-enriched Mg–B–O nanolayers (marked "L" in Fig. 5(a)), while those ones prepared at higher temperature (1050°C) contain separate Mg–B–O inclusions ("I" in Fig. 5(b)). It follows that the increase of reaction temperature causes O_2 segregation, thus resulting in a "cleaning" of the MgB_2 matrix from O_2. While some

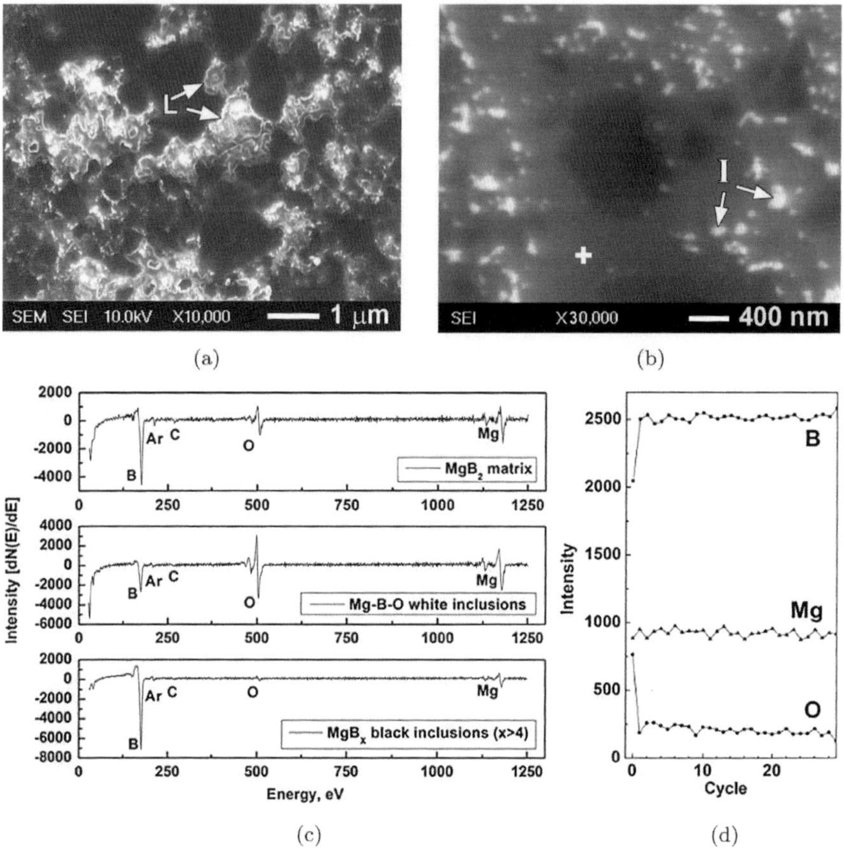

Fig. 5. (a) and (b): SEI images of MgB_2 samples from B(III) mixtures under 2 GPa, for 1 h at 800°C and 1050°C; "L" and "I" mark O_2-enriched Mg–B–O layers and inclusions; (c): Auger spectrum of the matrix: $MgB_{2.2-1.7}O_{0.4-0.6}$, white $MgB_{0.6-0.8}O_{0.8-0.9}$ and black ($MgB_{11-13}O_{0.2-0.3}$ inclusions prepared from B(III) at 2 GPa, 1050°C, shown in Fig. 5(b). The Ar peak is due to Ar gas used for etching. The C signal comes from impurities in B; (d) Depth profile of the O, B and Mg amounts in the MgB_2 matrix (at the position marked "+" in Fig. 5(b)) obtained from quantitative Auger analysis and multiple etching in Ar.[20]

O_2 is solved in MgB_2 matrices of the materials prepared at high temperature (1050°C), This kind of transformation is typical for samples prepared from Mg + 2B mixtures in the pressure range between 0.1 and 2 GPa. After synthesis at 0.1 MPa, 1050°C the composition of the matrix will rather correspond to MgB_4 or MgB_7

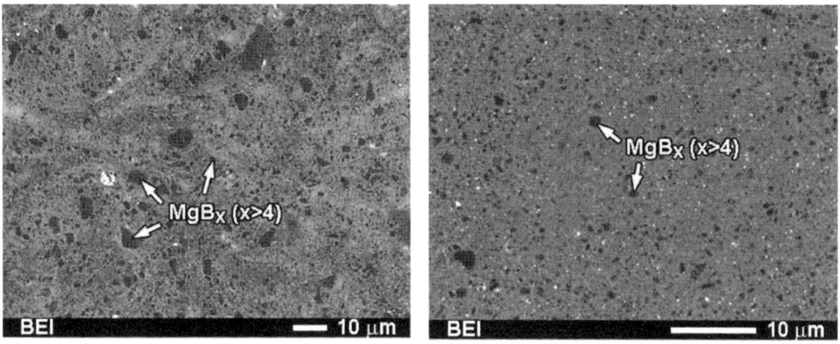

(a) (b)

Fig. 6. Structures obtained by SEM in BEI regime under the same magnification of MgB_2 prepared from B(III) under 2 GPa for 1 h (a) at 800°C and (b) at 1050°C.

(depending on time). As mentioned above, the general view of the structures prepared at 800 and 1050°C under 0.1 MPa will be similar to that one formed at 2 GPa. The second typical feature is the reduction of size and amount of the higher B inclusions MgB_x ($x = 4 - 25$: MgB_4, MgB_7, MgB_{12}, MgB_{17}, MgB_{20}, Mg_2B_{25}, etc., which appear darkest) with increasing temperature (Fig. 6). The observed structural variations are accompanied by a shift from grain boundary pinning toward point pinning, which may be attributed to inhomogeneities at the nanosize level.

The MgB_x inclusions are clearly visible in SEM (backscattering electron images) BEI, while O_2-enriched precipitations are practically "invisible," due to the similarity of the atomic numbers Z with MgB_2. But in spite of the different structures and properties, the MgB_x inclusions are practically "invisible" for the traditional X-ray diffraction analysis. This can be caused by their fine dispersion in the material structure and the large number of atoms in the unit cell of low-symmetry resulting in many "reflecting planes" which results in the low intensities of X-ray reflexes as compared to those of the MgB_2 matrix having a simple hexagonal unit cell.

The nanohardness of MgB_x ($x > 2$) inclusions is usually higher than that of the MgB_2 matrix, thus reflecting the crystallinity of

MgB_x. The Berkovich nanohardness and the Young's modulus of MgB_x ($x \sim 12$) inclusions under 10–60 mN load are 32.2 ± 1.7 and 385 ± 14 GPa, respectively. These values are even higher than those of sapphire: 31.1 ± 2.0 and 416 ± 22 GPA, respectively, and about twice the values for MgB_2. Usually x in small black looking MgB_x inclusions varied from 8 to 16. In materials prepared at 2 GPa, $x = 9$–14 and mainly $x \sim 12$ were observed. In the samples prepared by SPS, there was $x = 8$–16 (for small black inclusions) and $x = 4$–6 (for big dark gray inclusions) while in the samples prepared by hot pressing $x = 6$–8. Sometimes inclusions with a stoichiometry up to $x = 18$–25 or even pure B can be found in the samples. But they cannot influence the materials characteristics as a whole due to their random appearance.

The detailed study of MgB_2 structures by SEM and Auger showed that no MgO was found. Figure 5(c) presents a typical Auger spectrum obtained from the MgB_2 matrix as well as from inclusions of Mg–B–O and of MgB_x, which appear grey, white and black in Fig. 5(b), respectively. The quantitative Auger analysis allows an excellent localization of the electron beam (to analyze areas of 10 nm in diameter and about two united cell layers in depth) and Ar etching in the chamber during analysis, so the oxidized layer was removed from the surface of the material before the analysis and the presence of O_2 in the "white" layers "L" and inclusions "I" is clearly confirmed. It is rather difficult to specify the exact stoichiometry of the O_2-enriched layers or inclusions, but the amount of O_2 in their composition is high. On the other hand, O_2 is practically absent in higher magnesium borides MgB_x, but the matrix contains some O_2 as well. According to the quantitative Auger analysis, the approximate stoichiometry of the matrices of the materials shown in Figs. 5(a) and 5(b) was $MgB_{2.6-3.1}O_{1.6-2}$ and $MgB_{2.2-1.7}O_{0.4-0.6}$, respectively.[20] Figure 5(d) shows the results of a quantitative Auger depth profile: the amount of O, B and Mg was analyzed in the MgB_2 matrix (at the position marked "+" in Fig. 5(b)) after each of the multiple etching steps in Ar (30 etching cycles of 10's each were performed) and the composition of the matrix was analyzed to a depth of 200–300 nm. As shown in Fig. 5(d), the amount of O, B and Mg is practically unchanged indicating that O_2 was incorporated in the MgB_2 structure and formed a solid solution.[20]

The different Mg–B–O phases are not resolved by X-ray diffraction, where only peaks corresponding to MgB_2 or MgO are found. The inconsistency between the results of X-ray diffraction and Auger spectroscopy or SEM study can be explained as follows. Boron can be dissolved in MgO and O_2 in MgB_2 structures forming solid solutions which cannot be detected by traditional X-ray diffraction, due to the similarity of the scattering ability.[20] In the structure of the material which exhibited the highest B_{c2} value (No. 1 in Fig. 2(b)) prepared at 600°C, 2 GPa from B(II), i.e., using boron with a partial concentration of C (up to 3.5 wt.% C), the impurity O_2 is not localized and the average material composition according to SEM microprobe analysis was $MgB_{3.1}O_{0.3}$. The material exhibited a low connectivity of 18%, 17% porosity, 90% shielding fraction and 30% of MgO phase according to X-ray diffraction analysis (No. 5, Tables 2, 3). The small variation of both a and c parameters of MgB_2, as well as SEM and Auger analyses indicate that not only C, but also O_2 can be incorporated in the MgB_2 structure and influence the SC characteristics.

The presence of a large amount of higher magnesium boride inclusions and of "delocalized" impurity O_2 in the MgB_2 structure leads to high SC currents at high magnetic fields. This does not mean that we discard the role of C incorporated into the MgB_2 structure for the enhancement of B_{c2}, but the tendency of an increasing B_{c2} with decreasing synthesis temperature was also observed when other types of B with low carbon concentrations and different preparation methods (HP, HotP, SPS) were used. As mentioned above the addition of SiC leads to a sizeable enhancement of J_c, while Ti, Ta and Zr lead to a slight enhancement only. The same tendency has been observed for the variation of J_c by a variation of the reaction temperature. The positive effect of Ti and Zr on J_c reported in Refs. 14, 33, 34 could be explained by the formation of TiB_2 and ZrB_2 inclusions into a very thin layer, thus improving pinning. The positive effect of SiC addition is due to the C substitution in the MgB_2 structure after the decomposition of SiC to C and Mg, the latter forming Mg_2Si, while the SiC additive acts as a source of C. The authors of Refs. 2, 4, 17–19, 25 and 35 observed effects of O_2 segregation in the materials with Ti, Zr, Ta, SiC additions and the formation of higher amount of MgB_x inclusions in the materials with Ti.

Figures 7 and 8 show microstructures of MgB_2 with 10 wt.% Ti and SiC additions, respectively, prepared at 1050°C under 2 GPa. Their J_c values are shown in Figs. 1(c) and 1(d). The use of different SEI and BEI regimes (Figs. 8(a), 8(b), 8(c) and 8(d)) allows distinguishing the locations of Mg–B–O and of SiC inclusions in SiC-doped MgB_2, the locations of Ti containing phases (Fig. 7(a)) and of Mg–B–O inclusions (Fig. 7(b)) in Ti-doped MgB_2. The effects of SiC and Ti additions is somewhat different and is not totally understood yet.

In bulk MgB_2 reacted at 2 GPa, the rather coarse inclusions of the Ti based phases are randomly distributed (Fig. 7(a)). There was no diffusion of Ti into MgB_2. After reaction at low temperatures the TiH_2 inclusions are present but transform into TiB_2 at higher temperatures, as shown by X-ray diffraction and SEM analysis. The absorption of H by Ti is very useful, preventing the formation of MgH_2 and allowing the manufacture of large, dense MgB_2 blocks

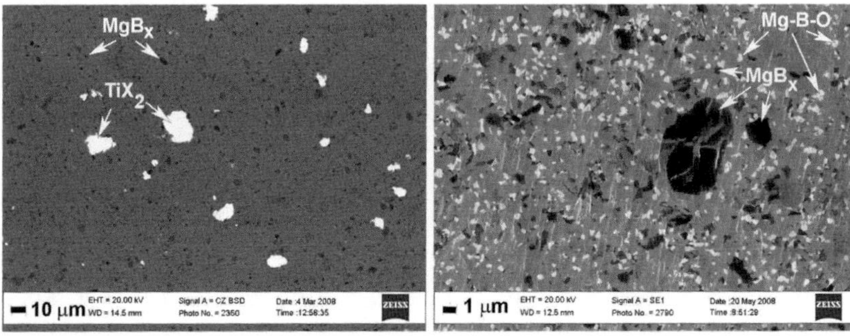

(a) (b)

Fig. 7. SEM images of the MgB_2 material prepared at 2 GPa at 1050°C from B(III) with 10% Ti), after etching in Ar under various magnifications and microscopic regimes: (a) BEI and (b) SEI image, showing an area where The Ti containing phase is absent. Brightest spots: Mg–B–O inclusions with high O_2 content. The darkest inclusions are MgB_x.

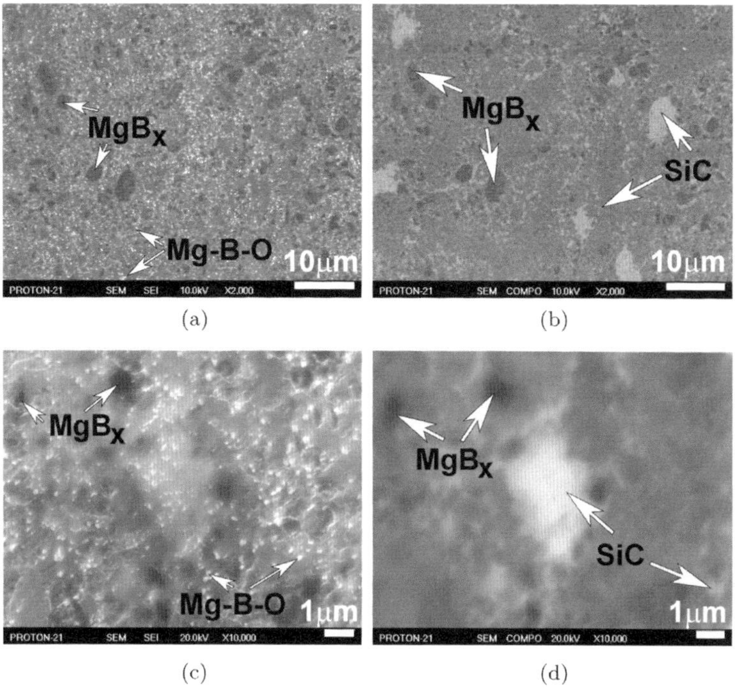

Fig. 8. SEM images of MgB_2 samples prepared under high pressure: 2 GPa at 1050°C from B(I) with 10% SiC, after etching in Ar. (a) and (b) show the same area under the same magnification as in SEI and BEI; (c) and (d) show the same area, but under higher magnification as compared to (a) and (b) as it is seen by SEI and BEI. The use of different microscopic regimes allows distinguishing the locations of Mg–B–O and SiC inclusions.

being free of cracks. For example, the addition of 10 wt.% TiH_2 powder instead of Ti results in a very porous MgB_2 material with extremely low J_c despite a high pressure reaction at 800°C. The absorption of H is not the only effect caused by Ti in MgB_2. Even more important in view of the J_c variation is the segregation of O_2 and thus the "cleaning" of the MgB_2 matrix from impurity O_2 (confirmed by SEM EDS analyses). In addition, Ti influences the formation of higher amounts of MgB_x ($x > 4$) inclusions in MgB_2. The materials with metallic additions were found to exhibit a higher hardness (due to the higher amount of harder higher boride inclusions) and fracture toughness, k_{1c}. The MgB_2 material with 10% Ta manufactured at 2 GPa had a values $k_{1c} = 7,6 \pm 2,0\,\mathrm{MPa\ m^{0.5}}$ (load of 149 N) and that one without additions $k_{1c} = 4,24 \pm 0,14\,\mathrm{MPa\ m^{0.5}}$ (estimated even under a much lower load (4,96 N). The Vickers microhardness of MgB_2 prepared at 2 GPa without additions was $H_v = 16,85 \pm 0,74\,\mathrm{GPa}$ under a load of 4,96 N.

The addition of SiC as well as O_2 segregation possibly lead to the formation of separate Mg–B–O inclusions instead of Mg–B–O layers at lower synthesis temperatures (in the material with 10 wt.% SiC, separate Mg–B–O inclusions are forming at 2 GPa and 800°C[37]).

2.3 *Application of bulk MgB_2 in electromotors and FCL*

Figure 9(a) demonstrates the general view of the zebra-type rotors of a SC motor of 1,300 W with MgB_2-10%Ti inserts reacted under high pressure (2 GPa), while Fig. 9(b) represents a 1600 W motor with an insert of melt textured MT–YBCO.[2] The motors were constructed and tested in the Moscow State Aviation Institute. Figure 9 also shows a cross-sectional view of the rotor as well as the working characteristics of the motors with these rotors at 15–20 K at a voltage of 210–215 V.

In spite of the much higher transition temperature of MT–YBCO (~93 K), and a smaller amount of SC material in the MgB_2 rotor both motors showed comparable performances. It is expected that electromotors with higher output power should have a much better performance. Taking into account that 1 kg of MT–YBCO costs about €2000, while 1 kg of high-pressure synthesized MgB_2 is available for about €700–800, one can estimate the competitive ability of MgB_2 based electromotors at an operation temperature of 20 K.

Bulk MgB_2 is very promising for application in MAGLEV transport and for manufacturing of magnetic bearings. The trapped magnetic fields estimated for MgB_2 blocks and items of different configurations are shown in Fig. 10. It should be mentioned that trapped magnetic field essentially depends on the dimensions of the SC part. Larger parts from a MgB_2 block of higher quality (with

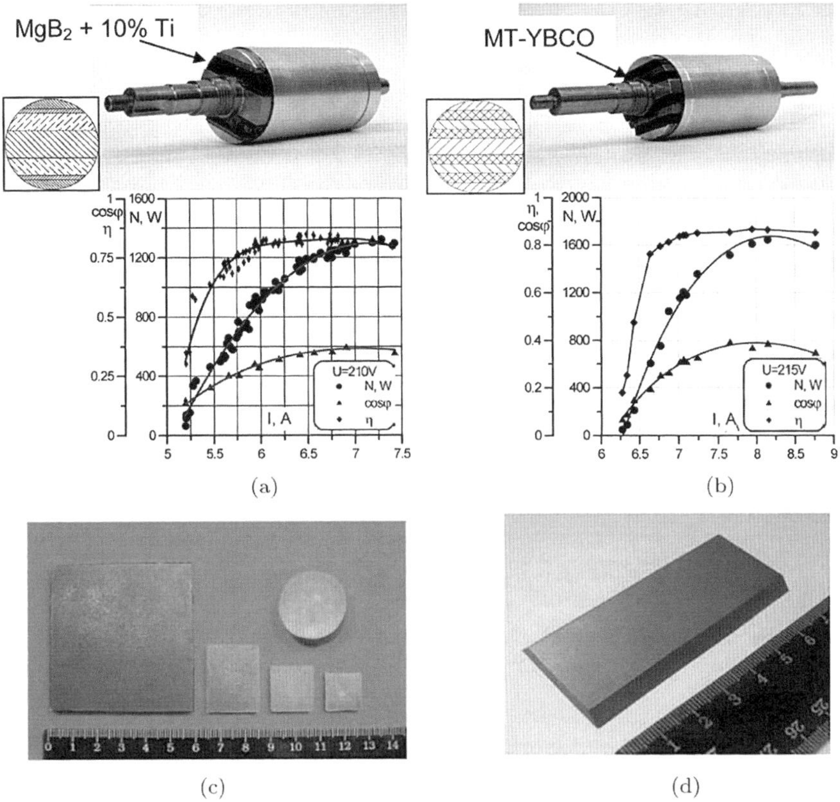

Fig. 9. General view of a zebra-type rotor of a SC motor with (a) MgB_2–Ti and (b) MT–YBCO inserts (layers). Also shown is the cross-section scheme of the rotors and the performance characteristics of the motors with rotors at 15–20 K and 210–215 V,[2] (c) General view of the MgB_2 bulk blocks produced under 2 GPa pressure and (d) View of the MgB_2 insert (or layer) from zebra-type rotor of SC motor.

higher J_c) can trap higher magnetic fields. Figure 10(c) demonstrates the effect of the rate of magnetic field variation on the trapped magnetic field of the ring cut from MgB_2 block produced from Mg:2B(III) with 10% SiC at 2 GPa, 1050°C, 1 h. For field variation rates was lower than 0.1 T/min, the maximum trapped field was above 1.5 T. Low quality MgB_2 is very sensitive to flux jumps, induced by external magnetic field variations. Since MgB_2 of higher

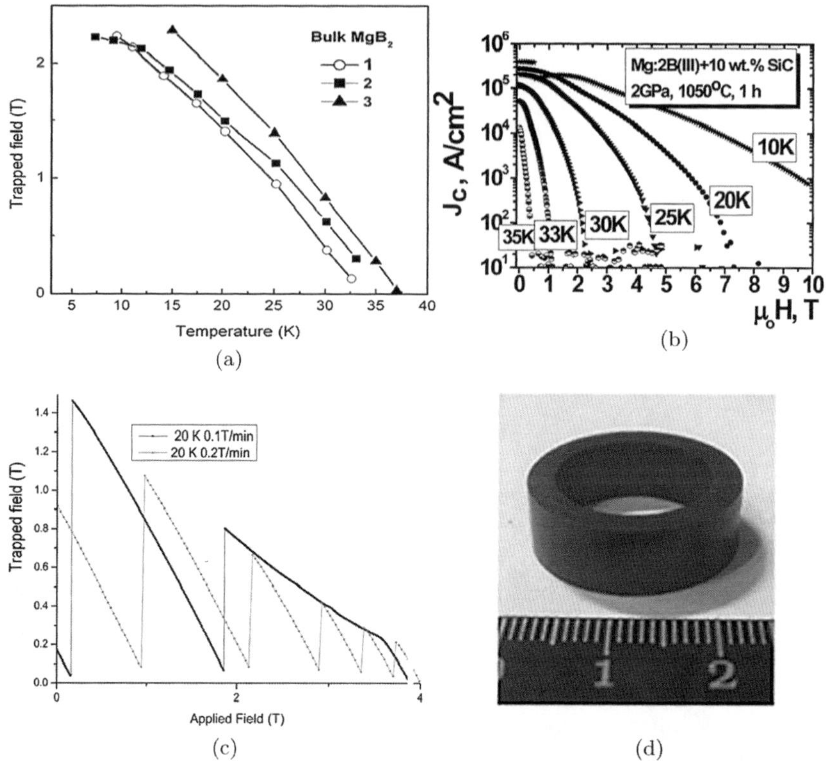

Fig. 10. (a) Trapped field at sample center. 1: from B(V), 10% Ti (sample size 14×14×4 mm), 2: from B(V), 5% Ta (Ø22.5×6 mm), 3: from B (V), 10% Ti (Ø30× 7.5 mm)[2]; (b) Trapped field at 20 K at center of the MgB_2 ring Ø24.3 × 17.9 mm, h = 7.8 mm, see Fig. 10(d). The measurements were performed by M. Eisterer (ATI Vienna), with different decreasing rates of outer magnetic field (0.1 T/min: solid line, 0.2 T/min: gray line with squares), (c) J_c versus magnetic field of the ring in Fig. 10(d).

quality is more stable against flux jumps, the stability of its SC characteristics in variable magnetic fields depends to high extent on the material manufacturing conditions.

The trapped magnetic field versus increasing temperature measured by Hall sensors above two MgB_2 disks (Ø30 mm $h = 10$ mm) and between such two MgB_2 disks[36] shown in the (Chapter 6b) by Yamamoto. The disks measured by Yamamoto *et al.*,[36] were prepared *in situ* by HIP: mixture of Mg and B was uniaxially pressed at

100 MPa, then sealed in a SS tube and heat treated at 850°C for 3 h under Ar atmosphere. The trapped field between two disks is usually higher than that one above a single disk; the trapped field measured 3 mm above such a bulk disk at 20 K just reached 1 T.[36]

The function of an inductive FCL is based on magnetic coupling between a SC element and a protected circuit. The inductive FCL consists of a primary normal metal coil coupled to a secondary short-circuited SC coil via a ferromagnetic core. Under normal conditions for the protected circuit, the secondary coil is in the SC state. The magnetic flux produced by the primary coil is compensated by the flux originated from induced screen currents in the short-circuited SC coil. The device impedance is determined only by the leakage flux in an air gap between the primary and secondary coils. Under short-circuit conditions, the secondary coil or its part passes to the resistive state due to the increase of the current above the critical value. If the resistance due to heat dissipation exceeds the inductive reactance of the coil, the induced current in the secondary coil is sharply reduced and the magnetic flux of the primary coil is no more compensated: the FCL impedance increases and limits the fault current.

Using the setup shown in Fig. 11, estimations were performed for the current of quenching, the quenching critical current density at 4.2 K and the AC losses of MgB_2 rings (with outer diameters from 20–110 mm, wall thicknesses of 3–6 mm and a height of 5–14 mm), manufactured under high pressure (2 GPa) and hot pressure (30 MPa) conditions. The critical current density of the ring materials determined using the magnetization loops obtained by VSM using the Bean Model varied from 2,24 to 5×10^5 A/cm^2 at 10 K in zero field. The AC activation current for these rings varied from 11 to 50 A (50 Hz) and a quenching current change in the range of 4, 5–24 $\times$ 10^3 A, corresponding to 1,9 to 2,47 $\times$ 10^4 A/cm^2 inductive or quenching critical current density. The thermal conductivity of MgB_2 (~1 W/K·m) is two orders of magnitude lower than that of BISCO ceramics, while recovering after quenching event occurs for about 1 s as required for applications of FCL for power systems. The estimated inductive critical current density is around 10–20 times

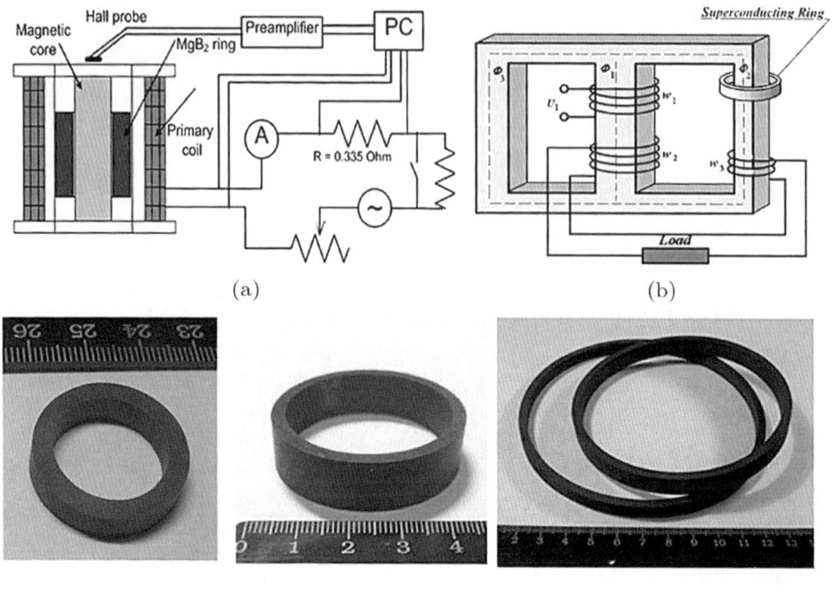

Fig. 11. (a) Setup for AC losses measurements by a contactless transformer method[5] designed in the Physics Department of Ben-Gurion University of the Negev, Israel (V. Sokolovsky, V. Meerovich)[5]; (b) Scheme of the current-limiting transformer: w_1 — primary winding; by, w_2 — secondary winding; w_3 — counter-connected winding in series. The ring performs as SC short-circuited winding, (c) The tested MgB_2 rings of various sizes.

lower than that estimated magnetically, which can be explained by the sensitivity of MgB_2 for quenching in the rapidly varying magnetic field induced by the AC current. The critical quenching current decreases very slowly for a temperature increase from 4.2 to 20 K, but sharply decreases in the temperature range from 20–34 K. The AC losses per period in MgB_2 rings before quenching were 12–17 J while the power of the losses was around 200 W. The investigations showed that all tested MgB_2 rings can be successfully used for fault current limitation. The advantage of MgB_2 over Low Temperature Superconductors (LTS) is a higher thermal stability, due to a higher difference between operational and critical temperature. A further advantage is the possibility to use liquid H instead of liquid He as a coolant (H is used for cooling in usual generators). The highest characteristics was obtained for a MgB_2 matrix with 12% Ti additives, synthetized by hot pressing.

References

1. L. K. Kovalev, K. V. Ilushin, V. T. Penkin, K. L. Kovalev, V. N. Poltavets, S. M. A. Koneyev, K. A. Modestov, W. Gawalek, T. A. Prikhna and I. I. Akimov, An experimental investigation of a reluctance electrical drive with bulk superconducting elements in the rotor at temperature below 20 K, *J. Phys. Conf. Ser.* **43**, pp. 792–795 (2006).
2. T. Prikhna, W. Gawalek, Ya. Savchuk, N. Sergienko, V. Moshchil, S. Dub, V. Sverdun, L. Kovalev, V. Penkin, M. Wendt, G. Fuchs, Ch. Grovenor, S. Haigh, V. Melnikov and P. Nagorny, High-pressure high-temperature synthesis of nanostructural magnesium diboride for electromotors and devices working at liquid hydrogen temperatures, *Adv. Sci. Technol.* **47**, pp. 25–30 (2006).
3. M. Tomsic, M. Rindfleisch, J. Yue, K. McFadden, J. Phillips, M. D. Sumption, M. Bhatia, S. Bohnenstiehl and E. W. Collings, Overview of MgB_2 superconductor applications, *Intl. J. Appl. Ceramic Technology* **4**(3), pp. 250–259 (2007).
4. T. A. Prikhna, W. Gawalek, Ya.M. Savchuk, N. V. Sergienko, V. E. Moshchil, V. Sokolovsky, J. Vajda, V. N. Tkach, F. Karau, H. Weber, M. Eisterer, A. Joulain, J. Rabier, X. Chaud, M. Wendt, J. Dellith, N. I. Danilenko, T. Habisreuther, S. N. Dub, V. Meerovich, D. Litzkendorf, P. A. Nagorny, L. K. Kovalev, Ch. Schmidt, V. S. Melnikov, A. P. Kozyrev, A. V. Sverdun, V. B. J. Kosa and A. V. Vlasenko, Nanostructural superconducting materials for fault current limiters and cryogenic electrical machines, *Acta Phys. Polonca A* **117**(1), pp. 7–14 (2010).
5. V. Meerovich, V. Sokolovsky, T. Prikhna, W. Gawalek and T. Habisreuther, AC losses in high pressure synthesized MgB_2 bulk rings measured by a transformer method, *Supercond. Sci. Technol* **26**, p. 035015 (p. 8) (2013).
6. T. Prikhna, J. Noudem, W. Gawalek, A. Mamalis, A. Soldatov, Ya. Savchuk, V. Moshchil, M. Eisterer, H. Weber, S. Dub, T. Habisreuther, J. Dellith, S. You, Ch. Schmidt, F. Karau, U. Dittrich, I. Vajda, N. Sergienko, D. Litzkendorf, X. Chaud, V. Sverdun and R. Kuznietsov, Spark plasma synthesis and sintering of superconducting MgB_2 based materials, *Mater. Sci. Forum* **721**, pp. 3–8 (2012).
7. C. Buzea and T. Yamashita, Review of superconducting properties of MgB_2, *Supercond. Sci. Technol.* **14**(11), p. R115 (2001).

8. E. W. Collings, M. D. Sumption, M. Bhatia, M. A. Susner and S. D. Bohnenstiehl, Prospects for improving the intrinsic and extrinsic properties of magnesium diboride superconducting strands, *Supercond. Sci. Technol.* **21**(10), pp. 103001 (2008).
9. S. X. Dou, O. Shcherbakova, W. K. Yeoh, J. H. Kim, S. Soltanian, X. L. Wang, C. Senatore, R. Flükiger, M. Dhalle, O. Husnjak and E. Babic, Mechanism of enhancement in electromagnetic properties of MgB_2 by nano SiC doping, *Phys. Rev. Lett.* **98**, p. 097002 (2007).
10. J. H. Kim, S. Oh, Y-Uk, Heo, Hata, S. Kumakura, H. Matsumoto, A. Mitsuhara, S. Choi, Y. Shimada, M. Maeda, J. L. MacManus-Driscoll, S. X. Dou, Microscopic role of carbon on MgB_2 wire for critical current density comparable to NbTi, *NPG Asia Materials* **4**, e3 (2012).
11. X. L. Wang, S. Soltanian, M. James, M. J. Qin, J. Horvat, Q. W. Yao, H. K. Liu and S. X. Dou, Significant enhancement of critical current density and flux pinning in MgB_2 with nano-SiC, Si, and C doping, *Physica C* **408–410**, pp. 63–67 (2004).
12. Y. B. Zhang, X. J. Shan, X. W. Bai, T. Y. Liu, H. M. Zhu and Ch.B. Cai, In situ synthesis and current-carrying characteristics of superconducting MgB_2–B_4C composites with MgB_2 fractions ranging from 18% to 85%, *Supercond. Sci. Technol.* **25**, p. 095003 (2012).
13. H. Yamada, N. Uchiyama, A. Matsumoto, H. Kitaguchi and H. Kumakura, The excellent superconducting properties of *in situ* powder-in-tube processed MgB_2 tapes with both ethyltoluene and SiC powder added, *Supercond. Sci. Technol.* **20**, p. L30 (2007).
14. Y. Zhao, D. X. Huang, Y. Feng, C. H. Cheng, T. Machi, N. Koshizuka and M. Murakami, Nanoparticle structure of MgB_2 with ultrathin TiB_2 grain boundaries, *Appl. Phys. Lett.* **80**, pp. 1640–1642 (2001).
15. N. Hörhager, M. Eisterer, H. W. Weber, T. Prikhna, T. Tajima and V. F. Nesterenko, Ti and Zr doped MgB2 bulk superconductors, *J. Phys.: Conf. Series* **43**, p. 500, (2006).
16. Y. Zhu, L. Wu, V. Volkov, Q. Li, G. Gu, A. R. Moodenbaugh, M. Malac, M. Suenaga and J. Tranquada, Microstructure and structural defects in MgB_2 superconductor, *Physica C* **356**(4), pp. 239–253 (2001).
17. T. A. Prikhna, W. Gawalek, Ya.M. Savchuk, V. E. Moshchil, N. V. Sergienko, A. B. Surzhenko, M. Wendt, S. N. Dub, V. S. Melnikov, Ch. Schmidt and P. A. Nagorny, High-pressure synthesis of a bulk superconductive MgB_2-based material, *Physica C* **386**, pp. 565–568 (2003).
18. T. A. Prikhna, W. Gawalek, Ya.M. Savchuk, V. E. Moshchil, N. V. Sergienko, T. Habisreuther, M. Wendt, R. Hergt, Ch. Schmidt, J. Dellith, V. S. Melnikov, A. Assmann, D. Litzkendorf and

P. A. Nagorny, High-pressure synthesis of MgB_2 with addition of Ti, *Physica C* **402**, pp. 223–233 (2004).

19. T. Prikhna, M. Eisterer, H. W. Weber, W. Gawalek, V. Kovylaev, M. Karpets, V. Moshchil, A. Kozyrev, T. Basyuk, X. Chaud, W. Goldacker, V. Sokolovsky, J. Noudem, A. Borimskiy, V. Sverdun and E. Prisyazhnaya, Temperature-pressure induced nano-structural inhomogenities for vortex pinning in bulk MgB_2 of different connectivity, *Physica C* **503**, pp. 109–112 (2014).
20. T. A. Prikhna, M. Eisterer, H. W. Weber, W. Gawalek, V. V. Kovylaev, M. V. Karpets, T. V. Basyuk and V. E. Moshchil, Nanostructural inhomogeneities acting as pinning centers in bulk MgB_2 with low and enhanced grain connectivity, *Supercond. Sci. Technol.* **27**(4), 044013 (9pp) (2014).
21. P. Kováč, I. Hušek, T. Melišek, J. C. Grivel, W. Pachla, V. Štrbík, R. Diduszko, J. Homeyer and N. H. Andersen, The role of MgO content in *ex situ* MgB_2 wires, *Supercond. Sci. Technol.* **17**(10), pp. L41–L46 (2004).
22. M. Vignolo, G. Romano, A. Malagoli, V. Braccini, M. Tropeano, E. Bellingeri, C. Fanciulli, C. Bernini, V. Honkimaki, M. Putti and C. Ferdeghini, Role of the grain oxidation in improving the in-field behavoir of MgB_2 *ex situ* tapes, *IEEE Trans. Appl. Supercond.* **19**, pp. 2718–2721 (2009).
23. C. B. Eom, M. K. Lee, J. H. Choi, L. J. Belenky, X. Song, L. D. Cooley, M. T. Naus, S. Patnaik, J. Jiang, M. Rikel, A. Polyanskii, A. Gurevich, X. Y. Cai, S. D. Bu, S. E. Babcock, E. E. Hellstrom, D. C. Larbalestier, N. Rogado, K. A. Regan, M. A. Hayward, T. He, J. S. Slusky, K. Inumaru, M. K. Haas and R. J. Cava, High critical current density and enhanced irreversibility field in superconducting MgB_2 thin films, *Nature* **411**, pp. 558–560 (2001).
24. R. F. Klie, J. C. Idrobo and N. D. Browning, Observation of coherent oxide precipitates in polycrystalline MgB2, *Appl. Phys. Lett.* **80**, p. 3970 (2002).
25. T. A. Prikhna, W. Gawalek, V. M. Tkach, N. I. Danilenko, Ya.M. Savchuk, S. N. Dub, V. E. Moshchil, A. V. Kozyrev, N. V. Sergienko, M. Wendt, V. S. Melnikov, J. Dellith, H. Weber, M. Eisterer, Ch. Schmidt, T. Habisreuther, D. Litzkendorf, I. Vajda, A. P. Shapovalov, V. Sokolovsky, P. A. Nagorny, V. B. Sverdun, J. Kosa, F. Karau and A. V. Starostina, Effect of higher borides and inhomogeneity of oxygen distribution on critical current density of undoped and doped magnesium diboride, *J. Physics: Conf. Series* **234**, p. 012031 (2010).

26. Thermal analysis, phase equilibria, and superconducting properties in MgB_2 and carbon doped MgB_2, Dissertation, Scot David Bohnenstiehl, the Ohio State University, 2012.
27. R. Flukiger, M. S. A. Hossain and C. Senatore, Strong enhancement of J_c and B_{irr} in binary *in situ* MgB_2 wires after cold high pressure densification, *Supercond. Sci. Technol.* **22**, p. 085002 (2009).
28. M. Eisterer, Calculation of the volume pinning force in MgB2 superconductors, *Phys. Rev. B* **77**, p. 144524 (2008).
29. T. Prikhna, W. Gawalek, Y. Savchuk, A. Soldatov, V. Sokolovsky, M. Eisterer, H. W. Weber, J. Noudem, M. Serga, V. Turkevich, M. Tompsic, V. Tkach, N. Danilenko, W. Goldacker, F. Karau, I. Fesenko, M. Rindfleisch, J. Dellith, M. Wendt, S. You, V. Meerovich, S. Dub, V. Moshchil, N. Sergienko, T. Habisreuther, C. Schmidt, D. Litzkendorf, P. Nagorny and V. Sverdun, Effects of high pressure on the physical properties of MgB_2, *J. Supercond. Nov. Magn.* **24**, pp. 137–150 (2011).
30. C. B. Bean, Magnetization of high field superconductors, *Rev. Mod. Phys.* **36**, pp. 31–36 (1964).
31. M. Eisterer, Magnetic properties and critical currents of MgB_2, Supercond. *Sci. Technol.* **20**, pp. R47–R73 (2007).
32. T. Prikhna, W. Gawalek, M. Eisterer, H. W. Weber, J. Noudem, V. Sokolovsky, X. Chaud, V. Moshchil, M. Karpets, V. Kovylaev, A. Borimskiy, V. Tkach, A. Kozyrev, R. Kuznietsov, J. Dellith, C. Schmidt, T. Basyuk, D. Litzkendorf, F. Karau, U. Dittrich and M. Tomsic, Superconductivity in multi-phase Mg-B-O compounds, *Physics Procedia* **36**, pp. 475–478 (2012).
33. Y. Feng, Y. Zhao, P. Sun, F. C. Liu, B. Q. Fu, L. Zhou, C. H. Cheng, N. Koshizuka and M. Murakami, Improvement of critical current density in MgB2 superconductor by Zr doping at ambient pressure. *Appl. Phys. Lett.* **79**, pp. 3983–3985 (2001).
34. Y. Zhao, Y. Feng, C. H. Cheng, L. Zhao, Y. Wu, T. Machi, N. Koshizuka and M. Murakami, High critical current density of MgB_2 bulk superconductor doped with Ti and sintered at ambient pressure, *Appl. Phys. Lett.* **79**(8), pp. 1154–1157 (2001).
35. T. Prikhna, N. Novikov, Y. Savchuk, W. Gawalek, N. Sergienko, V. Moshchil, M. Wendt, V. Melnikov, S. Dub, T. Habisreuther, Ch. Schmidt, J. Dellith and P. Nagorny, High-pressure synthesis of MgB_2-based material with high critical currents, *in "Innovative superhard materials and sustainable coatings for advanced manufacturing", Dordrecht; New York: Springer, Eds. by* J. Lee and N. V. Novikov, *NATO science series, Series II, Mathematics, physics and chemistry, Chapter 6* **200**, pp. 81–90 (2005).

36. A. Yamamoto, A. Ishihara, M. Tomita and K. Kishio, Permanent magnet with MgB_2 bulk superconductor, *Appl. Phys. Lett.* **105**, p. 032601 (2014).
37. T. Prikhna, M. Eisterer, W. Gawalek, A. Kozyrev, H. W. Weber, V. Sokolovsky, X. Chaud, J. Noudem, T. Habisreuther, V. Moshchil, M. Karpets, T. Basyuk, V. Kovylaev, J. Dellith, V. Sverdun, R. Kuznietsov, C. Shmidt, T. Vitovetskaya and L. Polikarpova, Synthesis pressure–temperature effect on pinning in MgB_2-based superconductors, *J. Supercond. Nov. Magn.* **26**, pp. 1569–1576 (2013).

Chapter 3b

The Reactive Liquid Infiltration (RLI) Technique for the Bulk Reaction to MgB_2

G. Giunchi

Materials Science Consultant
via Teodosio 8
20131 Milano, Italy
giovanni.giunchi@gmail.com

The densification of MgB_2 using a new processing technique, the Reactive Liquid Infiltration (RLI) is described, based on the liquid infiltration of Magnesium (Mg) in a Boron (B) powder preform. Bulk, polycrystalline MgB_2 material composites are produced, consisting of a superconducting MgB_2 matrix embedding normal conducting borides, identified by X-ray diffraction (XRD) as Mg_2B_{25}. The infiltration process is self-sustaining and its driving forces are described together with the reaction kinetics. The use of nanometer size B powders shortens the reaction time and improves the critical current density, but renders the superconductor thermally more instable. Nevertheless, the presence of normal conducting borides of several tens of μm size does not prevent the MgB_2 matrix to exhibit a sufficiently high critical current density, suitable for many medium field applications. The RLI processed MgB_2 material has a high density, about 90% of the theoretical value, similar to the material obtained by the hot pressing process. Therefore the process yields products exhibiting optimal mechanical strength, ideal for the manufacture of large superconducting bulk pieces for magnetic shielding purposes or for permanent superconducting magnets.

1. Introduction

The news about the discovery of superconductivity at 39 K in MgB_2 by Japanese researchers[1] at the beginning of 2001 was commented with curiosity in our research group on superconducting materials in EDISON R&D, but also with some initial misgiving for the relative "low T_c" of this material. At that time, our main interest was the electrodeposition of High Temperature Superconductors (HTS) cuprates (BSCCO-2212 and YBCO-123) on metallic substrates, the race for the HTS being our challenge. Nevertheless, the apparent simplicity of the new compound and some lucky events forced our curiosity to prevail and we decided to make some attempts to prepare MgB_2. My principal collaborator at that time was Sergio Ceresara, a physical metallurgist who in the eighties promoted the Italian industry of superconducting materials by installing various wire production plants for NbTi, Nb_3Sn and Nb_3Al at Europa Metalli, in Fornaci di Barga (Lucca). In the early days of 2001, one lucky circumstance was the presence of crystalline B lumps in our laboratory, this material being used for the fabrication of a special nickel based alloy, as a substrate for HTS. At that time, the delivery terms for MgB_2 commercial powders from an international supplier exceeded one month, which in the retrospective appears as another lucky event. Indeed, we thus started to prepare MgB_2 by ourselves, finding unexpected chemistry features in the Mg–B system and obtaining encouraging results on the superconducting properties of polycrystalline samples. We developed a new synthesis process, the Reactive Liquid Infiltration (RLI).

Today, the concern about the "low T_c" in MgB_2 is history and the progress on the material quality renders MgB_2 a real new "high T_c" superconductor which will be able to substitute the so-called Low Temperature Superconductors (LTS) in some large scale applications. Here, I am focusing on the bulk MgB_2 preparation by the RLI process and on the physical properties of the resulting products. Starting from the understanding of the infiltration mechanism, the advancements reached in the few last years and the material potentiality for superconducting bulk applications will be reviewed, with a particular attention to the fields where at present cuprates cannot easily be applied, due to their high cost and low mechanical characteristics.

2. Discovery of the Mg RLI Process

The first papers on the MgB_2 superconducting characteristics and material preparation appeared in January and February 2001 and described high pressure sintering (Hot Isostatic Pressing) of Mg and B powders mixtures in Japan[1] and the Mg vapor sintering of B fibers by American scientists.[2] Both techniques were able to produce well connected MgB_2 small samples with a sharp transition at the critical temperature, measured either by resistivity or by magnetic susceptibility techniques. In order to produce MgB_2 materials in our laboratory we filled a steel container with B powders and large chunks of Mg, welded the container and performed a thermal treatment of a few hours at about 900°C in a conventional oven. The morphology of the products resulting from one of the first attempts is reported in Fig. 1. After reaction the Mg disappeared and the B powders, still located in the same place where they were inserted, were transformed into MgB_2, as revealed by X-Ray Powder Diffraction (XRPD) analysis. This unexpected result, that to my knowledge had no counterpart in any ceramic sintering or in any inorganic material reaction, drew our attention.

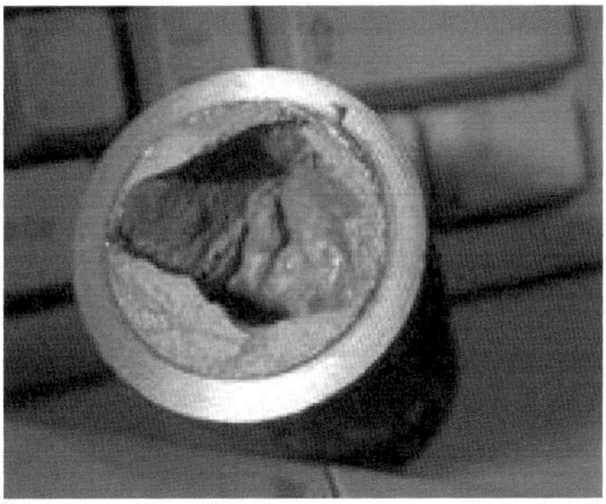

Fig. 1. Reacted MgB_2 material resulting from the liquid infiltration originated by a central Mg chunk.

Trying to understand the possible causes of these findings, it emerged that the liquid Mg was spontaneously attracted by the B powders and wetted them by percolation between the powder's interstices, simultaneously inducing their reaction into MgB_2. Other strange evidences were soon discovered, for example: (i) the transformed material was very dense, near to the theoretical density of MgB_2, (ii) the liquid Mg percolation proceeded also against gravity, probably helped by a kind of capillary phenomena. Looking at the morphology of the resulting product, the materials shows large fractured grains embedded in a matrix consisting of much finer grains, the fractures being filled by MgB_2. Figure 2 illustrates the typical morphology of one of our first MgB_2 products.

There was no more evidence of the original B powders in the XRPD spectrum, showing MgB_2, Mg, a faint peak of MgO and other impurity peaks, not related to the original beta-rhombohedral B. These very surprising initial results of the reaction remained for many months completely inexplicable to us. The superconducting analysis, performed on the first reacted product, confirmed that it was a good superconducting bulk material, with the same sharp transition as well sintered MgB_2 samples. The first experiments with this new and

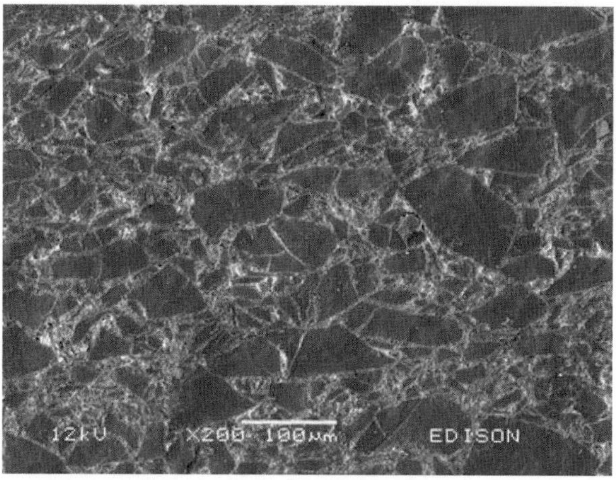

Fig. 2. Composite morphology, by Scanning electron microscope (SEM), of the MgB_2 by RLI, showing large borides grains glued by fine MgB_2 grains. Reproduced with permission from AIP Publishing LLC.

unexpected production method of dense bulk MgB_2 by Mg liquid infiltration, were the object of an Italian patent application (filed May 11, 2001[3])], published in 2003.[4]

3. MgB_2 Morphology and B Grain Size Effects

3.1 *The bulk RLI–MgB_2 peculiar morphology*

The morphology of liquid infiltration products is related to the original B powder size. We started with crystalline B powder (99.5% purity), which were crushed and sieved to particle sizes of $\leq 100\,\mu$m. An SEM image of the product (Fig. 4) reveals the presence of large fractures in the B grains, like channels. X-ray fluorescence (XRF) microanalysis inside these channels showed the presence of pure MgB_2.[5,6] The apparent B grains were borides, as shown by careful crystallographic and XRF analysis in the following years,[7] thus confirming that the matrix surrounding the large boride grains was a mixture of MgB_2 and Mg, with the presence of low amounts of MgO. For smaller B particle sizes, there was a reduction of the grain size of impurity borides, reaching an almost homogeneous morphology when B particles of the micrometer size were used, as shown in Fig. 5. The density of the final products was about 2.40 g/cm^3, 90%

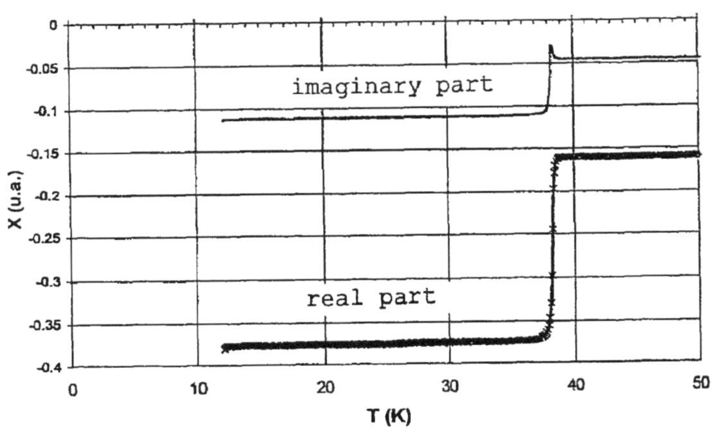

Fig. 3. AC susceptibility curves on early samples of MgB_2 by RLI process. Reproduced with permission from AIP Publishing LLC.

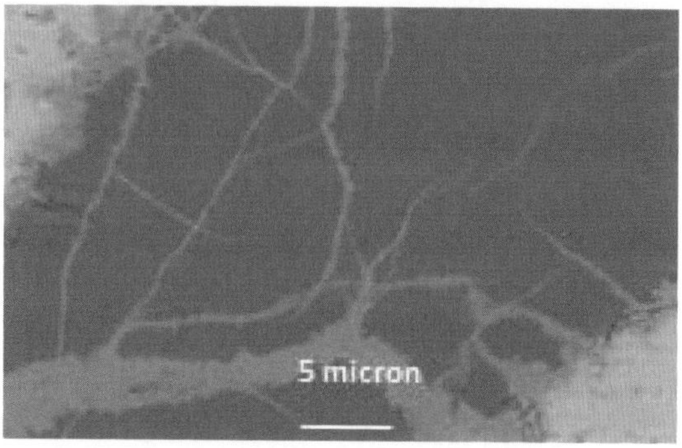

Fig. 4. SEM picture of the MgB_2 channels crossing the large borides grains. Reproduced with permission fro AIP Publishing LLC.

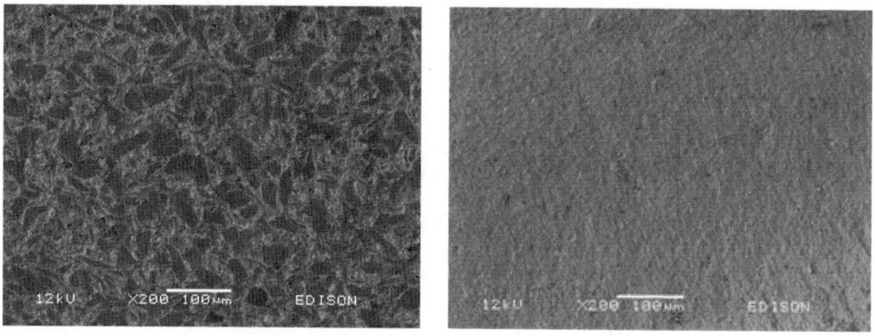

(a) (b)

Fig. 5. Comparison between morphologies of MgB_2 produced by RLI: (a) infiltration of boron; (b) infiltration of boron with micrometer size.

of the theoretical density of MgB_2 (2.63 g/cm^3). SEM, XRPD and energy dispersive X-ray spectrometry (EDS) analysis, in Fig. 6 and in Table 1, of the product obtained by infiltration of microcrystalline B (Starck AG, grade K2),with grains less than about 5 μm size, showed that some microcrystalline borides were still present in MgB_2 materials (less than a few %), even if their presence was not detectable by XRPD (Fig. 6(a)). The evaluation of the composition, made by EDS microanalysis in the positions showed in Fig. 6(b), is reported in Table 1. The compositions determined by this analysis should be only considered as qualitative, due to incertitude in the B evaluation.

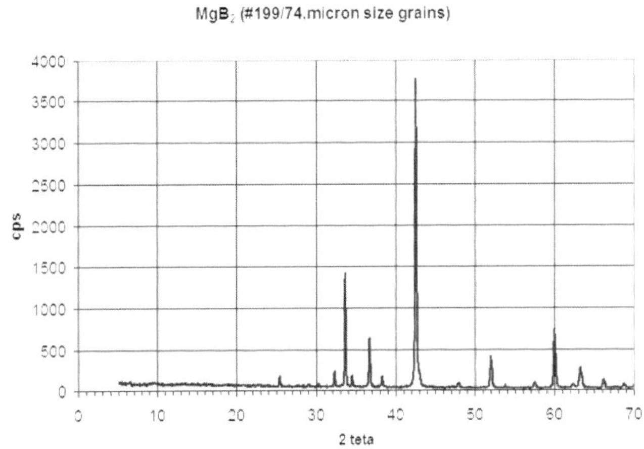

(a)

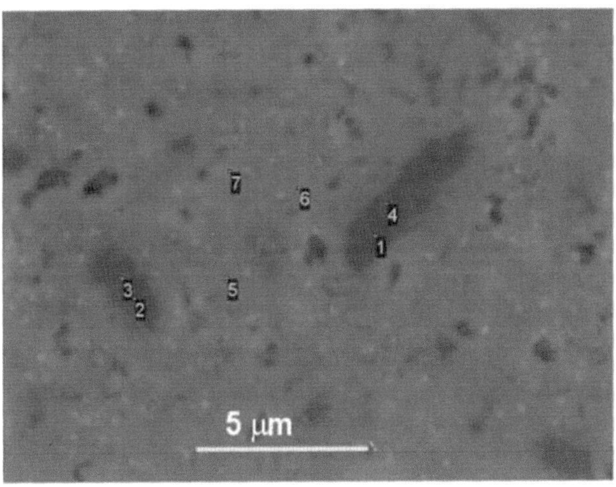

(b)

Fig. 6. Mg infiltration of microcrystalline beta-rhombohedral B: (a) XPRD spectrum, (b) The EDS microanalysis.

Table 1: EDS point-like microanalysis at the sites numbered in Fig. 6(b).[7]

	1	2	3	4	5	6	7
$(B/Mg)_{EDS}$	0.019	0.130	0.014	0.109	0.015	0.020	0.009
$(B/Mg)_{mol}$	2.330	6.094	2.000	5.581	2.070	2.390	1.604

3.2 *Early infiltration experiences from other laboratories*

Being the solid–liquid infiltration reaction a natural event for the B–Mg system, its occurrence was visible in any *in situ* experiment, even if the first experiments gave only a faint suggestion of its reality.

In 2001, in several labs the *in situ* reaction to produce MgB_2 from the elemental B and Mg was independently applied. At Iowa State University, a diffusion process was hypothesized for the reaction of a mixture of B powders and Mg lumps inside a Ta tube.[8] The experiment allowed measuring the isotope effect in MgB_2 for the first time. At Northwestern University, the production of a composite Mg–MgB_2 was attempted by a reaction promoted by gas pressure on liquid Mg surrounding a preform of B powders.[9] In this case, MgB_2 was formed only partially, probably due to the interposition of MgO powder layers between B and Mg. Nevertheless, the superconducting quality of the reacted products was excellent, as revealed by the magnetization measurements. At the Bhabha Atomic Research Center in Mumbai (India), the infiltration reaction was observed by mixing B powders with Mg flakes or powders.[10] In this case too, porosity and inhomogeneity of the product prevented a clear understanding of the infiltration reaction, even after observing that large MgB_2 grains were formed (up to about 2 mm). The Institute of Solid State in Moscow performed a high gas pressure synthesis from lump Mg layered on B powders of different granularity and observed that finer B powder particles were more difficult to be infiltrated by Mg.[11] It was concluded that, in this case, the reaction was limited at the surface layer of B due to the formation of a dense product at the reaction front that prevented further percolation of the melt. An alternative hypothesis was proposed, inspired by the closed tube geometry of the used crucible: stopping of the Mg melt percolation by a counter pressure of gas entrapped inside the finer powders.

4. The Infiltration Pathways to Bulk MgB_2 and the Discovery of a Precursor Boride Phase: Mg_2B_{25}

4.1 *The Reactive Infiltration description*

The use, from the beginning on, of beta-rhombohedral crystalline B powders (99.5% purity) of large grain size (up to 100 μm, hereafter called P100), turned out to be very instructive, both for a clear microscopic description of the infiltration process as well as for its understanding. The benefits of such LG size B powders, with an average grain size of about 40 μm, may be summarized as follows: (a) a high tap density of B in the preform (about 1.4 g/cm^3), higher than the partial B tap density for a theoretical stoichiometric reaction giving a 100% dense product (1.20 g/cm^3), (b) the B preform had a small amount of boron oxide at the surface as well as hydroxide species, (c) the porosity of the preform (about 50 vol%) was localized in relatively large inter-particles channels, where the melt can percolate without large liquid–solid interface tension effects, and (d) the crystalline B phase matched the crystalline phase of the first intermediate product of the reaction (a high-B Mg boride). The pure Mg melts at 650°C, therefore it is expected that the liquid Mg infiltration starts at this temperature, even if it is known that a gas–solid reaction between Mg and B can start at lower temperatures.[12] In Fig. 7, the two zones of the (P,T) phase diagram, by Liu *et al.*,[13] of interest for the MgB_2 production, are colored. The zone pertaining to the liquid infiltration in dark gray, that one pertaining to the thin film gas reaction in light gray.

The first unexpected behavior of the Mg melt was its capability to wet a B powder preform up to a depth of several cm, even against the gravity field. This behavior is well represented by the experiment reported in Ref. 5, where a long tube ($H > 200$ mm, $\Phi = 5$ mm) filled with crystalline B and with Mg at both ends was reacted in a vertical position. The resulting MgB_2 product was a separated double-rod infiltration of 110 mm height (bottom) and 75 mm depth (top),

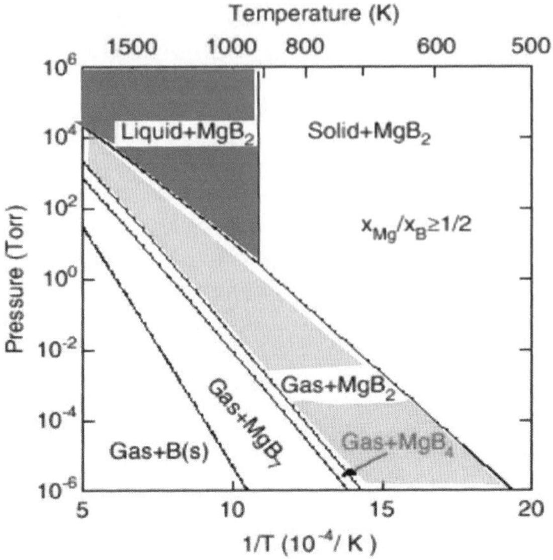

Fig. 7. The computed Mg–B phase diagram. The area pertaining to the liquid infiltration is marked in dark gray, the area pertaining to the vapor synthesis in light gray. (Reprinted with permission from Ref. 13. © 2001 AIP).

respectively, both separated by a central hole of 15 mm. The density of the rods was respectively 2.31 (bottom) and 2.09 g/cm^3 (top). In addition to the already mentioned peculiarities of the infiltration process, two additional features were observed: (a) the differences in the densities was attributed to an inhomogeneous filling of B powders in the tube (by hand pressing): the bottom part was more densely packed than the upper part, the higher tap density leading to a denser MgB_2 final product, i.e., a product with a smaller amount of residual pure Mg inside; (b) the formation of a hole between the two MgB_2 rods (the creation of such a 15 mm long hole during the filling can be excluded), showing that there is a sort of self-suction of the reacting species.

A further peculiar behavior of the infiltration is a marked similarity between the reacted product shape and the original B preform. This behavior suggested the possibility to use the RLI process for obtaining a semi-finished product directly during the reaction. This was a novelty with respect to the conventional ceramic

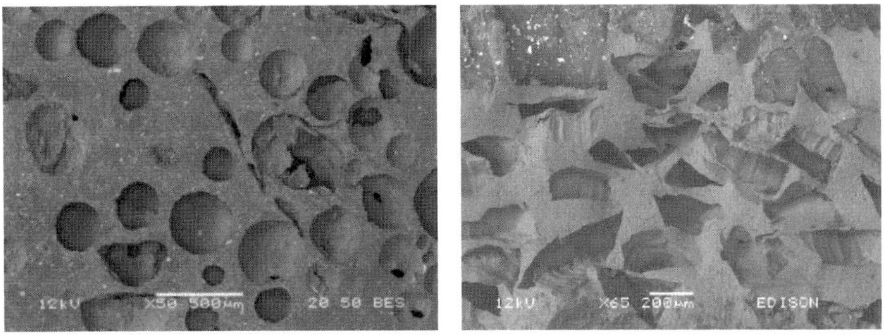

Fig. 8. A closed porosity is introduced in the MgB_2 by RLI by adding Mg in different shapes.

processes involving in a first phase, the production of the ceramic powders, followed by sintering methods to obtain a dense material.

Changing both Mg shape and disposition, the RLI process can create a faint hollow or porous object with holes that reproduce the Mg shape. For example, Fig. 8 shows some porous MgB_2 obtained with different shapes of the Mg lumps.

The voids produced by the RLI process were also applied to produce hollow superconducting MgB_2 fibers, patented in 2002[14] and published in the following year.[15]

4.2 *The RLI driving forces*

The peculiar RLI process with its self-sustaining behavior and spreading on macroscopic large volumes needs a rational understanding in term of driving forces. In my opinion, two main reasons may explain the peculiarities of the reaction mechanism[16]:

(a) — the molar volume reduction during the reaction $Mg + 2B \rightarrow MgB_2$, i.e., about 24%, and

(b) — the presence of a metastable crystalline boride phase, Mg_2B_{25}, which may be the first product of the chemical attack of liquid Mg against the crystalline beta-rhombohedral B.

The volume reduction causes a lowering of the pressure inside the reacting B preform, so that more liquid Mg is sucked inside the B powders. The suction prevails over the Mg vapor pressure as well as initial air atmosphere trapped in the B powders. As noted in Ref. 16, several couple of elements may exhibit this peculiar volume reduction, but Mg and B show to our knowledge the most effective reduction. One expects that infiltration on evacuated B preform could enhance the penetration depth, even if this procedure could complicate the process.

The second driving force is related to the peculiar crystallinity of the chosen B powders, i.e., beta-rhombohedral. Most commercially available B powders, either in lumps (99.5% purity) or in micrometer size (95–98% purity), are erroneously named "amorphous," the crystalline phase being still present. However, some low purity B (88% purity), or some new specialty nanosize B powders (99.99% purity) can be really amorphous. Therefore, we may anticipate that this second driving force can act only in presence of a B beta-rhombohedral crystallinity.

4.3 *The Mg boride species Mg_2B_{25}, as intermediate crystalline phase in the RLI process*

In the literature of Mg borides, there was a great debate on the thermo-dynamically stable phases in the range of the high B borides. In 1955 the existence of a phase MgB_{12} was proposed.[17] In 2003, several reflections of a Mg boride XRPD spectrum were indexed with a hexagonal unit cell; however, but the composition of the compound was inferred on the base of a XRF microanalysis and not from a well resolved crystallographic analysis.[18] In 2002, a crystallographic study revised the question: they detected a MgB_{20} phase and performed a full description of the crystalline structure.[19] In 2006, three papers described, almost simultaneously, the crystal structure of MgB_{12} (Orthorhombic, space group: Pnma),[20] Mg_2B_{25} (Rhombohedral, space group: R-3m)[7]and $MgB_{17.4}$ (Rhombohedral, space group: R-3m).[21] Different preparation procedures can explain the different compositions of the various phases. For example, both MgB_{12} and $MgB_{17.4}$ were obtained at very high temperature (1600°C), but with different Mg/B ratios of 1.5 and 1.333, respectively. The MgB_{20} phase was obtained at 1150°C with a ratio Mg/B = 0.08. In the RLI process the temperature is much lower and the Mg/B ratio, even if not precisely measured, should be much larger than the previous values, due to the abundance of liquid Mg. The Mg atoms occupy interstitial sites of the crystalline B framework and may also enter as a substitute of a B–B couple, as it happens for Mg_2B_{25}. In Fig. 9, the locations of the Mg atoms resulting from a Rietveld refinement are reported. These locations can be considered as statistical, the interatomic Mg–Mg distances being much lower than 0.25 nm, the usual length of the Mg–Mg bond. In Table 2 a review of the lattice unit cell (hexagonal setting) of the rhombohedral phases with formula MgB_x is reported, with $x > 12$, where a slight increase of the unit cell parameters and volumes with the Mg content is observed. A very

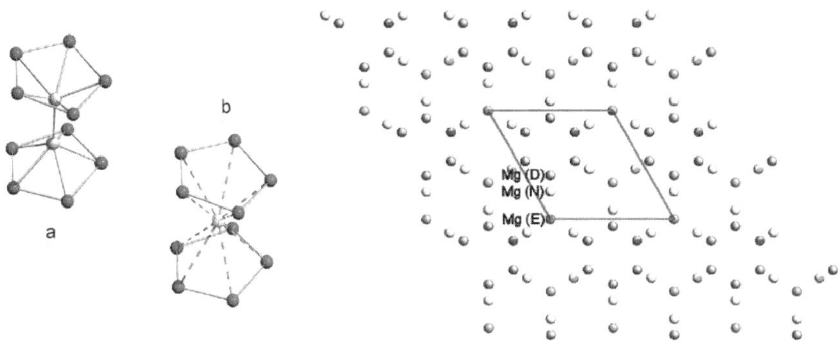

(a) (b)

Fig. 9. Details of the Mg_2B_2 atomic packing: (a) two possible Mg insertions between B icosahedra; (b) the interstitial sites (D,N,E), arranged in hexagonal configuration and occupied by Mg in the B lattice. (Reprinted from Ref. 6. © 2006 Elsevier).

Table 2: Lattice parameters of the hexagonal high B Mg borides MgB_x.

MgB_x	a(nm)	c(nm)	vol(nm^3)	Ref.
$x = 12.5$	1.10402 (2)	2.4198 (1)	2.55425	7
$= 17.4$	1.0991 (2)	2.4161 (4)	2.52767	21
$= 20$	1.0983 (4)	2.41561 (16)	2.52348	19
$= \infty$	1.09253 (1)	2.38103 (15)	2.46129	7

recent confirmation of the appearance of the Mg_2B_{25} phase in the RLI process was obtained by extracting a boride grain from a typical RLI MgB_2 product (obtained by powders P100). A single crystal X-ray diffraction (XRD) analysis confirmed that this grain was an Mg_2B_{25} single crystal.[22]

4.4 *The issues of the fine B powder infiltration*

The first attempts to infiltrate the "amorphous" commercial micrometer size B to obtain thick bulk samples in similar conditions as set for crystalline powders, were unsuccessful in extending the reaction to large depths.

To overcome this difficulty it was sufficient to evacuate or to fill the B preform in an inert atmosphere (Ar), without any previous treatment of the commercial B powders. By using the micrometer size B powders, one should in any case take care to maximize the tap density of the powders, in order to avoid large amounts of residual Mg in the reacted MgB_2 product. Today, the infiltration of conventional fine milled B produces a minimum level of residual Mg around 6 wt.%[23] as determined by quantitative XRPD. The mixture of fine and large B grains gives rise to intermediate values of the product characteristics. On the other hand, mixtures of B with MgB_2 powders may cause unsatisfactory results, the elimination of the fine MgB_2 powder aggregates that entrap Mg being quite difficult. The higher superconducting quality of bulk MgB_2 produced by micrometer size, high purity B powders could therefore promote their use in the most demanding applications.

5. Reaction Kinetics

5.1 *In situ MgB_2 formation kinetics*

Several articles have been dedicated to the description of the kinetics of the *in situ* MgB_2 reaction. The process is very rapid and strongly dependent on the kind of B powders. A first study[12] considered the MgB_2 formation by commercial "amorphous" B of low purity (90%) mixed to Mg powders. The paper reported isothermal growth at two

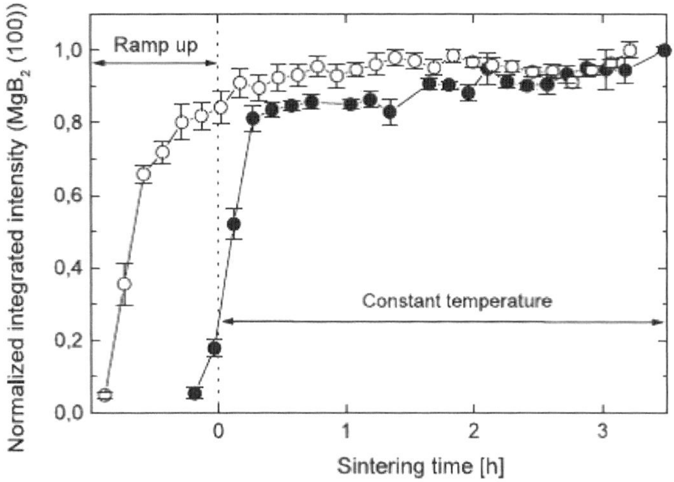

Fig. 10. Isothermal growth kinetics of MgB_2 at two temperatures (open symbol at 780°C, full symbol at 650°C). (Reprinted with permission from Ref. 12. © 2006 IOP).

target temperatures (650°C and 780°C) heated at a ramp rate of 0.055°C/s (see Fig. 10). The reaction starts at about 575°C, well below the melting temperature of Mg, due to the Mg evaporation, but is very rapid at $T = 650°C$, 80% of the final MgB_2 product being reacted in ~900 s.

5.2 *Kinetics of the RLI process*

A reaction kinetics similar to the RLI process, with a rapid temperature increase (1°C/s), was reported by the Northwestern University[24] and was monitored by XRD measurements of increasing amount of MgB_2 using synchrotron radiation. In Fig. 11, the isothermal growth of MgB_2 at various temperatures is reported in comparison with the infiltration reaction effects of commercial submicron B and specialty plasma synthesized ultrapure B. At lower temperatures, commercial powders had a different induction period than ultrapure powders, probably due to surface impurities. The reaction kinetics of the latter was faster than that one observed for the less pure powders. Indeed, the time to reach 80% of the final product was 900 s at the

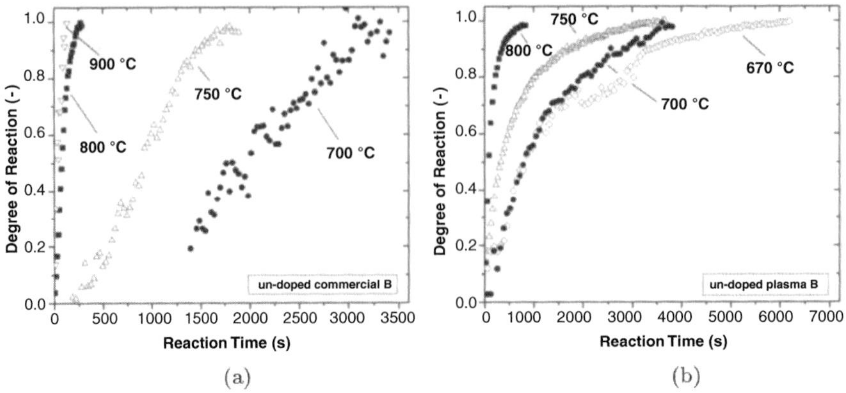

Fig. 11. Infiltration reaction kinetics, evaluated with synchrotron XRD at different temperatures and for different B powders: (a) undoped micrometer B powders, (b) undoped nanosize B. (Reprinted with permission from Ref. 24. © 2008 Elsevier).

temperature of 750°C. A measurement of the RLI process kinetics using much larger crystalline B powders (P100) (99.5% purity) is reported in Ref. 25.

The main results are illustrated in Fig. 12, where the composition of the three phases MgB_2, Mg_2B_{25} and Mg are computed by means of a Rietveld compositional analysis of the XRPD diagrams of the composite products. The compositions are reported as a function of the elapsed reaction time, starting from the isothermal condition. The experiment had been performed by packing the P100 powders at a tap density of $1.4\,g/cm^3$. Small deviations are observed for varying initial B tap density, but the substantial features of the reaction kinetics are maintained. Using LG crystalline B there is a rapid transformation of the B in Mg_2B_{25} boride (in less than 1 hour), followed by the transformation of this boride in MgB_2, at longer times.

This kinetics is fairly well reproduced by a simple log-linear model, for each phase, as:

$$\text{wt.\% phase } X = A_X(T) + \mathrm{B}_X(T)\mathrm{Log}(t), \tag{1}$$

where for each phase X, the A and B parameters, are linear functions of the variation $\Delta T = T - T_{\mathrm{m}}$ with respect to the Mg melting

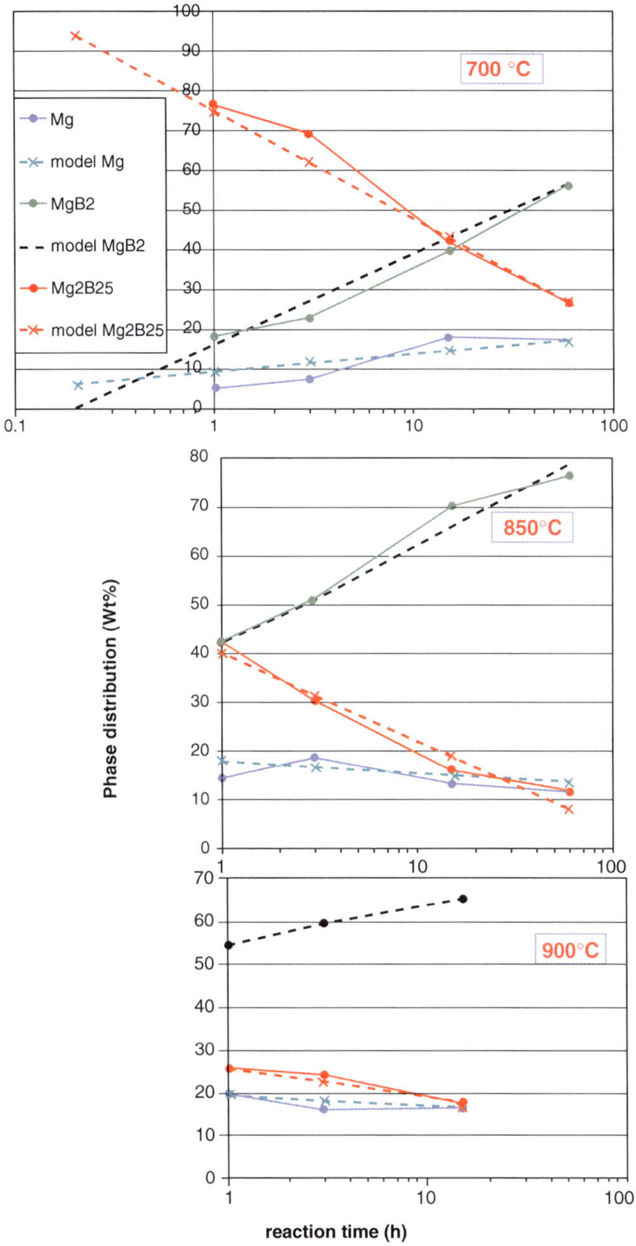

Fig. 12. Bulk MgB_2 reaction by RLI kinetics at 700°C, 850°C and 900°C, starting from LG B (distribution in wt.% phase: black = MgB_2, red = Mg_2B_{25}; blue = Mg).

temperature, $T_m = 650°C$:

$$A(T) = a_o + a_1 \Delta T \quad \text{and} \quad B(T) = b_o + b_1 \Delta T. \qquad (1\text{ bis})$$

The fitting quality of this simple log-linear model with the measured composition for MgB_2 and Mg_2B_{25}, respectively, is compared in Fig. 13. The model parameters values used in this fitting are given in Table 3. The model reproduces quite well the data in the intermediate time–temperature ranges and may be considered as qualitatively valid at the beginning of the reaction (up to 1 h) and at temperatures near to the Mg melting point.

When comparing with the previously mentioned kinetic studies, there are some similarities and some differences as well. Among

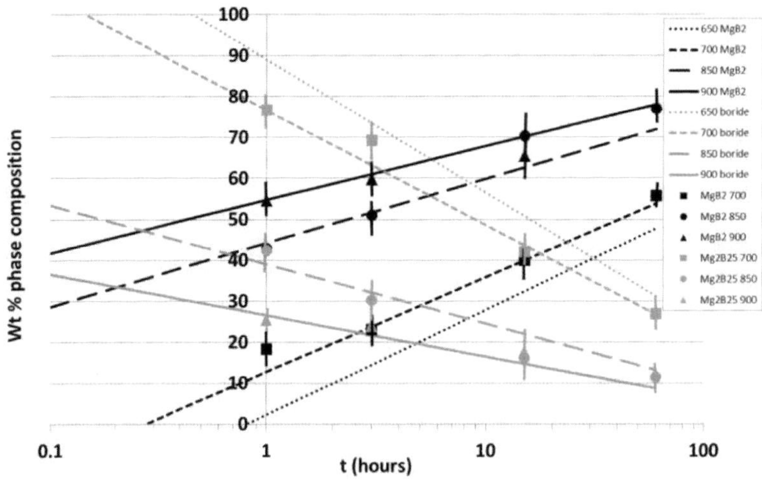

Fig. 13. Model kinetics of Eqs. (1) and (1 bis), with the parameters of Table 3, to interpolate the experimental data for the phases MgB_2 (*black*) and Mg_2B_{25} (*gray*).

Table 3: Parameters values of the log-linear kinetic model.

Phase	a_0	a_1	b_0	b_1
MgB_2	2.25	0.21	25.6	−0.05
Mg_2B_{25}	89	−0.25	−32.5	0.09

the similarities, one may consider the presence of an incubation period of similar duration with respect to the infiltration at 700°C of the less pure submicron B powders. Higher reductions of this duration are found at 850°C and at higher temperatures. Among the differences it is evident that large B powders substantially slow down the reaction rate. Of course, it may be expected that bulk MgB_2 obtained by micrometer size or finer B powders may be obtained by the fastest kinetics. Of course, more studies are necessary to evaluate the reaction kinetics in different percolation paths of the liquid Mg. These different paths will be due, in addition to the temperature gradient in the B preform, also to its porosity, to its chemical purity and to the gravitational force fields acting on the Mg melt.

6. The Infiltration Process: Practical Details

6.1 *Production of a MgB_2 piece*

As with many other ceramic forming processes, the RLI process for the production of bulk MgB_2, gives only satisfying results if all processing variables are under control. Today, the main steps being under control, when fabricating large MgB_2 pieces, are:

(1) Selection of the B powders of appropriate granularity to form a very homogeneous powder preform of tap density $> 1.25\,g/cm^3$. When using micron size B an additional degassing step is necessary.
(2) Design of an appropriate metallic container of the B preform, taking into account the differential thermal expansion/contraction between MgB_2 and the metallic container.
(3) Size definition of the semi-finished product, assuming a differential linear expansion of $\leq$2–3% with respect to the size of the initial B preform.
(4) Careful choice of the initial location of the Mg constituent in order to avoid large gravitational movements of the melt, which can locally compromise the stoichiometry of reactants. This problem may arise, Mg being present in large excess with respect to MgB_2 stoichiometry.

Fig. 14. Bulk MgB_2 prototypes produced by RLI.

(5) The container should be carefully welded, followed by a check for tightness.
(6) Post manufacturing surface grinding by spark erosion machines or diamond tooling. The typical brittleness of ceramics may be an issue for very large and thin pieces.

During the last years, a laboratory production included many shape prototypes being of interest for superconducting applications; some of them are displayed in Fig. 14. The RLI process is advantageous with respect to hot pressing in case of fabrication of bulk objects with large size. Today, the experience is limited to lengths of some tens of centimeters, but the technology appears to be scalable to larger sizes.

6.2 *Assembling parts by joining*

The RLI process being a joining process at the atomic level, it was not a great surprise that the process also allowed to realize superconducting joints between already synthesized MgB_2 parts.[26] A MgB_2 superconducting disc made by RLI with P100 B powder was cut into four pieces which were enclosed in a container with

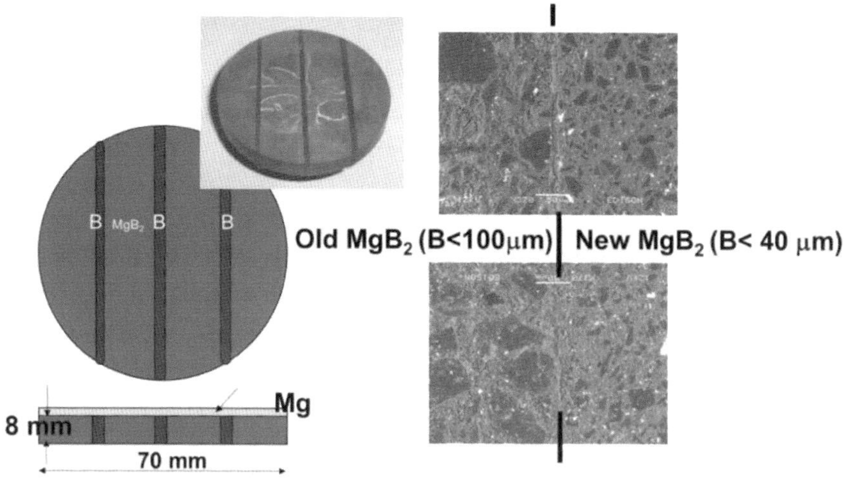

Fig. 15. Superconducting joining of a cut MgB_2 disc, with new MgB_2 material.

pressed fine B powders (<40 μm, for a better analysis of the joint) between them and were covered by a Mg disc, as shown in Fig. 15.

After repeating the thermal treatment and removing the Mg excess, the disc appeared again mechanically resistant and no discontinuity could be detected by scanning microscopy between the old and the new MgB_2 material. To confirm the superconductivity of the joint, we verified that the levitation characteristics of the disc were restored, after the joining process.

7. Some Mechanical Properties of the Bulk MgB_2 by RLI at 300 K

The mechanical performances of well densified MgB_2 are better than those of HTS materials, and their noticeable properties are expected to promote their use under high stress conditions as encountered in high power electromechanical applications. Due to its ceramic nature, bulk MgB_2 material is brittle, the quality of the mechanical performances must thus be judged with respect to the class of ceramics and not compared to ductile metals. The large morphological difference between the RLI MgB_2 bulk pieces produced using B with different particle size also causes differences in their mechanical properties.

Table 4: Mechanical properties of MgB_2 produced by RLI with SG or LG B powders.

RLI–MgB_2 bulk	LG	SG
Buckling Young modulus (GPa)	125 ± 5	85 ± 5
Buckling strength (Mpa)	360 ± 15	330 ± 100
Flex Yield strength (Mpa)	259 ± 28	278 ± 43
Buckling strain at rupt. (%)	0.14	0.27
Vickers microhardness (Gpa)	25–28	7
K_{1c} (Mpa $m^{0.5}$)	1.8–2.6	2.8

Let us consider two typical products: (a) Large grain (LG) samples made by crystalline B powders P100 (<100 μm) with composition MgB_2 = 45 wt.%, Mg_2B_{25} = 40 wt.%, Mg = 15 wt.% and (b) Small grain (SG) samples made by microcrystalline B powders (<5 μm), with the composition MgB_2 = 70 wt.%, Mg = 30 wt.%. Vickers microhardness, fracture toughness and elastic properties have been measured in two different ways.[27] For beams about 18÷40 mm length, 4-points flexural tests and beam buckling tests were performed. A summary of the results is given in Table 4.

The most evident differences between LG and SG materials emerged from the buckling experiment, where the larger toughness ($K_{1c} = 2.8$ MPa $m^{1/2}$) of the SG material causes an almost double strain at rupture ($\varepsilon = 0.27\%$) and a larger strength ($\sigma = 550$ MPa). As expected, the Young's modulus of the two materials tends towards opposite directions, with the larger value for the harder LG material ($E = 125$ GPa). The Vickers microhardness of the LG material varies from 38–2.8 GPa. More homogeneous values (7 GPa) are observed for the SG material. It is important to note that the measured samples were reacted at 850°C for only 1 h, therefore one could expect to find different values for the mechanical properties of materials with larger amounts of MgB_2 phase and less residual Mg and Mg_2B_{25}.

It is instructive to compare the elastic and inelastic behavior of RLI MgB_2 materials with that of a highly dense, pure MgB_2 material obtained by hot pressed MgB_2 powders. Measurements by resonant internal friction at high frequencies (1600 ÷ 2050 kHz) of the hot pressed material in the 4–300 K range[28] showed a low monotonic damping (about 1.5%) of the shear modulus and of the bulk modulus of 115–117 GPa and 142–145 GPa, respectively. A small dumping peak of resonant modes in the 20–60 K range was also observed and connected to anharmonic instabilities at T_c. On the contrary, measurements by inelastic spectrometry at lower frequencies ($6 \div 70$ kHz) on one of the first RLI–MgB_2 bulk samples[29] did not present a monotonic behavior of the Young's modulus in the low temperature range, but had a high dumping effect and yielded a room temperature Young modulus of 167 GPa. After the current microscopic description of the RLI MgB_2 samples, it is possible to hypothesize that the inelastic features and the high dumping are partially due to metallic Mg residues but mainly to the presence of the large Mg boride species.

8. A Review of the Superconducting Properties of Bulk MgB_2 by RLI

Initially, the superconducting quality of the RLI MgB_2 bulk materials was also tested with magneto-optical (MO) imaging[30] to verify possible grain boundary effects due to the presence of large boride grains. It was a surprise to see that the Meissner magnetic field exclusion effect involved a wide surface area of the sample (several mm^2 in size), regardless of the normal conducting areas of less than 100 μm size. Figure 16(c) shows the MO images of a RLI MgB_2 plate, taken at 11 K, under a magnetic field of 120 mT (ZFC mode).

Comparing the MO image with the microscopic morphologies (Optical and SEM) of the same plate (see Figs. 16(a) and 16(b) where the Mg boride grains are displayed) it is clearly visible that the magnetic field enters only in correspondence with a crack. Therefore, the normal conducting borides are embedded in a fully connected superconducting MgB_2 matrix. The microscopic details of the interior borides shown in Fig. 4 display a net of MgB_2 channels that helps to distribute the currents in a fairly "homogeneous" way in the RLI–MgB_2 samples, even if they consist of a superconducting/normal conducting composite. Looking at

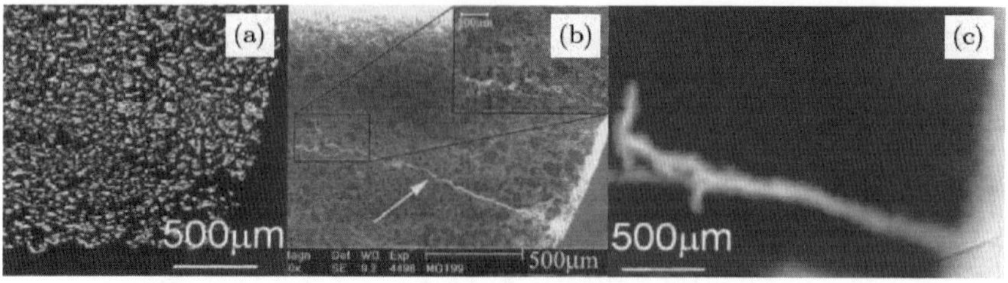

Fig. 16. MO imaging of bulk MgB_2 produced by RLI using LG B, around a crack: (a) Optical image, (b) SEM image with inset magnification, (c) MO image in ZFC and 120 mT applied field, at 11 K (Reprinted from Ref. 30. © 2002 Taylor & Francis).

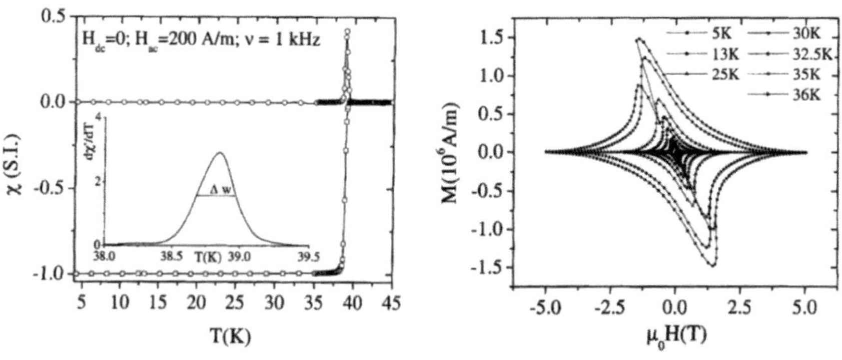

Fig. 17. AC susceptibility and magnetization cycle as a function of the temperature (Reprinted from Ref. 30. © 2002 Taylor & Francis).

the AC susceptibility measurements and magnetization cycles at different temperatures[30] shown in Fig. 17, several peculiar features of the RLI MgB_2 material are already evident: (a) a very sharp T_c, indicating pure superconducting material, even in presence of external impurities (Mg and borides); (b) only a single flux jump is visible in the magnetization cycle at 5 K, indicating that the sample has an intrinsic thermal stability, even for a thickness of 0.34 mm, in contrast to the usual behavior of LTS bulk materials. Of course, the good thermal stability, in this case, is also favored by the relatively low critical current density of the whole sample. It should be mentioned that the presence of a large amount of normal conducting inclusions has only a limited lowering effect on the critical current density. This may be due to the presence of pure MgB_2 channels, crossing the large boride grains.

8.1 *Critical current density of RLI MgB_2 bulk materials in magnetic field*

The current transport characteristics of RLI–MgB_2 bulk samples in a magnetic field were mainly evaluated by means of magnetization cycles at different temperatures, but also by transport measurements using the four point technique, allowing the determination of the exponential n values of the $V(I) \approx I^n$ curves.[31]

The morphological differences among bulk RLI–MgB_2 samples obtained with various B particle sizes also affect the critical current density. The micrometer size B produces the best transport characteristics, even being less thermally stable at low temperature and low magnetic fields: many flux jumps were observed, in contrast to a small number of flux jumps in MgB_2 samples produced using P100 B powders. Figure 18 shows the comparison between the J_c(B,T) curves of two typical RLI-MgB_2 bulk samples (identified on the base of the starting boron powders). The experimental data, measured by

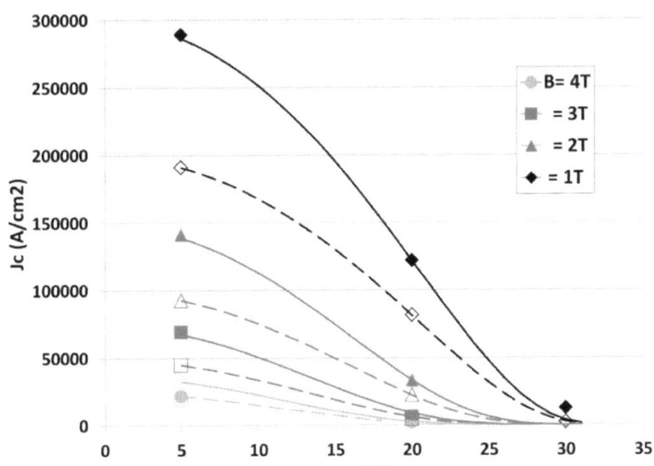

Fig. 18. $J_c(B,T)$ experimental data of bulk RLI–MgB_2 samples interpolated by Eq. 2, for LG (dashed) and SG (continuous) materials, respectively.

magnetization cycles, are well fitted by the analytical formula:

$$J_c(B,T) = J_c(0)(1 - (T/T_c)^2)\exp(-\gamma B/(1 - (T/T_c)^2)^2), \quad (2)$$

where $J_c(0) = 6 \times 10^5$ A/cm^2 and 4×10^5 A/cm^2, for SG and LG samples, respectively, the parameters $\gamma = 0.7(1/T)$ and $T_c = 39$ K being valid for both samples.

The SG type bulk samples made using "amorphous" B show $J_c(B,T)$ lower values than the RLI–MgB_2 hollow wires made from the same powders.

Typically at 4.2 K and in self-field, the SG bulk samples exhibit $J_c(B,T)$ values 2.5 to 3 times lower than the corresponding values for the wires.[15] This may be due to a more compact B preform of the wires, owing to the drawing process, and consequently to a smaller amount of residual Mg among the grains of the infiltrated product.

Regarding the exponential n value, a wide range of values has been measured, going from $n = 10$–100. The behavior of n(B,T) at low temperatures and high fields is well represented by the formula:

$$n(B,T) = n_o \exp(-\alpha B/(1 - (T/T_c)^\beta). \quad (3)$$

For a LG sample the fitting parameters are: $n_o = 103$, $\alpha = 0.3(1/T)$, $\beta = 1.33$, $T_c = 39$ K. The β value gives a very similar temperature trend of the exponential part in Eqs. (2) and (3) at $T < 15$ K. The n parameter may be related to the flux creep activation energy U(B,T) assuming a logarithmic dependence: $n = \mathrm{U}(B,T)/K_B T$.[32] The activation energy may have values ranging from tens to several hundred meV. The values of U(B,T) were also determined by means of AC susceptibility on a similar LG sample (U = 20 meV), but the comparison with Eq. (3) is only qualitative: the experiments yield $n = 7.5$ at 30 K and B = $2T$), while Eq. (3) yields $n = 1.5$. The discrepancy may be due to the low accuracy of the model in the operation range of temperatures and fields. The good transport performances of the bulk RLI–MgB_2 materials are mainly due to their high grain connectivity, which is between 60% and 89%, according to the Rowell analysis of the resistivity behavior at high temperatures.[33] The connectivity is defined as

$$\text{Connectivity} = (\rho^{PC}_{300K} - \rho^{PC}_{40K})/(\rho_{300K} - \rho_{40K}),$$

where $\Delta\rho^{PC} = 8.5\,\mu\Omega\text{cm}$ is the limit polycrystalline $\Delta\rho$. Our typical connectivity values are: 61%, 89%, 73% for RLI–MgB_2 samples based on P100, P38 and P_{am} B powders, respectively.[34]

8.2 *Trapped magnetic fields and levitation*

Bulk superconductors may be charged with persistent currents, giving rise to a persistent magnetization. Magnetic charging thus leads to powerful permanent magnets (PM) with a magnetic energy density up to thousand times larger than that of conventional PM. The magnetic trapping characteristics of a bulk superconductor are demonstrated by YBCO pellets that, in the textured form, have very high critical current densities and are able to trap very high magnetic fields, up to 17 T.[35] However, this HTS breaks up due the mechanical stresses of these extreme magnetic conditions, and bulk MgB_2 may thus represent a valid alternative, even if the trapped fields are considerably lower. MgB_2 superconductors have the clear advantage that, sufficiently high current densities for realizing trapped fields of practical interest can be obtained in the polycrystalline form. Indeed, large superconducting MgB_2 PM with intermediate trapped fields of the order of 4 to 5 T may be applicable in various electro-technical devices. Recently, high trapped fields have been reached for bulk MgB_2 densified by hot pressing procedures,[36] therefore we feel confident that these values can also be achieved for bulk MgB_2 materials produced by the RLI process, in particular when using high purity submicron boron powders with high tap densities (1.3 to 1.4 g/cm^3).[37] The present trapped field values for several shapes of bulk RLI–MgB_2 objects (discs, hollow cylinders or rings), all based on P100 B powders are reported in Table 5. From the measured central trapped field, one can evaluate the total current I, the single spire inductance L, thus allowing the calculation of the total energy, 0.5 L I^2, stored in the superconducting magnet.

Table 5: Trapped magnetic field in several bulks RLI–MgB_2 prototypes with different shapes.

MgB_2 Bulk	OD/ID; H (mm)	Central trapped field (T)	*Temp* (K)	J_c (A/mm^2)	$L(\mu H)$	Energy (J)	En. Density (kJ/m^3)	Ref.
Ring	42/25; 1	0.52	4.2	1300	1.24	463	279000 (*)	38
"	155/100; 8	0.40	4.2	185	2.01	6640	44000 (*)	39
Hollow cylinder	48/44; 31	0.82	13	590	0.07	180	110000 (*)	40
"	65/56; 40	1.55	4.2	500	0.09	1466	3210 (**)	41
Disc	40; 8	2.00(**)	4.2	250	0.05	157	17500	42
"	55; 15	1.27(**)	15	100	0.05	169	4750	37

(*)External volume of the bulk; (**) Field at 1 mm distance from the surface.

The performances of the MgB_2 bulk samples depend on their shape, and their flux jump behavior is thus related to the details of the magnetic field map on their surface. Since the shape with the highest energy density is a thin ring, it could be interesting to investigate the geometry of a bulk superconducting magnet assembled by adding many thin rings.

The levitation characteristics of such superconducting magnets are generally tested measuring their repulsive/attractive forces with respect to classical PM, a condition that only partially exploits the possibility of superconductors. Nevertheless, it is a condition of practical interest, due to the stable equilibrium that can be established between superconducting and PM. The levitation characteristics of typical RLI–MgB_2 discs produced using P100 B were important to consider how the forces change after a loop, depending on the temperature of the SC: indeed, the appearance of the measured when a PM (i.e., SmCo) was approaching and going away from the SC, in ZFC condition. The axial forces between superconducting and PM are reported as a function of their distance in Fig. 19.[43]

A hysteretic behavior is related to the beginning of flux trapping in the bulk. In this example, flux trapping starts above about 25 K. Below this temperature no hysteresis is detected: when the bulk

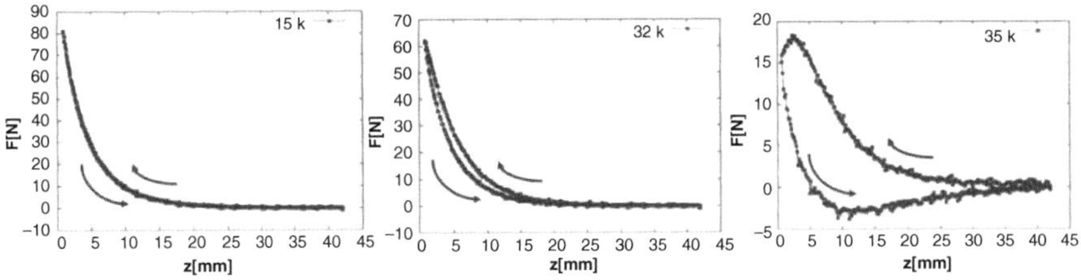

Fig. 19. Axial forces measured in ZFC levitation experiments of a permanent magnet on RLI MgB_2 discs, at different temperatures. The large hysteretic behavior at higher temperature is due to the effect of the trapped vortices. (Reprinted from Ref. 43. © 2005 AIP).

MgB_2 approaches the PM, the gradient dB(z)/dz induces superconducting currents in the superconducting magnet and the generated magnetic dipole repels the permanent dipole. The repulsion of the two magnetic dipoles is highest near the superconducting magnet surface (in this case the minimum distance between superconducting and PM was 1 mm). When the PM is moved in opposite direction, the induced currents flow in the opposite direction and the repulsion forces tend to become zero. Above 25 K, the mentioned behavior is superimposed to the force generated by the magnetic vortices trapped inside the MgB_2. At higher temperatures, a larger fraction of the magnetic induction of the PM enters steadily in the SC; this trapped field reduces the maximum axial repulsion force and is responsible of the minimum attractive force, appearing when the PM is going forward. Indeed near T_c, the trapped field is so high and the induced repulsive axial forces are so low that, during the PM going forward, there is a point where the total axial force starts to change from repulsive to attractive, reaching an intermediate position (height) of maximal attraction, corresponding to the levitation equilibrium point. Such a maximum attractive force is always present in the FC condition, when the superconducting magnet is cooled in a position very close to the PM, which is the usual way to perform a levitation test.

In Fig. 20, the axial forces of the FC levitation experiment are reported at two temperatures, 26 and 36 K.[44] The initial distance between PM and SC was 1 mm. In this position the trapped flux

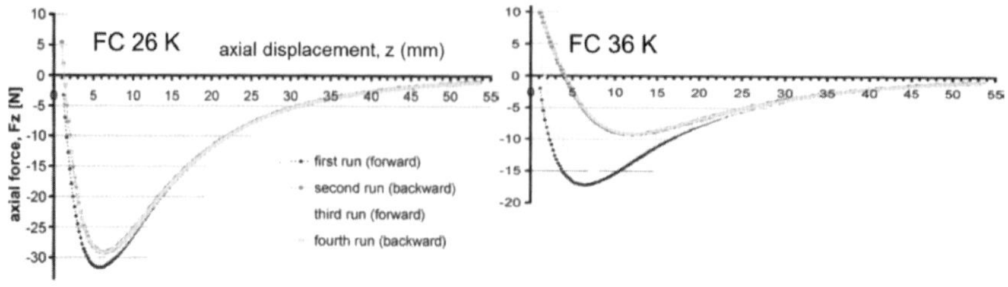

Fig. 20. Axial levitation of a PM on a RLI–MgB_2 disc, in FC experiments at different temperatures.

was exactly equal to the PM induction field, with zero mutual forces. When the PM is moved forward it is subjected to the restoring attractive forces similar to the forces between two magnets. This attractive force is again superimposed to the repulsive forces generated by the induced supercurrents by the field gradient and the minimum corresponds to the point of maximal total force (attractive). This force represents the useful levitation forces of the system. The levitation force in backward PM runs lower than in forward runs because the compensating currents are generated above the previous currents. Even in the FC case the effect of higher temperature is to increase the hysteresis effect: at higher temperatures the forward and backward currents are lower and more spread apart than at low temperatures, giving rise to a low attractive effect.

From a practical point of view, the important parameters for the levitation are the maximum attractive force and the stiffness, represented by the second derivative of the force curve. Of course, at low temperatures and without hysteresis, there is an optimal levitation, with maximum force and stiffness. A practical estimation of the RLI–MgB_2 — PM levitation effects can be made: at 26.5 K and at 4 mm distance, a maximum levitation force of about 65 N over a planar stack of PM was measured, giving a levitation pressure of 0.7 atm and a stiffness of about 65 N/mm. A larger stiffness, about 120 N/mm, was measured on a cylindrical stack of three PM inside a MgB_2 hollow cylinder, to simulate a superconducting bearing system.[45] Alternatively, assuming an interaction between two powerful superconducting magnets, the levitation characteristics will be substantially improved. However, a fully superconducting levitating system, even if feasible, will present at least two practical problems: (a) the magnet charging operation should be performed by a high-field superconducting magnet or by pulsed field techniques[46] and (b) cooling of the two magnets will be anyway complex.

8.3 *Passive magnetic field shielding and shaping*

The capability of a bulk SC to sustain persistent currents can be further utilized in a passive magnetic shielding systems, either for drastically reducing the stray field of powerful superconducting magnets or to protect sensitive devices in the presence of external magnetic fields.[47] The same persistent currents can be used in a different bulk configuration, consisting of a thick hollow cylinder with a longitudinal slit aperture. Such a superconducting system modifies the magnetic flux lines and represents a lens able to concentrate on the magnetic field.[48] For these applications large cylinders are generally necessary, therefore bulk RLI–MgB_2 appears as preferred choice[49] with respect to HTS materials, combining workability with good mechanical resistance, high critical current density and sufficient thermal stability. To date the larger prototype bulk RLI–MgB_2 manufacts, produced at lab scale, are displayed in Fig. 21.

(a) (b) (c)

Fig. 21. Large RLI–MgB_2 cylinders and a ring: (a) OD/ID = 90.9/76.7 mm, $H = 152.5$ mm, $m = 680$ g. (b) OD/ID = 240/220 mm, $H = 45$ mm, $m = 780$ g, (c) OD/ID = 152/140 mm, $H = 120$ mm, $m = 790$ g.

Table 6: List of RLI–MgB_2 shields, and their shielded AC magnetic fields.[50]

Sample (OD/ID–H) mm	B_o (T)	B_{AC}(T)	B_{sh} (T)	Freq (Hz)
cylinder (16/7–7)	0.35	0.39	0.74	7
cylinder (16/7 –7)	0.58	0.18	0.76	10
cylinder (25/15–15)	0.58	0.56	1.14	5
cylinder (25/15–15)	0.88	0.38	1.26	5
cylinder (25/15–15)	0.24	0.18	0.42	61
Tube (12.5/8.5–43.5)	0.24	0.30	0.54	11
Tube (12.5/8.5–43.5)	0.24	0.18	0.42	21
Tube (12.5/8.5–43.5)	0.24	0.05	0.29	51
Tube (9/7–60)	0.24	0.40	0.64	7
Tube (9/7–60)	0.24	0.30	0.54	11
Tube (9/7–60)	0.24	0.05	0.29	21

Today the magnetic shielding evaluation of the RLI–MgB_2 pieces is limited to cylinders being small enough to be inserted in the bore (about 8 cm) of a superconducting laboratory magnet. The shielding performances of RLI–MgB_2 cylinders and tubes of various sizes were tested in AC axially applied magnetic fields.[50] At 4.2 K and for a maximum frequency of 61 Hz of the AC fields component, the maximum measured shielded field is $B_{sh} = B_o + B_{AC}$. Table 6 summarizes the most interesting shielding characteristics of different MgB_2 shapes and RLI preparations.

The low frequency AC magnetic field shielding is of particular interest, as no classical system is known to be working under these conditions. The environmental earth magnetic field and the magnetic noises due to the common 50 Hz AC devices are particularly disturbing for Superconducting quantum interference device (SQUID) magnetometers, giving rise to 1÷100 nT field amplitudes, which have to be shielded. A MgB_2 cup+lid (ID = 37, H = 35 mm) was manufactured and tested for this purpose at 3 K[51] and the field gradient noises were reduced to 5 pT/(cm $\sqrt{\text{Hz}}$). The higher T_c of MgB_2 and its low density render these superconducting cups very competitive to equivalent LTS shielding materials as Pb and Nb, particularly in space applications.

Larger RLI–MgB_2 hollow cylinders were produced and tested at 4.2 K in axial applied fields[52] as well as in transversal fields.[53]

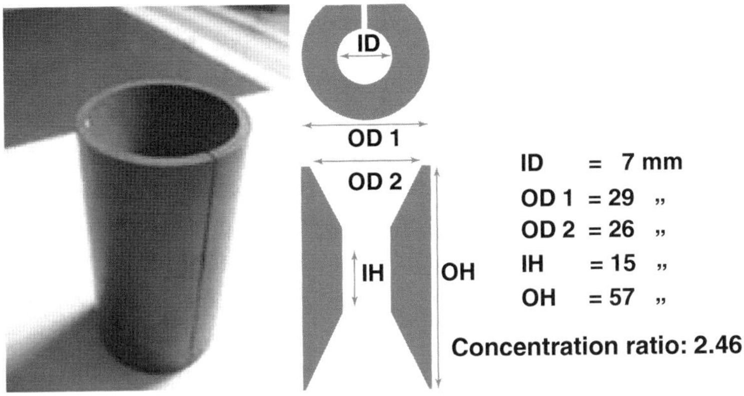

Fig. 22. MgB_2 conical cylinder with slit aperture, utilized for magnetic field concentration (Reprinted from Ref. 54. © 2012 IOP).

For applied axial field (ramp rate 0.6 T min^{-1}), the fully shielded field at the center of a RLI–MgB_2 cylinder of dimensions OD/ID = 17.75/9 mm and H = 70 mm was 2T, with the Hall sensor resolution of 10 μT. No flux jumps were detected and no measurable trapped field was observed after ramp down. By applying a transversal field (same ramp rate) on a cylinder of RLI–MgB_2 of larger size (OD/ID = 65/56, H = 40 mm), the field measured inside the cylinder, near the wall, was almost null up to about 0.6 T, while at 1.5 T it was less than 30% of the applied field.

The magnetic field shaping was tested for large superconducting MgB_2 conical cylinders with a longitudinal slit aperture (see Fig. 22). The magnetic field concentrations were measured at 4.2 K, their ratio ranging from 2.5–2, depending on the temperature and on the background magnetic field.[54] This ratio compares well the modeling concentration ratio of 2.46, based on the size of the conical cylinder. During the experiments, some flux jumps occurred in the MgB_2 cylinders again, the RLI–MgB_2 material with P100 B was more thermally stable, with less flux jumps.

8.4 *Magnetic relaxation characteristics*

All bulk superconductors undergo an intrinsic relaxation of their magnetization, due to thermal instabilities of the vortices, which influence the persistent current distribution. These instabilities have different dynamics, mainly depending on $\Delta T_c = T - T_c$ and are especially severe for the HTS materials,[55] where grain boundaries have a great influence on the degradation of the superconducting currents. The MgB_2 material has much less grain boundary barriers with respect to cuprates and even in the polycrystalline state its current intensity and persistency are sufficient, so that the corresponding bulk manufacts could practically be considered as superconducting PM. After current charging, several RLI–MgB_2 manufacts were tested for their magnetic relaxation rate, described by the parameter S:

$$S = -1/\mathrm{M}(\Delta\mathrm{M}/\Delta\log\mathrm{t}) \tag{4}$$

where ΔM/M is the normalized magnetization loss in the time interval Δ log t.

Typical S values for textured YBCO pellets from 77–20 K shift from 8% to 2%.[55] For the LTS materials, like NbTi, S is more than one order of magnitude lower (0.1%). Referring to a RLI–MgB_2 cylinder, used as magnetic lens, the measured rate S has values similar to the LTS materials (S = 0.106% at 4.2 K at 2÷3T)[54] (see Fig. 23). This S rate value corresponds to a magnetization decrease of 0.5% in about 16 hours.

8.5 *Radiofrequency behavior*

The microwave (MW) absorption energy of the bulk RLI–MgB_2 has been checked to test the potentiality of the material to build superconducting cavities or to be used in MW device applications, i.e., as filters for telecommunications.[56] MgB_2 thin films have already shown encouraging low MW surface resistance, of the order of tens of $\mu\Omega$ at 4.2 K, which are not far from the pure Nb at low temperature.[57] We measured the MW surface resistance and the quality factors Q, for different cavities and for different type of B powders. The fine grain B is far better than the P100 powders. A comparison for these materials has been reported for plate like samples inside a cylindrical Nb cavity at 9.4 GHz.[58] In three MgB_2 samples prepared with P40 and P100 B powders the surface resistance at 4.2 K was $< 0.5\,\mathrm{m}\Omega$,

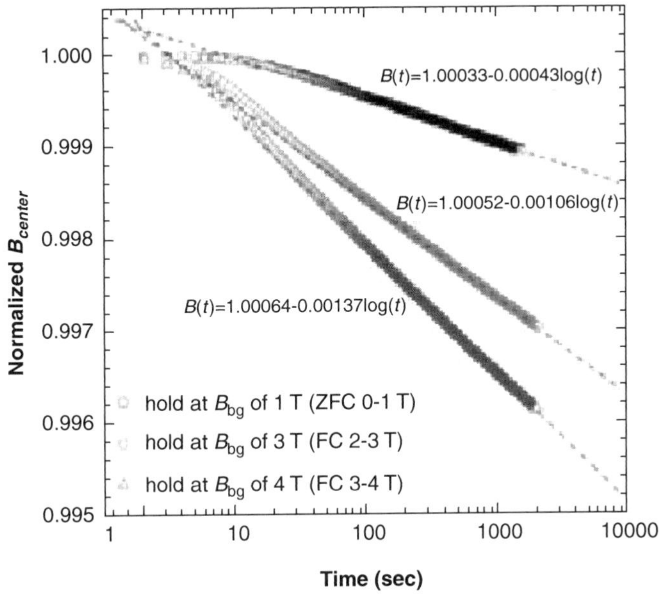

Fig. 23. Magnetic relaxation of the trapped fields of the MgB_2 magnetic lens, at different applied fields (Reprinted from Ref. 54. © 2012 IOP).

4 mΩ and 7.5 mΩ. More recently, a Cu coaxial cylindrical cavity with an inner MgB_2 rod (prepared with P38 powders, dimensions: d = 3.8 mm, L = 96 mm) exhibited the frequency dependence of the surface resistance shown in Fig. 24.[59]

The unloaded and loaded quality factors, Q^U (intrinsic of the cavity) and Q^L (modified by the additional losses due to coupling of the excitation and detection lines), were measured at different temperatures for two different cavities. The first cavity, a P100 based RLI–MgB_2 hollow cylinder with lids (OD/ID = 48/40, H = 42.5 mm), was evaluated in the frequency response interval of 5 to 13 GHz. The best performing mode was the TE_{011} (9.79 GHz) mode, $Q^U = 2.2 \times 10^5$ at 4.2 K, with very stable values with respect to the input power level (see Fig. 25).[60]

The second cavity was a coaxial one and consisted of a P38 based RLI–MgB_2 tube (OD/ID = 20/12.8, H = 60 mm) with 2 mm thick lids and a central MgB_2 rod (D = 3.8, L = 45 mm), (Fig. 26).[61] The cavity was tested in the 1 ÷ 13 GHz frequency interval and the

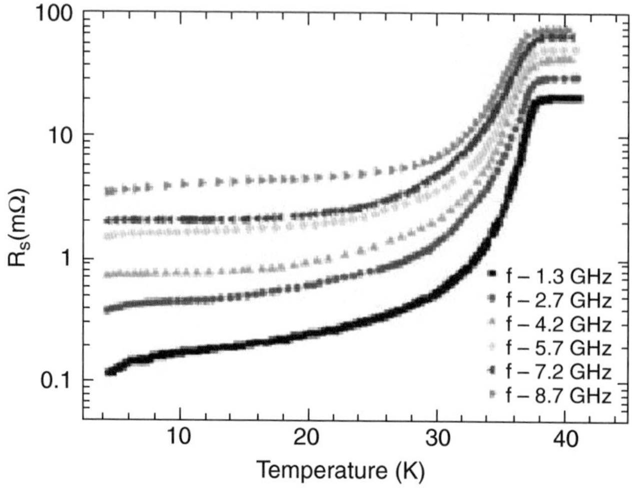

Fig. 24. Surface resistance of a MgB_2 rod prepared using P38 B, measured in a coaxial Cu cavity at different frequencies. (Reprinted from Ref. 59. © 2014 Elsevier).

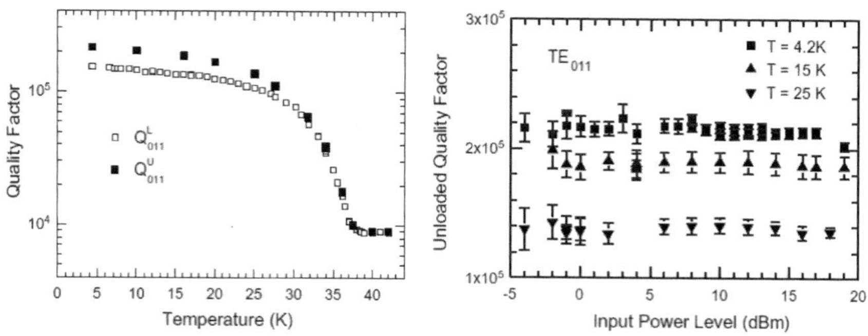

Fig. 25. MgB_2 coaxial cavity quality factors, at 9.79 GHz (Reprinted from Ref. 60. © 2008 Elsevier).

best performing mode was the fundamental TEM mode (2.6 GHz), characterized by radial electric field lines. Magnetic field lines, wound around the inner rod and, at both ends of the inner conductor, the electric field was highest and the magnetic field was zero.

At 4.2 K, the Q^U of the fundamental TEM mode was 65,000 for the full MgB_2 system (cavity, rod and lids), and 50,000 for the hybrid system (cavity and rod in MgB_2 and brass lids). From the

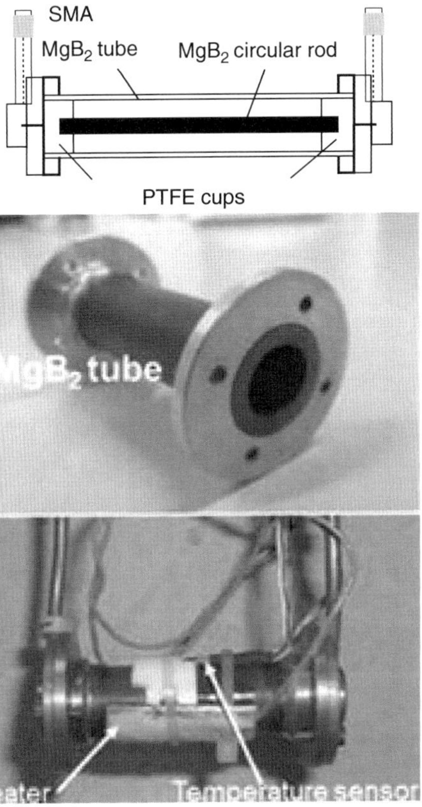

Fig. 26. MgB_2 coaxial cavity, and the measuring setup.

measured Q^U it was possible to compare the MW surface resistance of the MgB_2 rod at 4.2 K with that of a similar Bi-2223 type central rod of a previous analogous experiment at 77 K,[62] the values being 0.6 mΩ and 8 mΩ, respectively. As a final comment about the bulk RLI–MgB_2 applications to MW systems, it is possible to anticipate substantial lowering of the MW surface resistance by using finer and purer B powders.

9. The Doping Efficiency by the RLI Process

It is well known that the low critical fields of the MgB_2 material can be substantially improved by doping. The most studied doping mechanism is the partial substitution of B atoms by C atoms. In the *in situ* reaction process, the C may be substituted into the MgB_2 structure using various procedures: by adding pure fine C (carbon black or graphene), by adding nanosize SiC or B_4C powders which release C after being decomposed by the Mg melt or by adding various hydrocarbons or organic compounds. All these doping types act in a similar way: the highest J_c values are found for a C amount of 5÷10 wt.% with respect to B, while T_c shows a decrease of several degrees. At the same time, the critical fields, B_{irr} and B_{c2} are enhanced by 2–4 T. The critical current increase can be one order of magnitude at high fields, in contrast to the behavior at lower fields (<3T), where the C substitution leads to a decrease of J_c with respect to binary MgB_2. The C-doping creates a point-like contraction in the B layers of the MgB_2 structure (reduction of the cell parameter a, detectable by XRD), and the presence of these defects, together with the critical fields increase, correlates with the higher critical current density at high-fields. The debate on the mechanism of action of the C defects is open.[63] Other not carbonaceous dopings have been proposed (mainly oxides),[64–66] with not evident presence of lattice distortion.

The RLI *in situ* process contributes to an improved efficiency of the doping in two ways: by increasing the connectivity between grains and by a better dispersion of the doping agents among the grains. At some extent, the C-doping decreases the self-infiltration mechanism and, in such case, the infiltration depth may be expected to be limited to few cm. The increase of the critical magnetic fields, B_{c2} and B_{irr}, gives a measure of the doping efficiency. The practical critical field to be monitored is the irreversibility field, B_{irr}, which represents the limit below which a stable superconducting current can flow. The usual way to measure the critical fields is: (a) to perform a $\rho(T)$ curve to control the $\Delta\rho(T_c)$ at different applied fields, (b) to evaluate the temperature corresponding to a 10% $\Delta\rho$, which is used to evaluate B_{irr} and the temperature corresponding to a 90% $\Delta\rho$, which is used to evaluate B_{c2}. Sometimes, for the wires, some authors use a different criterion to evaluate B_{irr}: the field corresponding at a critical current density of 100 A/cm^2. This criterion gives higher B_{irr} values.

Table 7: Irreversibility fields for various MgB_2 samples prepared using the RLI process.

	B_{irr} (T)				
MgB_2	4.2 K	10 K	20 K	25 K	30 K
Pure by P100 B	8.3	7.2	4.6	3.3	1.7
Pure by micronic B	12*	8.5*	5.5*	3.6	1.9
Doped SiC 10wt.%	15.5*	12*	7*	4.5	2.4
Doped C 6wt.%	16*	12.5*	7.2*	4.6	2.15

(*)extrapolated value.

Assuming the quite linear behavior of the B_{irr} (T) at $T < T_c/2$, we summarize in Table 7 the B_{irr} values, measured by resistivity, at the temperature interval $25 \div 30$ K. The low temperature values were extrapolated by a linear function. The measurements were performed on two pure RLI–MgB_2 samples with different B powders and on two doped typical RLI–MgB_2 samples based on micrometer size B, doped with 10 wt.% SiC and 6 wt.% C.[67–69]

10. Infiltration Process Variants (from inserts to hollow wires)

The infiltration process can be applied in many large variety of arrangements of the B and Mg as well as by doping constituents, according to the needs. Two settlings, very different from the bulk manufacts, were patented and may have practical applications.[70,71]

10.1 *Inserts*

Manufacts named "inserts"[72] are characterized by a metallic substrate having surface grooves that contain the densified MgB_2 material. The geometry of the grooves may be chosen in such a way to form close or open loops, in order to realize either a PM or a winding magnet that can be optionally charged by an external current supply. The choice of the metallic substrate, magnetic or not, should be based on the physical application of the insert and should be selected with a thermal expansion similar to the one of polycrystalline MgB_2 (about 7 ppm/K), to avoid large strains which could damage the brittle MgB_2 during the cooling process from the reaction temperature. The cross-section of the grooves may be chosen in such a way to minimize the thermal instability, typical of the bulk materials, known to be related to the SC thickness.[73] Acceptable thickness values are smaller than a critical thickness L, which may be evaluated, for each temperature and critical current density, as:

$$L = \sqrt{12\delta C(T)(T_c - T)/\mu_0 J_c^2} \tag{5}$$

where δ is the density and $C(T)$ is the specific heat of the MgB_2 at the temperature T. The resulting critical thickness L displayed in Fig. 27 has been determined from the MgB_2 experimental values of $C(T)$[74] and typical J_c values for bulk MgB_2. The mean thickness for avoiding flux jumps must be about 100 μm at T $<$ 15 K and may reach 1 mm at $T > 20$ K, within certain limits of J_c. The use of SG MgB_2, instead of LG leads to a decrease of thermal stability of the system, due to the higher critical current density.

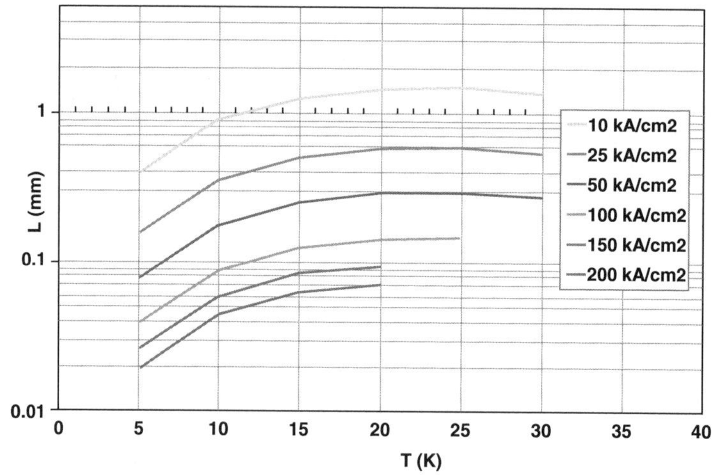

Fig. 27. The maximum thickness (L) of a MgB_2 bar for avoiding thermal instability by flux jumps, for various critical current densities.

Fig. 28. MgB_2 inserts on metallic substrates.

Examples of MgB_2 inserts on metallic substrates are given in Fig. 28. The superconducting properties of some of these inserts have been evaluated with respect to their capability to trap permanently a magnetic field. The larger measured trapped field, on the surface of the sample, reached the value of 1.12 T, at 12 K, for a double side 16 turn helix, with grooves of cross-section $1.5 \times 2.5\,mm^2$. The magnetic relaxation, measured after 13 hours after reaching a magnetization of about 1 T, yields $S = 0.16\%$, in line with our MgB_2 bulk manufacts. The very limited amount of superconducting material and the robustness of these superconducting macrocomposites are very promising for their application in electrical motors or other permanent magnets applications. For example, plates with radial MgB_2 inserts have been proposed for a synchronous axial motor and are under test.[75]

10.2 *Hollow wires*

The infiltration process was initially applied to obtain bulk MgB_2 objects having the shape of plates, cylinders or tubes. Later on, it was attempted to scale down the process to wires, inserting a Mg rod, surrounded by B micrometer powders in a metallic tube and drawing the tube to a wire.[76]

A first review which compared the RLI–MgB_2 wires with the other MgB_2 wires, obtained by different processes, is reported in Ref. 76. The application of the infiltration process to a wire configuration was initially affected by the doubt that an HCP metal, like Mg, should have brittleness problems during the drawing process, but the experience quickly overcame the doubts. Probably the B powders act as a lubricating layer preventing the initiation of microcracks. By drawing, it was possible to reduce the diameter of the tube at least 60 times, the cylindrical shape of the Mg and B being maintained during the drawing process. This isomorphic reduction gives the opportunity to maintain the local composition unchanged, a condition for optimized properties after the reaction. Further doubt of movement of the liquid Mg along the wire, during the reaction, was clarified by experimental evidences. When the B internal diameter is small enough, i.e., less than about $400\,\mu m$, the interfacial tension between liquid and powders prevents the liquid movement. The resulting RLI–MgB_2 hollow wires have a very dense superconducting layer and, even in presence of the central hole, present a quite high fill factor (about 0.23) and show high engineering current densities.[77] More recently, some labs have reproduced the infiltration process, defining it after minor changes as "Internal Magnesium Diffusion" or as "Advanced Internal Magnesium Infiltration", obtaining the best critical current densities of the MgB_2 wires.[78–80] Details of these more recent MgB_2 wire developments are described in the following chapters of this book.

11. Conclusions

The mechanical and superconducting performances of the bulk, polycrystalline MgB_2 material have been optimized by the introduction of an innovative reaction-sintering process, the RLI, which can be considered an easy alternative to the more complex hot-pressing techniques. Main features of the RLI process are the realization of a very high mass density and high grain connectivity with the use of conventional ovens, avoiding the intrinsic MgB_2

structure decomposition arising at elevated temperatures (>1000°C) and normal pressure.

The peculiar novelty of the RLI process is its self-propagating characteristics, without any trigger. The driving mechanism is so powerful that the liquid Mg is even pushed against the gravitational field, reaching depths up to several centimeters inside the B preform. A complete mathematical description of the reaction dynamics is still lacking, due to the complexity of the system. The liquid metal percolates into the B preform porosity and, at the same time, reacts with the B surfaces forming an intermediate phase, Mg_2B_{25}, before giving rise to the final MgB_2. It is experimentally confirmed that the crystalline beta-rhombohedral boron helps to speed up the reaction kinetics, also with B grains as large as 100 μm, however the amorphous micrometer size B can be infiltrated as well. In the case of micro or nanosize B, the powder purity may be a critical issue because adsorbed species on B surfaces can drastically reduce the reaction kinetics.

The RLI process, being an *in situ* type process, favors the doping of MgB_2 with a large variety of dopants, which are presumably better dispersed by the presence of the liquid phase.

In my opinion, the most practical aspect of the RLI process is its capability to produce MgB_2 objects of large dimensions and good mechanical strength. Up to now, prototype rings of more than 200 mm diameter have been realized and, in principle, there is no limitation to the manufacture of larger products. Among the high temperature superconductors, only BSCCO-2212 is adequate to produce large objects by the melt texturing process,[81] but the mechanical strength of this material is much lower than that of the MgB_2.

In the current LTS applications at 4.2 K, the use of bulk thick MgB_2 superconductors poses the issue of flux jumping, in particular when high currents are fed in the most performing fine grain material, in a way similar to all LTS bulk superconductors, e.g., Nb_3Sn. Nevertheless, MgB_2 pieces produced by the RLI process may be formed as a composite material, with different shorter percolating current paths minimizing the flux jumping problems. A clear example of this behavior is the MgB_2 material being thermally more stable, being composed by large normal conducting boride grains, glued by fine grain MgB_2. The LG, even with a reduced critical current density, maintain sufficiently high J_c values for medium field applications. Of course, once applications up to 20÷25 K will have a consolidated market, this issue will be no more important, due to the higher specific heat of the material at these temperatures.

Finally, giving a look to the prospected superconducting application that can be promoted by the development of bulk MgB_2, one should mention all applications defined as "passive," i.e., where the material is exposed to external magnetic flux variations, without any current supply. Examples of these passive applications are: magnetic shielding, magnetic field concentrators, levitation for transport systems and many superconducting electro-technical devices, as electrical motors, where bulk RLI–MgB_2 materials can act as powerful PM. In these applications, MgB_2 enters into a strong competition with bulk HTS cuprates: MgB_2 devices may be somewhat less performing than cuprates, but there is space for alternatives, thanks to their considerably lower processing cost.

From the applicative point of view, it seems possible to anticipate that the arrival of MgB_2 bulk superconductors will move superconductivity towards the design and realization of practical and cheap cryogenic systems at intermediate temperatures (produced by one-stage compact and reliable cryocoolers), the performance and low cost of MgB_2 material being a major issue.

References

1. J. Nagamatsu, N. Nakagawa, T. Muranaka, Y. Zenitani and J. Akimitsu, *Nature* **410**, p. 63 (2001).
2. P. C. Canfield, D. K. Finnemore, S. L. Bud'ko, J. E. Ostenson, G. Lapertot, C. E. Cunningham and C. Petrovic, Superconductivity in Dense MgB_2 Wires, *Phys. Rev. Lett.* **86**, pp. 2423–2426 (2001).
3. G. Giunchi and S. Ceresara, Method for the preparation of highly densified superconductor massive bodies of MgB_2, relevant solid end-products and their use, Edison Italian *Pat. Appl. N.* MI2001A0978 (May 11, 2001).

4. G. Giunchi, High density MgB2 obtained by reactive liquid Mg infiltration, *Int. J. Mod. Phys. B* **17**, pp. 453–460 (2003).
5. G. Giunchi, S. Ceresara, G. Ripamonti, S. Chiarelli and M. Spadoni, MgB_2 sintering from the elements, *IEEE T. Appl. Supercond.* **13(2)**, pp. 3060–3063 (2003).
6. G. Giunchi, C. Orecchia, L. Malpezzi and N. Masciocchi, Analysis of the minority crystalline phases in bulk superconducting MgB_2 obtained by reactive liquid Mg infiltration, *Physica C* **433**, pp. 182–188 (2006).
7. G. Giunchi, L. Malpezzi and N. Masciocchi, A new crystalline phase of the boron-rich metal-boride family: the Mg_2B_{25} species, *Solid State Sciences* **10**, pp. 1202–1208 (2006).
8. S. L. Bud'ko, G. Lapertot, C. Petrovic, C. E. Cunningham, N. Anderson and P. C. Canfield, Boron Isotope Effect in Superconducting MgB_2, *PRL* **86**, N. 9, pp. 1877–1880 (2001).
9. David C. Dunand, Synthesis of superconducting Mg/MgB_2 composites, *App. Phys. Lett.* **79**, p. 25 (2001).
10. D. K. Aswal, K. P. Muthe, A. Singh, S. Sen, K. Shah and L. C. Gupta, Synthesis and Characterization of MgB_2 superconductor, *Physica C* **363**, pp. 149–154 (2001).
11. N. N. Kolesnikov and M. P. Kulakov, Synthesis of MgB_2 from Elements *Physica C* **363**, pp. 166–169 (2001).
12. J.-C. Grivel, R. Pinholt, N. H. Andersen, P. Kovac, I. Husek and J. Homeyer, *In situ* investigations of phase transformations in Fe-sheathed MgB_2 wires, *Supercond. Sci. Technol.* **19**, pp. 96–101 (2006).
13. Z. K. Liu, D. G. Scholm, Q. Li and X. X. Xi, Thermodynamics of the Mg-B system: Implications for the deposition of MgB_2 thin films, *Appl. Phys. Lett.* **78**, pp. 3678–3780 (2001).
14. G. Giunchi and S. Ceresara, Edison Italian Patent Application No. MI2002A1004 (May 10, 2002).
15. G. Giunchi, S. Ceresara, G. Ripamonti, A. Di Zenobio, S. Rossi, S. Chiarelli, M. Spadoni, R. Wesche and P. L. Bruzzone, High performance new MgB_2 superconducting hollow wires, *Supercon. Sci. Tech.*, **16**, pp. 285–291 (Feb 2003).
16. G. Giunchi, A. Figini, L. Albisetti, L. Malpezzi, L. Saglietti and E. Perini, High Performance Bulk MgB_2 obtained by the Reactive Mg-Liquid Infiltration Process: New Advancements and Solutions, *Proc. 2011 IEEE Int. Conf. on Applied Superconductivity and Electromagnetic Devices*, pp. 14–16 Sydney, Australia (2011).
17. L. Y. Markowski, Y. D. Kondrashev and G. V. Kaputovskaya, *J. Gen. Chem. URSS* **15**, p. 409 (1955).
18. T. Wenzel, K. G. Nickel, J. Glaser, H. J. Meyer, D. Eyidi and O. Eibl, Electron Probe Microanalysis of Mg-B compounds: stoichiometry and

heterogeneity of superconductors, *Phys. Stat. Sol. A* **198**, pp. 374–386 (2003).
19. S. Brutti, M. Colapietro, G. Calducci, L. Barba, P. Manfrinetti and A. Palenzona, *Intermetallics* **10**, p. 811 (2002).
20. V. Adasch, K.-U. Hess, T. Ludwig, N. Vojteer and H. Hillebrecht, Synthesis and crystal structure of MgB_{12}, *J. Solid State Chemistry* **179**, pp. 2916–2926 (2006).
21. V. Adasch, K.-U. Hess, T. Ludwig, N. Vojteer and H. Hillebrecht, Synthesis, crystal growth and structure of Mg containing β-rhombohedral boron: $MgB_{17.4}$, *J. Solid State Chemistry* **179**, pp. 2900–2907 (2006).
22. L. Bindi, A. Figini Albisetti, G. Giunchi, L. Malpezzi and N. Masciocchi, Redetermination of Mg_2B_{25} based on single-crystal X-ray data, *Acta Crystallographica, E* **68**, p. i50 (2012).
23. L. Saglietti, E. Perini, G. Ripamonti, E. Bassani, G. Carcano and G. Giunchi, Boron Purity Effects on Structural Properties of the MgB_2 Obtained by Mg-Reactive Liquid Infiltration, *IEEE TAS*, **19**, pp. 2739–2742 (2009).
24. J. D. DeFouw, J. P. Quintana and D. C. Dunand, In situ X-ray synchrotron diffraction study of MgB_2 synthesis from elemental powders, *Acta Materialia* **56**, pp. 1680–1688 (2008).
25. A. Figini Albisetti, L. Saglietti, E. Perini, C. Schiavone, G. Ripamonti and G. Giunchi, The Mg_2B_{25} formation and its role in the preparation of bulk MgB_2 superconductors, *Solid State Sciences* **14**, pp. 1632–1635 (2012).
26. G. Giunchi, L. Saglietti, G. Ripamonti, A. Figini Albisetti, E. Bassani and E. Perini, Superconducting Joints Between MgB_2 Wires and Bulks, *IEEE T. App. Supercond*, **20**, pp. 1524–1527 (2010).
27. G. Giunchi, T. Cavallin, P. Bassani and S. Guicciardi, The mechanical properties of the MgB_2 bulk materials obtained by Reactive Liquid Mg Infiltration, *Advances in Cryogenic Engineering: Transaction of the ICMC*, **54**, pp. 396–404 (2008).
28. U. Harms, A. Serquis, R. B. Schwarz and V. F. Nesterenko, Low Temperature Elastic Constants of Polycrystalline MgB_2, *J. Supercond. Nov. Magn.*, **16**, pp. 941–944 (2003).
29. F. Cordero, R. Cantelli, G. Giunchi and S. Ceresara, Search for incipient lattice instabilities in MgB_2 by inelastic spectroscopy, *Phys. Rev. B* **64**, p. 132503 (2001).
30. L. Gozzelino, F. Laviano, D. Botta, A. Chiodoni, R. Gerbaldo, G. Ghigo, E. Mezzetti, G. Giunchi, S. Ceresara and G. Ripamonti, Critical state analysis in bulk MgB_2 by means of a quantitative magneto-optical technique, *Phil. Magazine Part B* **82**, pp. 1–11 (2002).

31. G. Giunchi, S. Raineri, R. Wesche and P. L. Bruzzone, The voltage–current relations for MgB_2 obtained by reactive liquid infiltration, *Physica C* **401**, pp. 310–315 (2004).
32. R. Wesche, Temperature dependence of critical currents in superconducting Bi-2212/Ag wires, *Physica C* **246**, pp. 186–194 (1995).
33. J. M. Rowell, *Supercond. Sci. Technol.* **16**, R17 (2003).
34. A. M. Awasthi, S. Bhardwaj, V. P. S. Awana, A. Figini Albisetti, G. Giunchi and A. V. Narlikar, Carrier transport in magnesium diboride: role of nano-Inclusions, *Appl. Phys. Lett.* **103**, p. 112601 (2013).
35. M. Tomita and M. Murakami, *Nature* **421**, p. 517 (2003).
36. G. Fuchs, W. Häßler, K. Nenkov, J. Scheiter, O. Perner, A. Handstein, T. Kanai, L. Schultz and B. Holzapfel, High trapped fields in bulk MgB_2 prepared by hot-pressing of ball-milled precursor powder, *Supercond. Sci. Technol.* **26**, p. 122002 (2013).
37. E. Perini, G. Giunchi, L. Saglietti, A. Figini Albisetti, A. Matrone and V. Cavaliere, Magnetic Field Trapping in MgB_2 Bulks and Inserts, *IEEE TAS* **21**, p. 2690 (2011).
38. A. Matrone, Private Communication (2005).
39. G. Giunchi, E. Perini, T. Cavallin, E. Bassani, R. Quarantiello, V. Cavaliere and A. Matrone, Superconducting Permanent Magnet made by a bulk MgB_2 ring, *IEEE TAS* **18**, pp. 1216–1219 (2008).
40. E. Perini, A. Matrone, R. Quarantiello, V. Cavaliere, U. Gambardella, S. Pace, T. Cavallin and G. Giunchi, The magnetic characterization of a MgB_2 superconducting hollow cylinder, at variable temperatures up to Tc, *IEEE TAS* **17**, pp. 2730–2733 (2007).
41. G. Giunchi, A. Figini Albisetti, G. Montenero and A. Ballarino, Magnetic shielding and field trapping of a superconducting MgB_2 cylinder in a transversal field (2015).
42. A. Di Zenobio, S Chiarelli and M. Spadoni, Private Communication (2002).
43. M. Zeisberger, W. Gawalek and G. Giunchi, Magnetic levitation using magnesium diboride, *J. Appl. Phys.*, **98**, p. 023905 (2005).
44. E. Perini, G. Giunchi, M. Geri and A. Morandi, Experimental and Numerical Investigation of the Levitation Force between Bulk Permanent Magnet and MgB_2 Disk, *IEEE TAS* **19**, pp. 2124–2128 (2009).
45. E. Perini and G. Giunchi, Field Cooling of a MgB_2 cylinder around a Permanent Magnets stack: axial Forces and stiffness for a prototype magnetic bearing, *Supercond. Sci. Technol.* **22**, p. 045021 (2009).
46. H. Fujishiro, T. Ujiie, T. Naito, A. Figini Albisetti and G. Giunchi, Trapped magnetic field in double stacked MgB_2 bulks magnetized by pulsed field, *J. Phys. Conference Series* **507**, p. 032016 (2014).

47. G. Giunchi, MgB_2 bulk cylinders as magnetic shields for physical instrumentation, *Proc. 20th IMEKO-TC4 Intern. Symposium*, Benevento (Italy), pp. 1033–1037 (Sept. 2014).
48. Z. Y. Zhang, S. Choi, S. Matsumoto, R. Teranishi, G. Giunchi, A. Figini Albisetti and T. Kiyoshi, Magnetic Amplifiers Using Different MgB_2 Bulk Superconductors, *Supercond. Sci. Technol.* **25**, p. 025009, (2012).
49. G. Giunchi, G. Ripamonti, T. Cavallin and E. Bassani, The reactive liquid Mg infiltration process to produce large superconducting bulk MgB_2 manufacts, *Cryogenics* **46**, pp. 237–242 (2006).
50. T. Cavallin, R. Quarantiello, A. Matrone and G. Giunchi, Magnetic shielding of MgB_2 tubes in applied DC and AC field, IOP: Conf. Series **43**, pp. 1015–1018 (2006).
51. G. Giunchi, E. Bassani, J. Hassel, P. Helistö and J. Penttilä, MgB_2 cylindrical cup for magnetic shielding in pulse tube cryostat, *Adv. Cryo. Eng. Transaction CEC*, **55**, pp. 1636–1640 (2010).
52. J. J. Rabbers, M. P. Oomen, E. Bassani, G. Ripamonti and G. Giunchi "Magnetic shielding capability of MgB2 cylinders" *Supercond. Sci. Technol.* **23**, p. 125003 (2010).
53. P. Arpaia, A. Ballarino, G. Giunchi and G. Montenero, MgB_2 cylindrical superconducting shielding for cryogenic measurement applications: a case study on DC current transformers, *J. Instrumentation* **9**, p. P04020 (2014).
54. Z. Y. Zhang, S. Choi, S. Matsumoto, R. Teranishi, G. Giunchi, A. F. Albisetti and T. Kiyoshi, Magnetic Amplifiers Using Different MgB_2 Bulk Superconductors, *Supercond. Sci. Technol.* **25**, p. 025009 (2012).
55. G. Krabbes *et al.*, *High Temperature Superconductor Bulk Materials*, WILEY-VCH (2006).
56. M. J. Lancaster, *Passive Microwave Device Applications of High-Temperature Superconductors*. Cambridge University Press (1997).
57. B. P. Xiao, X. Zhao, J. Spradlin, C. E. Reece, M. J. Kelley, T. Tan and X. X. Xi, Surface impedance measurements of single crystal MgB_2 films for radiofrequency superconductivity applications, *Supercond. Sci. Technol.* **25**, p. 095006 (2012).
58. A. A. Gallitto, G. Bonsignore, G. Giunchi, M. Li Vigni, Yu. A. Nefyodov, *Microwave response of bulk MgB_2 samples of different granularity*, IOP: Conf. Series **43**, pp. 480–483 (2006).
59. A. Agliolo Gallitto, P. Camarda, M. Li Vigni, A. F. Albisetti and G. Giunchi, Frequency dependence of the microwave surface resistance of MgB_2 by coaxial cavity resonator, *Physica C* **503**, pp. 150–153 (2014).

60. A. A. Gallito, G. Bonsignore, M. Li Vigni, G. Giunchi and Yu. A. Nefyodov, Microwave response of a cylindrical cavity made of bulk MgB_2 superconductor, *Physica C* **468**, pp. 66–71 (2008).
61. A. A. Gallitto, P. Camarda, M. Li Vigni, A. Figini Albisetti, L. Saglietti and G. Giunchi, Microwave Response of Coaxial Cavities Made of Bulk Magnesium Diboride. *IEEE TAS*, **24**, p. 1500109 (Feb 2014).
62. A. A. Gallitto, G. Bonsignore, M. Li Vigni and A. Maccarone, Tunable coaxial cavity resonator for linear and nonlinear microwave characterization of superconducting wires, *Supercond. Sci. Technol.* **24**, pp. 095008-1–095008-8, (2011).
63. R. Flükiger, private communication (2015).
64. C. B. Eom, *et al.*, High critical current density and enhanced irreversibility field in superconducting MgB_2 thin films, Nature **411**, 558–560 (2001).
65. W. K. Yeoh *et al.*, On the Role of Graphene Oxide Doping for Enhanced Supercurrent in MgB_2, *Cond. Mat. arxiv* 1405.4619 (2014).
66. Guangze Li, M. D. Sumption, M. A. Rindfleisch, C. J. Thong, M. J. Tomsic and E. W. Collings, Enhanced higher temperature (20–30 K) transport properties and irreversibility fields in nano-Dy_2O_3 doped advanced internal Mg infiltration processed MgB_2 composites, *Appl. Phys. Lett.* **105**, p. 112603 (2014).
67. G. Giunchi, E. Perini, C. Orecchia, G. Ripamonti, G. Carcano, G. De Marzi, R. Viola and S. Turtù, Analysis of various dopants on the MgB_2 superconducting properties, *IEEE TAS* **19**, pp. 2802–2806 (2009).
68. L. Gozzelino, B. Minetti, R. Gerbaldo, F. Laviano, G. Lopardo, G. Giunchi, E. Perini, T.Cavallin, A. Agostino, M. Panetta and E. Mezzetti, Effects of Nanoparticle Doping on Electrical Properties of MgB_2 Bulks and Wires Obtained by Reactive Mg Liquid Infiltration Technique, *IEEE TAS* **17(2)**, pp. 2726–2729 (2007).
69. L. Gozzelino, R. Gerbaldo, G. Ghigo, F. Laviano, G. Lopardo, G. A. Ummarino, G. Giunchi, E. Perini, E. Bassani, A. Agostino and B. Minetti, Effects of annealing and nanoparticle doping on electrical properties of MgB_2 bulks grown by Reactive Mg Liquid Infiltration Technique, *IEEE TAS* **19(3)**, pp. 3524–3528 (2009).
70. G. Giunchi and S. Ceresara, Method for the production of superconductive wires based on hollow filaments made of MgB_2, Edison Italian *Patent Application No.* MI2002A1004 (May 10, 2002).
71. G. Giunchi, Superconductive Element and relative preparation process, Edison Italian *Patent Application No.* MI2009A 001078 (June 18, 2009).
72. G. Giunchi, MgB_2 Superconductive Inserts: Products between Bulk and Wires, *IEEE Trans. on Appl. Supercond.* **21(3)**, pp. 1564–1567 (2011).

73. M. Wilson, *Superconducting Magnets*, Oxford Sci. P. Ed. (1983).
74. Y. Wang, T. Plackowski and A. Junod, *Physica C* **355** p. 179 (2001).
75. F. Marignetti *et al.*, Use of MgB_2 Superconductors for Excitation Field in Synchronous Machines — Part II: Inserts, *IEEE TAS* **23**, p. 8002606 (2013).
76. G. Giunchi, S. Ceresara, G. Ripamonti, A. Di Zenobio, S. Rossi, S. Chiarelli, M. Spadoni, R. Wesche and P. L. Bruzzone, High performance new MgB_2 superconducting hollow wires, *Supercond. Sci. Technology* **16**, pp. 285–291 (2003).
77. G. Giunchi, L. Saglietti, A. F. Albisetti, E. Perini, C. Schiavone, G. Ripamonti, D. Uglietti and P. L. Bruzzone, MgB_2 Hollow Wires and Cables Embedded in a Mg Alloy Matrix, *IEEE TAS*, **23**, p. 6200605 (2013).
78. G. Nishijima, S. J. Ye, A. Matsumoto, K. Togano, H. Kumakura, H. Kitaguchi and H. Oguro, Mechanical properties of MgB_2 superconducting wires fabricated by internal Mg diffusion process, *Supercond. Sci. Technol.* **25** p. 054012 (2012).
79. G. Z. Li, Y. Yang, M. A. Susner, M. D. Sumption and E. W. Collings, Critical current densities and n-values of MgB_2 strands over a wide range of temperatures and fields, *Supercond. Sci. Technol.* **25**, p. 025001 (2012).
80. P. Kovác, I. Hušek, T. Melišek, L. Kopera and J. Kovác, Critical currents, *Ic*-anisotropy and stress tolerance of MgB_2 wires made by internal magnesium diffusion, *Supercond. Sci. Technol.* **27**, p. 065003 (2014)).
81. J. Bock, H. Bestgen, S. Elchner and E. Preisler, *IEEE TAS* **3**, p. 1659 (1993).

Chapter 4a

Processing of Amorphous Nanosize Boron Powder

M. S. Somer*,‡, S. Acar† and I. Kokal†

*Koç University, Chemistry Department
34450 Sariyer/Istanbul, Turkey
†Pavezyum Kimya, Ulu Sohak No. 3, Tuzla, Turkey
‡msomer@ku.edu.tr

This article presents natural resources, properties, allotropes and the different synthesis methods for elemental amorphous boron (B) in small and large quantities. The industrial production of elemental amorphous B is mainly based on Moissan process while the gas phase pyrolysis and Chemical vapor deposition (CVD) techniques are more or less restricted to semi-commercial use and research applications. Among all, only thermolysis of B hydrides performed in the temperature range of 500–1000°C yields fully amorphous nanosize B powders. The properties the commercially amorphous B samples of different purity grades and sources were characterized by using powder X-ray diffraction, SEM, TEM and chemical analysis. The results show that nanosize B has superior properties with respect to purity, particle size, amorphousness and chemical reactivity. The use of reactive nanosize B powder as precursor decreases not only the reaction temperature but leads also to the formation of nanosized MgB_2. These attitudes are well compatible with the outstanding performance reported for MgB_2 wires manufactured from fully amorphous B powder (B-A9999).

1. Introduction

The development of the synthesis of elemental Boron (B) is a process which endured almost two centuries. Until 2001, to the year of the discovery of the superconductivity of MgB_2,[1] the use of elemental B was mainly restricted to the area of energetic materials such as igniters for airbags, solid fuel for rockets and missiles as well as for pyrotechnics. These applications are closely related to B's high affinity to O_2 coupled with the high gravimetric heat of combustion value which is the third highest of all elements. Another important application field of elemental B is its use as thermal neutron absorbers in nuclear plants, consisting of boron carbide (B_4C) control rods.[2] Over the years, the industrial production of elemental B has reached a certain level of balance in quantity and quality, meaning that the amount and purity grades of the B powders in the market met the demands of the consumers to a large extent.

Until very recently the highest purity grade for a commercially available amorphous B powder was 95–97 wt.%. Unlike the amorphous form, crystalline B can be obtained in ultra-high purity grades (99.999%) but has two major disadvantages. It is very expensive (~100 €/g) and compared to the amorphous B, chemically less reactive.[3,4] Ultra-high pure crystalline B is commercially available in comparatively small quantities and is used as a dopant in the semiconductor industry. It is produced by pyrolysis of diborane or B halides and hydrogen at high temperatures and is then further purified either by zone melting or by Czochralski processes.[5] For all other purposes such as airbags, military and nuclear applications, the fabrication of metal-based compounds and others, amorphous B with ≤95–97% purity has been the material of choice for the scientists. The demand for ultra-pure amorphous B (99.99%, metal impurities) in larger quantities started practically with the discovery of superconductivity in magnesium diboride (MgB_2) in 2001.

The compound MgB_2 is easily accessible via a solid state reaction of the elements in the temperature range 650–850°C under Ar atmosphere. It turned out that MgB_2 samples prepared from a special B powder called B-A9999 (meaning 99.99% purity, based on trace metal analysis) which was supplied by the company Alfa Aesar had superior properties compared to specimen synthesized from other commercial powders of inferior purity. Besides the lower reaction temperature which allowed a synthesis under easier conditions, the most striking advantage of the B-A9999 based MgB_2 was the increase of the critical current density J_c by more than one order of magnitude[6–9] compared to those obtained from other commercial powders. However, by the end of the last decade its availability was severely reduced due to military applications and limited resources[10] and Alfa Aesar discontinued the production of B-A9999 in 2010.

The significant increase in J_c achieved by using B-A9999 powder has been decisive in the development of scientific research in the system MgB_2 in the years 2001–2009, demonstrating the crucial role and importance of the quality of the precursor B powder.

In this chapter a short history of the synthesis of pure elemental B will be presented and the difficulties related to its high reactivity with respect to O_2 as well as to other elements, e.g., C, N and most of the metals will be described. Novel approaches for the production of sustainable, high-purity and economic amorphous B powder will be discussed, as an important precursor for high performance MgB_2 wire.

2. Elemental B

2.1 *Natural sources*

B cannot be found in the elemental form in nature, due to its high affinity to O_2. Thus, B ores are basically borate minerals based on B and O_2, but also containing H, alkali and earth alkaline elements.[11] Highly concentrated, economically feasible deposits of B minerals are found in regions with a history of volcanism or hydrothermal activity (Mojave desert in the southwest of U.S., Mountains Zone in South America and South–Center Asia–Alpine zone).[12] The largest amount of the proven B mineral deposits, 71.3%, are located in Turkey.[13] Table 1 depicts the B deposits distribution all over the world. Today, 90% of the B mining activities in the world are performed in Turkey and the USA.[14]

Among almost 250 different known B minerals, the commercially most important ones are borax ($Na_2B_4O_7{\cdot}10H_2O$), collemanite ($Ca_2B_6O_{11}{\cdot}5H_2O$) and ulexite ($NaCaB_5O_9{\cdot}8H_2O$).

B minerals are processed to obtain commercial B products which find more than 250 application areas, the commercially most important refined B chemical being boric acid, H_3BO_3.[15] Globally the annual consumption of B compounds was estimated to four million

Table 1: B deposits in the world.

Country	Total reserve (1000 tons)	Total reserve (% B_2O_3)
Turkey	885.000	71.3
Kazakhstan	102.000	8.2
U.S.A	80.000	6.5
China	47.000	3.8
Chile	41.000	3.3
Russia	35.000	2.8
Peru	22.000	1.8
Rest	29.000	2.3
Total	**1.241.000**	**100**

tons of B_2O_3 in 2012,[16] while the production and consumption of elemental B occurs in much smaller quantities. To date, there are worldwide four companies producing elemental amorphous B on an industrial scale: H.C. Starck (Germany), Tronox and S.B. Boron (both U.S.) and Pavezyum (Turkey). The dominating airbag industry consumed 2012 between 80 and 100 tons[17,a] of elemental amorphous boron.

2.2 *The production of elemental boron: historical retrospective*

The historical review of the elemental B production is summarized in two comprehensive articles[10,18] describing the most important milestones in the development of the elemental B synthesis (see also Fig. 1). It is historically proven that elemental B was first produced in crude form by Sir Humphry Davy in 1807.[19] In 1808 Joseph Gay-Lussac and Louis Thenárd announced independently and almost simultaneously the preparation of elemental B through reduction of B_2O_3 with elemental potassium.[20] In 1809 Davy electrolyzed molten boric acid and borax to B, a method which has been improved and employed by several scientist in the following years.[21–23] However,

[a]The estimation is based on more than 90 million airbags globally produced in 2011 and the use of ca. 1 g elemental B igniter per airbag.

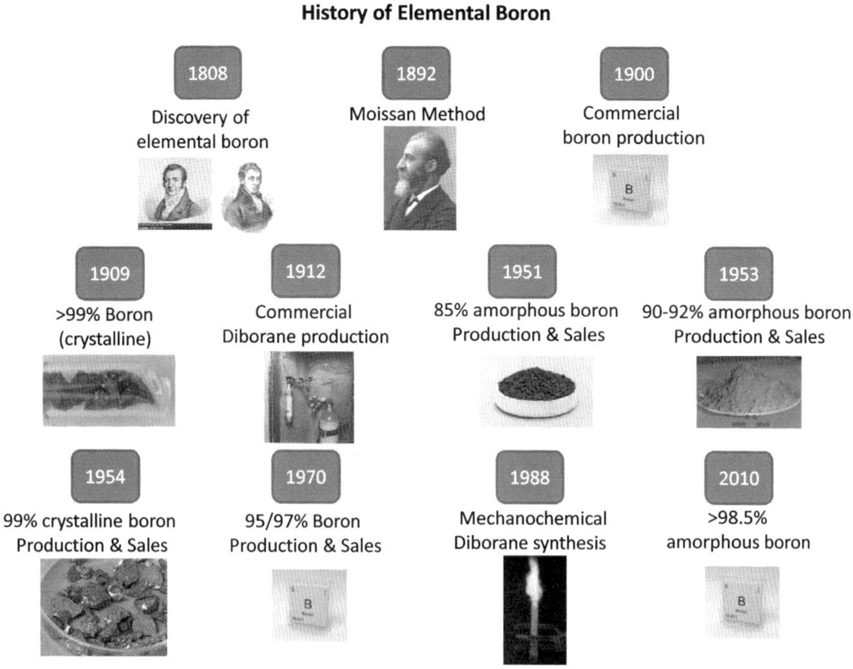

Fig. 1. Timeline for boron processing developments.

the purity of these elemental forms remained in all cases well below 70 wt.% B.[24] In 1892, Henri Moissan produced amorphous B by reducing B_2O_3 with magnesium, the most efficient reducing agent among other alkaline and earth alkaline metals.[25,26] The purity of amorphous B obtained with this method can be increased up to 95–97% by several chemical procedures being mainly based upon acid treatment of the reaction products.[27–31] Reduction of B_2O_3 with magnesium metal, known as metallothermal — or magnesiothermal — reaction, is called "*Moissan Method*" on behalf of its inventor. The Moissan method is still the most widely used process for the commercial production of amorphous B.

In addition to the employment of boric acid or boron oxide as precursor for the reduction process, there have been several others processes for production of elemental B. Worth mentioning are passing of B trichloride and H gas in an electric arc, deposition on hot metal filaments and finally pyrolysis of volatile B compounds such as boranes and BI_3 (see Secs. 3.2.2 and 3.3).

2.3 The Allotropes of B and their physical properties

Elemental B exists in amorphous as well as in crystalline forms. Up to now, 16 crystalline B modifications have been well characterized and reported in literature.[32–35] Among these 16 crystalline polymorphs most of them were reported to be B-rich compounds, like YB_{66}, NaB_{15}, $B_{50}N_2$ etc.[32] The preparation and conversion of pure B are performed at elevated temperatures and it is well-known that contamination of the base element even in parts per million (ppm) level can have an influence on the electronic structure of inorganic solids and result in dramatic changes in material properties. Two crystalline B phases exist at ambient conditions, α and β, both rhombohedral: β is thermodynamically stable at high temperatures only, while α B is a low temperature phase.[25,32–34]

There are two more crystalline B modifications, named α and β tetragonal B, both being high temperature, metastable phases. However, since the two tetragonal modifications were reported to be stabilized by foreign atoms (N and C), it is still a subject of discussion whether they are real allotropes of B.[25] The recently published γ-B_{28} form[32,35] is a high pressure modification of β-rhombohedral B, consisting of isolated $[B{-}B]^{2+}$ cations and $[B_{12}]^{2-}$ cluster anions. The most important physical properties of elemental B are depicted in Table 2.

Table 2: The most important physical properties of elemental B.

	Amorphous	α rhombohedral	β rhombohedral	γ orthorhombic
Melting Point (K)	2573	2453		—
Boiling Point — Sublimes (K)	2923	3923		—
Heat of Combustion (kJ/mol)	630	—	—	—
Density (g/cm^3)	2.35	2.45	2.35	2.52
Appearance	Brown	Red-brown	Black	—

3. Industrial Relevant Production Methods for Elemental B

The usual oxidation state of B in his minerals and compounds is +3. For conversion to element it must be chemically reduced by suitable agents or thermally decomposed (pyrolysis). This can be achieved by four basic processes as was proposed by Newkirk.[36]

3.1 *Electrochemical reduction of molten borates and fluoroborates*

The electrolysis of molten B oxide is probably the oldest preparation method (Sir Davy, 1808) for elemental B. It is based on the deposition of the element by electrolysis of fused metal borates or fluoroborates as B source. For the electrolysis, pure B_2O_3 and metal borates as base material with high B oxide content have the important disadvantage of high viscosity and low electrical conductivity. Additives such as K_2O, KF, KBF_4 and KCl increase the mobility and the current density[37,38] by the formation of fluoroborate complexes. B generated by electrolysis from molten KBF_4–B_2O_3 with 2–15% boron oxide content was claimed to have a purity of 98 wt.%.[39] Later studies, however, showed that the purity was lower. Similar results were reported for electrolysis from KBF_4–KCl melts (mole ratio 5:1) using graphite as electrodes.[40] The main contamination introduced boron during the process, carbon, was prevented using boron electrodes.[41] Makyta *et al.*[42] reported the cathodic reduction of $[BF_4]^-$ obtained from a molten B_2O_3–LiF–KF bath resulting in the formation of elemental boron.[42] According to Polykova,[43] trivalent boron is reduced to elemental boron in a single irreversible step in fused LiF–NaF–KF for concentrations of KBF_4 inferior to $5.7 \cdot 10^{-2}$ mol.%. In 2010 J. C. Gomez[44] presented a new electrolyte system consisting of a molten MgF_2–NaF–LiF eutectic mixture with 4–20% B_2O_3 as B source. The process yielded elemental boron with a purity of ≤ 90 wt.%.

3.2 *Chemical reduction of B compounds*

The B precursors in the most cases are B oxide and rarely B halides.

3.2.1 *Reduction of B oxide*

(*a*) *with Al, Ca, Zn and Na*

The reduction of B oxide with Al yields Al borides such as AlB_{10} and AlB_{12}.[45] In case of Ca, main product is CaB_6.[46] Na metal reacts with B_2O_3 to form elemental boron of low purity (~86%) which can be further purified by chlorination.[47] In the same manner potassium reduces B oxide to elemental B as was reported by Gay-Lussac and Thénard.[20]

(*b*) *with Mg (Metallothemal Reduction) or "Moissan Method"*

This technique, also known as Moissan method, is based on the reduction of B oxide powder with Mg in the form of powder or chips. The technical details of the process, first reported in Refs. 48 and 49, were later presented by Karau at ISBB in 2011.[50] It is the most important and widely used process for the industrial production of amorphous B. The stoichiometric reaction is represented by Eqs. (1)–(3) being side reactions leading to borate byproducts:

$$B_2O_3 + 3Mg \rightarrow 2B + 3MgO, \tag{1}$$

$$2B_2O_3 + 3Mg \rightarrow 2B + Mg_3(BO_3)_2, \tag{2}$$

$$4B_2O_3 + 3Mg \rightarrow 2B + 3Mg(BO_2)_2, \tag{3}$$

The highly exothermic process is called Self Propagating High Temperature Synthesis (SHS). It is a typical batch reaction occurring in large steel containers which are placed in a cooling water bath. Once the reaction mixture is ignited the reaction proceeds extremely fast ($T_{max} > 2500°C$ if no heat lost.[48] Since amorphous B is very sensitive to air and moisture at elevated temperatures, the reaction must be performed under protective atmosphere such as Argon. The product obtained after the reaction is a black mass containing the main products B and MgO I addition to the byproducts: magnesium borates ($Mg_3(BO_3)_2$, $Mg(BO_2)_2$), magnesium borides (MgB_4, Mg_2B_{25}), subborates ($MgB_{30}O_{10}$), B suboxides ($B_{12}O_2$ and B_7O) and BN.[51] The removal of MgO and the borates is performed by hot acid leaching with HCl, followed by filtration, washing and drying. The final yield of the amorphous B powder is 86% to 88%,

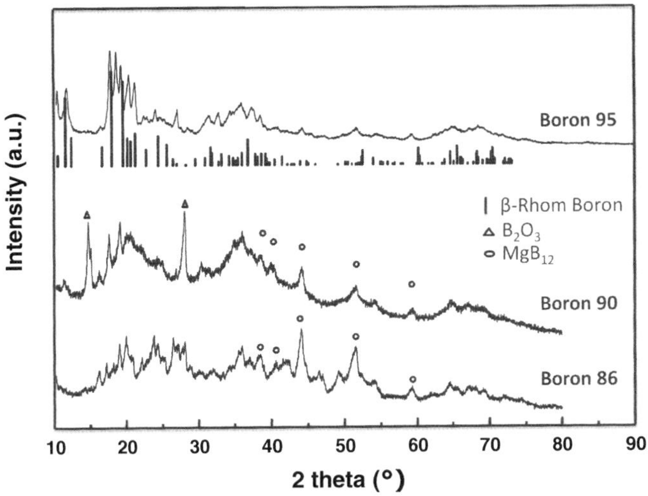

Fig. 2. PXRD of amorphous B powders of different purity grades produced by the Moissan method.

still containing magnesium (10–12%) in boride, B oxide and suboxide impurities. The maximum purity that can be achieved by the "Moissan method" is limited to 90–92%. The rest consists of Mg (4–6%) and B oxides. To achieve higher purity grades such as 95–97% — the optimum standard that can be reached so far by industrial processes, further treatment with elemental chlorine[47] or KBF_4, HF or KF[52] is necessary.

According the PXRD, SEM and TEM results (Fig. 2), B powders produced by the Moissan method are not fully amorphous but contain a considerable amount of crystalline fractions, as was repeatedly reported.[6,7,53] The characteristic properties of the powders are listed in Table 3. Corresponding SEM and TEM pictures are shown in Fig. 3.

3.2.2 *Reduction of B halides*

(a) by metals and metal compounds

Elementary zinc is suitable for reduction of BCl_3 at temperatures between 873 and 1123 K, resulting in elemental B of 99.5 wt.% purity. The product contains amorphous and α-rhombohedral B and a zinc

Table 3: Particle size and surface area values of different amorphous B grades.

Material	Boron (wt.%)	Magnesium (wt.%)	Particle Size (nm)	Surface area (m^2/g)
nano-B	>98.5	—	50–300	>35
B 95	95–97	<0.8	<1000	>10
B 90	90–92	<6.0	<800	>18
B 86	86–88	<12.0	<1400	>5

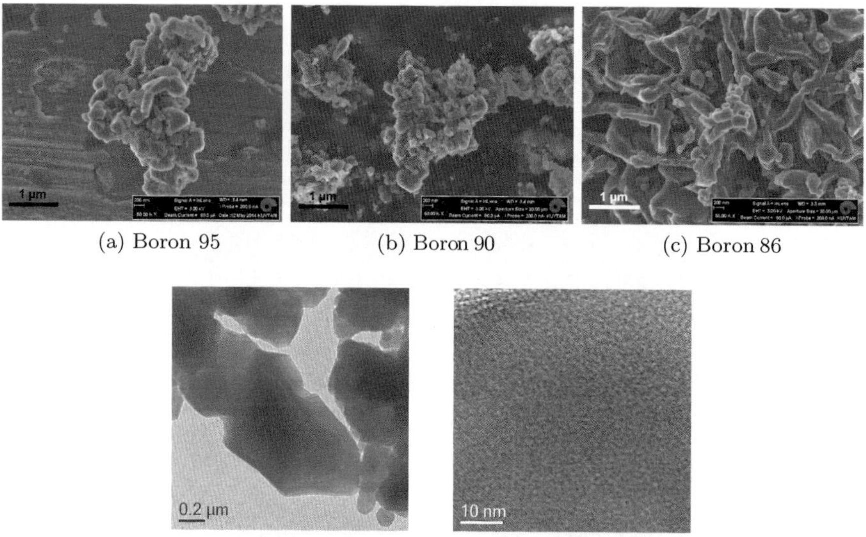

(a) Boron 95 (b) Boron 90 (c) Boron 86

(d) Boron 95 (e) Boron 95

Fig. 3. SEM pictures of B 95 (a), B 90 (b), B 86 (c), TEM micrograph of B 95(d), lattice fringes and Electron Diffroctagram of B 95 (e).

containing B phase ($B_{50}Zn$).[54] Sodium can be used in the gas phase reduction of B halides around 753 K.[55] Tetragonal B (98% B) is formed by the reduction of BCl_3 with tin at 1323 K, followed by the transformation to β rhombohedral structure after annealing at 1273 K.[56]

According to Ref. 48, B halides are thermochemically reduced by binary borides, such as MgB_2, AlB_2, AlB_{12} and SiB_6 to form elemental B (in crystalline and amorphous forms) and metal halides.

(b) by hydrogen

In this case volatile B halides other than BF_3 (requires higher reaction temperatures) can be reacted in the gas phase with H gas in the temperature range of 1123–1723 K in an arc or on the surface of a suitable substrate (CVD, Chemical Vapor Deposition) to form elemental amorphous B and hydrogen halides[57–60]:

$$BX_3(g) + 1.5H_2(g) \rightarrow B(s) + 3HX(g), \quad \text{where X} = \text{Cl, Br,} \tag{4}$$

$$BX_3(g) + 1.5H_2(g) + \text{Substrate} \rightarrow B(\text{on substrate}) + 3HX(g). \tag{5}$$

The Tungsten, tantalum, molybdenum wires, graphite, BN, and also B rods can be used as substrate materials. The gas phase reduction of B halides with hydrogen is by far the most commonly employed preparative route for pure (>99.9%) elemental B — both in semi-amorphous or crystalline form — in smallest to commercial quantities. Almost 70 articles and reviews were published between 1960–1977[61] and more than 20 patents are available on this topic to date.

The U.S. company SMI developed this technique by injecting a BCl_3–H_2 mixture into an inductively coupled plasma, The particle size of the resulting boron powder is in the range of 5–200 nm. The powder is not fully amorphous revealing also fractions of crystalline material.[62]

3.3 *Pyrolysis (thermal decomposition)*

(a) Pyrolysis of gaseous B halides on hot substrate surface

The formation of pure elements by transport reaction of their volatile halides followed by pyrolysis on hot (1500–1700 K) micrometer size single filament W wire were first reported by A. E. van Arkel and J. H. de Boer in 1925.[63] In the same manner gaseous B halides (preferably BI_3) can be decomposed on hot substrate surfaces, e.g., B, tantalum, tungsten, molybdenum or vitreous BN according to Eq. (6)[64–67]:

$$BX_3 + \text{substrate B} \leftrightarrows (\text{onsubstrate}) + 1.5\,X_2, \quad \text{where X} = \text{Cl, Br, I}). \tag{6}$$

(b) Pyrolysis of B hydrides (boranes)

According to Ref. 68 the direct thermal decomposition of B compounds to high purity B is limited only to halides and hydrides (boranes). Due to their lower thermal stability — compared to halides — the thermolysis of boranes takes place in the range of 873–1173 K in which the formation of amorphous form of B is favored. Highly pure B powder of predominantly amorphous nature is obtained by thermal decomposition of diborane, pentaborane or decaborane, usually diluted with hydrogen, by passing through a heated quartz tube and by the pyrolysis of solid B hydrides.[48,69–71]

The most important precursor for the borane pyrolysis is diborane, B_2H_6, first isolated and characterized by Stock in 1912.[72,73] It can be prepared in small to industrial quantities. The major industrial process is based on the reaction of NaH with BF_3 developed in 1950's by Schlesinger and Brown.[74] It is technically an important reagent with wide range of applications.[75] Diborane is a highly reactive and toxic gas which can be thermally decomposed in the range of 180–300°C[75,76] to higher boranes (e.g., B_5H_9, B_5H_{11}, $B_{10}H_{14}$, etc.). Further pyrolysis starting from 600°C up to 800°C results in complete dissociation into H and elemental B.[77]

Depending on the temperature range, high purity amorphous (<1000°C) or semi-crystalline B (>1000°C)[78] is produced by the gas phase pyrolysis of diborane on substrates according to the reaction:

$$B_2H_6 + \text{substrate} \leftrightarrow 2B\ (\text{on substrate}) + 3H_2. \qquad (7)$$

The substrate consists of a thin wire of tantalum or tungsten and B rods of 250 mm length and 10 mm diameter can be produced. All impurities except C can be easily removed by zone melting. C free diborane is a key factor for avoiding the formation of stable, volatile carbonyl–borane, according to $B_2H_6 + CO \rightarrow 2\ BH_3CO$.[48]

During the disposal period (1982–1996) of unused U.S. pentaborane, B_5H_9, stocks of approximately 1800 lb — once produced as a fuel additive to power advanced high-speed bombers, missiles and rockets of the U.S. Army and Navy — a small quantity (25 cylinders) of the material was converted to amorphous B by pyrolysis in 1995.[79,80] But after only one year, the production had to be stopped due to several accidents during the treatment of still bottoms generated during the thermal decomposition.[79] Pentaborane, B_5H_9, is an exceptionally reactive and hazardous liquid,[81] accessible via controlled thermolysis of diborane at 200°C.[72] After multiple incidents even of fatal nature during its production and the later disposal period, pentaborane has been completely banned as a viable fuel since the 70's. At present it is commercially available only in small quantities, mainly for use in research.

Decaborane, $B_{10}H_{14}$, is a higher derivative of diborane which was — as pentaborane — once developed as an additive to special high-performance rocket fuel and was produced in more than 1,000 tons per year.[82] It is a flammable white solid accessible via thermal decomposition of diborane or pentaborane above 240°C. Similar to diborane, decaborane undergoes a thermal decomposition in the range of 500–1000°C to yield pure elemental B powder and H. Due to its easier handling compared to diborane, decaborane is a superior elemental B source for research and small scale applications.[83] In this manner, the pyrolysis of decaborane has been successfully employed for generating B thin films,[83] manufacturing semiconductors,[84,85] as well as for synthesizing nanosize B particles.[82,86]

(c) Thermal decomposition of metal borohydrides

Several attempts have been undertaken to synthesize pure elemental B by decomposition of metal borohydrides, since their crystal structure comprises "$[BH_4]^-$ anions and consequently "BH_3," units can be pyrolyzed to yield B. Until very recently, the method was considered as not very effective, like the above mentioned decomposition reactions,[10] mainly because most of these materials exhibit good to excellent hydrogen storage properties. These materials can thus act as "hydrogen sources"[87,88] for H batteries and fuel cells but only very few are suitable to generate borane[89–91] and consequently B by thermolysis.

The pyrolysis path of borohydrides is quite different, depending predominantly on the nature of the metal present in the compound. For instance, while the ultimate decomposition for the monovalent alkali metal hydrides such as $NaBH_4$ yields H_2, amorphous B and

NaH,[92,93] the products obtained from borohydrides with bi-, tri- or tetravalent cations are — besides hydrogen gas — MB_2 (e.g., M = Mg)[94,95] and MB_2 and diborane) (e.g., M = Ti, Zr), respectively.[96,97] Due to formation of the reactive hydrides (i.e., LiH, NaH, etc.), the pyrolysis of alkali metals is not suitable for the B synthesis. The borohydrides of tri- and specifically tetravalent cations are unstable, very hygroscopic liquids which are available in small amounts only. Bivalent borohydrides $M(BH_4)_2$ are the most promising candidates, due to their higher thermal stability and easier handling, as was reported in Refs. 89–91.

4. Nanoscaled Boron Powders

To date, B particles of 10–150 nm size have been successfully produced by various techniques, e.g., arc decomposition of diborane gas,[83,98] gas phase pyrolysis of decaborane, $B_{10}H_{14}$,[81] reduction of BCl_3 by H_2 in an inductively coupled plasma[62,99] or in thermal plasma,[100] solution reduction of BBr_3[101] and ball milling of oleic acid[102] or dopamine functionalized B powders.[103] The major disadvantage of these methods resides in the fact that they either yield semi-crystalline products or cannot be produced economically and suitably for a large scale production, the latter being necessary for developing industrial applications of the MgB_2 superconductor technology.

4.1 *The "other" approach: diborane "on demand"*

The pyrolysis of diborane allows the most efficient and straightforward access to commercial production of high purity amorphous B. The important disadvantages related to its handling and storage can be overcome by "on demand" generation and "in-situ" conversion of B_2H_6 to amorphous B, as explained below.

The possibility of the controlled production and storage of diborane gas by vacuum decomposition of mechanochemically synthesized $Sn[BH_4]_2$ was first reported in 1999[89] and later in Ref. 91. But a successful commercialization of the process has not yet been achieved.

In 2007, mechanochemically activated solid mixtures of $SnCl_2$, $ZnCl_2$, $CuCl_2$ and $NiCl_2$ with alkaline metal borohydrides were investigated with respect to borane gas evolution characteristics.[104] Hereby, a metathesis reaction takes place where alkali metal hydrides are converted to the bivalent borohydrides.[89–91,105–107] In the following, the borohydrides can be decomposed through gentle heating to diborane according to Refs. 89–91 and 105:

$$MCl_2 + 2M_ABH_4 \rightarrow M(BH_4)_2 + 2M_ACl, \tag{8}$$

where M : Sn, Zn, Cu, Ni ; M_A : Li, Na, K), and

$$M(BH_4)_2 \rightarrow M + B_2H_{6(g)} + H_{2(g)}. \tag{9}$$

The theory behind these reactions is the use of a metal chloride where the metal has a high electronegativity.[87,108,109] As a result of the metathesis (see Eq. (7)), the anions are mutually substituted to form alkali metal chloride and metal borohydride. The tetrahedral complex anion $[BH_4]^-$ is stabilized by one delocalized electron. If the $[BH_4]^-$ group is combined with electrons withdrawing counter ions, i.e., highly "electronegative" metal cations such as Sn^{2+}, Cu^{2+}, etc., the electronic structure of the tetrahydridoborate group $[BH_4]^-$ will be markedly destabilized, decomposing elemental B and hydrogen or "BH_3" and H. The monomer BH_3 molecules dimerize at low temperatures and form B_2H_6. Due to the high Pauling electronegativity of Sn, Zn and Cu (1.96, 1.65 and 1.90, respectively), complex borohydrides of these metals have positive heat of formations,[108,109] they are thus thermodynamically unstable (metastable) and decompose according to Eq. (8) to generate B_2H_6 at relatively low temperatures (between 300 and 373 K).

One can obtain B_2H_6 "on demand" by using one of the solid state reactions described above. When heated above 873 K, B_2H_6 is pyrolyzed to yield solid elemental B and H gas (Eq. (8)).

$$B_2H_{6(g)} \rightarrow 2B_{(s)} + 3H_{2(g)} \quad (T > 873\,K). \tag{9}$$

Under optimized conditions[85] the pyrolysis of "on demand" generated diborane gas produces nanosize elemental boron powder of relatively high purity (>98.5 wt.% B) in entirely amorphous form, similarly to the previously mentioned commercial B-A9999 powder (Fig. 4).

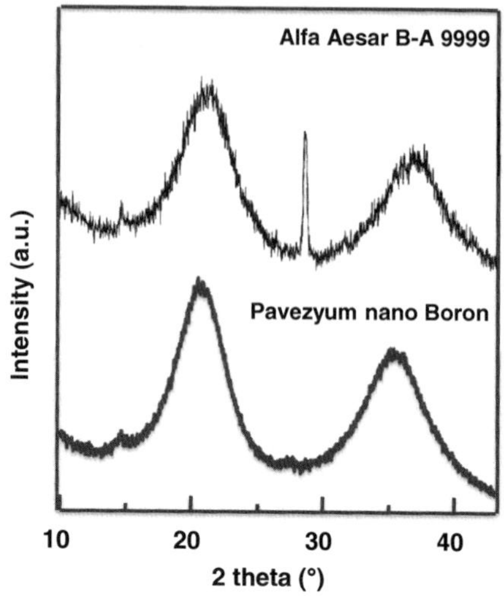

Fig. 4. Comparison of the PXRD of B powders from Alfa Aesar (B-A 999) and Pavezyum (nanosize B).

4.2 *Characterization of nanosize B powders*

4.2.1 *Physical properties, morphology and purity*

Nanosize B powder samples were characterized by means of p-XRD, ICP-OES, EDX, SEM, TEM, TG/DTA/MS, Combustion Calorimetry, DLS and BET with the focus on nanosize B of Pavezyum Kimya. The amorphous B samples of 86, 90 and 95 wt.% purity used for comparison and denoted as B 86, B 90 and B 95, respectively, are commercial powders. The results can be summarized as follows:

> The XRD diagram of nanosize B is very similar to that of A-B9999 (Fig. 4). The electron diffraction picture (Fig. 5) with related TEM micrograph are depicted in Fig. 5, revealing the characteristic halos which confirm that the powder is completely amorphous.

The B content of samples was determined by both acid–base titration (mannitol method) and ICP–OES analyses. The average value found for B was 98.5 wt.%. Other impurities are O_2, N, C and H. The amount of metallic contaminations such as Na, Si and Fe was in the ppm region.

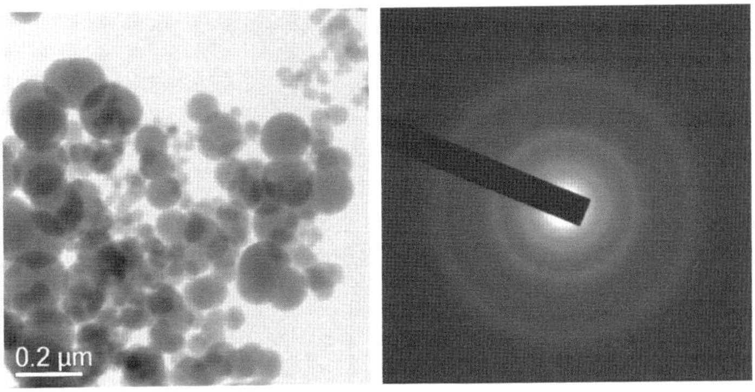

(a) TEM micrograph (b) Electron Diffractogram

Fig. 5. Nanosize B powder: (a) TEM micrograph, (b) Electron Diffractogram.

The powder particle size was determined by means of DLS and SEM/TEM measurements. The result of the DLS recorded from samples pyrolized at 973 K reveals for the average size distribution (d_{50}) a value of 190 nm. Rising the decomposition temperature enhances the size of the nanosize B particles. This phenomenon is well-known[110,111] and is based on the fact that the powder particles are better sintered at high pyrolysis temperatures nucleating to larger spherical cores, and *vice versa*. Another important indicator in context with the particle size is the polydispersity index value, PDI. PDI values close to 0.1 indicate a narrow particle size distribution, indicating that the majority of particles have sizes close to the average value and possess identical shapes. Alternatively, TEM and SEM measurements were performed to observe the distribution of the particle size and their shape (see Fig. 6).

TEM images reveal that amorphous B particles have spherical shapes (aspect ratio >0.9) their size being ≤ 300 nm. For further clarification complementary SEM imaging was performed. Figure 6 shows the SEM images of the nanosize B powder samples.

Although it is difficult to depict an average size value from the SEM images, it is clear that the powder particles are in the range of 50–300 nm, in good agreement with the results from TEM and DLS measurements. The surface area (BET) of nanosize B powder particles

Fig. 6. SEM image of nanosize B particles.

Table 4: Combustion temperatures, specific surface areas (BET) and heats of combustion of different B powders.

Sample	Nanosize B	B 95	B 90	B 86
BET (m^2/g)	35	10	18	5
T_{comb} (°C)	696	815	796	875
ΔH^*_{comb}(kJ/g)	1847*	1526*	1528*	1752*

*ΔH^*_{comb} (kJ/g) obtained from powder mixtures of B:KNO_3 = mass ratio 1:3.

was determined to as 35 m^2/g, which is almost twice the surface of any other commercially available amorphous B powder. Table 4 gives the comparative particle size and surface area values of nanoamorphous B and commercial amorphous B.

4.2.2 *Thermal properties and reactivity*

Nanosize particles coupled with a high specific surface are important premises for high chemical reactivity. To verify this DTA/TG and calorimetric measurements were performed on the four B samples

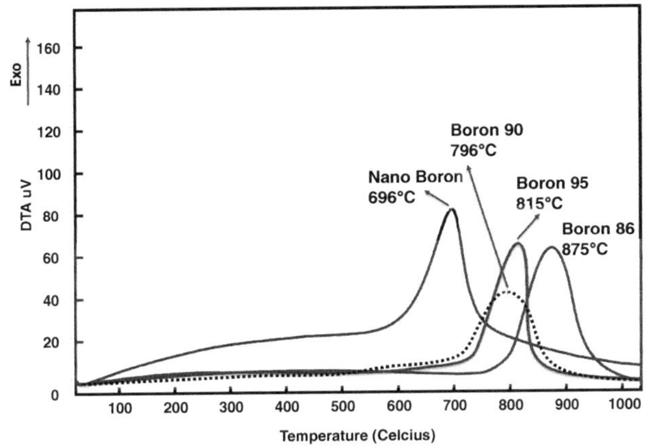

Fig. 7. DTA curves for the combustion of amorphous B nanosize powder samples of various purity grades under air flow.

depicted in Table 4. For thermal analysis (DTA), the specimen were heated (combusted) under air flow from 300–1273 K (see Fig. 7).

Based on the peak maxima, the DTA curves show that the ignition temperature for nanosize B is 100–180°C lower than that of the other powders. The heat of combustion values (kJ/g) of the different B powders were determined from combustion of B–K nitrate (KNO_3) mixtures (mass ratio 1 : 3) — a blend industrially used for airbag ignitors (BPN) — in bomb calorimeter under Ar.[112] Due to very high heat of combustion of B, the results of the calorimetric experiments of pure B samples and specifically that of nanosize B under O_2 atmosphere show strong deviations and, hence, were not accurate enough for quantitative analysis. To use blends where KNO_3 acts as internal oxidizer has the advantage of having a complete, controlled combustion under "diluted" (mass ratio 1:3) conditions. The results are summarized in Table 4.

A comparative evaluation of the data shows that the heat of combustion values for nanosize B–KNO_3 (1847 kcal/g) blend exceeds those of B 95 (1526 kcal/g) and B 90 (1528 kcal/g) by almost 18%, as was expected from the morphology and high specific area data. The high heat of combustion for B 86 blend (1752 kcal/g) is directly correlated to the fairly high Mg content of the B sample (10–12%)

resulting in partial formation of thermodynamically stable Mg borates as combustion products. In case of B-rich B 90, B 95 and nano B powders the combustion yields almost entirely KBO_2.[92]

An interesting detail in this context, is the comparatively high heat of combustion value and the low ignition temperature of B 90 indicating that the relevant property for describing the reactivity of the related B powders is given by the surface morphology and the specific surface area rather than by the purity grade.

A further reactivity test was furnished by analyzing the effect of the four different B powders on the formation temperature of MgB_2. The reaction of the stoichiometric mixtures of the powdered reactants (pellets of 1 g, $T_{\text{max}} = 650°\text{C}$, 3 h., steel ampoule, Ar) were monitored by p-XRD (Fig. 8) yielding in all cases MgB_2 as the main product.

For the sake of completeness, it should be mentioned that B particles obtained from plasma synthesis of SMI[62] and pyrolysis of decaborane at 500–700° C[82] reveal significant similarities with those of the relevant nanosize powder with respect to purity, particle size (5–200 nm), shape (quasi-spherical) and combustion behavior.[82] The main deference between the two powders is that plasma synthesized

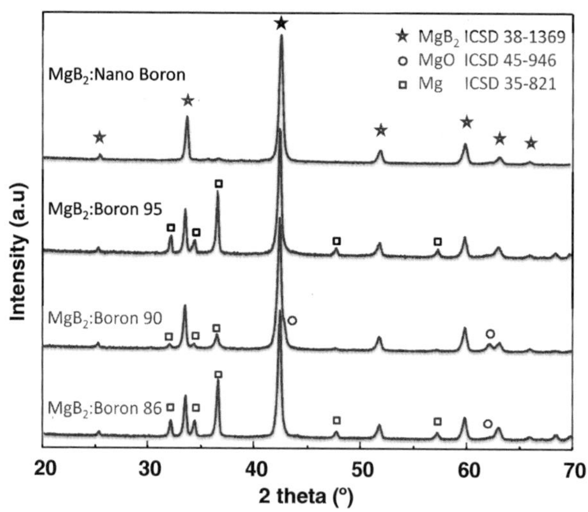

Fig. 8. PXRD diagram of MgB_2 samples prepared at 700°C by using different amorphous B precursor Nano B (top), B 95, B 90, B 86 (bottom).

B is not entirely amorphous but showing large portions of both amorphous and β-rhombohedral crystalline material.[62] The material reported in Ref. 82 is practically identical with the nanosize powder since the "feedstock" of both powders are boranes (B_2H_6 and $B_{10}H_{14}$).

A further criterion for reactivity is the amount of unreacted Mg. As shown in Fig. 8, under the given conditions, only nanosize B fully reacts, followed by amorphous B of purity of 90, 95 and 86%. The corresponding Jc vs. B values in wires are shown in Fig. 9. An important purity indicator for MgB_2 is in general the content of MgO, present in all MgB_2 samples, mainly due to the intrinsic O_2 (at the particle surface or in bulk B_2O_3) of the used B sources. Based on the XRD patterns, nanosize B based MgB_2 shows no detectable MgO, in contrast to the other powders (B 90 exhibited the highest level of MgO impurity). It is known as the most oxygen rich material among all commercially available B powders.[93]

As was reported for the B powder B-A9999 and plasma synthesized material from SMI, the critical current density J_c values measured for nanosize B based specimen are by the critical current density J_c values measured for nanosize B based specimen are by several order of magnitudes higher than those for magnesiothermal produced B powders. This phenomenon has been discussed at various occasions and publications[7,94] but was not fully understood. It is known that in solid materials, the g-factor of an electron spin, obtained from electron paramagnetic spectra (EPR), varies from the free electron g-factor (g = 2.0023). Differences from this value originate from orbital angular momentum induced by the neighboring ions by the spin. For a localized defect (here vacancy), the surrounding atoms have filled orbitals which makes the magnitude of the orbital angular momentum small. These systems have therefore similar or very near g-values to free electrons which mean $g \sim 2.0$. On the other hand, delocalized electrons, weakly bound to surface, can have strong orbital momentum components which result significant deviations of g-factor from the free electron value. A very recent EPR study[95] on MgB_2 powders synthesized from nanosize B and B-95 powder particles shows that using nanosize B as precursor generates reactive nanosize MgB_2 particles in the range of 50–250 nm. By considering the g-factors, EPR spectra

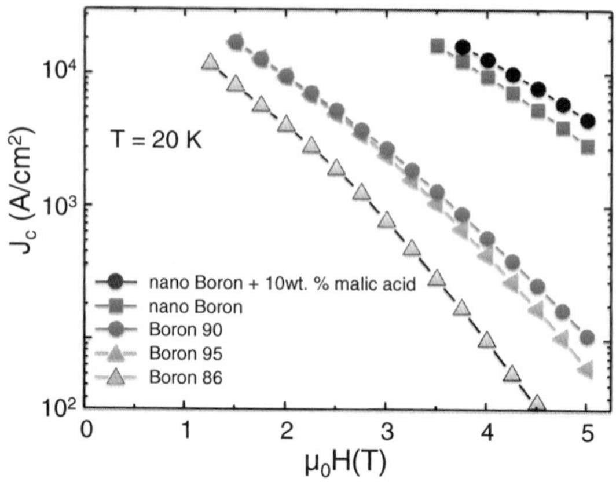

Fig. 9. Transport measurements at 20 K on *in situ* prepared MgB_2 wires using different purity amorphous B precursors performed at ISEM, University of Wollongong.

revealed the presence of defect centers in nanosize MgB_2 which are delocalized electrons due to enhancement of surface defects and defect centers on the bulk MgB_2 (based on B-95) which are localized arising from the core of MgB_2. From the point of view of defect structure, the EPR signal of bulk MgB_2 is related to oxygen vacancies, whereas one of the samples based on nanosize B is compatible with the vacancy of Mg atoms, indicating an enhancement of non-metallic conductivity due to localization of the unpaired electrons. This observation may offer a new insight for giving additional functionalities to MgB_2 at the nanoscale for industrial application in relevant superconducting technology area.

5. Nano B Production at Semi-Commercial Scale

The characterization experiments on nanosize B have confirmed that this material has unique properties such as complete amorphousness, higher purity, homogeneous nanometric particle size distribution, high surface area and high chemical reactivity compared to the other commercial amorphous B powders. These features have been once attributed to the legendary B-A9999 of Alfa Aesar, which is not produced anymore. The comparison of the PXRD diagrams of the two powders in Fig. 4 shows that nanosize B exhibits a higher purity with respect to the B_2O_3 content, too.

In 2012, Pavezyum Kimya (Turkey) constructed a pilot plant for industrial production of nanosize B based on the "on demand diborane" method developed at Koç University, Inorganic Lab.[85] During the development of the production facility, several technical problems regarding safe handling of diborane and process parameters (temperature, flow rates) were engineered in view of a continuous nanosize B powder production of 1 kg/day. It should be mentioned that today amorphous nanosize B powder is commercially also available from SMI (U.S.). As repeatedly mentioned, the production process is somewhat different, involving plasma pyrolysis of BCl_3 in present of hydrogen gas.[62,99,117] According to Ref. 117 this powder contains large portions of crystalline material.

6. Conclusion

In this chapter the different synthesis routes of elemental B and the resulting products were discussed in context with their purity, crystallinity, amorphousness, morphology, particle size and reactivity. The most widely used commercial production is the magnesiothermal process (Moissan method) yielding semi-amorphous powders of maximum 97% purity with a particle size range of 0.8–1.2 μm. Higher purity (>99% wt.) and smaller particles size can be reached either by pyrolysis of B halide–hydrogen mixtures or volatile boranes in arc or on substrate at elevated temperatures. The limiting reaction temperature of B processing is roughly T = 1000°C, below which the B powders are entirely amorphous. Due to their higher thermal stability, the thermolysis of B halides is carried out in a range of 1100–1500°C, where the formation of semi-crystalline or crystalline B is favored. The U.S. manufacturer SMI Inc. uses this technique in an inductively coupled plasma system yielding nanosized B particles in the range of 5–200 nm. However, the powder is not fully amorphous containing some small crystalline fractions. Compared with this, boranes can be pyrolized at more moderate conditions and temperatures (600–800°C), resulting in highly pure and entirely amorphous B powders. The most common borane precursor is diborane gas which is commercially available in small quantities. Beside the high price, handling, toxicity and storage related problems have been severe hurdles for the employment and application of diborane on large scale. These handicaps can be overcome by "on demand" production and "*in situ*" pyrolysis of diborane, a technique which was developed at Koc University (Istanbul) in 2010 and later commercialized by Pavezyum Kimya in 2012. The material obtained from thermolysis of "on demand" generated diborane is a loose dark brown powder whose PXRD shows a perfect match with that of the "legendary" B-A9999 of Alfa Aesar. Further investigations by means of chemical analysis, DTA–TG, DLS, SEM, TEM and bomb calorimeter confirmed that the nanosize material had superior features compared to other commercially available B powders, such as high purity (>99.9%wt. B) fully amorphousness, nanometric particle size range (50–300 nm), higher chemical reactivity and lower reaction and combustion temperatures.

The transport measurements conducted on *in situ* prepared MgB_2 wires from B powders of different origins and purity show enhanced J_c values in presence of magnetic fields for wires synthesized from nanosize B. The results are directly comparable with those previously reported for *B*-A9999- and plasma synthesized B-based (SMI Inc.) MgB_2. The outstanding J_c performance is very likely related to nanometric particle size of MgB_2 obtained from the synthesis of nanosized B.

References

1. J. Nagamatsu, N. Nakagawa, T. Muranaka, Y. Zenitani and J. Akimitsu, *Nature* **410**, pp. 63–64 (2001).
2. J. R. Lamarsh, Introduction to Nuclear Engineering,*Addison-Wesley Series in Nuclear Science and Engineering*, 2nd Edition, 1983.
3. A. W. Laubengayer, D. T. Hurd, A. E. Newkirk and J. L. Hoard, *J. Am. Chem. Soc.* **65**, pp. 1924–1931 (1943).
4. M. S. Sethi and P. S. Raghavan, *Concepts and problems in Inorganic Chemistry*, Discovery Publishing House, New-Delhi, 1998.
5. L. I. Berger, *Semiconductor Materials*, CRC Press, 1996, pp. 37–43.

6. J. T. Tobin, *Superconductivity Research Development*, Nova Science Publishers, Inc., 2008.
7. S. K. Chen, K. A. Yates, M. G. Blamire and J. L. MacManus-Driscoll, *Supercond. Sci. Tech.* **18**, pp. 1473–1477 (2005).
8. S. K. Chen, K. P. Lim, K. B. Tan and S. A. Halim, *Solid State Science and Technology* **17**, pp. 173–181 (2009).
9. J. Jiang, B. J. Senkowicz, D. C. Larbalestier and E. E. Hellstrom, *Supercond. Sci. Technol.* **19**, pp. L33–L36 (2006).
10. N. R. Neelameggham, *J. Manufacturing Science and Production* **12**, pp. 155–160 (2012).
11. C. Helvaci, A Mineralogical approaching to the mining, storing and the marketing problems, Geological Engineering Department of Dokuz Eylul University, Izmir, 2006, pp. 5–17.
12. W. G. Woods, *An Introduction to Boron: History, Sources, Uses, and Chemistry*, Office of Environmental Health and Safety, University of California: Riverside, California (USA), 2000.
13. R. I. Report, *Boron: global industry markets & outlook*, 2010, Roskill: London.
14. Y. Kar, N. Şen and A. Demirbaş, *Minerals & Energy — Raw Materials Report* **20**, pp. 2–10 (2006).
15. N. Acarkan, "Boron Products and Their Uses" in *Proc. First Intl. Boron Symposium,* ed.K. Erarslan, Dunlupinar University Press, 2002, pp. 1–5.
16. Global reserves chart, retrieved August 14, 2010.
17. "An Industry Report on Airbag Industry", Outsource, India, 2010.
18. J. W. Mellor, *A comprehensive Treatise on Inorganic and Theoretical Chemistry*, Vol. V, 1967 edition.
19. H. Davy, *Philosophical Transactions of the Royal Society of London* **99**, pp. 33–104 (1809).
20. J. L. Gay-Lussac and L. J. Thénard, 1808, *Ann. Chim.* **68**, pp. 169–174 (1808).
21. R. D. Thompson, *Phil. Mag.* **2**, p. 410 (1831).
22. F. Wohler and H.St.C. Deville, *Compt. Rend.* **43**, p. 1088 (1856).
23. N. P. Nies, *J. Electrochem. Soc.* **107**, pp. 817–820 (1960).
24. H. C. Starck, GmbH, *Boron — from eye drops to rocket propellants*, 2001.
25. B. Albert and H. Hillebrecht, *Angew. Chem. Int. Ed.* **48**, pp. 8640–8668 (2009).
26. H. Moissan, *Ann. Chim. Phys.* **6**, pp. 296–304 (1895).
27. E. Amberger, "Elemental Boron" *Gmelin Handbook of Inorganic Chemistry, Boron, Volume Suppl. 2*, Berlin: Springer-Verlag, 1981, pp. 1–112.

28. T. Onak, *Gmelin Handbook of Inorganic Chemistry, Boron Compounds* ed. K.-C. Bushbeck. Vol. 3–1, Berlin: Springer-Verlag, 1987.
29. L. H. Jansen, *Boron, Elemental in Kirk-Othmer Encyclopedia of Chemical Technology* 2001, pp. 132–138.
30. K. Beeker, *Gmelin Handbuch der Anorganishen Chemie, Bor-Halogen-Verbindungen*, Ed. K. Niedenzu, Berlin: Springer-Verlag, 1978.
31. L. Barton, *Gmelin Handbook of Inorganic and Organometallic Chemistry, Boron Compounds*, Ed. J. Faust. Vol. 4-1a, Berlin: Springer-Verlag, 1994.
32. A. R. Oganov and V. L. Solozhenko *J. Superhard Mater* **31**, pp. 285–291 (2009).
33. T. Ogitsu, F. Gygi, J. Reed, Y. Motome, E. Schwegler and G. Galli, *J. Am. Chem. Soc.* **131**, pp. 1903–1909 (2009).
34. R. H. Wentorf Jr.: *Refractory Boron Compounds, Kirk-Othmer Encyclopedia of Chemical Technology*, John Wiley & Sons, New York, 2000.
35. A. R. Oganov, J. Chen, C. Gatti, Y.-Z. Ma, Y.-M. Ma, C. W. Glass, Z. Liu, T. Yu, O. O. Kurakevych and V. L. Solozhenko, *Nature* **457**, pp. 863–867 (2009).
36. A. E. Newkirk, 'Elemental Boron' in "*Boron, Metallo-boron Compounds and Boranes*", Interscience Publishers, 1964.
37. Z. N. P. Nies, U.S. Borax and Chemical Corp., U.S. patent 2 909 471, 1959.
38. D. Z. Hobbs, T. T. Campbell and F. E. Block, U.S. Bur. Mines Rept. Invest. No. 5893, 1961.
39. H. S. Cooper, W. M. Weil, U.S. 2 984 605, 1961.
40. J. J. Rameau and M.-J. Rameau, *Rev. Intern. Hautes Temp. Refract.* **8**, pp. 59–70 (1971).
41. A. Uchiyama, D. R. Stern and Q. H. McKenna, American Potash & Chemical Corp. U.S. 2 940 911, 1960.
42. M. Makyta, K. Matiasovsky and P. Fell, *Electrochim. Acta.* **29**, pp. 1653–1657 (1984).
43. L. P. Polyakova, G. A. Bukatova and E. G. Polyakov, *J. Electrochem. Soc.* **143**, pp. 3178–3186 (1996).
44. P. Taylor and J. Gomez, U.S. 20 100294 670, 2010.
45. W. Dietze, M. Miller and E. Amberger, *Electron Technol.* **3**, pp. 73–79 (1970).
46. Ö. Balcı, D. Ağaoğulları, I. Duman and M. L. Öveçoğlu, *Powder Technology* **225**, pp. 136–142 (2012).
47. R. W. Baier and J. C. Schuhmacher, U.S. 3 001 855 A, 1961.
48. A. F. Zhigach and D. Stasinevich, "Methods of Preparation for Amorphous Boron" in "*Boron and Refractory Borides*" ed. V. I. Markovich, Springer-Verlag, Berlin-Heidelberg, pp. 214–226 (1977).

49. La. Ya. Markovski and M. P. Belov, *J. Appl. Chem.* [USSR] **37** pp. 1650–1656 (1964).
50. F. Karau, *Industrial Synthesis of Boron and Borides*, International Symposium on Boron, Borides and Related Materials, Istanbul, 2012.
51. La. Ya. Markovski and M. P. Belov, *J. Appl. Chem.* [USSR] **40**, pp. 706–712 (1967).
52. J. Yannacakis and N. P. Nies, *Boron. Syn. Struct. Properties. Proc. Conf.* Asbury Park, N. J., 1959, pp. 38–41.
53. Z. Dou, T. Zhang, G. Shi, C. Peng, M. Wen and J. He, *Nonferrous Met. Soc. China* **24**, pp. 1446–1451 (2014).
54. M. Picon and P. Pichat, *J. Cuelleron, Comp. Rend. C* **258**, pp. 5877–5880 (1964).
55. B. Kamenar, *Z. Anorg. Allg. Chem.* **342**, pp. 108–110, (1966).
56. P. Pichat, *J. Cuelleron, Comp. Rend. C* **264**, pp. 1121–1123 (1967).
57. L. B. Thomas Onak, *Gmelin Handbook of Inorganic Chemistry, Boron Compounds* Volume 3-1. 8 ed, Ed. K.-C. Bushbeck. Vol. 3-1, Berlin: Springer-Verlag, 1987.
58. T. O. Lawrence Barton, *Gmelin Handbook of Inorganic and Organometallic Chemistry, Boron Compounds* Volume 4-1a. 8, Ed. J. Faust. Vol. 4-1a, Berlin: Springer-Verlag, 1994.
59. A. E. Hultquist and M. E. Siber, "*Glow Discharge Deposition of Boron, Chemical Reactions in Electrical Discharges*", 1st ed., *Am. Chem. Soc.*, Educational Library, Washington, DC, 1969, pp. 182–97.
60. D. Agaogullari, Ö. Balci, I. Duman and M. L. Övecoglu, *Met. Mat. Trans. B* **42B**, pp. 568–574 (2011).
61. E. Amberger, "Elemental Boron" *Gmelin Handbook of Inorganic Chemistry, Boron, Volume Suppl. 2*, Berlin: Springer-Verlag, 1981, pp. 2–14.
62. J. V. Marzik, R. C. Lewis, M. E. Tillman, Y. Q. Wu, D. K. Finnemore, M. Rindfleisch, M. Tomsic, J. Yue and W. J. Croft, *Mater. Res. Soc. Symp. Proc.* **1148**, pp. 12–01, Materials Research Society, Warrendale, Pennsylvania, 2009.
63. A. E. van Arkel and J. H. de Boer, *Z. Anorg. Allg. Chem.* **148**, pp. 345–350 (1925).
64. W. Dietze and E. Amberger, *Angew. Chem.* **79**, p. 278 (1967).
65. J. Cueilleron and J. C. Viala, *J. Less Common Met.* **65**, pp. 167–173 (1979).
66. E. Amberger, W. Dietze, M. Druminski and K. Ploog, *Electron Technol.* **3**, pp. 133–141 (1970).
67. J.-C. Viala, *Thesis* Univ. Claude Bernard Lyon, 1977.
68. Kirk-Othmer, Encyclopedia of Chemical Technology, Third Ed. Vol IV, p. 65.

69. E. Wiberg and E. Amberger, *Hydrides of Elements of main Groups I–IV*, Elsevier, Amsterdam, 1971.
70. S. H. Bauer, *Chem. Rev.* **31**, pp. 43–75 (1942).
71. A. J. Owen, *J. Appl. Chem.* **12**, pp. 483–493 (1960).
72. A. Stock, *Hydrides of Boron and Silicon*, Cornell University Press, Ithaca, New York, 1933.
73. P. Laszlo, *A Diborane Story, Angew. Chem. Int. Ed.* **39**, pp. 2071–2072 (2000).
74. H. C. Brown and B. C. Subba Rao, *J. Am. Chem. Soc.* **81**, pp. 6423–6437 (1959).
75. N. N. Greenwood and R. Greatrex, *Pure & Appl. Chem.* **59**, pp. 857–868 (1987).
76. C. E. Housecroft and A. G. Sharpe, *Inorganic Chemistry*, 1st Ed., Pearson Education, 2001, p. 273.
77. A. F. Holleman, E. Wiberg and N. Wiberg, *Lehrbuch der Anorganischen Chemie*, 101 Auflage, Berlin-New York, de Gruyter, 1995, p. 1002.
78. P. Jash and M. Trenary, *16th International Symposium on Boron, Borides and Related Materials. IOP Publishing. J. Physics: Conference Series* **176**, p. 012011 (2008).
79. J. W. Gold, C. Militscher and D. D. Slausen, Pentaborane Disposal: taming the dragon, *Integrated Environmetal Services*, 2009.
80. U.S. Army Environmental Center, *Environmental Monitor* **5** pp. 1–6 (2000).
81. Acute Exposure Guideline Levels for Pentaborane, *Interim* 2008, pp. 3–49.
82. B. J. Bellott, W. Noh, R. G. Nuzzo and G. S. Girolami, *Chem. Commun.* **22**, pp. 3214–3215 (2009).
83. Y. Kumashiro, *Electric Refractory Materials*, Markel Dekker Inc., New York Basel, 2005, p. 657.
84. Y.-S. Sohn, S.-J. Joo, U.S. 6 753 230 B2, 2004.
85. R. A. Rosenberg, F. K. Perkins, D. C. Mancini, G. R. Harp and B. P. Tonner, *Appl. Phys. Lett.* **58**, pp. 607–609 (1991).
86. A. H. Robert and J. Ibrahim, U.S. 5 013 604, 1991.
87. Y. Nakamori, H. Li, K. Miwa, S. Towata and S. Orimo, *Materials Transactions* **47**, p. 18981901 (2006).
88. H.-W. Li, Y. Yan, S. Orimo, A. Züttel and C. M. Jensen, *Energies* **4**, pp. 185–214 (2011).
89. V. V. Volkov and K. G. Myakishev, *Inorg. Chem. Acta.* **289**, pp. 51–57 (1999).
90. E. Jeon and Y. W. Cho, *J. Alloys Comp.* **422** pp. 273–275 (2006).
91. K. G. Myakishev and V. V. Volkov, *Chemistry for Sustainable Development* **14**, pp. 375–378 (2006).

92. J. Urgnani, F. J. Torres, M. Palumbo and M. Baricco, *Int. J. Hydrogen Energy* **33**, pp. 3111–3115 (2008).
93. M. Somer, S. Acar, C. Koz, I. Kokal, P. Hoehn, R. Cardoso-Gil, U. Aydemir, L. Akselrud, *J. Alloys Comp.* **491**, pp. 98–105 (2010).
94. T. Matsunaga, F. Buchter, K. Miwa, S. Towata, S. Orimo and A. Züttel, *Renewable Energy* **33**, pp. 193–196 (2008).
95. I. Saldan, I. Llamas-Jansa, S. Hino, C. Frommen and B. C. Hauback, *Mater. Sci. Eng.* **77**, p. 012041 (2015).
96. G. L. Soloveichik, *Material Matters* **2**, pp. 11–14 (2007).
97. F. C. Gennari, L. F. Albanesi and I. J. Rios, *Inorg. Chim. Acta.* **362**, pp. 3731–3737 (2009).
98. P. Z. Si, M. Zhang, C. Y. You, D. Y. Geng, J. H. Du, X. G. Zhao, X. L. Ma and Z. D. Zhang, *J. Mater. Sci.* **38**, pp. 689–692 (2003).
99. J. V. Marzik, R. J. Suplinskas, R. H. T. Wilke, P. C. Canfield, D. K. Finnemore, M. Rindfleisch, J. Margolies and S. T. Hannahs, *Physica C* **423**, pp. 83–88 (2005).
100. W. G. Shin, S. Calder, O. Ugurlu and S. L. Girschick, *J. Nanopart. Res.* **13**, pp. 7187–7191 (2011).
101. A. L. Pickering, C. Mitterbauer, N. D. Browning, S. M. Kauzlarich and P. P. Power, *Chem. Commun.* **6**, pp. 580–582 (2007).
102. B. van Devener, J. P. L. Perez, J. Jankovich and S. L. Anderson, *Energ. Fuel* **23**, pp. 6111–6120 (2009).
103. G. Zhe, N. I. Walton, A. Malugin, H. Ghandehari and I. Zharov, *J. Mater. Chem.* **22**, pp. 877–882 (2012).
104. V. Vladimir and K. G. M. Volkov, *Chemistry for Sustainable Development* **10**, pp. 221–233 (2002).
105. S. Acar, Ph.D. *Thesis*, Koç Universitesi Istanbul, 2012.
106. H. Hagemann and R. Cerny, *Dalton Trans.* **39**, pp. 6006–6012 (2010).
107. J. Huot, D. B. Ravnsbæk, J. Zhang, F. Cuevas, M. Latroche and T. R. Jensen, *Prog. Mater Sci.* **58**, pp. 30–75 (2013).
108. K. Nakamori, A. Miwa, H. Ninomiya, N. Li, S. Ohba, A. Towata and S. Züttel, Orimo, *Phys. Rev. B* **74**, p. 045126 (2006).
109. P. Błoński and Z. Łodziana, *Phys. Rev. B* **90**, p. 054114 (2014).
110. M. T. Swihart, *Curr. Opin. Colloid Interface Sci.* **8** pp. 127–133 (2003).
111. F. E. Kruis, H. Fissan and A. Peled, *J. Aerosol Sci.* **29**, pp. 511–535 (1998).
112. Bomb calorimeter measurements were performed in Laboratories of Roketsan A.Ş., Ankara-Turkey.
113. MIL-P-46994B, Military Specification: Pellets/Granules Boron/ Potassium Nitrate, 1982.

114. Certificates of Analysis for elemental amorphous boron powders of the commercial Producers, H. C. Starck, Tronox, S. B. Boron and Pavezyum Kimya.
115. Y. Zhang, C. Lu, S. Zhou and J. Joo, *J. Nanosci. Nanotech.* **9**, pp. 7402–7406 (2009).
116. A. Bateni, S. Repp, R. Thomann, S. Acar, E. Erdem and M. Somer, *Appl. Phys. Lett.* **105**, p. 202605 (2014).
117. J. V. Marzik, R. C. Lewis, M. R. Nickles, D. K. Finnemore, J. Yue, M. Tomsic, M. Rindfleisch and M. D. Sumption, *AIP Conf. Proc.* **1219**, pp. 295–301 (2010).

Chapter 4b

MgB_2 Wires Fabricated Using the *Ex Situ* Technique

Andrea Malagoli* and Valeria Braccini

C.N.R.–SPIN
Corso Perrone 24, 16152 Genova, Italy
**andrea.malagoli@spin.cnr.it*

In this chapter the *ex situ* powder-in-tube (PIT) technique used for the fabrication of MgB_2 conductors will be presented. A thorough analysis of the various procedures developed at CNR–SPIN will be given, from the synthesis of the precursor powders to the analysis of the cold working processes and their influence on the conductor properties. We will describe how the pinning properties of the wires were enhanced, either through grain boundary (GB) oxidation or by reduction of the grain size via high energy ball milling, following a procedure for the home-made synthesis of nanosize Boron (B). All these steps were performed with or without the presence of dopants. The conductor development and its optimization is described including the fabrication of the first multifilamentary tapes on a km scale, and of multifilamentary wires with square or round shape. The description includes the fabrication of wires with a twist pitch and with very fine filaments. Finally, a brief overview of industrial manufacturing and of present commercial wires is given.

1. Introduction

Various processing routes have been introduced in view of the successful fabrication of polycrystalline MgB_2 conductors, their transport critical current density J_c being not substantially limited by the grain boundary (GB) problem.[1–3] Amongst these routes, the powder-in-tube (PIT) method showed the largest potential in terms of performance as well as of scalability, reproducibility and flexibility: it takes advantage of the low material costs and of the relative simplicity of the employed deformation techniques.

Since the beginning, researchers have tried both the *in situ* and the *ex situ* approaches, driven by different aims and getting different results. Most groups focused on the *in situ* technique, its peculiarities being well described in Chapters 4c, 4e and 4f of this book: this approach has been industrialized by HyperTech at Columbus, Ohio (U.S.). A few groups performed a substantial development of the *ex situ* technique: we report here about the most significant results achieved on MgB_2 wires, especially at CNR–SPIN, leading to a technology transfer towards Columbus Superconductors in Genova (Italy) for fabrication and commercialization.

The *in situ* route presents several advantages as low costs, high fill factors, high speed processing, low temperature reaction processes and relatively more straightforward MgB_2 nanoparticle doping. The *ex situ* technique, however, still appears at present to be more suitable for the development of long wires with complex multifilamentary configuration; it allows a better control of the granulometry and purity degree of the powders and also leads to more robust wires that can be readily be employed to realize magnets by the Wind and React process.[4,5]

2. The *Ex Situ* PIT Method

The *ex situ* PIT process can be subdivided into the following steps:

(a) MgB_2 synthesis from precursors,
(b) Various powder treatments (mechanical, thermal, oxidation),
(c) Filling of the metallic tube with MgB_2 powder, followed by tube sealing,
(d) Mechanical deformation to obtain tapes, wires or cables,
(e) Final sintering process of the conductor.

2.1 *The ex situ technique in detail*

MgB_2 powders are prepared from amorphous B (95–97% purity) and Mg (99% purity), which are mixed and submitted to a heat treatment at various synthesis temperatures between 740 and 910°C in controlled atmosphere. After cooling, the powders are extracted and treated either in air or in a controlled atmosphere. A better choice for reducing a possible uncontrolled oxidation consists in using MgH_2 instead of metallic Mg as a precursor.[6–8] Indeed, the formation of H_2 by MgH_2 decomposition produces a reducing atmosphere which hinders O_2 contamination. The processes involved in the MgB_2 synthesis can be summarized by the following main reactions:

$$\mathrm{MgH}_{2(\mathrm{s})} = \mathrm{Mg}_{(\mathrm{s})} + H_{2(\mathrm{g})} \text{ at T} = 450^\circ\mathrm{C},$$

$$\mathrm{Mg}_{(\mathrm{s})} = \mathrm{Mg}_{(\mathrm{l})} \text{ at T} = 650^\circ\mathrm{C},$$

$$\mathrm{Mg}_{(\mathrm{l})} + 2\mathrm{B}_{(\mathrm{s})} = \mathrm{MgB}_{2(\mathrm{s})}.$$

Detailed *in situ* high energy synchrotron X-ray diffraction (HEXRD) investigations[9] on the phase formation was performed; one of the main results is shown in Fig. 1, representing the phase evolution as mass percentage versus reaction time for MgB_2 synthesis at 760°C. Metallic Mg is already present (7%) at the beginning of the reaction, probably due to an incomplete hydrogenation or to a decomposition reaction of MgH_2 during its storage. The formation of MgB_2 takes place at very low temperatures (below 300°C), about 30 min after the beginning of the thermal treatment. This formation

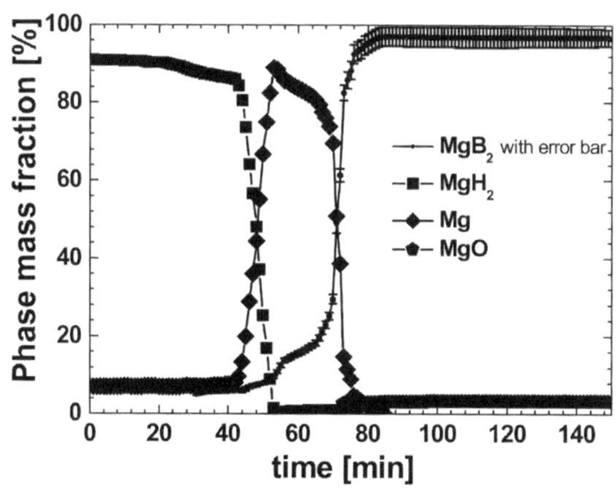

Fig. 1. Evolution of the mass fraction as a function of time during the synthesis of MgB_2.[9]

at exceptionally low temperatures is probably due to a solid–solid reaction between MgH_2 and B: indeed, the MgH_2 amount slightly decreases after 30 min, whereas the amount of metallic Mg remains constant. The solid state reaction between Mg and B was already reported earlier, but was located at about 500°C.[10–13]

The lower reaction temperature may be due to a catalytic effect of H_2, even if the evidence has not been verified yet.

Once the powders are synthesized, the basic steps of the following cold working process toward the wires are shown in Fig. 2. The MgB_2 powders are loaded into a metallic tube which has to be ductile and malleable. The entire loading procedure can be carried out in controlled atmosphere, and both the tube ends are tightly closed with, for example, tin plugs. Such a tube is drawn or groove-rolled in

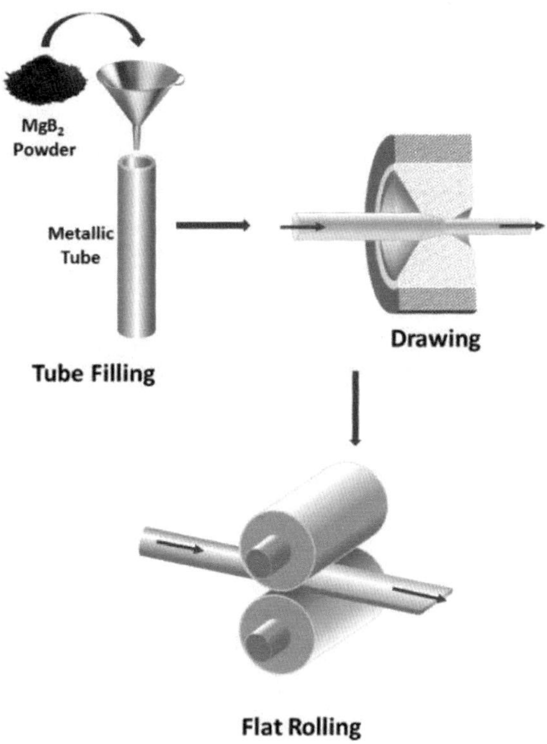

Fig. 2. Schematic of the steps of the cold working procedure.

several steps in order to reduce its cross-section to a wire of the final dimensions. This wire is subsequently flat-rolled in following steps of thickness reduction in order to obtain a flat tape. Several parameters have to be carefully controlled and optimized such as the percentage of wire diameter and tape thickness reduction and the drawing and rolling speed. As a final step a sintering heat treatment is needed, mainly in order to obtain a useful connectivity between the grains. As it will be described in the following, the choice of the heat treatment parameters, such as temperature and duration, strongly depends on the components of the wire i.e., sheath, powder granulometry, etc., and has a crucial impact on the final performances of the conductor.

2.2 *Effects of processing on the wire and its properties*

First attempts to produce practical MgB_2 conductors followed a preparation route without applying any sintering process, to give much less restrictions to the conductor fabrication route and design[14,15]: a typical transverse cross-section of a Ni-sheathed unsintered tape with low filling factor is shown in Fig. 3.

Such unsintered Ni-sheathed tapes showed J_c values of $10^5\,A/cm^2$, comparable with those achieved by more complex high pressure and high temperature sintering of the same material, but still about one order of magnitude lower than the intragrain MgB_2 powder current density. This is because the pressure applied to the superconducting phase during each deformation step of the cold working process is relevant and actually increases with the toughness of the sheath material and its proportion within the tape.

Despite the promising results obtained on unsintered MgB_2 tapes, improved transport properties were subsequently achieved and reported on Fe- and Ni-sheathed tapes subjected to a final thermal

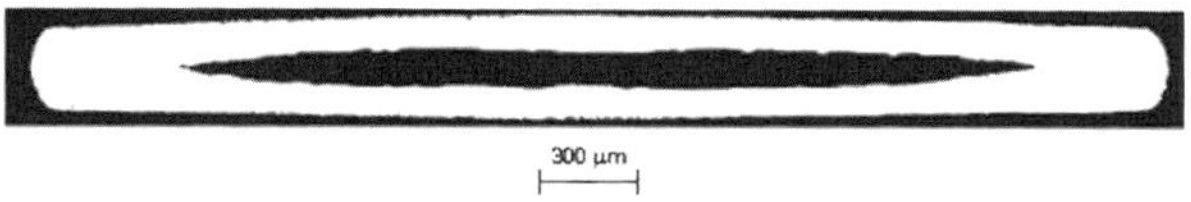

Fig. 3. Transverse cross-section of a Ni sheathed unsintered MgB_2 tape.

treatment.[2,16–18] A thermal treatment after mechanical deformation was found to be useful to recover the high critical temperature T_c value of the MgB_2 phase as well as to partly sinter the grains between them. On the other hand, an increased heat treatment temperature causes both, the growth of too large grains leading to a reduction of the GB pinning and inducing a detrimental reaction between the MgB_2 core and the metallic sheath, particularly in the case of Ni.[19]

In order to understand the effects of mechanical and thermal treatment on the microstructure properties and to establish possible correlations with the superconducting properties, different samples obtained from a tape at different stage of preparation were analyzed by means of neutron diffraction.[20] The experiment described here was performed by using the D1A high resolution diffractometer at ILL (Grenoble) on samples at different stages of deformation and annealed at various temperatures between 400°C and 1000°C. Particular attention was paid to the formation of the reaction layer between the Ni sheath and the MgB_2 core: as already reported in other works,[19] its thickness was estimated to be between 10 μm and 20 μm depending on the temperature (Fig. 4). Moreover, from the neutron diffraction pattern, we identified the ternary phase $Mg_2B_4Ni_5$ to be the main constituent of this layer, its relative amount being calculated by the refined scale factor. The ternary phase

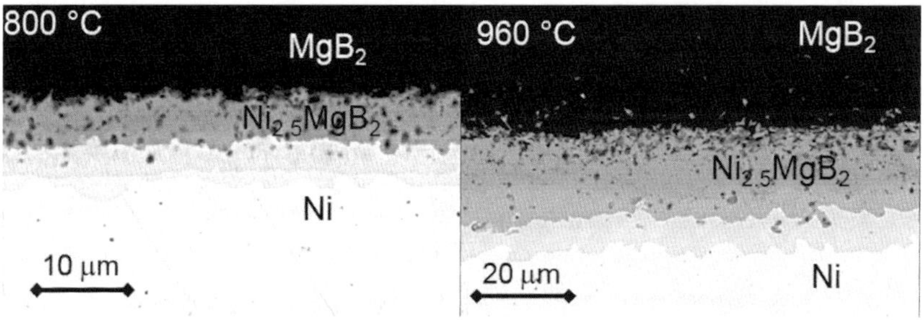

(a) (b)

Fig. 4. Back-scattered electron images of Ni-sheathed MgB_2 tapes annealed at (a) 800°C and (b) 960°C. Note the different scales.

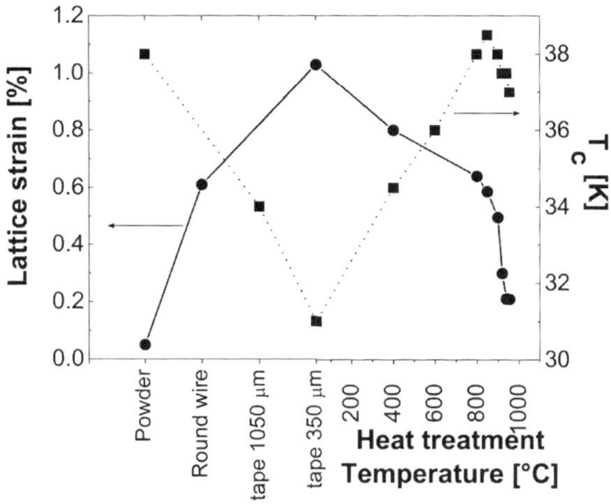

Fig. 5. MgB_2 lattice strain and T_c versus deformation steps (left side of abscissa) and heat treatment temperature (right side of abscissa) (from Ref. 20. © 2003 by IOP).

reaction layer is detected in the samples treated above 800°C, its thickness increasing monotonically with the annealing temperature. For the highest temperatures reached in this work (940–960°C), a progressive diffusion of Ni and a consequent formation of the ternary compound deeply into the superconducting core was observed, with a clearly negative effect on the superconducting properties.

The deformation induces a large residual strain (Fig. 5) up to 1.1% that is partially relaxed during the thermal treatment at low temperatures: an almost complete lattice strain relaxation is reached for thermal treatment at temperatures higher than 900°C only. The opposite behavior of T_c versus the deformation steps and the annealing temperature clearly shows that T_c is strongly affected by the lattice strain and decreases down to 31 K for a residual strain of 1.1% in the rolled tape before any thermal treatment. The strain is relaxed after annealing, T_c is recovered and a value close to 39 K is obtained for the tape annealed at 920°C. Simultaneously with the strain relaxing due to the heat treatment, the grain size is found to increase, as shown in Fig. 6. For higher annealing temperatures a decrease in T_c is again observed, linked to the detrimental effect

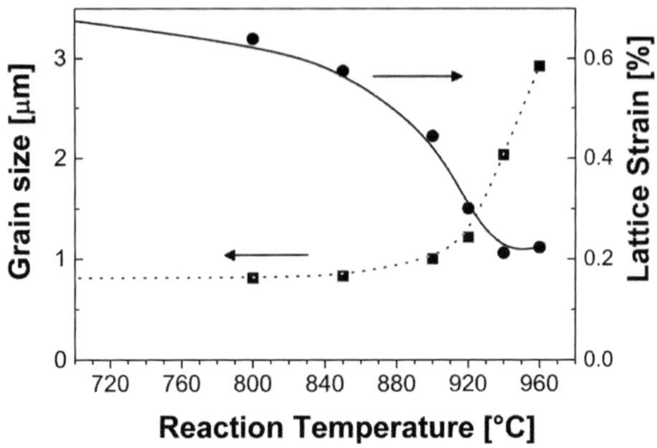

Fig. 6. MgB_2 lattice strain and grain size versus annealing temperature.

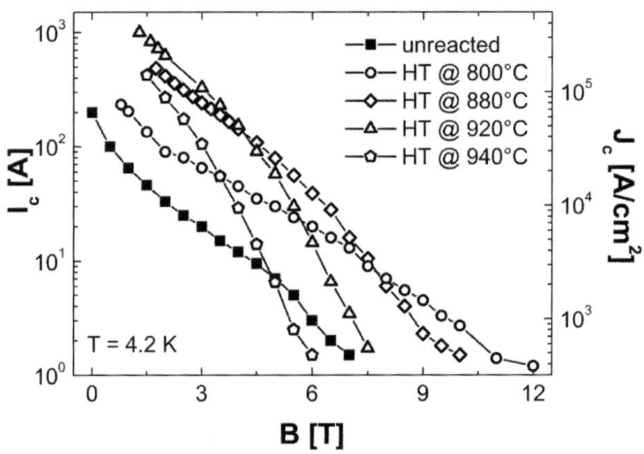

Fig. 7. Critical current of MgB_2 tapes treated at various temperatures as measured by V–I characteristics versus the applied field B, applied perpendicularly to the tape surface (from Ref. 21 © 2003 by IOP).

of the Ni diffusion and the secondary phase appearance in the superconducting core.

In Fig. 7, the values of J_c as a function of the applied field up to 12 T at 4.2 K are shown for different annealing temperatures, the field being applied perpendicular to the tape surface. As can be seen, the heat treatment temperature clearly influences both the values of the critical current density in the low-field range and the slope of the magnetic field dependence. Optimized conductors can carry critical currents I_c larger than 1000 A in a magnetic field of 1 T, which corresponds to a current density of about 4×10^5 A/cm^2.[21] The optimal heat treatment temperature in view of the low-field behavior of J_c for the different wires, is about 920°C, but is clearly lower if the high-field behavior is considered.

Furthermore, it has be noticed that the conductors treated at temperatures below 920°C are still able to carry appreciable critical currents at magnetic fields clearly larger than the expected irreversibility field B^* of the MgB_2 phase, which typically ranges between 6 to 8 T for bulk polycrystalline sample at 4.2 K. In particular, the wire treated at 880°C carries about 40 A at 6 T. In this sample a large lattice strain is still present, indicating a high density of dislocation. On the other hand, these defects, detrimental for the low-field superconducting properties such as J_c and T_c could act as effective pinning centers for flux lines at higher magnetic fields, resulting in larger J_c values. We can therefore state that a direct correlation exists between the MgB_2 lattice strain, induced by cold deformation, and B^* and T_c. In particular, a certain amount of defects seems to have a positive effect on the flux pinning.

3. How to Enhance Pinning Properties

MgB_2 presents reasonably high and potentially appealing upper critical field (B_{c2}) values, as shown on thin films, where extrapolated values of 70 T in the direction parallel to the *ab* planes were reported,[22] either through introducing disorder[23] or after C-doping. One of the main tasks in order to improve the applicative potential is to further raise B^* — which approximately corresponds to $B_{c2}^{\perp}$ because of the strong effect of B^* — over the whole $J_c(B,T)$ curve. The measured J_c represents a complex balance between connectivity, B_{c2} and flux pinning induced by grain boundaries and precipitates.[24] Most of the data reported in the literature in terms of enhanced B_{c2} and J_c have been taken on *in situ* wires, and the highest B_{c2} values and best performances of the current-carrying capability were obtained after doping with C in different forms, i.e., C, SiC, C nanotubes or carbohydrates. However, attempts to increase B^*, J_c and the pinning properties of the tapes have been done through the *ex situ* process as well. In particular, we have analyzed the effect of mastering the powder grain size s by lowering the synthesis temperature and by high energy ball milling with or without additives. A decisive role was found to be played by the presence of an amorphous layer (AL) at the grain boundaries (GB).

The evidence of a thin AL at the surface of MgB_2 crystals was shown by TEM analysis[25] as shown in Fig. 8: regardless of the synthesis conditions, its thickness was quite constant, and consisted of a 5 nm thick amorphous phase with O, Mg and B and a 5 nm thick crystalline MgO layer. The surface of MgB_2 rapidly reacts with O or atmospheric moisture leading to the following phase sequence, from the surface inward:

(a) External layer constituted of MgO nanoparticles;
(b) Intermediate AL composed of Mg, B and O;
(c) Crystalline MgB_2 core.

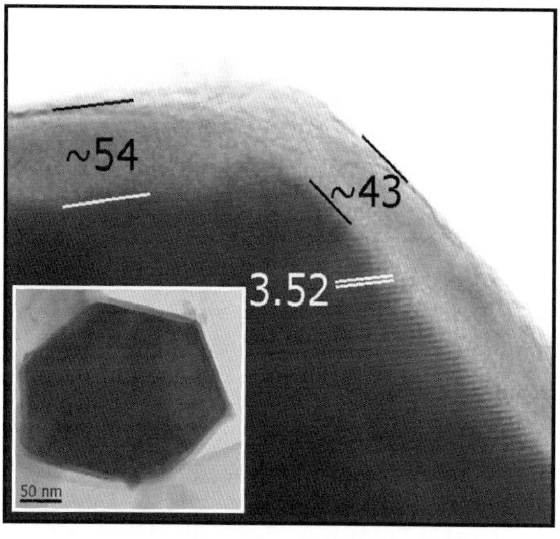

Fig. 8. HR–TEM image of a MgB_2 crystal showing 001 lattice fringes ([100] zone axis) and an amorphous continuous layer at its surface (distances are reported in Å). The inset shows a low magnification image of the same crystal exhibiting a hexagonal habit.

In order to study the effect of the AL as GB on the pinning centers, we tried to lower the grain size, both by reducing the synthesis temperature of the powders and through high energy ball milling. We showed that MgO actually serves as pinning center, while its removal improves the grains connectivity, leading to a doubling of J_c at low fields.

3.1 *Role of grain size and oxidation*

The role of the whole layer (simply defined in the following as MgO layer) in view of the superconducting properties was investigated by comparing the in-field behavior of two series of monofilamentary MgB_2 tapes prepared with powders synthesized at several temperatures in different atmospheres: Inert Atmosphere (IA) — powders extracted and processed in controlled atmosphere under Ar flow — and (OA) Oxidizing Atmosphere — powders extracted at 300°C and processed in air.[26] In Fig. 9 the transport J_c is shown for two IA and OA tapes prepared with powders synthesized at 760°C, representative of several couples of samples synthesized various

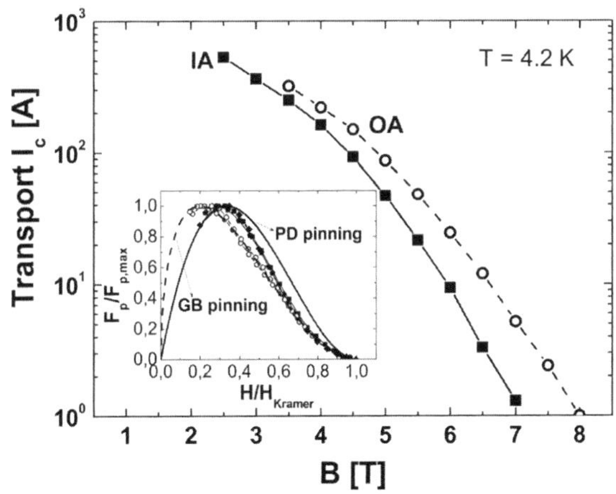

Fig. 9. Transport I_c measurements at 4.2 K performed on tapes processed in Inert IA and OA starting from MgB_2 powders synthesized at 760°C.[26] In the inset, normalized volume pinning force curves versus normalized magnetic field for various OA and IA samples GB and Point Defect pinning models are also plotted as dashed and continuous lines.

temperatures between 700°C and 920°C. While the magnetic field dependence of J_c is more favorable for the OA tape which therefore shows a higher B^*, the IA tape shows a higher J_c exhibits a higher J_c in the low-field region.

The differences between the samples prepared with IA and OA powders can be summarized as follows:

(a) The crystallite size is the same in both samples, while the grains (crystallite agglomerates) are larger in IA case;
(b) T_c and B_{c2} and are similar in the two samples;
(c) B^* increases in OA samples;
(d) GB pinning dominates in OA samples, while point defect pinning is evident in IA samples, as shown in the inset of Fig. 9.

Therefore T_c and B_{c2}, which are related to the crystallite properties, are not affected by oxidation, while pinning force turns out to be related to the larger scale morphology, being improved by the presence of the MgO at the grain surface. In fact, the IA samples not exhibiting a MgO layer have a reduced GB pinning: this allows the manifestation of mechanisms which are ordinarily less effective, such as Point Defect pinning.

3.2 *High energy ball milling of undoped and C-doped powders*

Another way to refine the grain size and thus to enhance the GB density consists in ball milling the starting MgB_2 powders, which was first reported as responsible for enhancing B^* and J_c, especially in high fields.[27] We thoroughly explored the high energy ball milling of the pre-reacted powders, which was conducted in a planetary ball mill with tungsten carbide (WC) jar and media for different times[5,28] and/or milling revolution speed.[25,29] Increasing time and/or speed, an improvement was observed until a maximum value was reached, followed by a decrease for higher milling times and/or speeds.

The variation of J_c versus B at 5 K reported in Fig. 10 was extracted from the M–H loops with the field perpendicular to the tape surface: even if J_c is not much affected at lower fields, a

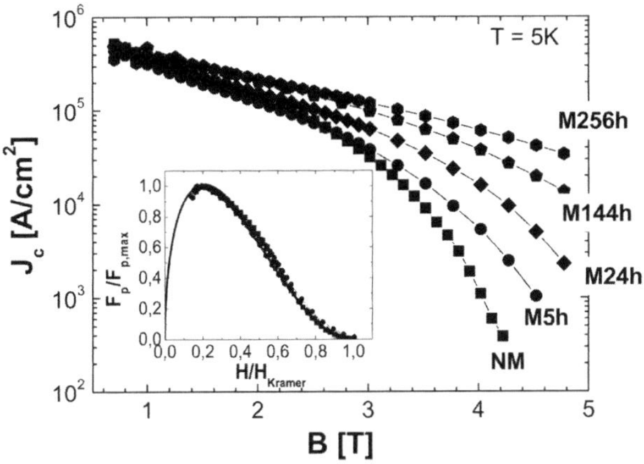

Fig. 10. Magnetic J_c versus B at 5 K in perpendicular orientation for tapes fabricated with MgB_2 powders without milling (NM) and milled (M) during various times.[5] In the inset, normalized pinning force versus normalized magnetic field, compared with the GB pinning model $F_P = b^{0.5}(1-b)^2$ (continuous line).

substantial improvement is obtained at high fields by increasing the milling time up to 144 h, while upon further increasing the milling time the in-field behavior was found to worsen. In the inset we show the pinning force F_p at 5 K normalized by its maximum value as a function of the normalized field $B/B_{\text{Kramer}} = B/B^* = b$ (B^* was determined with a linear extrapolation of the Kramer function $F_k = J_c^{0.5}B^{0.25}$). The F_p behavior is well described by the GB pinning model $F_p = b^{0.5}(1-b)^2$ for all milling conditions and rises from 3.8 GN/m^3 for the original powders to 4.5 GN/m^3 for the tape prepared with powders milled for 144 h, suggesting that the ball milling process lowers the grain size and thus increases the density of grain boundaries, without introducing additional pinning mechanisms on milling: the enhanced J_c in magnetic field and the higher B^* values are therefore probably due to increased perpendicular B_{c2} values.[28]

A deep analysis has been performed with the aim to understand the competing influence of connectivity, B_{c2} and flux pinning after lowering the size of the starting undoped MgB_2 precursor powders[28] by milling. The powder milling process produces a grain refinement; at the same time the calculated connectivity decreases from about

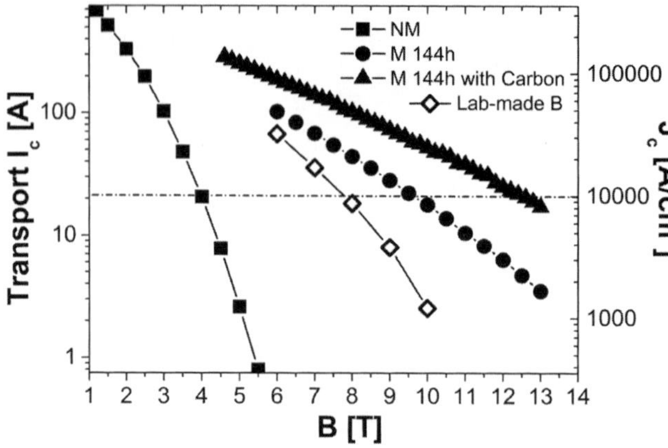

Fig. 11. Transport I_c and corresponding calculated J_c as measured up to 13 T at GHMFL in He bath for the tapes with powders without milling and after 144 h milling, without and with C. J_c is reported also for a wire prepared with laboratory made B (see 3.3).

12–8%[30] and the anisotropy of both J_c and B_{c2} disappears. The increased GB density increases the pinning force at the same time substantially increasing B_{c2} since the crystallite size is comparable to the electronic mean free path.

The milling time was optimized in our system to obtain the best in-field performance, the transport J_c reaching the value of 10^4 A/cm^2 at 4.2 K and 10 T (see Fig. 11), B^* of about 14 T at 4.2 K, and B_{c2} of 9 T at 25 K, all high values for undoped MgB_2. Similar samples were prepared as well through adding small percentages of C nanoparticles (below 1 wt.% to the starting MgB_2 powders) and ball milling them for the same number of times. As shown in the same Fig. 11, the effect of milling in the presence of C was even stronger and led to higher transport I_c values: a I_c value of 20 A, corresponding to $J_c \sim 10^4$ A/cm^2, was reached at 13 T, leading to B^* values as high as 17 T.[5,31] Similar values have been reached in samples prepared following similar approaches: J_c of 10^4 A/cm^2 at 14.3 T and 4.2 K was reported by combining mechanical alloying on powders and C-doping in *in situ* PIT tapes,[32] while grain refinement induced by high energy ball milling C doped powders was indicated as the reason for the increased $B_{c2}^{\perp}$ and therefore B^* — (up to 17.2 T at 4.2 K) — in ball-milled and C doped *ex situ* bulk samples[33] which attained 10^4 A/cm^2 at 14 T.

3.3 *Nanosize B synthesis procedure*

Starting from very fine MgB_2 powders is crucial for several reasons, either to enhance the GB density, as shown in 3.2, or to develop multifilamentary wires, especially for the case of very fine filaments, as will be explained in 4.3. The milling process gave the best results and could in principle be scaled to long lengths multifilamentary tapes for the fabrication of industrial quantities with the same improved performances. However, it is not so straightforward to scale up the milling process, in particular to replicate the same conditions of transfer of energy working with big (and expensive) WC jars and suitable balls.

A new method has been therefore developed at SPIN for the preparation of B powders, based on the *magnesiothermic* reaction, adding a supplementary initial step resulting in B powder with nanometric grain size.[34] At the beginning of the process, B_2O_3 was dissolved in hot water, cryogenic freezed in liquid nitrogen and then freeze dried. The precursor of B was then reduced following the Moissan's process, i.e., the magnesiothermic reduction of B_2O_3 using elemental Mg followed by several acid leaching steps to remove the secondary product MgO. B powders prepared this way were around 100 nm in size but their amorphous nature caused difficulties in powder packaging and in friction phenomena occurring during cold rolling, thus limiting grain connectivity and causing failure and cracking of the external tube. In order to avoid these problems a final crystallization treatment at high temperatures was applied (1200 and 1700°C), non-stoichiometric MgB_2 was synthesized and Mg was used as lubricant.[35,36] The J_c behavior of a sample prepared starting from B purified and treated at 1200°C in an Fe boat for 120 min under Ar flow[36] is shown in Fig. 11, and compared with the other wires prepared with milled powders. This process is very interesting and deserves further optimization, being able to produce nanometric powders leading to wires with improved J_c performances with respect to wires prepared with commercial powders; furthermore, large scale production is possible and has already been demonstrated for up to 100 grams.[37]

4. Conductor Development and its Optimization

4.1 *Multifilamentary tapes and long length conductors in km lengths*

When thinking about applications, it is necessary to develop multifilamentary tapes: despite a more complex fabrication, they allow to reduce the fluctuation of the superconducting cross-section along the tape length, to improve the mechanical properties and to reduce the AC losses, their configuration consisting of several separated superconducting filaments.

Multifilamentary tapes are fabricated through the PIT method as well but, at the end of the drawing step performed on the monofilamentary rod, a further step is introduced: the monocore rod undergoes a further drawing step by a hexagonal draw-plate to obtain a wire with a hexagonal cross-section form. This wire is cut in several pieces that are stacked in a second tube, followed by the mechanical deformation process. The number of filaments is calculated in such a way as to make the filling factor as high as possible and actually, for this reason, only a few configurations are suitable. In Fig. 12 a typical cross-section of a multifilamentary tape is shown.

The multifilamentary tape development is based on various sheath materials, as Fe[38] or Ni,[5] the latter being more appropriate for industrial commercialization. For this reason, the present description is limited to the issues encountered in the Ni-sheathed multifilamentary tape development.

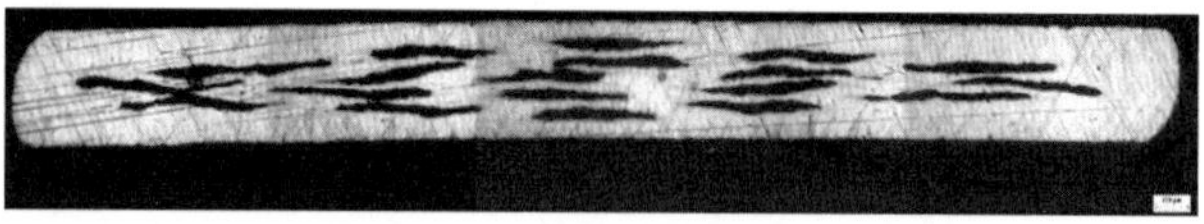

Fig. 12. Typical cross-section of a 19-filament multifilamentary MgB_2 tape, where Ni is used for both internal and external sheaths.

In Sec. 2.2, it was shown that an undesired reaction layer appears as a consequence of the heat treatment. In the case of multifilamentary tapes, the thickness of the filaments can be as low as 30–40 μm while the thickness of the reaction layer can be about 10–20 μm in a tape heat treated at 920°C for 2 h. Under these conditions the effective superconducting cross-section is drastically reduced. This issue is even more important in the case of long length wires requiring a large furnace for the heat treatment: such furnaces cannot work using fast temperature ramps; consequently the total heat treatment time is much longer, thus allowing the formation of a quite thick reaction layer. A deeper investigation on the heat treatment parameters has been carried out for optimizing heating time and temperature ramp speed. For instance, using a much faster temperature ramp and a quite short heat treatment time (about 20 min), higher values of I_c have been obtained keeping the temperature at 920°C. In such a short time the growth of the reaction layer between MgB_2 and Ni is very limited. This result led to the development of a continuous process for the final heat treatment, consisting in passing the tape trough a furnace kept at high temperature: the speed of the tape is set in such a way as to keep the tape at high temperature for a few minutes only. This method does not need further procedures; on the contrary, it can be easily included at the end of the rolling process without handling the conductor. In tapes heat treated by in-line method, the reaction layer thickness is almost negligible and has only a little influence on the effective superconducting cross-section.

In addition, especially when different and more reactive alloys are used as sheath material (i.e., Cu–Ni alloys), it is possible to insert a chemical barrier between the MgB_2 and the sheath material before the powder filling. The barrier material has obviously to be ductile and has thus to be softer than the sheath material. Furthermore, it should not react with MgB_2: good candidates are Nb, Fe, Ta. When establishing the right size of the barrier, in particular its thickness, the possible Ni diffusion into the barrier during the heat treatment and the mechanical deformation that the barrier undergoes have to be taken into account. Indeed, for instance, the Ni can diffuse through the metallic barrier and reach MgB_2 if the latter is too thin. Furthermore the strong mechanical deformation can break the barrier if it is not thick enough. On the other side, too thick a barrier limits the transfer of the deformation stress from Ni-sheath to MgB_2 acting as a kind of damper.

The multifilamentary configuration allows to obtain tapes showing J_c values similar to those of monocore ones, a reasonably lower fluctuation of the transverse superconducting cross-section along their length and the aptitude of sustaining a larger bending strain. The various envisaged industrial applications though require further precautions and protective measures. One of the most significant problems concerns the appropriate protection of a winding made with superconducting tapes, in case the superconducting state is interrupted for any reason, for example a loss of coolant, even if only momentarily. In this transient case it is essential to have an immediately available additional electrical path, being as much as possible parallel to the superconducting one and with a very low electrical resistance, in order to protect the superconducting compound from the current flow when the latter is temporarily in the normal state. We have approached this problem by introducing an oxygen-free high conductivity (OFHC) cu core in the center of the conductor, properly coated by a thin barrier for preventing chemical diffusion. The resulting composite is incorporated in the central part of a superconducting wire comprising a plurality of MgB_2 filaments arranged in a ring around the core. The barrier consists typically of a thin tube introduced (in our case an Fe tube) around the copper core.

In Fig. 13, a typical transverse cross-section of this type of multifilamentary conductor is reported. The first example of long

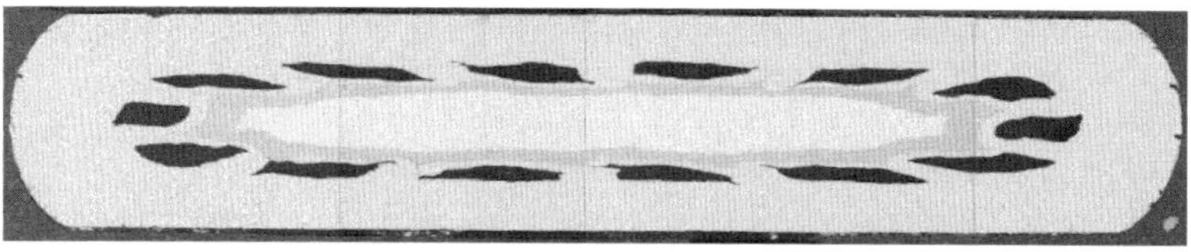

Fig. 13. Cross-section of a stabilized multifilamentary MgB_2 tape commercialized by Columbus Superconductors. In this case the barrier material is iron. The typical cross-section size is 3,5 × 0,65 mm^2 and the fill factor is about 10%.

lengths (1.53 km) of this type of tape was achieved in April 2005 by our institute and was the first tape commercialized by Columbus Superconductors SpA.

4.2 *Multifilamentary squared- and round-shaped wires*

The shape and the architecture of the conductor are designed depending on the particular application it needs to serve. For some applications square- or round-shaped conductors with a larger number of filaments and higher fill factors are preferable. The drawing process gives the desired round shape to a wire, but cannot transfer high stresses and high compaction to the powder filaments. In order to obtain square or round wires with good superconducting properties the so-called groove-rolling schematically shown in Fig. 14 (a), was introduced of drawing to deform the second stacked tube. In case of round wires, a further deformation process was added, performed by a swaging machine (Fig. 14(b)).

In Fig. 15 we show as an example various composite wires, all with 19-filaments, prepared by suitably varying the fill factor, the hardness of the sheath material and the thermal and electric

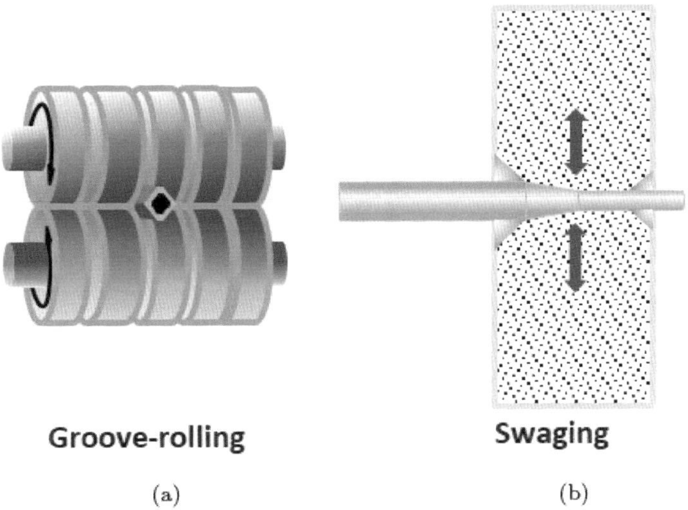

Fig. 14. Schematic representation of (a) groove-rolling and (b) swaging processes.

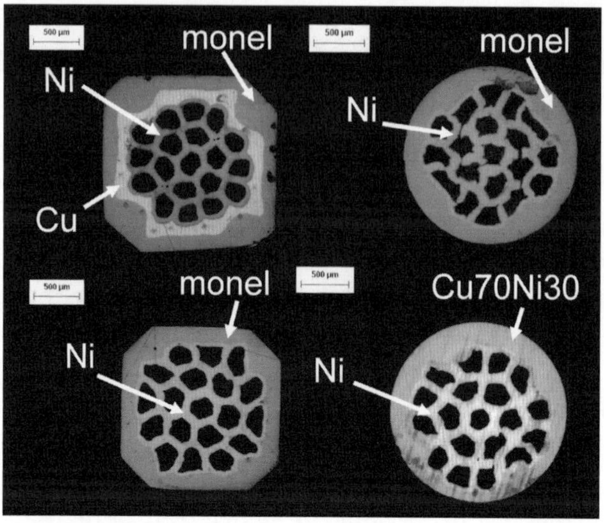

Fig. 15. 19-filaments multifilamentary wires: (a) stabilized Monel/Cu/Ni, f.f. (fill factor) = 21%; (b) Monel/Ni, f.f. = 25%; (c) Monel/Ni, f.f. = 28%; (d) Cu70Ni30/Ni, f.f. = 26%.

stabilization.[25] Cu–Ni alloys (monel), Fe and SS were chosen for the external sheath. In Fig. 16 we compare the J_c of the first standard commercial Columbus flat tape with recent square and round wires. Taking into account the different fill factors, these wires have reached higher values at fields above 3 T, even after fewer deformation steps.

4.3 *Twisting and fine filaments*

The rapid development and improvement of MgB2 wires and tapes have allowed the employment of this superconducting material in industrial DC applications such as low-field MRI magnets.[39] Several papers have also been published on the applicability of the MgB2 conductors in AC devices such as SFCL[40] and transformers.[41] Nevertheless, further efforts are still needed to develop and optimize these conductors in terms of their performance in the AC regime, in particular in facing the losses issue. Therefore, the research work should be focused on the development of multifilamentary wires or

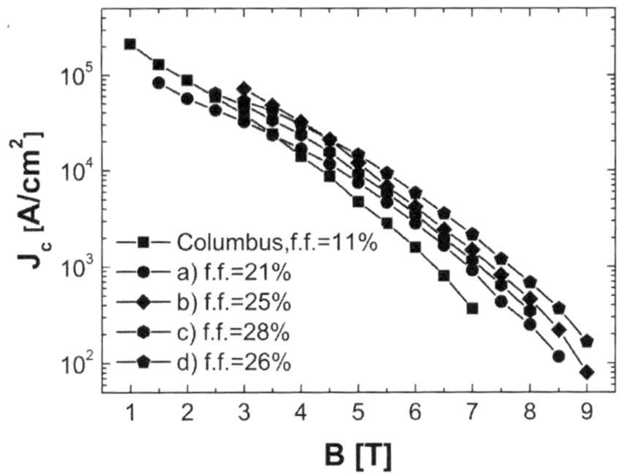

Fig. 16. Transport J_c measurements versus applied field at 4.2 K for multifilamentary MgB_2 wires compared with the Columbus tape.[25]

tapes with a large number of fine and possibly twisted filaments. Furthermore, a non-magnetic and high resistivity metal sheath is preferable in order to minimize the contribution to the AC losses of the magnetic hysteresis and the eddy currents in the matrix. Some publications treating this kind of MgB2 wire development have appeared,[42–44] while other groups worked on the fabrication of multistrand MgB2 cables.[45–47] Conductors with a high number of fine filaments have been reported in.[42,43] In all these cases, the *in situ* PIT technique was employed to realize multifilamentary MgB2 conductors. *Ex situ* twisted tapes and the related twisting technique are described in Ref. 48.

Concerning the feasibility of *ex situ* wires with very thin filaments down to 30 μm,[49] we will describe the relationship between the starting granulometry and the final single filament size in terms of transport properties. We have previously shown the positive effect of a finer MgB_2 granulometry on the superconducting properties of *ex situ* monocore tapes,[28] which was demonstrated for multifilamentary tapes as well, for example in Ref. 50. Here we report about an experiment where MgB_2 multifilamentary wires were realized using ball-milled (M, milling time = 144 h) and un-milled (NM) MgB_2 powders with average grain sizes estimated to 450 nm and 1.5 μm respectively. For each powder, three multifilamentary wires were

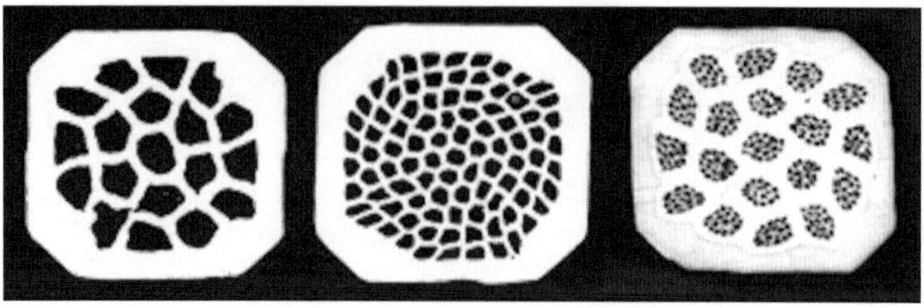

Fig. 17. Cross-section of the 19-, 91- 361-filament wires described in the text.

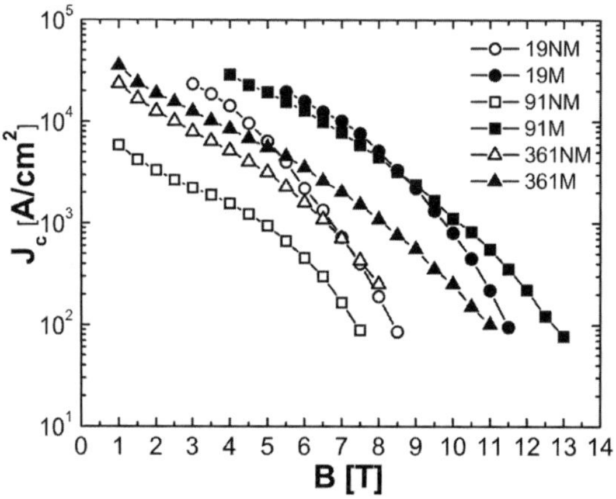

Fig. 18. Transport J_c as measured up to 13 T at GHMFL in He bath for all samples. (From Ref. 49; © 2010 by IOP).

prepared with 19-, 91- and 361-filaments. Figure 17 shows the cross-sections for the three configurations with average filament sizes of 270, 110 and 30 μm.

The measured J_c values for all the square wires are reported in Fig. 18. J_c was found to improve with milling for all wires: focusing on the behavior of the un-milled powders, passing from 19- to 91-filaments revealed a remarkable J_c degradation, which was partially recovered going to 361 filaments. On the contrary the wires based on milled powders 19M and 91 M had almost identical J_c values a slightly better behavior in-field being observed for 91 M. When increasing the number of filaments to 361, J_c decreased, but still staying above the value of wire 361 NM.

However, such a J_c behavior in presence of field cannot be explained and well understood by milling effects alone. In these complex wires several factors have to be taken into account, having an effect on the transport properties: the starting granulometry of the MgB_2 powders, the pressures transmitted during cold deformation and the final filament size. As already mentioned, the effect of the strong cold deformation of *ex situ* wires not only leads to a compacting of the powders inside the sheath, thus improving the connectivity but has a crashing effect and further reduces the average grain size.[28] On the other hand, when the sheath is particularly hard, the cold deformation could cause microcracks or filaments sausaging, acting as obstacles on the percolative current paths. In this context it is clear that the flow of the MgB_2 grains or agglomerates inside the sheath during drawing or rolling also depends on the starting granulometry and on the final individual filament size. Therefore, in the final analysis, the capability of these conductors to transport high currents crucially depends on a proper balance of all these parameters. Considering these arguments, the reason for the considerably lower J_c value of wire 91 NM with respect to wire 19 NM resides in the larger grain size of its starting powder, thus rendering the grain flow inside the sheath more difficult and causing microcracks.

On the other hand the stronger cold deformation undergone by the 361 NM wire — thanks to the double restacking resulting in a higher powder compaction and therefore in a partial recovery of J_c. This value is limited by microcracks and/or filament sausaging inevitably occurring during the deformation to such fine filaments. Analysing the behavior of the wires with milled powders it seems that the J_c enhancement in the 91-filament wire is just due to the finer grains of the used powder. For the 361 M wire we still observe a J_c increase with respect to 361 NM thanks to the finer grains, even if the effect is again limited by microcracks: the thinner the filaments, the more obstructive are the cracks on the percolative current paths. A finer MgB_2 granulometry seems to be needed to realize very thin filaments (10–30 μm) with high J_c, although the critical current measured on the 361-filament sample together with the advantages given by the high number of 30 μm thin filaments is promising for low-field applications.

5. Industrial Manufacturing and Present Commercial Wires at Columbus Superconductors SpA

The production of industrial quantities of wires based on a new superconducting compound at the desired performance level and at a competitive price is a very challenging objective. In addition to that, the implementation of a superconducting technology into practical devices has to cope with a number of hurdles, including adequate mechanical strength, weldability and durability with respect to time and operating conditions. The achievement of all these characteristics into a single wire technology represents a major technological breakthrough.

The production of long lengths of *ex situ* MgB_2 wire with a multifilamentary architecture has been already achieved in early 2006.[51] The first conductor produced in long lengths was a tape with 14 superconducting filaments, as described in 4.1 and shown in Fig. 13.

In Fig. 19, a wire evolution is reported, in which the filament number has been reduced to 12, whilst a higher filling factor has been achieved. The better understanding of wire workability, the optimization of the manufacturing methods and of the sintering conditions has led to a more than two-fold improvement of the superconducting performance of the wire without impacting on the wire cost or the processing complexity. At 20 K, in a background magnetic field of 1 T the I_c improvement has been from 200–450 A in a conductor with identical external size and architecture. In

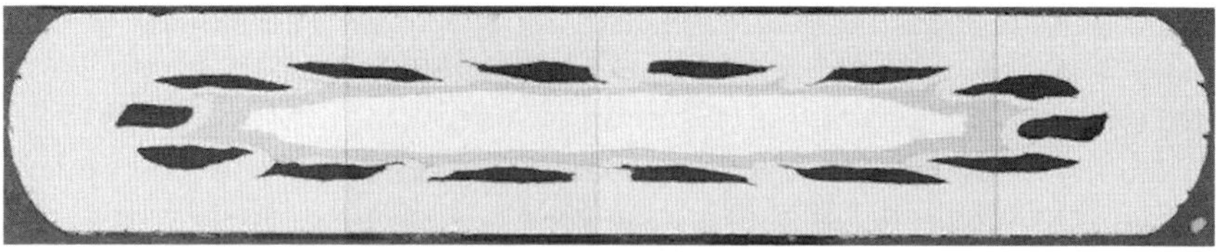

Fig. 19. Transverse cross-section of MgB_2 *ex situ* wires produced regularly in 4 km unit length.

Table 1: Wire components.

Material	MgB_2	Ni	Fe	Cu	Total	Dimensions
Area (mm^2)	0.23	1.55	0.32	0.36	2.37	3.5 × 0.65
%	10	65	10	15	100	—

Table 2: Contents of different materials in the wire cross-sections shown in Fig. 20.

Configurations	30 filaments + Ni–Cu core	36 filaments + Ni core	37 filaments
Materials %			
Cu	7	—	—
MgB_2	11	16	17
Nb	11	12	21
N-Monel matrix	71	72	62

parallel, the increase of the initial billet size from 21.3 mm to 48.2 mm has led to the stable manufacturing of 4 km long unit piece lengths on a regular basis. Since 2006, this tape conductor has been produced steadily in a volume of 100–200 Km/year. No major intrinsic limitations to lengths longer than 4 km unit piece lengths have been detected so far.

In parallel, an important number of wire architectures has been developed and scaled up to long lengths. Round *ex situ* MgB_2 wires are very promising for high current cable applications and have been recently implemented in various successful cable demonstrators, particularly at CERN in view of the high current leads for the future HL-LHC accelerator.[52] The optimization of round *ex situ* MgB_2 wires has been lengthy, as metallurgical processes are less favorable for increasing MgB_2 powder packing density in round geometry. Nevertheless, different wire cross-sections have been produced in km-long unit piece lengths with J_c values similar to those of tapes: some of the most common architectures are reported in Fig. 20. The J_c value in the SC is slightly dependent on the sheath material choice: the harder the sheath the higher the wire performance. Critical

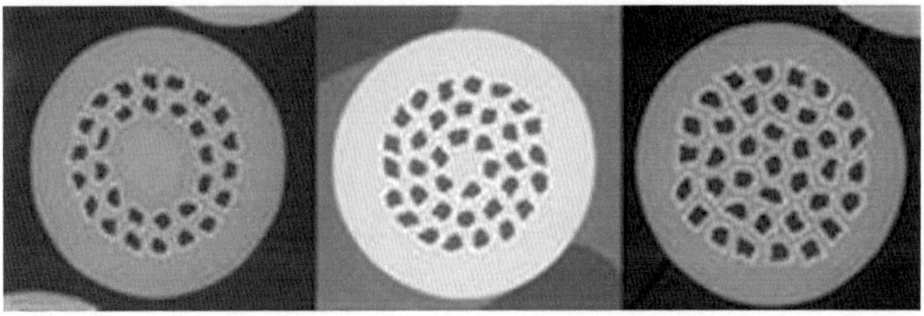

Fig. 20. Industrial *ex situ* multifilamentary round MgB_2 wires of 1 mm diameter: (a) 30 filaments +Ni–Cu core, (b) 36 filaments + Ni core, (c) 37 filaments. Courtesy of Columbus Superconductors, Genova (Italy).

Table 3: J_c in the SC and J_e values for the configurations shown in Fig. 20.

Configuration ($\Phi = 1$ mm)	J_c at 20 K, 0.4T (A/mm^2)	J_e at 20 K, 0.4T (A/mm^2)
30-filament +Ni–Cu core	3000	460
36-filament +Ni core	3200	500
37-filament	3600	600

current data at 20 K at a background field of 0.4 T are reported in the Table 3. The confirmation that even round wires can sustain reasonable mechanical handling is visible in the next graph, were the critical current is reported for a 1 mm round wire subjected to bending on different diameters. No sign of degradation of the wire performance is observed for a bending diameter down to 150 mm, as shown in Fig. 21.

A significant wire production exceeding several hundreds km/year is now in place at Columbus Superconductors SpA (Italy), based on the *ex situ* MgB_2 processing route. This process has proven to produce very homogeneous wires. Nevertheless, significant effort is still ongoing in order to transfer the know-how developed on a laboratory scale to industrial manufacturing of long wire lengths with enhanced wire performance. High energy ball milling, specific MgB_2 powder doping technologies involving organic substances, substances,

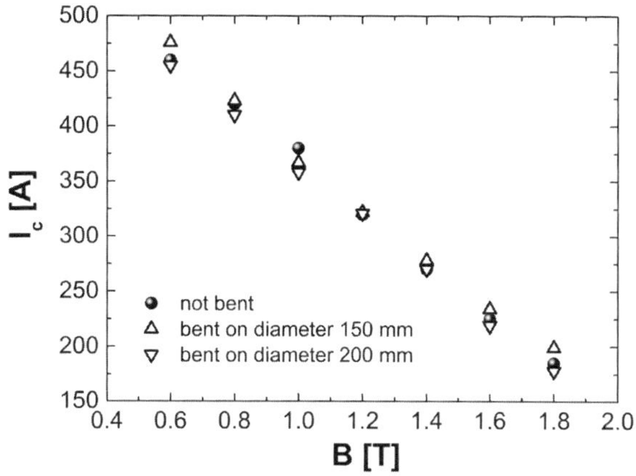

Fig. 21. I_c at 20 K for 1 mm round MgB_2 wires after bending to a diameter of 150 and 200 mm.

surface passivation control, to cite a few processes, are of more complex scalability and still require alternative routes and concepts in order to be brought at industrial level. There is still a noticeable factor to be gained in wire performance and efforts are undertaken in this direction.

Acknowledgments

We are thankful to G. Grasso from Columbus Superconductors SpA for the contribution on the industrial manufacturing and present commercial wires, and to all the former and present colleagues at CNR–SPIN who have been working on MgB_2.

References

1. P. C. Canfield, D. K. Finnemore, S. L. Bud'ko, J. E. Ostenson, G. Lapertot, C. E. Cunningham and C. Petrovic, Superconductivity in dense MgB_2 wires, *Phys. Rev. Lett.* **86**, pp. 2423–2426 (2001).
2. S. Jin, H. Mavoori, C. Bower and R. B. van Dover, High critical currents in iron-clad superconducting MgB_2 wires, *Nature* **411**, pp. 563–565 (2001).
3. S. Soltanian, X. L. Wang, I. Kusevic, E. Babic, A. H. Li, H. K. Liu, E. W. Collings and S. X. Dou, High-transport critical current density

above 30 K in pure Fe-clad MgB_2 tape, *Physica C* **361**, pp. 84–90 (2001).
4. R. Musenich, P. Fabbricatore, S. Farinon, C. Ferdeghini, G. Grasso, M. Greco, A. Malagoli, R. Marabotto, M. Modica, D. Nardelli, A. S. Siri, M. Tassisto and A. Tumino, Behavior of MgB_2 react and wind coils above 10 K, *IEEE Trans. Appl. Supercond.* **15**, pp. 1452–1456 (2005).
5. V. Braccini, A. Malagoli, A. Tumino, M. Vignolo, C. Bernini, C. Fanciulli, G. Romano, M. Tropeano, A. S. Siri and G. Grasso, Improvement of magnetic field behavior of *ex situ* processed Magnesium Diboride tapes, *IEEE Trans. Appl. Supercond.* **17**, pp. 2766–2769 (2007).
6. T. Nakane, H. Fujii, A. Matsumoto, H. Kitaguchi and H. Kumakura, The improvement of *in situ* powder in tube MgB_2 tapes by mixing MgB_2 to the starting powder of MgH_2 and B, *Physica C* **426–431**, pp. 1238–1243 (2005).
7. T. Nakane, H. Kitaguchi, H. Fujii and H. Kumakura, Improvement of *ex situ* MgB_2 powder in tube processed tapes using MgH_2 powder *Physica C* **445–448**, pp. 784–787 (2006).
8. C. H. Jiang, H. Hatakeyama and H. Kumakura, Preparation of MgB_2/Fe tapes with improved J_c property using MgH_2 powder and a short pre-annealing and intermediate rolling process, *Supercond. Sci. Technol.* **18**, pp. L17–L22 (2005).
9. M. Vignolo, G. Romano, E. Bellingeri, A. Martinelli, D. Nardelli, A. Bitchkov, C. Bernini, A. Malagoli, V. Braccini and C. Ferdeghini, *In situ* high-energy synchrotron X-ray diffraction investigation of phase formation and sintering in MgB_2 tapes, *Supercond. Sci. Technol.* **24**, p. 065014 (2011).
10. Y. C. Liu, Q. Z. Shi, Q. Zhao and Z. Q. Ma, Kinetics analysis for the sintering of bulk MgB_2 superconductor, *J. Mater. Sci, Mater. Electron.* **18**, pp. 855–861 (2007).
11. S. C. Yan, G. Yan, C. F. Liu, Y. F. Lu and L. Zhou, Experimental study on the phase formation for the Mg–B system in Ar atmosphere, *J. Alloys Compounds* **437**, pp. 298–301 (2007).
12. Z. Ma, Y. C. Liu and J. Huo, Influence of ball-milled amorphous B powders on the sintering process and superconductive properties of MgB_2, *Supercond. Sci. Technol.* **22**, p. 125006 (2009).
13. B. Q. Ning, H. Jiang and Z. S. Yan, Sintering mechanism of Ag-doped MgB_2 superconductor from low temperature to high temperature, *J. Mater. Sci., Mater. Electron.* **21**, pp. 666–670 (2010).
14. G. Grasso, A. Malagoli, C. Ferdeghini, S. Roncallo, V. Braccini, M. R. Cimberle and A. S. Siri, Large transport critical currents in unsintered MgB_2 superconducting tapes, *Appl. Phys. Lett.* **79**, p. 230 (2001).

15. G. Grasso, A. Malagoli, D. Marré, E. Bellingeri, V. Braccini, S. Roncallo, N. Scati and A. S. Siri, Transport properties of powder-in-tube processed MgB_2 tapes, *Physica C* **378–381**, pp. 899–902 (2002).
16. M. Dhallé, P. Toulemonde, C. Beneduce, N. Musolino, M. Decroux and R. Flükiger, Transport and inductive critical current densities in superconducting MgB_2, *Physica C* **363**, pp. 155–165 (2001).
17. B. A. Glowacki, M. Majoros, M. Vickers, J. E. Evetts, Y. Shi and I. McDougall, Superconductivity of powder-in-tube MgB_2 wires, *Supercond. Sci. Technol.* **14**, pp. 193–199 (2001).
18. W. Goldacker, S. I. Schlachter, S. Zimmer and H. Reiner, High transport currents in mechanically reinforced MgB_2 wires, *Superconcond. Sci. Technol.* **14**, p. 787 (2001).
19. H. L. Suo, C. Beneduce, M. Dhallé, N. Musolino, J. Y. Genoud and R. Flükiger, Large transport critical currents in dense Fe- and Ni-clad MgB_2 superconducting tapes, *Appl. Phys. Lett.* **79**, p. 3116 (2001).
20. E. Bellingeri, A. Malagoli, M. Modica, V. Braccini, A. S. Siri and G. Grasso, Neutron scattering studies of superconducting MgB_2 tapes, *Supercond. Sci. Technol.* **16**, pp. 276–280 (2003).
21. G. Grasso, A. Malagoli, M. Modica, A. Tumino, C. Ferdeghini, A. S. Siri, C. Vignola, L. Martini, V. Previtali and G. Volpini, Fabrication and properties of monofilamentary MgB_2 superconducting tapes, *Supercond. Sci. Technol.* **16**, pp. 271–275 (2003).
22. V. Braccini, *et al.*, High-field superconductivity in alloyed MgB_2 thin films, *Phys. Rev. B* **71**, p. 012504 (2005).
23. V. Ferrando, P. Manfrinetti, D. Marré, M. Putti, I. Sheikin, C. Tarantini and C. Ferdeghini, Effect of two bands on critical fields in thin films with various resistivity values, *Phys. Rev. B* **68**, p. 094517 (2003).
24. A. Matsumoto, H. Kumakura, H. Kitaguchi, B. J. Senkowicz, M. C. Jewell, E. E. Hellstrom, Y. Zhu, P. M. Voyles and D. C. Larbalestier, Evaluation of connectivity, flux pinning, and upper critical field contributions to the critical current density of bulk pure and SiC-alloyed MgB_2, *Appl. Phys. Lett.* **89**, p. 132508 (2006).
25. M. Vignolo, G. Romano, A. Malagoli, V. Braccini, C. Bernini, M. Tropeano, A. Martinelli, V. Cubeda, A. Tumino, M. Putti, C. Ferdeghini and A. S. Siri, Development of MgB_2 powders and study of the properties and architecture of *ex situ* PIT wires, *IEEE Trans. Appl. Supercond.* **18**, pp. 1175–1178 (2008).
26. M. Vignolo, G. Romano, A. Malagoli, V. Braccini, M. Tropeano, E. Bellingeri, C. Fanciulli, C. Bernini, V. Honkimäki, M. Putti and C. Ferdeghini, Role of the grain oxidation in improving the in-field behavior of MgB_2 *ex-situ* tapes, *IEEE Trans. Appl. Supercond.* **19**, pp. 2718–2721 (2009).

27. P. Lezza, V. Abächerli, N. Clayton, C. Senatore, D. Uglietti, H. L. Suo and R. Flükiger, Transport properties and exponential n-values of Fe/MgB_2 tapes with various MgB_2 particle sizes, *Physica C* **401**, pp. 305–309 (2004).
28. A. Malagoli, V. Braccini, M. Tropeano, M. Vignolo, C. Bernini, C. Fanciulli, G. Romano, M. Putti, C. Ferdeghini, E. Mossang, A. Polyanskii and D. C. Larbalestier, Effect of grain refinement on enhancing critical current density and upper critical field in undoped MgB_2 *ex situ* tapes, *J. Appl. Phys.* **104**, p. 103908 (2008).
29. G. Romano, M. Vignolo, V. Braccini, A. Malagoli, C. Bernini, M. Tropeano, C. Fanciulli, M. Putti and C. Ferdeghini, High-energy ball milling and synthesis temperature study to improve superconducting properties of MgB_2 *ex-situ* tapes and wires, *IEEE Trans. Appl. Supercond.* **19**, pp. 2706–2709 (2009).
30. J. M. Rowell, The widely variable resistivity of MgB_2 samples, *Supercond. Sci. Technol.* **16**, p. R17 (2003).
31. V. Braccini, D. Nardelli, R. Penco and G. Grasso, Development of *ex situ* processed MgB_2 wires and their applications to magnets, *Physica C* **456**, pp. 209–217 (2007).
32. M. Herrmann, W. Haessler, C. Rodig, W. Gruner, B. Holzapfel and L. Schultz, Touching the properties of NbTi by carbon doped tapes with mechanically alloyed MgB_2, *Appl. Phys. Lett.* **91**, p. 082507 (2007).
33. B. J. Senkowicz, J. E. Giencke, S. Patnaik, C. B. Eom, E. E. Hellstrom and D. C. Larbalestier, Improved upper critical field in bulk-form magnesium diboride by mechanical alloying with carbon, *Appl. Phys. Lett.* **86**, pp. 202502–202504 (2005).
34. M. Vignolo, G. Romano, A. Martinelli, C. Bernini and A. S. Siri, A novel process to produce amorphous nanosized boron useful for MgB_2 synthesis, *IEEE Trans. Appl. Supercond.* **22**, p. 6200606 (2012).
35. G. Bovone, M. Vignolo, G. Romano, A. Palenzona, A. Malagoli and A. S. Siri, Improved performances of MgB_2 conductor by using of innovative amorphous and nano-structured boron, *IEEE Trans. Appl. Supercond.* **23**, p. 6200504 (2013).
36. M. Vignolo, G. Bovone, C. Bernini, A. Palenzona, S. Kawale, G. Romano and A. S. Siri, High temperature heat treatment on boron precursor and PIT process optimization to improve the J_c performance of MgB_2-based conductors, *Supercond. Sci. Technol.* **26**, p. 105022 (2013).
37. M. Vignolo, G. Bovone, D. Matera, D. Nardelli, C. Bernini and A. S. Siri, Nano-sized boron synthesis process towards the large scale production, *Chem. Engineering. J.* **256**, pp. 32–38 (2014).

38. H. L. Suo, C. Beneduce, X. D. Su and R. Flükiger, Fabrication and transport critical currents of multifilamentary MgB_2/Fe wires and tapes, *Supercond. Sci. Technol.* **15**, pp. 1058–1062 (2002).
39. M. Razeti, S. Angius, L. Bertora, D. Damiani, R. Marabotto, M. Modica, D. Nardelli, M. Perrella and M. Tassisto, Construction and operation of cryogen free MgB_2, magnets for open MRI systems, *IEEE Trans. Appl. Supercond.* **18**, pp. 882–886 (2008).
40. R. D'Alessandro, M. Bocchi, V. Rossi and L. Martini, Test results on 500 kVA-class MgB_2-based fault current limiter Prototypes, *IEEE Trans. Appl. Supercond.* **17**, pp. 1776–1779 (2007).
41. Y. S. Hascicek, Y. Akin, T. W. Baldwin, M. M. Rindfleisch, J. Yue, M. D. Sumption and M. Tomsic, A MgB_2 12.5 kVA superconductor transformer, *Supercond. Sci. Technol.* **22**, p. 065002 (2009).
42. S. I. Schlachter, A. Frank, W. Goldacker, B. Liu, H. Orschulko, B. Ringsdorf and A. Will, Filament Microstructure and Superconducting Properties of MgB_2 Wires With Thin Filaments, *IEEE Trans. Appl. Supercond.* **17**, pp. 2842–2845 (2007).
43. M. D. Sumption, M. A. Susner, M. Bhatia, M. A. Rindfleisch, M. J. Tomsic, K. J. McFadden and E. W. Collings, High Critical Current Density Multifilamentary MgB_2 Strands, *IEEE Trans. Appl. Supercond.* **17**, pp. 2838–2841 (2007).
44. K. Funaki, T. Sueyoshi, K. Kajikawa, M. Iwakuma, K. Tanaka, M. Okada, H. Kumakura and H. Hayashi, AC loss properties of Cu-sheathed MgB_2 Composites with Nb barriers, *J. Phys.: Conf. Ser.* **97**, p. 012316 (2008).
45. T. Holùbek, S. I. Schlachter and W. Goldacker, Fabrication and transport properties of superconducting MgB_2 cables, *Supercond. Sci. Technol.* **22**, p. 055011 (2009).
46. R. Musenich, M. Greco, M. Razeti and G. Tavilla, Electrical characterization of a multi-strand MgB_2 cable, *Supercond. Sci. Technol.* **20**, pp. 235–238 (2007).
47. P. Kovac, I. Husek and T. Melisek, MgB_2 cable made from two-axially rolled wires, *Supercond. Sci. Technol.* **21**, p. 125003 (2008).
48. A., Malagoli, C., Bernini, V., Braccini, C., Fanciulli, G. Romano and M. Vignolo, Fabrication and superconducting properties of multifilamentary MgB_2 conductors for AC purposes: twisted tapes and wires with very thin filaments, *Supercond. Sci. Technol.* **22**, p. 105017 (2009).
49. A. Malagoli, V. Braccini, C. Bernini, G. Romano, M. Vignolo, M. Putti and C. Ferdeghini, Study of the MgB_2 grain size role in *ex situ* multifilamentary wires with thin filaments, *Supercond. Sci. Technol.* **23**, p. 025032 (2010).

50. R. Flükiger, P. Lezza, C. Beneduce, N. Musolino and H. L. Suo, Improved transport critical current and irreversibility fields in mono- and multifilamentary Fe/MgB_2 tapes and wires using fine powders, *Supercond. Sci. Technol.* **16**, pp. 264–270 (2003).
51. A. Malagoli, G. Grasso, M. Vignolo, A. Tumino, V. Braccini, C. Bernini, M. Tropeano, A. S. Siri, D. Nardelli and M. Modica, Long length MgB_2 conductors for industrial applications, *Adv. Sci. Tech.* **47**, pp. 238–245 (2006).
52. A. Ballarino, Design of an MgB_2 feeder system to connect groups of superconducting magnets to remote power converters, *J. Phys. Conf. Series* **234**, p. 032003 (2010).

Chapter 4c

MgB_2 Wires by *In Situ* Technique, Mechanical Alloying

Wolfgang Hässler

Leibniz Institute for Solid State and
Materials Research (IFW) Dresden
PO-Box 270116, D01171 Dresden, Germany
w.haessler@ifw-dresden.de

In this chapter the *in situ* method for the preparation of MgB_2 wires is described. The precursor powders were treated by high energy milling. For the improvement of the superconducting properties, the Mg + B powders were additionally doped by Carbon (C). The multicore wires were manufactured by the powder-in-tube (PIT) technique. The wire properties are discussed as a function of the milling energy applied on the precursor, of the C-doping level and the boron (B) powder quality. The achieved critical current densities are comparable to those of NbTi at 4.2 K.

1. Introduction

Three different technologies are used for the preparation of MgB_2 multifilamentary wires: *ex situ*, *in situ* and IMD (internal Mg diffusion). The present chapter gives a detailed description of the *in situ* processing, where the whole deformation is performed on wires containing a mixture of B and Mg powders, in contrast to the *ex situ* processing (Chapter 4b) where the starting material consists of already reacted MgB_2 powders. The reaction to MgB_2 takes place after the deformation to wires at their final diameter. The internal Mg diffusion technique discussed in Chapter 4e, starts with a central Mg rod which is surrounded by densified B powder: at a given temperature, Mg first diffuses into B before the compound MgB_2 is formed.

In this chapter we will concentrate on the *in situ* method for MgB_2 wire processing, with a particular attention on the details of the precursor powder preparation and of the wire production following the powder-in-tube (PIT) technique. The optimization of the superconducting properties of *in situ* MgB_2 wires is discussed with respect to various factors, e.g., powder size and quality, doping, sheath material and deformation/annealing parameters.

2. Precursor Powder

2.1 *Powder preparation — mechanical alloying*

The precursors are commercially available Mg and B powders of various qualities. The Mg powder purity was better than 99%, the particle size $\leq 250\,\mu m$. The corresponding data for B are 95% and $\leq 10\,\mu m$. The B powder should be either amorphous or should contain a small crystalline fraction only, due to the lower reactivity of crystalline B with Mg. Besides the metallic impurities given by the retailers, the amount of C and O_2 is important. Depending on the powder preparation technique, a large variety of particle sizes is available, from about 10 mm down to about 10 nm, the latter being generally agglomerated. Due to the high reactivity of Mg powder (especially fine grained) with O_2, the entire preparation process has to be carried out in inert atmosphere (Ar or N_2), i.e., the whole powder process has to be operated in a glove box.

The first preparation step consists of mixing the Mg and B powders, which can be performed in a mortar or in a mill, for instance in a planetary ball mill at low milling energies. Investigations have shown that the superconducting properties of the wires can be improved by milling the mixed powders. Two processes take place during milling: the decrease of Mg and B particle size and — above a certain value of milling energy — the partial reaction of both elements, this process being called *mechanical alloying*. Mechanical alloying of Mg and B was first reported by Gümbel *et al.*[1] in 2002.

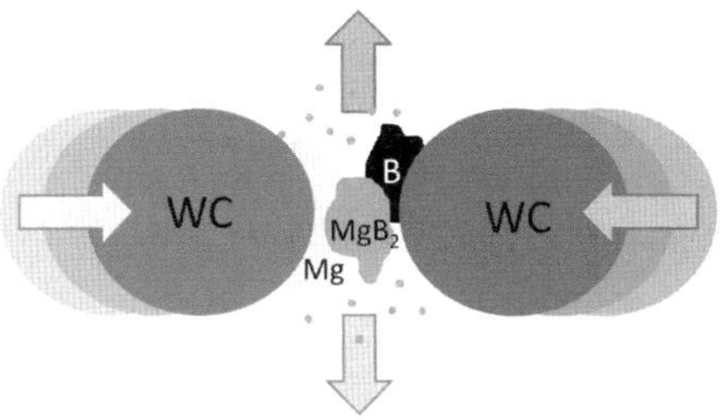

Fig. 1. Mechanical alloying during the high-energy milling process in a planetary ball mill with Tungsten carbide balls.

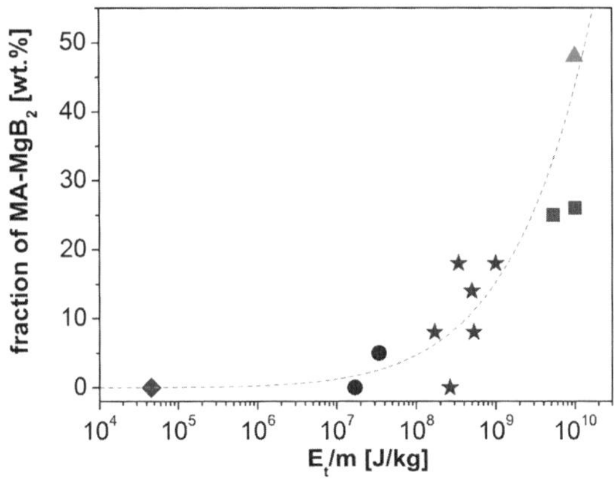

Fig. 2. MgB_2 phase fraction formed during high energy milling as a function of milling energy, calculated by Rietveld analysis of X-ray diffraction patterns[2] ().

In Fig. 1, a schematic illustration of this process is shown where Mg and B powder particles are crushed, partly reacting to MgB_2 after being hit by two milling balls. High milling energies are achieved by a combination of high rotation speeds of the planetary mill at various milling times as well as by appropriate milling tools (balls and vessels), e.g., WC, having a high mass density.

From X-ray powder diffraction patterns it can be seen that the reaction of Mg and B to MgB_2 starts above milling energies of the order of 10^5 J/kg (Fig. 2). Below this energy only grinding takes place. At milling energies of 10^9 J/kg, around 30 wt.% of the milled powder already consists of MgB_2. The dependence of the milling

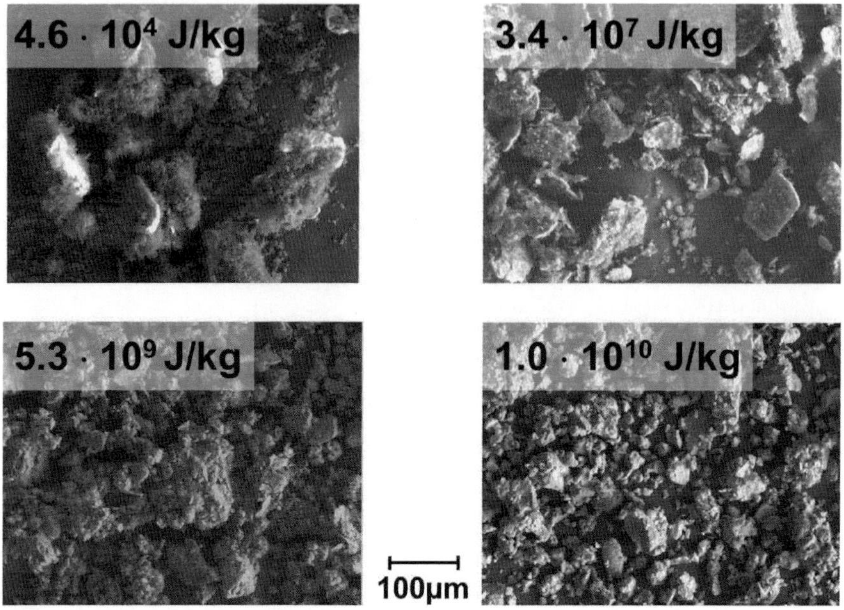

Fig. 3. SEM images of precursor powders milled at different milling energies[2] (© IOP Publishing. Reproduced with permission. All rights reserved.).

energy on the milling parameters of a planetary ball mill is calculated in Ref. 2. The crushing effect, i.e., the decrease of grain size during milling, is illustrated in Fig. 3, where SEM images of powders milled at different energies are shown. At low energies the Mg flakes of $\sim 100\,\mu m$ are decorated with fine B particles: With higher milling energy finer powders are obtained: only agglomerates are visible in the images. A Rietveld size-strain analysis of the X-ray data revealed a distinct reduction of the primary grain size (size of coherent diffracted regions) down to ~ 10 nm at milling energies of $\sim 10^9$ J/kg.

The higher reactivity of mechanically alloyed powders can be demonstrated by DSC analysis.[3] With increasing milling time with respect to energy, the enthalpy peak correlated with the MgB_2 formation decreases and is shifted to lower temperatures. The occurrence of smaller peak heights confirms the XRD result of a higher amount of partial reaction after higher milling energies, the residual part (which reacts during the DSC analysis) being smaller.

The shift of the DSC peak to lower temperatures can be explained by a smaller grain size of the powders and by a higher mechanical stress in the particles, thus resulting in a higher reactivity of the powders. The easy creation of this nanocrystalline microstructure by this technique[2] can provide a higher grain boundary pinning, thus increasing the critical current density J_c.

A very interesting effect of high energy milling of precursor powders was observed when dealing with powder storage in an O_2 containing atmosphere at room temperature and enhanced temperatures.[4] Indeed, it was surprisingly found that monocore Ti-sheathed wires prepared with high energy milled precursor powders stored for 48 h in air exhibited no degradation of J_c. Knowing that standard mixed *in situ* powders are very sensitive to O_2, this could simplify the handling of the precursor powder in industrial wire production.

The effect of mechanical alloying on the superconducting properties of MgB_2 samples will be discussed later in the wire section.

2.2 *Doping*

It is well known that the relatively low upper critical field, B_{c2}, of pure MgB_2 can be enhanced by C-doping. Most other possible doping elements do not lead to a marked improvement of the properties of MgB_2, as can be seen from a survey of different doping agents by Collings *et al.*[5] It follows that C partially substitutes B in the MgB_2 lattice, thus modifying the electronic structure of MgB_2, in particular the electronic mean free path. As a result, the upper critical field is remarkably enhanced.[6] The amount of C introduced in the lattice can be controlled by the change of the lattice constant a. The maximum concentration which can be incorporated in the lattice structure depends on the preparation technique. In single crystals, Lee *et al.*[7] reported C contents up to $x = 0.3$ in the formula $MgB_{2-x}C_x$. He found a linear dependence of the lattice constant a as a function of concentration x. In bulk samples prepared with mechanically alloyed powder doped with nanosize carbon powder, a maximum C content of $x = 0.13$ was found, which corresponds to $\sim$3.5 wt.%.[8,9]

The *in situ* process has in this context the advantage of a good mixture of the reacting agents Mg, B and C or C compounds. The incorporation of C in the lattice takes place during the reaction of Mg and B to MgB_2. Possible C sources are: pure C in the form of nanocarbon powder,[8,10,11] C nanotubes[12] or various C compounds, like carbides and organic compounds.[13–16] These compounds comprise SiC as well as various hydrocarbons in solid or liquid form. Some suppliers of B powder have recently started producing nanosize C coated B powder. In the literature it is often not clear why such huge number of organic substances had to be tested as dopants for MgB_2, the mechanism being always the same: decomposition of the C compound during annealing, followed by the incorporation of a certain amount of C in the lattice. It is important that the dopant is nanocrystalline and that it decomposes below the annealing temperature. A small O_2 content or even better a fully O_2 free additive would be desirable. The first major work on this subject was published by Dou *et al.*[13] with the investigation of SiC as doping substance for *in situ* powders.

The effect of various C compounds or pure C powders can be studied on hot pressed bulk samples. As an example, nanosize C was milled together with Mg and B powders. The precursor powders were hot pressed at 700°C and 640 MPa to bulk cylinders with 10 mm diameter. The resulting C content was calculated from the shift of the *a*-axis. Measurements of the electrical resistivity as a function of temperature and applied magnetic field allowed estimating the upper critical field B_{c2}. In Fig. 4, B_{c2} as a function of temperature is shown for two bulk samples, one undoped, the other one doped with $x = 0.13$. The upper critical field is markedly increased by C-doping: from 17–27 T at 10 K. The extrapolated B_{c2} value at 4 K is 32 T for the doped sample.

Chen *et al.*[19] have added small concentrations of around 1 wt.% of a C free compound, namely Dy_2O_3 powder, to a mixture of Mg and B. At bulk samples annealed at 900°C a remarkable increase in the inductively measured value of J_c was observed, not only in the high-field range but in low-fields as well. Nanoscale precipitates of

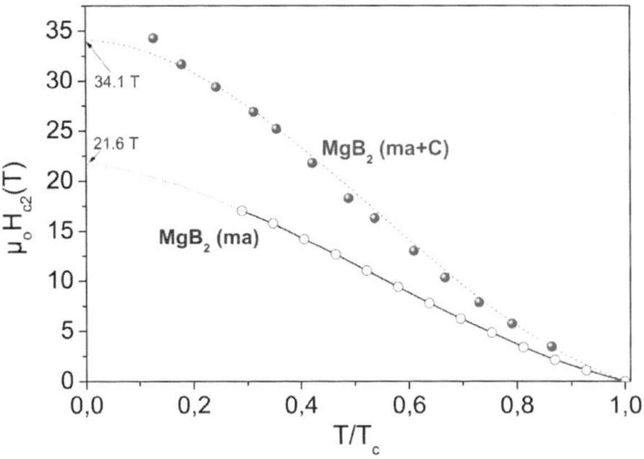

Fig. 4. Comparison between the temperature dependence of the upper critical field of two bulk samples prepared with undoped and C-doped precursor (x = 0.13 in $MgB_{2-x}C_x$) measured in a pulse field magnet.[17] The values at 0 K were extrapolated using the Gurevich model.[18]

DyB_4 and MgO were detected, which were assumed as being pinning centers.

3. Wires and Tapes

3.1 *Wire preparation*

Wires and tapes with MgB_2 filaments were fabricated by the well-known PIT method. This technique for the fabrication of superconducting wires was reviewed by Flükiger *et al.*[20] The precursor powder is filled in a ductile metallic sheath, the tube is sealed and then the composite is deformed to a wire by swaging and drawing or by groove-rolling. Another possible deformation technique is two-axial rolling (TAR).[21] The density in the filaments can be increased when the powders are not directly filled in the tube but if they are first densified to a rod by cold isostatic pressing (CIP) prior to be inserted in the tube. After deforming the wire to its final diameter, it is heat treated for a complete reaction of the precursor powder to MgB_2 and for getting a good connectivity between the MgB_2 grains by sintering. For *in situ* precursors the annealing temperature is in the region of the melting temperature of Mg, i.e., around 650°C. Mechanically alloyed powder can be already reacted at lower temperatures, around 600°C. This is much lower than the annealing temperature of *ex situ* wires, which is around 900°C.

At the beginning of the development of MgB_2 wires in 2001 or 2002, Fe or Cu were used as first sheath materials.[22] However, Cu is a good thermal and electrical conductor and has therefore good stabilization properties, but is not appropriate as a sheath material for MgB_2. It reacts strongly with Mg to $MgCu_2$ during the heat treatment. The reactivity of Fe with B to FeB_2 is relatively low, but the reaction layer between Fe and MgB_2 becomes a problem in case of fine filaments in multifilamentary wires.[22] Taking this into account the idea of a protecting barrier around the filament came up. Three different metallic barrier materials came into consideration, Nb, Ta[23] and Ti.[24] For all these metals high purity is necessary to get a good deformation behavior. Nb works well as a diffusion barrier for annealing temperatures up to 650°C. For higher annealing temperatures, B is found to diffuse into the Nb.[25] For these temperatures Ti is a much better choice.[26] In addition, the roughness of the interface between barrier and filament is smaller for wires with a Ti barrier.

A four-core wire with Ti barriers and a Cu stabilizer deformed by TAR[27] is shown in Fig. 5. For thermal stability of the wires

Fig. 5. Cross-section of a four-core wire 1,2 × 1,2 mm^2 with Ti-barrier (Ti/Cu/Monel sheath).[48]

(a) (b)

Fig. 6. Metallographic cross-sections of a monocore wire with (a) Nb/Glidcop sheath and (b) of a multifilamentary wire with an outer monel sheath.

the Nb or Ti tubes were surrounded by a second tube consisting of Cu or Glidcop (oxide dispersion strengthened Cu). This double sheath system is normally used for the preparation of monocore wires. Images of the cross-sections of MgB_2 wires MgB_2 powder as barrier material[28] are shown in Fig. 6. The preparation technique of this "ceramic barrier", described by Kario *et al.*[29] has the advantage to be superconducting in the designed application temperature range of MgB_2 wires, being not limited to 9 K as in the case of Nb. But this technique was so far not applied to long wires involving multiple deformation steps.

An alternative route, reducing the number of deformation steps during the fabrication of monocore wires was described by Tomsic *et al.*:[30] the *continuous tube filling and forming* (CTFF). In the *in situ* version, Mg and B powders are directly filled in a foil consisting of the barrier material, mostly Nb. The overlap-closed filled tube is then inserted in a tube of stabilizing material like Glidcop Cu and drawn to the final wire size.

For the preparation of multifilamentary wires, the deformed monocore wires are bundled and inserted in an outer sheath of Monel or SS. This composite is then deformed again by swaging and drawing, groove-rolling or TAR. To improve the mechanical properties of the wires, materials with a sufficient tensile strength are used for the outer sheath, e.g., Glidcop Cu or Monel (Fig. 6). For long wires a large number of deformation steps is necessary. A possibility to reduce the number of deformation steps for both, monocore and multifilamentary wires was described by Pachla *et al.*,[31] who used hydrostatical extrusion, having the beneficial effect of a core densification.

For a further densification of the filament the wire can be rolled to a tape. But the advantage of denser filaments in the tape is unfortunately connected with anisotropic superconducting properties in an external magnetic field. Densification of the filaments without introducing anisotropy can be obtained by cold hydrostatic pressure densification (CHPD) of the wires before annealing.[32]

3.2 *Superconducting properties of the wires*

3.2.1 *Influence of milling energy*

The transport critical current density J_c of wires can be enhanced by using precursor powder milled at higher energies.[2,33,34] In Fig. 7, the variation of J_c of monocore wires with a Cu/Nb- sheath is

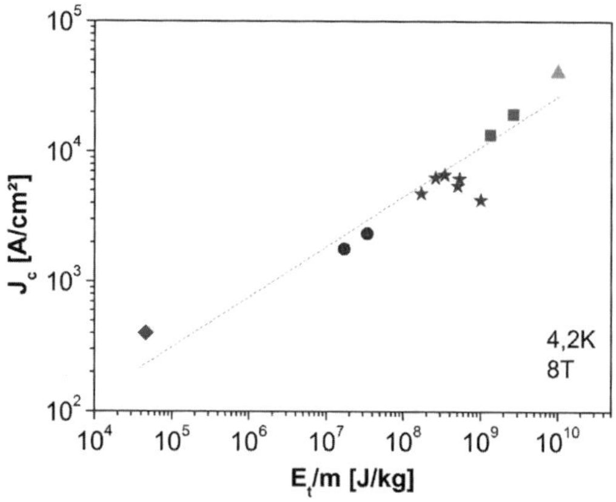

Fig. 7. Transport critical current density of monocore wires prepared with precursor powders milled at different milling energies (Fe-sheath)[2] (© IOP Publishing. Reproduced with permission. All rights reserved).

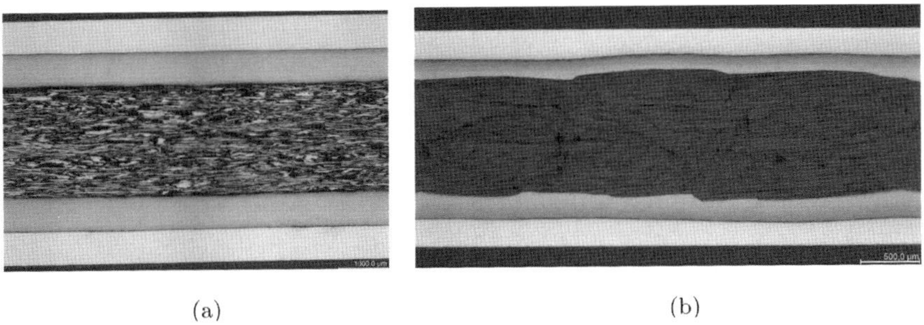

Fig. 8. Longitudinal sections of two monocore wires prepared with mixed precursor powder (a) and with high energy milled powders (b).

plotted as a function of the milling energy of the precursor powder. It demonstrates that J_c is shifted to higher values for higher milling energies. The same behavior is encountered in multifilamentary wires. The filaments of these wires are characterized by a denser microstructure in comparison to wires prepared with mixed powder (Fig. 8). Due to the smaller primary grain size, grainboundary pinning plays a dominant role in wires based on high energy milled precursor powders.

An additional advantage of high energy milled precursor powders is the higher reactivity, allowing to reduce the reaction temperature to values clearly below the melting temperature of Mg. The increase of milling energy, however, is limited by the deformation properties of the wires. Using very fine powder, shear bands in the filaments appear (Fig. 8). Therefore it is necessary to find a reasonable compromise of the preparation parameters between current density and deformation behavior.

3.2.2 *Influence of B quality*

The nominal purity of the B powder specified by the suppliers considers metallic impurities only, which is obviously not sufficient for their characterization. Indeed, it is also important to know the amount of O_2 impurities (usually in the range of 2–4 wt.%) as well as grain size and morphology.[35–37] Predominantly amorphous and nanosize B powders are more reactive, thus rendering lower reaction temperatures possible, yielding the highest values of J_c.

3.2.3 *Influence of C-doping*

The influence of C-doping on the transport J_c is illustrated in Fig. 9, where J_c is plotted as a function of applied magnetic field for monocore MgB_2 wires based on precursor powders with C contents of $x = 0$, 0.03 and 0.12, respectively. It can be seen that the field dependence of J_c becomes weaker, its values being enhanced at higher fields. For the undoped wire J_c was approximately 10 kA/cm^2 for B < 12 T, for the doped sample with $x = 0.13$ the corresponding value was B < 14 T. It is evident that *in situ* wires produced in view of applications at higher fields should be based on C-doped precursor powders. In recent papers the pure C as a doping agent was replaced by malic acid[16] or hydrocarbon.[38]

C-doping has another positive effect, the observed anisotropy in J_c being significantly decreased.[39] This is illustrated in Fig. 10, where the critical current of an undoped tape is compared to that of tape with a carbocontent of $x = 0.13$ in $Mg(C_xB_{2-x})$ as function the angle between the tape normal and the field direction at 7.5 T and 4.2 K (90° corresponds to the parallel field direction).

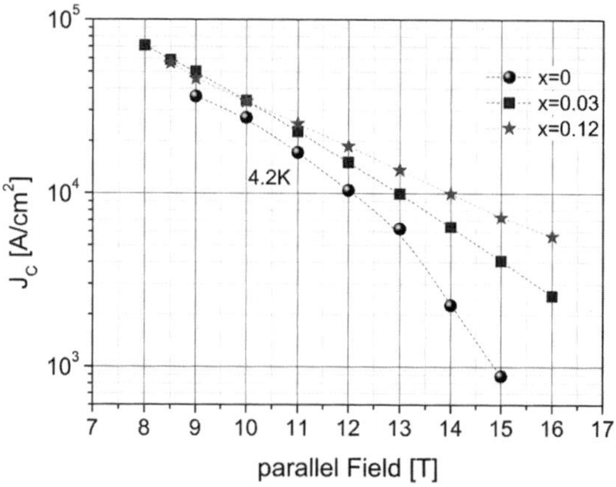

Fig. 9. Critical current densities of monocore tapes $Fe/MgB_{2-x}C_x$ with C-doped precursor powder at 4.2 K ($x = 0$–0.13).

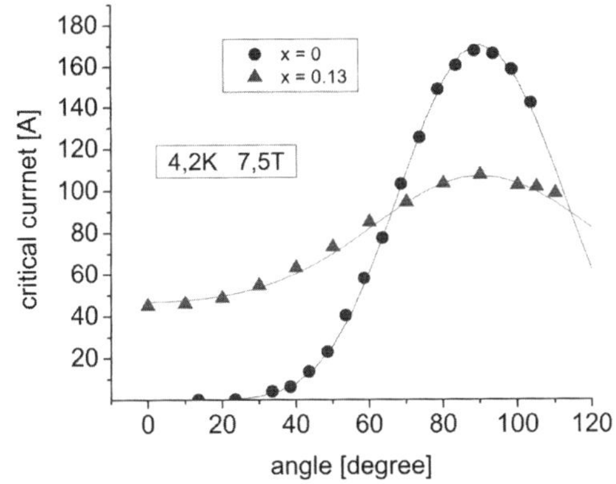

Fig. 10. Critical current I_c at 4.2 K and 7.5 T as a function of the field orientation angle in $Mg(C_xB_{2-x})$ in both cases with mechanically alloyed precursor powders[39] (© IOP Publishing. Reproduced with permission. All rights reserved.).

Besides the large variety of C and C-based additives, Cu and Ag additions should also be mentioned.[40–42] Both materials influence the formation kinetics of MgB_2 in the filaments, reducing the reaction temperature of Ag or Cu doped wires by at least 50 K, due to the formation of a low temperature eutectic liquid with Mg.

4. State of the Art

At present, multifilamentary *in situ* wires are prepared as follows. The starting materials Mg and B, with ≥99% purity, are chosen with an O_2 content as low as possible. The grain size of B should be in the submicron range and mainly amorphous. Several suppliers are meanwhile on the market. For high-field application, nanosize C or for instance a hydrocarbon like malic acid is added. Nanosize SiC comes into consideration as well. Nb or Ti are used for the barrier, while Glidcop Cu is applied for the second sheath of the monocore wire. Additional mechanical strength is usually obtained by an outer monel sheath, or stainless steel. The wire were deformed either by wire drawing or by groove-rolling. The CTFF processing is used as an alternative deformation route for wires in km length. The reaction takes place in inert atmosphere at temperatures in the range of 600–650°C for high energy milled precursors, in contrast to 700°C for wires with mixed precursors.

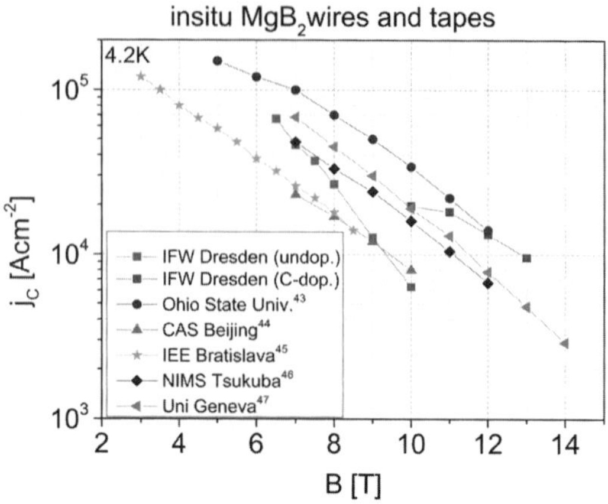

Fig. 11. Comparison between the critical current densities of multifilamentary C-doped MgB_2 *in situ* wires prepared by different groups — state of the art.[43–47]

The engineering critical current densities J_e achieved so far are in the range of 30 kA/cm^2 at 3 T and 4.2 K, which is comparable to the values for NbTi. For high-field applications 10 kA/cm^2 at 8 T are achievable with the *in situ* technology. An overview of J_c versus B characteristics of *in situ* wires prepared by different groups,[43–47] representing the state of art is shown in Fig. 11.

Summarizing, one has to keep in mind that an essential point of the *in situ* wire technology is a high density of the MgB_2 filaments, which has a direct influence on the connectivity between the grains in the filaments. It can be influenced by the powder properties (initial grain size, milling type, additives), by the deformation technique and the annealing parameters. The properties achieved up to now in *in situ* MgB_2 wires are promising for the potential application in magnetic coils but a further improvement by optimizing the processing parameters is necessary, particularly in view of an extension of the application possibilities.

References

1. A. Gümbel, J. Eckert, G. Fuchs, K. Nenkov, K.-H. Müller and L. Schultz, *Appl. Phys. Lett.* **80**, p. 2725 (2002).

2. W. Häßler, H. Hermann, M. Herrmann, C. Rodig, A. Aubele, L. Schmolinga, B. Sailer and B. Holzapfel, *Supercond. Sci. Technol.* **26**, p. 025005 (2013).
3. J. Eckert, O. Perner, G. Fuchs, K. Nenkov, K.-H. Müller, W. Häßler, C. Fischer, B. Holzapfel and L. Schultz, *Adv. Solid State Physics* **403**. Berlin, Springer (2003).
4. P. Kovaĉ, M. Kulich, W. Haessler, M. Hermann, T. Melisek and M. Reissner, *Physica C* **477**, p. 20 (2012).
5. E. W. Collings *et al.*, *Supercond. Sci. Technol.* **21**, p. 103001 (2008).
6. M. Eisterer, *Supercond. Sci. Technol.* **20**, R47 (2007).
7. S. Lee, T. Masui, A. Yamamoto, H. Uchiyama and S. Tajima, *Physica C* **397**, p. 7 (2003).
8. W. Gruner, M. Herrmann, A. Nilsson, H. Hermann, W. Haessler and B. Holzapfel, *Supercond. Sci. Technol.* **20**, p. 601 (2007).
9. M. Herrmann, W. Haessler, C. Mickel, W. Gruner, B. Holzapfel and L. Schultz, *Supercond. Sci. Technol.* **20**, p. 1108 (2007).
10. T. Takenobu, T. Ito, D. Chi, K. Prassides and Y. Iwasa, *Phys. Rev. B* **64**, p. 134513 (2001).
11. B. J. Senkowicz, A. Polyanskii, R. J. Mungall, Y. Zhu, J. E. Giencke, P. M. Voyles, C. B. Eom, E. E. Hellstrom and D. C. Larbalestier, *Supercond. Sci. Technol.* **20**, p. 650 (2007).
12. S. X. Dou, W. K. Yeoh, J. Horvat and M. Ionescu, *Appl. Phys. Lett.* **83**, p. 4996 (2003).
13. S. X. Dou, S. Soltanian, J. Horvat, X. L. Wang, S. H. Zhou, M. Ionescu, H. K. Liu, P. Munroe and M. Tomsic, *Appl. Phys. Lett.* **81**, p. 3419 (2002).
14. H. Yamada, M. Hirakawa, H. Kumakura and H. Kitaguchi, *Supercond. Sci. Technol.* **19**, p. 175 (2006).
15. D. L. Wang, X. P. Zhang, Z. S. Gao, L. Wang, Y. W. Ma, S. Awaji, G. Nishijima, K. Watanabe and E. Mossang, *Physica C* **469**, p. 23 (2009).
16. J. H. Kim, S. Zhou, M. S. Hossain, A. V. Pan and S. X. Dou, *Appl. Phys. Lett.* **89**, p. 142505 (2006).
17. N. Kozlova and G. Fuchs (IFW Dresden), private communication.
18. A. Gurevich, *Phys. Rev. B* **67**, p. 184515 (2003).
19. S. K. Chen, M. Wei and L. Mac-Manus-Driscoll, *Appl. Phys. Lett.* **88**, p. 192512 (2006).
20. R. Flükiger, C. Senatore, M. Cesaretti, F. Buta, D. Uglietti and B. Seeber, *Supercond. Sci. Technol.* **21**, p. 054015 (2008).
21. P. Kováč, I. Hušek, E. Dobročka, T. Melišek, W. Haessler and M. Herrmann, *Supercond. Sci. Technol.* **21**, p. 015004 (2008).

22. C. Fischer, W. Häßler, C. Rodig, O. Perner, G. Behr, M. Schubert, K. Nenkov, J. Eckert, B. Holzapfel and L. Schultz, *Physica C* **406**, p. 121 (2004).
23. W. Goldacker, S. I. Schlachter, S. Zimmer and H. Reiner, *Supercond. Sci. Technol.* **14**, p. 787 (2001).
24. M. Alessandrini, H. Fang, M. Hanna, P. Putman, Y. X. Zhou and K. Salama, *Supercond. Sci. Technol.* **19**, p. 129 (2006).
25. Y. Y. Sun, P. X. Zhang, Q. Y. Wang, Q. Ming, Y. Fang, F. J. Gao, Q. L. Guo, A. Sulpice and Y. Guo, *J. Supercond. Nov. Magn.* **25**, p. 943 (2012).
26. P. Kováč, I. Hušek, L. Kopera, T. Melišek, A. Rosová and E. Dobročka, *Supercond. Sci. Technol.* **26**, p. 025007 (2013).
27. P. Kováč, I. Hušek, T. Melišek and T. Holúbek, *Supercond. Sci. Technol.* **20**, p. 771 (2007).
28. B. A. Glowacki and A. Morawski, patent WO/2008/122802 (2008).
29. A. Kario, A. Morawski, W. Häßler, M. Herrmann, C. Rodig, M. Schubert, K. Nenkov, B. Holzapfel, L. Schultz, B. A. Glowacki and S. C. Hopkins, *Supercond. Sci. Technol.* **23**, p. 025018 (2010).
30. M. Tomsic, M. Rindfleisch, J. Yue, K. McFadden, D. Doll, J. Phillips, M. D. Sumption, M. Bhatia, S. Bohnenstiel and E. W. Collings, *Physica C* **456**, p. 203 (2007).
31. W. Pachla, A. Morawski, P. Kovac, I. Husek, A. Mazur, T. Lada, R. Diduszko, T. Melisek, V. Strbik and M. Kulczyk, *Supercond. Sci. Technol.* **19**, p. 1 (2006).
32. R. Flükiger, M. S. Hossain and C. Senatore, *Supercond. Sci. Technol.* **22**, p. 085002 (2009).
33. M. Takahashi, M. Okada, T. Naskane and H. Kumakura, *Supercond. Sci. Technol.* **22**, p. 125017 (2009).
34. C. Wang, Y. Ma, X. Zhang, D. Wang, Z. Gao, C. Yao, C. Wang, H. Oguro, S. Awaji and K. Watanabe, *Supercond. Sci. Technol.* **25**, p. 035018 (2012).
35. W. Häßler, B. Birajdar, M. Herrmann, O. Perner, W. Gruner, C. Rodig, M. Schubert, O. Eibl, B. Holzapfel and L. Schultz, *Supercond. Sci. Technol.* **19**, p. 512 (2006).
36. J. H. Kim, Y. U. Heo, A. Matsumoto, H. Kumakura, M. Rindfleisch, M. Tomsic and S. X. Dou, *Supercond. Sci. Technol.* **23**, p. 075014 (2010).
37. S. Safran, A. Kilic, E. Asikuzun, E. Kiicarslan, O. Ozturk and A. Gencer, *J. Mater Sci : Mater Electron* **25**, p. 2737 (2014).
38. H. Yamada, N. Uchiyama, H. Kumakura, H. Kitaguchi and A. Matsumoto, *IEEE Trans. Appl. Supercond* **17**, p. 2850 (2007).

39. W. Haessler, P. Kovac, M. Eisterer, A. B. Abrahamsen, M. Herrmann, C. Rodig, K. Nenkov, B. Holzapfel, T. Melisek, M. Kulich, M. von Zimmermann, J. Bednarcik and J. C. Grivel, *Supercond. Sci. Technol.* **23**, p. 65011 (2010).
40. J. Shimoyama, K. Hanafusa, A. Yamamoto, Y. Katsura, S. Horii, K. Kishio and H. Kumakura, *Supercond. Sci. Technol.* **20**, p. 307 (2007).
41. J. C. Grivel, A. Abrahamsen and J. Bednarčik, *Supercond. Sci. Technol.* **21**, p. 035006 (2008).
42. Z. Q. Ma and Y. C. Liu, *Int. Mat. Rev.* **56**, p. 267 (2011).
43. G. Z. Li, K. M. Reddy, J. B. Zwayer, M. A. Kuldell, M. A. Susner, Y. Yang, M. D. Sumption, J. J. Yue, M. A. Rindfleisch, M. J. Tomsic, C. J. Thong and E. W. Collings, *IEEE Trans. Appl. Supercond.* **23**, p. 6200204 (2013).
44. D. Wang, C. Wang, X. Zhang, Z. Gao, C. Yao, Y. Ma, M. Kanazawa, Y. Yamada, H. Oguro, S. Awaji and K. Watanabe, *Supercond. Sci. Technol.* **25**, p. 065013 (2012).
45. P. Kováč, T. Mellšek, L. Kopera, J. Kováč and I. Hušek, *Supercond. Sci. Technol.* **25**, p. 095008 (2012).
46. A. Matsumoto, H. Kitaguchi and H. Kumakura, *Supercond. Sci. Technol.* **21**, p. 065007 (2008).
47. M. S. Hossain, C. Senatore, R. Flükiger, M. A. Rindfleisch, M. J. Tomsic, J. H. Kim and S. X. Dou, *Supercond. Sci. Technol.* **22**, p. 095004 (2009).
48. P. Kovač, private communication.

Chapter 4d

Pressure Effects on J_c of *In Situ* and *Ex Situ* Processed MgB_2 Wires

Shahriar M. A. Hossain*,‡ and René Flükiger†

*ISEM, University of Wollongong, Northfields Av. Wollongong, New South Wales, 2522 Australia

†Department of Quantum Matter Physics, University of Geneva, 24, Quai Ernest Ansermet, 1211 Geneva 4, Switzerland

‡shahriar@uow.edu.au

The effects of Cold High Pressure Densification (CHPD) on square shaped MgB_2 wires are described. Pressures up to 2 GPa are built up by four hard steel anvils acting simultaneously on the wire, thus resulting in a higher mass density in the MgB_2 filaments. The effects of CHPD on the critical current density, J_c, are presented for binary and alloyed mono and multifilamentary MgB_2 wires prepared by both *in situ* and *ex situ* processing. On short alloyed wires, a pressure of 1.5 GPa enhanced J_c by factors of ~2 and ~4 at 4.2 K/8 T and 20 K/4 T, respectively. An automatic machine was constructed allowing a sequential press/release/advance operation on industrial wire lengths. The J_c enhancements obtained with the short sample device on *in situ* MgB_2 wires were confirmed for lengths exceeding 10 m. Using an improved anvil geometry, the wall friction was reduced, thus yielding the same J_c enhancements, but at considerably lower applied pressures: the same value of $J_{c,max}$ was already obtained at 0.85 GPA, in contrast to 1.3 GPA for the short sample device. A recent device for hot pressing of long MgB_2 wires is briefly discussed.

1. Introduction

In situ and *ex situ* are two powder metallurgical processes currently used for the industrial fabrication of multifilamentary MgB_2 wires.

There are characteristical differences between both methods:

(a) Different initial powder mixtures: Mg and B powders for *in situ* and already reacted MgB_2 powders for *ex situ* wires;
(b) Different final heat treatments: at 650–700°C for *in situ* wires (reaction to MgB_2) and at 850–950°C for *ex situ* wires (sintering between neighbor grains).

The mass density in the MgB_2 filaments of the wires after the final heat treatment is quite low with respect to theoretical density (2.36 g/cm^3). Values of ~50% and ~70% have been reported for *in situ* and *ex situ* wires, respectively.[1,2] In both cases, the presence of voids in the final wires strongly limits the critical current density, which can be correlated to the grain connectivity, according to Rowell.[3] Several attempts have been undertaken to enhance the mass density of *in situ* filaments by applying high pressure during the reaction heat treatment. Serquis *et al.*[4] performed the reaction under HIP conditions at 0.2 GPa in both *in situ* and *ex situ* wires and reported at 4.2 K an enhancement of $B(10^4)$ up to 1 T B (10^4) is defined as the field at which the critical current density J_c reaches the value of 1×10^4 A/cm^2). Prikhna *et al.*[5] produced very dense bulk samples of Ti alloyed MgB_2 after reaction at $T > 1000$°C in a multianvil device under Ar pressures of 2 GPa. In bulk samples the mass density was enhanced to values close to 100%, yielding B $(10^4) = 11$ T at 4.2 K.[5]

After hot pressing of SiC alloyed *in situ* tapes at 630°C under 100 MPa, Yamada *et al.*[6] found enhanced J_c values, the extrapolated value of $B(10^4)^{//}$ being close to 14 T. The large difference in J_c between hot pressed bulk samples[5] and the hot pressed tapes[6] is due to the deformation of tapes by rolling, inducing a partial texturing in the filaments, as shown by Abrahamsen *et al.*[7] Cold densification was applied as an alternative to high pressures at high temperatures. Matsushita *et al.*[8] analyzed the changes of J_c and of the electrical residual resistivity $\rho_{40\,K}$ on *in situ* MgB_2 pellets. Pachla *et al.*[9] studied the effect of cold isostatic pressure up to 1.5 GPa in *ex situ* MgB_2. Finally, hydrostatic extrusion followed by two-axial rolling (TAR) was reported by Kovac *et al.*[10] and Pachla *et al.*[11]

However, none of these high pressure proceedings has the potential to be applied on long wire lengths, the wire length being limited to several centimeters. Thus, new ways had to be investigated for the production of MgB_2 wires of km lengths with enhanced J_c values.

Very high mass densities in MgB_2 wires (up to 100%) can be produced without external pressure using the *infiltration method*, originally proposed by Giunchi *et al.*[12] (see Chapter 3b). Hur *et al.*[13] and Togano *et al.*[14] modified this technique (they called it Internal Mg Diffusion, or IMD) and applied it to wires. They found very high critical current densities in ring shaped MgB_2 filaments (see Chapter 4e). The hollow center of these filaments, however, introduces new homogeneity problems which require further developments (see Chapters 4e and 4f).

In the following, the cold densification method developed in Geneva[15,16] for enhancing the J_c values in multifilamentary *in situ* MgB_2 wires and its application to long wire lengths will be discussed in detail.

In order to get a complete overview of pressure effects on *in situ* MgB_2 wires, an alternative hot pressing method for long wire lengths recently developed at the High Pressure Physics Institute in Warsaw (Poland) will briefly be mentioned.[17]

2. Devices for Cold High Pressure Densification

The first cold high pressure densification (CHPD)[15] experiments were performed with a prototype cell (Fig. 1) at room temperature, with hard steel 39 mm long anvils acting simultaneously on all four sides of a square-shaped *in situ* MgB_2 wire.

The development was undertaken at the University of Geneva[15,16] on binary and alloyed MgB_2 wires, respectively. After reaction, a considerable increase of the mass density was observed on a CHPD treated monofilamentary binary MgB_2 wire. At the same time, the J_c value by a factor of 2 at 4.2 K/8 T was observed, the enhancement at 20 K/4 T reaching even a factor of 4.[15,16] Since no degradation of J_c was detected at the overlapping pressure zones between two neighbor regions when compared to short wire lengths, we extended

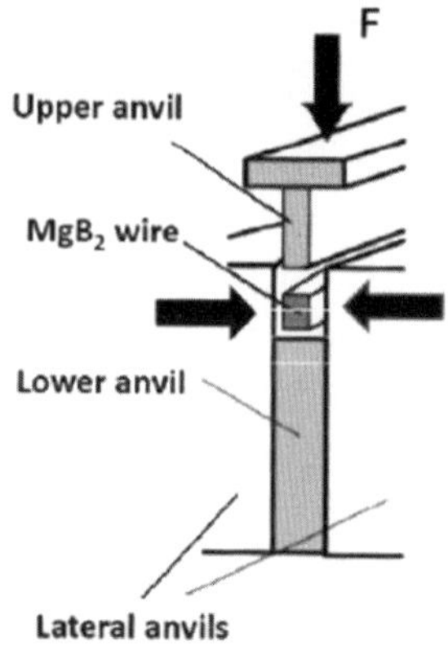

Fig. 1. Schematic drawing of the short sample pressure cell, with 39 mm anvil length. After densification, the upper and the lateral anvils are removed, in order to allow transport measurements on the wire (Reproduced from Ref. 17. © Springer-Verlag GmbH).

the CHPD method to lengths exceeding 10 m, a first step towards industrial lengths.[17]

2.1 *The prototype pressure cell*

The prototype pressure cell used in earlier publications[15–19] for short wires designed at the University of Geneva for the application of high pressures at room temperature is schematically drawn in Fig. 1. It consists of hard steel anvils acting simultaneously on all four sides of a square wire (shaped by means of a turks head). The square geometry was chosen as a first step of development in view of a systematical study of the densification effects under precise pressure conditions. The application of pressure on the wire is followed by a pressure release, allowing a recovery of the wire without damage. In Fig. 1, the upper anvil transmits the force F exerted by a hydraulic press to the sample. For most experiments, the horizontal anvil width was 1 mm. A width of 2 mm was chosen for the preparation of samples for specific heat and electrical resistivity measurements[18] (see Chapter 5a). Little or no wire elongation was observed after densification: for wires with initial square shape, the conditions of pressure inside the pressure cell are thus supposed to be nearly homogeneous.

2.1.1 *The effect of pressure on mass density*

After densification the wire was recovered without damage and was submitted to the final heat treatment. This initial solution was very useful as a first step towards an automatic device[17] for continuous CHPD processing of long wires (see Fig. 3). The quasi-hydrostatical densification step has the effect of enhancing the mass density of the unreacted Mg + B filament, letting the final degree of texturing unchanged. This is in contrast to the effect of pressing of a tape between two walls, where the tape flows in the direction parallel to the walls, thus enhancing the degree of texturing but letting the mass density essentially unchanged.[16]

The highest possible pressure applied on the four-wall cell was 6.5 GPa, after which the mass density of the unreacted Mg + B powder mixture reached 96%, while the corresponding value in the reacted MgB_2 filament increased up to ~73% of the theoretical value. For practical reasons, however, the pressure in most experiments was much lower. After applying 2.5 GPa,[15] the density of binary monofilamentary MgB_2 wires prior to reaction was enhanced from 0.44–0.54% of the theoretical value. As seen later, this corresponds to an enhancement of J_c by factors of 2 and 4 at 4.2 K/8 T and 20 K/4 T, respectively.

The enhancement of mass density in binary monofilamentary Fe/MgB_2 wires having a square shape as a function of applied pressure exerted by four anvil press is represented in Fig. 1. The effect was also observed on alloyed monofilamentary MgB_2 wires. A careful series of experiments on a large number of wires allowed to determine the enhancement of mass density of the MgB_2 filaments as a function of the applied pressure. On densified wires, a marked decrease of electrical resistance was observed, reflecting an improvement in connectivity.[15,16]

The highest possible applied pressure is limited by the properties of the anvil material. Even at the highest applied pressure of 6.5 GPa, the mass density inside the filaments does not exceed 75% (see Fig. 2). Considerably higher values of J_c are expected for a further enhancement of mass density. This is the case for MgB_2 wires

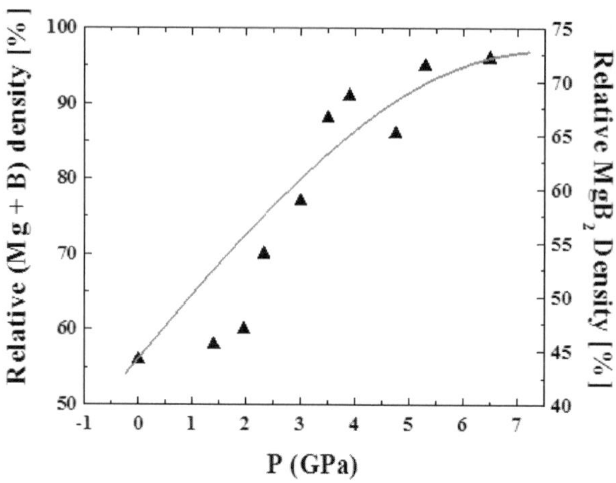

Fig. 2. Mass density of monofilamentary Fe/MgB_2 wires. Left scale: relative density of the unreacted (Mg + B) mixture. Right scale: relative mass density of reacted MgB_2 filaments. (From Flükiger *et al.*[15] Reproduced with permission IOP Publishing).

Fig. 3. Densification machine for long MgB_2 wires now operational at University of Geneva. (1) Vertical 40 t press, (2): Horizontal 16 t press. The arrow indicates the forward direction of the wire. (Reproduced with permission from Flükiger *et al.*,[17] Reproduced with permission from © Springer Science + Business Media New York 1972).

produced by the infiltration processes IMD[21] and AIMI[23] wires (see Chapters 4e and 4f), where the mass density approaches the theoretical maximum.

2.2 *Device for the cold densification of long wire lengths*

Based on the experiments with the short sample prototype device, a new machine with an automatic press/release/advance mechanism was built,[17] the declared goal being the densification of longer wire lengths.

The new machine was designed for a fast change between the various pressing and releasing operations, combined with a forward movement of the wire. Since ordinary swaging machines are designed for obtaining an elongation of the working piece by squeezing, first tests failed in producing substantially higher mass densities. It was soon recognized that the principle of swaging with an "open" sample room cannot be applied, the material flow at the borders for higher pressures becoming a major problem. An operational version of an automatic densification machine was recently installed in Geneva (see Fig. 3). The automatic monitoring of both independent hydraulic systems enables a press/release/travel work sequence, at pressures $p \leq 1.5\,\mathrm{GPa}$.

3. Results of CHPD Applied on Short *In Situ* MgB_2 Wires

3.1 *CHPD applied on in situ multifilament Fe/MgB_2 wires*

The first experiments with CHPD were performed on monofilamentary *in situ* tapes, first on Fe/MgB_2 tapes, then on MgB_2 tapes alloyed with $C_4H_6O_5$ (malic acid), with a Monel sheath and a Nb barrier. In both cases, a strong enhancement of J_c was observed[15]: the value of $J_c^{//}$ at 4.2 K and 10 T was raised by 100%!

The next step consisted in applying CHPD on wires, first with monofilamentary configurations: the results on MgB_2 wires alloyed with 10 wt.% $C_4H_6O_5$ (malic acid) confirmed the strong enhancement of J_c observed in the tapes. For a monofilamentary wire, the value of $B(10^4)^{//}$ at the pressure $p = 1.48$ GPa was enhanced by 1.9 T, from 11.3 to 13.2 T. The J_c values of the present wires at 8 T and 4.2 K are approaching those of industrial Nb–Ti wires (1.2×10^5 A/cm^2). It should be noted that the highest J_c recently reported for the C-doped second generation MgB_2 wire (AIMI method) was 10^5 A/cm^2 at 4.2 K and 10 T,[21] well exceeding those of commercial Nb–Ti wires.

The experiments were then extended to multifilamentary wire configurations. The results for a binary Fe/MgB_2 wire with 18-filaments are shown in Fig. 4 for $T = 4.2$ K. It is seen that J_c is enhanced over the whole field range. The enhancement factor of J_c is 2.1 at 10 T and decreases gradually to 1.4 at 5 T, reflecting that CHPD not only enhances the mass density, but also the irreversibility field, B_{irr}. It is also interesting that the anisotropy of J_c in the wires for parallel and perpendicular field orientation is always below a factor 1.1.

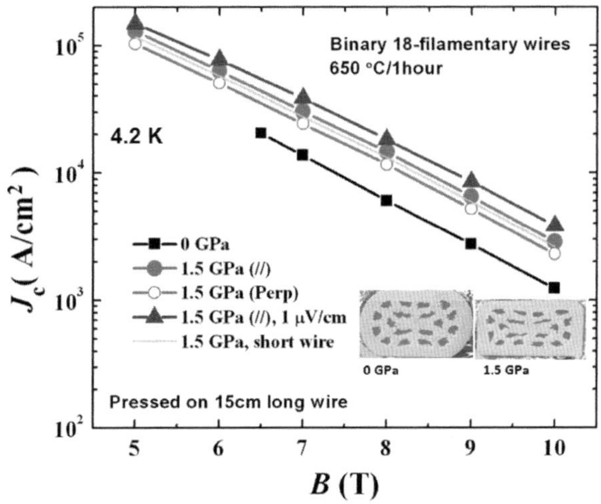

Fig. 4. J_c versus B at 4.2 K for binary 18-filament Fe/MgB_2 wires, at $p = 0$ and after CHPD at 1.5 GPa. The B (10^4) values are 8.48 and 8.18 T for // and $\perp$ field orientation with respect to the wider surface, respectively. Criterion for J_c: 0.1 μV/cm (triangles: 1 μV/cm). (Reproduced from Hossain *et al.*[22] Reproduced with permission from © IOP Publishing).

The next goal was to extend the CHPD processing to multifilamentary MgB_2 wires of longer wire lengths.[22] Sequential densification was applied to binary 18-filament wires of 150 mm total length, with overlapping pressure regions, leading again to a strong enhancement of the transport J_c, as shown in Fig. 4. The field $B\ (10^4)^{//}$ at which J_c (4.2 K) reaches 1×10^4 A/cm^2 for // field direction was determined to 8.5 T (0.1 μV/cm criterion). The observed enhancement of J_c by a factor 2.3 at 4.2 K at all applied fields up to 10 T in densified samples is directly correlated to the observed enhancement of mass density due to the filament volume contraction.

3.2 *Pressure effects on wires for application at 20 K and above*

The effect of densification at 20 and 25 K is particularly interesting in view of cryogen-free MRI applications. It was found that the increase of J_c of mono and multifilament wires after densification is considerably higher at 20 K than at 4.2 K. For the monofilamentary C-doped wire, J_c at 5, 6 and 7 T is enhanced by a factor 3.5, 5.6 and 10, respectively. After densification at 2 GPa, the value of $B(10^4)$ at 20 K was found to increase by 1.2 T, which is a promising result in view of the fabrication of magnets operating at 20 K. For the multifilament binary sample, after pressing at 1.5 GPa, the value of $J_c^{//}$ at 20 and 25 K exhibited a strong increase by a factor 4.5 in the field range between 4.5 and 5 T, the enhancement at 25 K and 3 T being close to 6. For comparison, the corresponding enhancement at 4.2 K is only 2.3 (Fig. 5). The increase of J_c at 20 K in Fig. 5 appears to be much stronger than the one at 4.2 K (Fig. 4), but is in reality due to the fact that the measurements at 4.2 and 20 K are performed at very different values of the reduced field $B/B_{irr}(T)$. Even for the MgB_2 cables, the I_c values of CHPD treated cable show 60 A at 4 T which is clearly higher than the cables deformed by other mechanical deformation techniques.[24] CHPD processing causes an enhancement of $B\ (10^4)$ compared to other mechanical deformation techniques. For cables subjected to CHPD, $B\ (10^4)$ at 20 K is found to increase from 4.2 to 5.0 T compared to cables without any mechanical

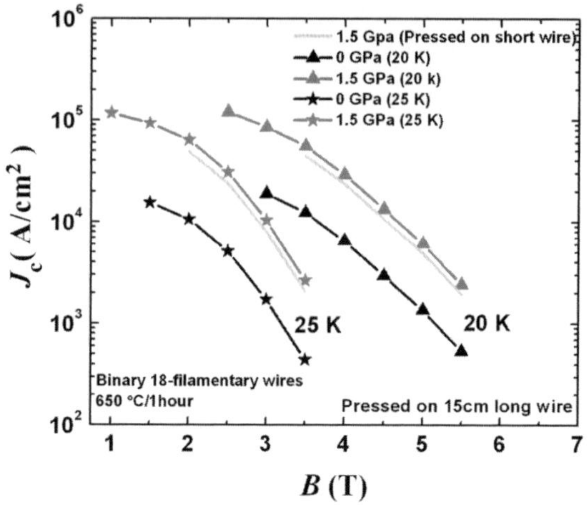

Fig. 5. J_c versus B at 20 and 25 K for binary 18-filament MgB_2 wires, at $p = 0$ and after CHPD at 1.5 GPa (Reproduced from Ref. Hossain *et al.*[22] Reproduced with permission from IOP Publishing).

deformation. These results are very promising to use this CHPD treated cables in cryogen-free or LH_2 cooled power applications.[26]

3.3 *Effect of CHPD on the exponential n factor*

In view of MRI or medium field NMR applications, the variation of the exponential n factor with applied field plays an important role. The increased homogeneity of the transport properties after CHPD is also reflected by the current–voltage (I–V) characteristics exhibiting a higher n value for the densified wires. At 4.2 K and low fields, n of MgB_2 wires is known to reach very high values.

It is well-known that the strong decrease of n with field is a limiting factor when envisaging applications at higher fields. As shown in Figs. 6(a) and 6(b) densification leads to an increase of the n factor in both mono and multifilamentary wires. The increase of n after densification at 1.5 GPa is around 50%. An argument of industrial importance is that the field at which n takes values of the order of 30 is higher in densified wires, the average increase for the binary MgB_2 being ~1 T at 4.2 K and ~0.5 T at 20 K. It should be noted that the enhancement for the monofilamentary malic added

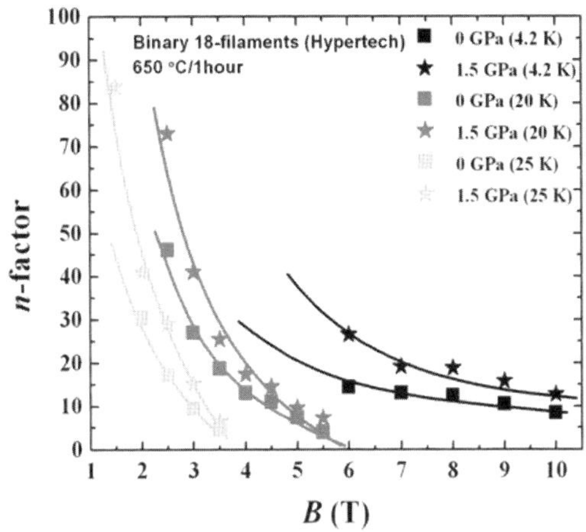

Fig. 6. Variation of the exponential n factor (criterion 0.1 μV/cm) as a function of applied field, at 4.2, 20 and 25 K for binary 18-filament MgB_2 wires after CHPD at 1.5 GPa. The value of n increases at all temperatures and fields (Reproduced from Hossain *et al.*[16] Reproduced with permission from IOP Publishing).

wires is considerably higher: ~2 T at 4.2 K and ~1 T at 20 K.[19] This difference may be due to the fact that the malic alloyed MgB_2 wire was monofilamentary, with a considerably thicker filament surface (0.092 mm^2) than for the multifilamentary wires with a filament surface of 0.005 mm^2 (we take here the surface, the filament shapes being irregular). Although n is an empirical factor depending on a variety of effects which are not all well-defined, it reflects in a certain way the homogeneity inside the filaments. The present enhancement of n can thus be interpreted as an improved homogeneity of the densified filament (this is also confirmed by the analysis of the electrical resistivity presented in the following).

3.4 *Effect of CHPD on anisotropy of in situ MgB_2 wires*

The variation of the anisotropy factor, $\Gamma = J_c^{//}/J_c^{\perp}$, versus. B at 4.2 K and 20 K for binary and malic doped tapes before and after CHPD processing is shown in Figs. 7(a) and 7(b). After CHPD, Γ of the binary tape increased with increasing aspect ratio from

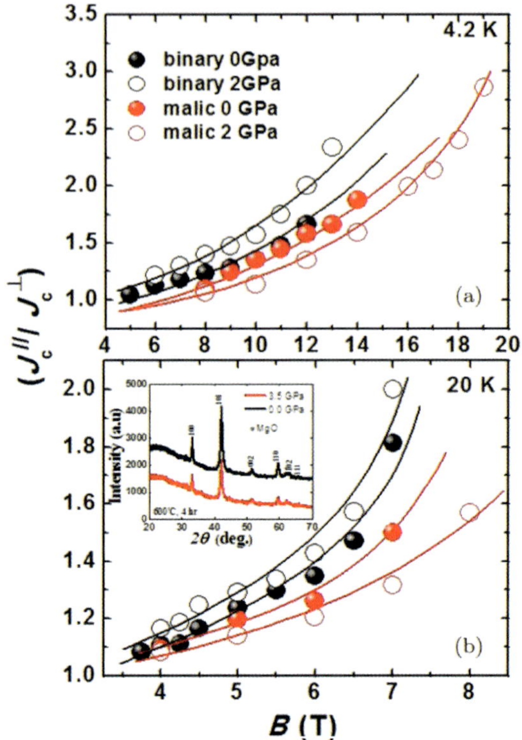

Fig. 7. Variation of $\Gamma(J_c^{//}/J_c^{\perp})$ versus magnetic fields at 4.2 and 20 K for binary and $C_4H_6O_5$ alloyed MgB_2 tapes (Monel sheath, Nb barrier) at 0 and 2 GPa, under 600°C/4 h.

3.87 to 6.96, in agreement with Kovac *et al.*[27] In malic acid doped tapes the value of Γ decreased after CHPD, even for aspect ratios between 4.34 and 7.60. All these tapes were submitted to the same deformation conditions, heat treatments and sheath materials. This difference may arise from the combination of two effects influencing the J_c anisotropy, i.e., C substitution and texture.

As expected, the J_c anisotropy after CHPD, is more pronounced in binary tapes where the texture effect is more visible. In the *in situ* process, the texture is introduced by plastic deformation of Mg, the higher current densities at 4.2 K and 20 K[7,28] being due to effective C-doping. The texture in tapes is affected by impurity scattering due to C-doping and hence, the anisotropy decreases in malic acid doped tapes compared to binary tapes, in good agreement with other works.[27,29] After pressing at 2 GPa, the malic acid-doped tape exhibits at both temperatures a further reduction of anisotropy. Since the aspect ratio is enhanced after CHPD at 2 GPa in malic acid doped tapes, a stronger texture in the filament is expected, leading to a higher anisotropy factor. But the interesting point is that malic acid tapes pressed with 2 GPa exhibit lower T_c and a slightly higher C substitution (the a lattice parameter is slightly decreased) than those without pressure, which is certainly the cause for the reduced anisotropy.

3.5 *CHPD applied on ex situ wires: A comparison*

CHPD was applied on *ex situ* MgB_2 wires,[25] too. However, a substantially different behavior was observed when compared to the one observed in *in situ* wires. In contrast to *in situ* wires, where densification at 1.5 GPa on binary and ternary alloyed MgB_2 wires causes a marked increase of mass density and a strong enhancement of J_c, *ex situ* wires up to 2 GPa show a negligible increase of the MgB_2 mass density only, while a considerable enhancement of J_c is still observed. In both cases, this reflects an enhancement of grain connectivity; in *ex situ* wires, though, the enhancement of J_c is connected to the partial disruption of the oxide layer around each MgB_2 powder particle. It is well known that J_c of *ex situ* wires decreases after longer exposition times when exposed to air prior to annealing; after several months J_c falls to values $\leq 50\%$ of the original value. A lower electrical resistivity is found on pressed *ex situ* MgB_2 pressed *ex situ* MgB_2 wires, which confirms the improvement of grain connectivity due to the breakage of the oxide layers.[25]

4. CHPD Applied on Long *in situ* MgB_2 Wires and Cables

4.1 *CHPD applied on long in situ wires*

As reported earlier,[15,16,19,22] no microcracks were detected in the transition region between pressed and unpressed filaments, neither by optical nor by SEM microscopy. This was also the case after repeated pressing steps on the same overlapping regions. This forgiving behavior of the Mg + B core opens the way to the densification of much longer, industrial wire lengths. In view of the densification of long wire lengths, the fact to work well below 1 GPa is a very positive point: indeed, the strong reduction of applied pressures will strongly prolongate the lifetime of the anvils. So far, wire lengths up to 10 meters were densified without any difficulties. We found that J_c reaches a maximum at 0.85 GPa and then decreases for higher pressures, thus confirming earlier observations on wires pressed with the short sample device.[15] At the present state, no explanation has been found for this behavior. Although earlier investigations failed in detecting microcracks,[15] their presence in these new wires densified with another anvil configuration cannot be excluded and is presently the object of further investigations.

4.2 *CHPD applied on cables based on in situ MgB_2 wires*

After the successful application of CHPD on mono and multifilament wires, it was tempting to apply the CHPD technique on MgB_2 cables[24] to improve their critical current density values. Multifilamentary and multistrand conductors with current capacities in the kA range and higher n factors are the key requirements for many large scale applications. In this study, the binary *in situ* MgB_2 wire with the highest performance was selected for the cable construction. The variation of the I_c, J_c and J_e values for the present binary 6-strand MgB_2 cables before and after CHPD with other mechanical deformations at 4.2 K are presented in Fig. 7 and compared with the values for the single wire. The highest critical current density (10^4 A/cm^2 at 10 T) was found in the cable after deformation by CHPD, due to higher filament mass density compared to the cables deformed by swaging, groove-rolling and TAR.

The I_c values of a CHPD treated cable show 60 A at 4 T, i.e., clearly higher than the cables deformed by other mechanical deformation techniques.[24] CHPD causes an enhancement of B (10^4)

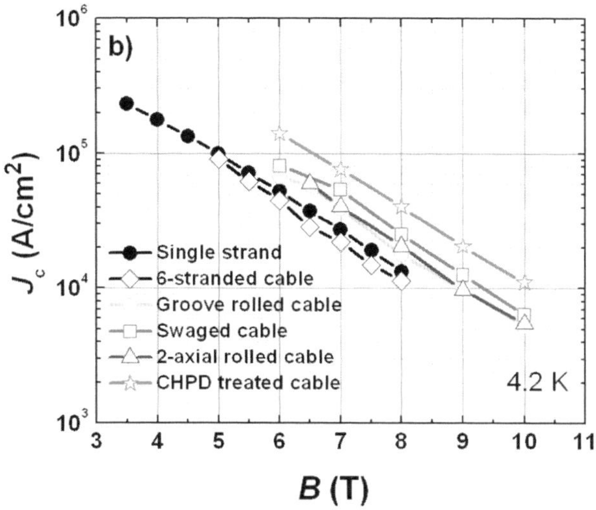

Fig. 8. J_c versus B of single monofilament wires and cables with and without additional mechanical deformations at 4.2 K (From Hossain *et al.*[24] Reproduced with permission from IOP Publishing).

compared to the other techniques of mechanical deformation. For cables subjected to CHPD, B (10^4) at 20 K is found to increase from 4.2 to 5.0 T compared to cables without any mechanical deformation (Fig. 8). These results are very promising to use there CHPD treated cables in cryogen-free or LH_2 cooled power applications.[25]

CHPD treated C-doped MgB_2 wires were placed in a toroidal molten salt based HIP chamber for further densification during reaction. The wire length was 1 m, the reaction conditions were 700°C/20 minutes under 1.4 GPa. The J_c performance of CHPD and HIP co-treated PIT MgB_2 wires in our work is close to that of IMD MgB_2 wires reported in previous studies. Although the AIMI wires exhibit so far the highest J_c values among all MgB_2 wires,[21] there is still some development to be performed on these second generation wires. Indeed, Mg diffuses into the B layer during the AIMI reaction, thus forming a hollow wire and a thin layer of fully dense MgB_2 inside the sheath tube, thus lowering the filling factor (for more details see Chapter 4f). On the contrary, the core of CHPD and HIP co-treated PIT MgB_2 wires in the recent molten salt wires (Morawski *et al.*[20]) exhibit a very high filling factor. One can imagine that this structure in our wires possesses higher J_e as well as better mechanical properties than IMD MgB_2 wires.

5. Alternative Method for the Densification of Long Wire Lengths: The Toroidal Molten Salt Based HIP Chamber

HIP is not suitable for the production of wires of industrial length, due to several reasons: (a) the possibility of Ar gas trapped inside the filaments and (b) the limitation of applied gas pressure to 0.2 GPa and (c) the size of the pressure chamber being too small for reacting wires in km lengths. In order to avoid these difficulties a new type of High Pressure device was recently constructed by Morawski *et al.*[20] at the Institute of High Pressure Physics in Warsaw (Poland), replacing the gas by a liquid, thus allowing pressures up to 1.4 GPa. In this device (Fig. 10) the pressing medium consists of molten salt specific eutectic mixtures of salts (i.e., NaCl and KCl) along with Boron Nitride (BN). BN is used for its high chemical stability and plasticity.

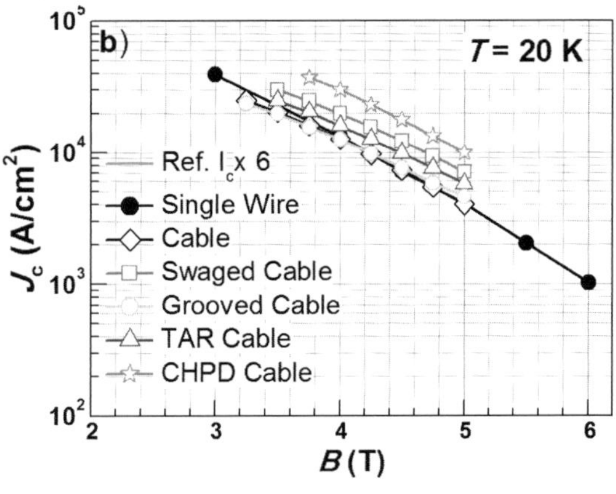

Fig. 9. Critical current density, J_c versus B, of single monofilament wire and cables with and without additional mechanical deformations at 20 K; all sintered at 650°C/20 min. (From Hossain *et al.*[22] Reproduced with permission from IOP Publishing).

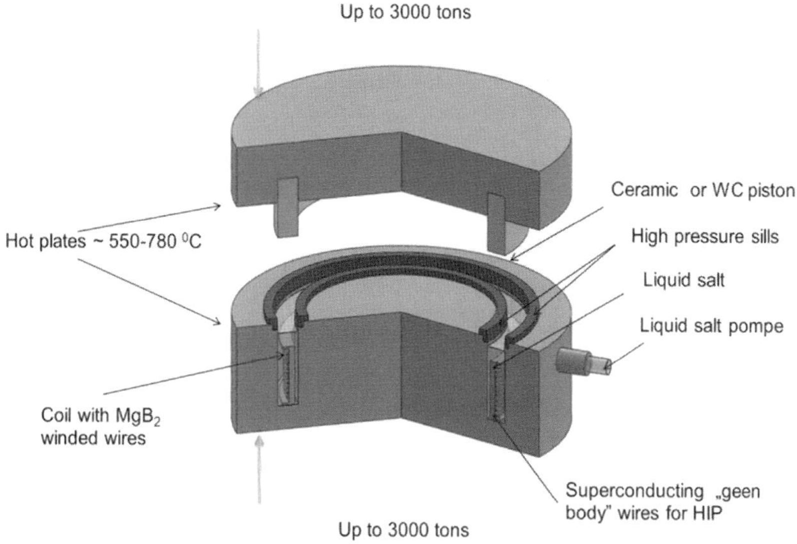

Fig. 10. Schematic drawing of the eutectic liquid salt based HIP system. CHPD treated long wire was further densified by this technique. (Courtesy from A. Morawski, Institute of High Pressure Physics in Warsaw, Poland).

In this process, a long wire sample is kept inside the salt bed in a sealed high pressure chamber. At the desired temperature the solid salt mixture melts in the high pressure chamber, which is tightly closed. External forces onto the upper cover are directly transmitted to the pressure chamber before to reach the reaction temperature of 650–700° C.

This new toroid type isostatic hot press enables the production of very long lengths of MgB_2 wire. The pictorial schematic design of the first prototype of liquid salt based HIP machine has been shown in Fig. 4. The second stage densification after CHPD using toroid HIP chamber has recently been applied on optimized CHPD treated wires under the collaboration between University of Wollongong, Australia and Institute for High Pressure Physics, Unipress, Poland. The highest J_c values have been obtained for a *in situ* PIT wires: as will be shown later in this work (Fig. 12), the results are well comparable with the highest known ones at the present day: second generation MgB_2 wires produced by the Advanced Internal Mg

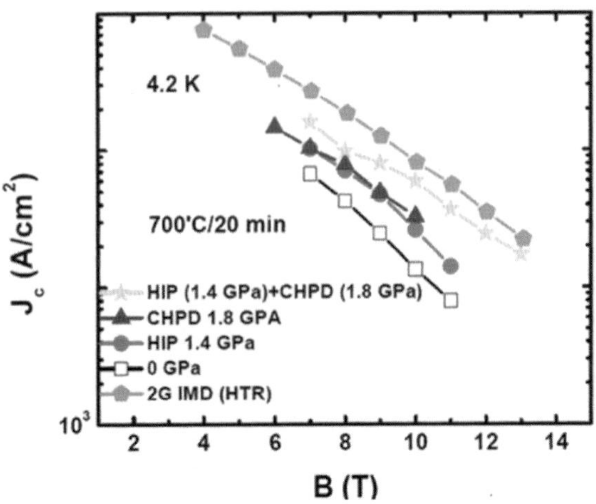

Fig. 11. Molten salt based HIP applied on CHPD treated pre-C-doped MgB_2 wire, compared to undoped, only CHPD and HIP treated separately and also compared with the best AIMI second generation wire.

Infiltration technique (AIMI) reported by Hyper Tech, Research, USA (see Chapter 4f).

6. Microstructure and Physical Properties of CHPD *in situ* MgB_2 wires

6.1 *Microstructure*

The insets in the Figs. 5 and 6 show the cross-section of mono and multifilament wires and cables, before and after applying CHPD. The change of the aspect ratio b/a and the reduction of the filament cross-section are clearly visible. The transition zone between the unpressed region and the one pressed at 1.9 GPa and reacted at 650°C for 1 h is shown in Fig. 12(a). The microstructure of conventionally treated MgB_2 filaments shows always a certain amount of voids (see Fig. 12(b)), due to volume shrinking during the reaction process. Most voids are not observed anymore after densification, as illustrated by Fig. 12 after densification at a pressure of 1.9 GPa. In doped wires, CHPD makes the C substitution more effective, leading to a reduction of T_c and of anisotropy.

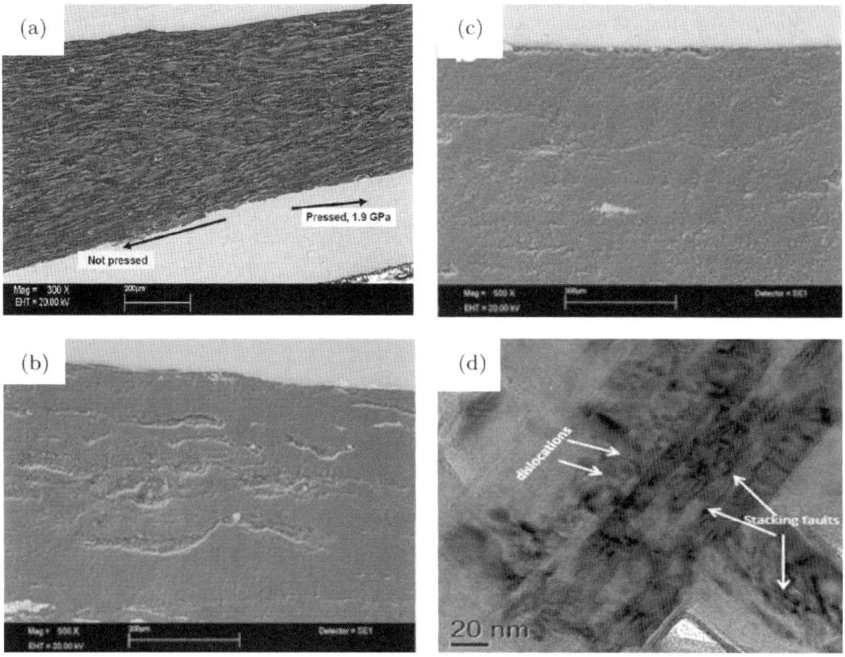

Fig. 12. (a) filament of a binary Fe/MgB_2 wire at the border of a region pressed at 1.9 GPa, (b) and (c) high resolution SEM and (d) dark-field STEM investigations revealing reduced porosity, stacking faults and dislocations due to the typical combination of C-doping and cold pressure.

This observation is further supported by the TEM shown in Fig. 13. The observed improvement of J_c after applying CHPD in malic-doped wires[16] is due to the reduced amount of voids (as seen by SEM). High resolution STEM investigations revealed that the subgrain lattice is randomly oriented and many crystal lattice defects such as stacking faults and dislocations are found in subgrains due to the typical combination of C-doping and cold pressure.

6.2 *Effect of CHPD on AC loss, intrawire resistance and strain dependence of critical current*

Zhou *et al.*[30] investigated and analyzed the impact of CHPD on the intrawire resistance, AC loss, and strain dependence of the critical

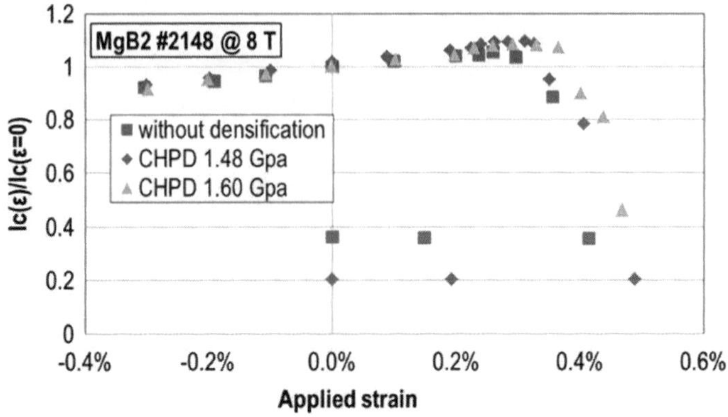

Fig. 13. Normalized I_c of a Fe/MgB_2 wire (No. 2148), as a function of axial strain, measured at 4.2 K and 8 T using a U-spring (Adapted from Ref. 30).

current for the first time. Increased critical current and AC loss, and a reduced intrawire resistance, are found after CHPD.

It is also found that the densified MgB_2 wires show less strain sensitivity and a higher irreversible strain limit, which is likely due to the enhanced connectivity and percolation of MgB_2 filaments. As indicated in Fig. 13, the irreversible strain is around 0.28%, 0.31% and 0.33% in the cases without CHPD, CHPD at 1.48 GPa and 1.60 GPa, respectively. In the high tensile strain region (above 0.3%), the densified wires also show better performance than those without densification. In addition to the reduced interfilament resistance, this is another advantage of the MgB_2 conductors with CHPD, although the effect is marginal.

6.3 *Electrical resistivity*

Grain connectivity and the pinning strength at grain boundaries are essential factors for determining the critical current density of polycrystalline MgB_2. Eisterer[31] and Matsushita *et al.*[11] used a percolation model for explaining the correlation between higher connectivity and the enhancement of J_c. Pachla *et al.*[12] evaluated the electrical connectivity on a series of bulk samples submitted to various pressures before reaction and found that J_c is approximately

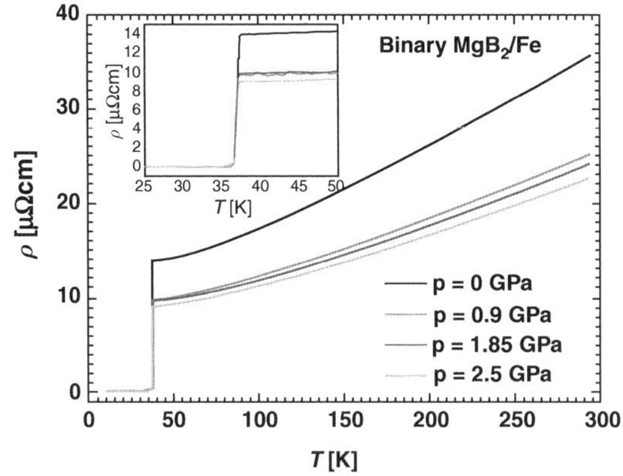

Fig. 14. Variation of the electrical resistivity for a binary MgB_2 wire before and after various pressures. The numbers are in GPa (adapted from Senatore *et al.*[18]).

proportional to the product of connectivity K, the elementary pinning force of grain boundaries and the reciprocal grain size, The connectivity $K = \Delta\rho_g/\Delta\rho$ was defined by Rowell,[1] $\Delta\rho_g$ being the value for a single crystal and $\Delta\rho = \rho_{300\,\mathrm{K}} - \rho_{40\,\mathrm{K}}$. These quantities are more difficult to determine for filamentary samples, due to the presence of a metallic sheath, e.g., Fe, Nb + Monel, . . . , but the same general tendency is followed.

The present measurements on a binary MgB_2 wire (Fig. 14) confirm the observation of Matsushita *et al.*[11] that the electrical resistivity $\rho_{40\,\mathrm{K}}$ of the filaments decreases for densified wires. The decrease of $\rho_{40\,\mathrm{K}}$ with pressure represents a strong evidence of the improvement of the grain connectivity. From the resistivity versus temperature measurements, it follows that after densification the width of the resistive transition for a given probing current density J is reduced. This reduction of the transition width also was observed by low temperature specific heat measurements.[18] Combining the analysis of the T_c distribution with the results of the $\rho(T)$ measurements, it is possible to obtain an insight of the influence of CHPD on percolation. By integrating the T_c distributions in Fig. 15 between the onset T_c and T ($\rho = 0$), Senatore *et al.*[18]

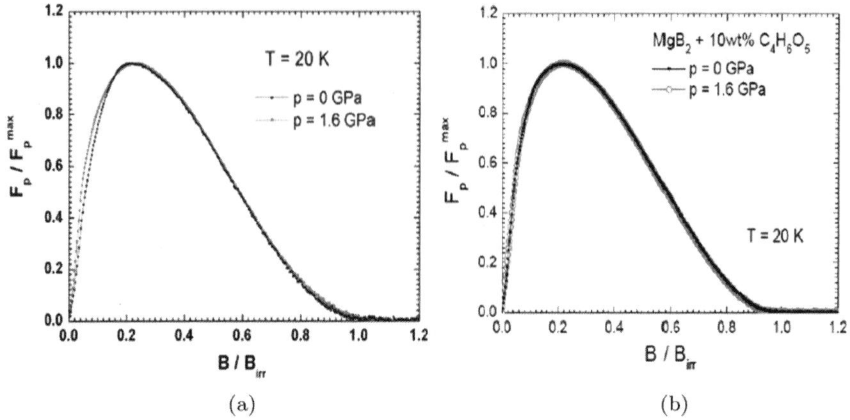

Fig. 15. Magnetization measurements for the same (a) binary and (b) C4H6O5 alloyed MgB_2 wire, conventional ($p = 0$) and after densification at $p = 1.6\,\mathrm{GPa}$ (adapted from Ref. 18).

determined the minimum superconducting volume fraction needed for a full percolation path with zero resistivity (see Chapter 5a in this book). For the non-densified binary sample, 14.9% of the filament volume needs to become superconducting in order to get percolation with zero resistivity. This fraction decreases to 8.4% after CHPD. For the percolation path in a malic acid-doped wire, the value of 25.3%, was determined.[18] The marked effect of CHPD on the percolation properties of MgB_2 binary and alloyed wires is a direct consequence of the improved connectivity between the MgB_2 grains.

6.4 *Pinning effects*

A definitive argument excluding effects on the pinning mechanism is given by the normalized pinning force $F_p/F_{p,\max}$ obtained by magnetization.

It is found that the plot $F_p/F_{p,\max}$ versus B/B_{irr} remains unchanged (Figs. 15(a) and 15(b)) at $B/B_{\mathrm{irr}} > 0.2$ after densification at 1.6 GPa for both binary and malic acid-doped samples.[15] It follows that the only effect of densification is to enhance the number of pinning centers.

7. Conclusions

Multifilamentary and multistrand conductors with critical currents of the kA class and higher n factors are the key requirements for many large scale applications. In 2008, a new kind of treatment was introduced at GAP in Geneva, the CHPD technique.[15,16] The relative mass density inside MgB_2 filaments of *in situ* mono and multifilamentary wires was enhanced from 48 to >60% for pressures well above 1 GPa, thus leading to strongly enhanced values of J_c. The J_c at 4.2 K and 8 T was enhanced to $\sim 10^5$ Acm^{-2} for a malic acid doped square conductor by CHPD. The highest J_c values reported for the AIMI second generation MgB_2 wire was 10^5 Acm^{-2} at 9.5 T,[20] well exceeding those commercial Nb–Ti wires. However, it appears that serious progress has to be achieved on the way to a successful scaling-up of this method, in particular concerning the homogeneity of filament cross-sections and reliability. It is expected that the alternative processing methods for MgB_2 wires will also undergo sizeable improvements. The further improvement of MgB_2 wires involves the following main factors:

(a) Availability of amorphous, high purity B powder with particle sizes in the 100 nm range,
(b) Optimization of mass density in MgB_2 filaments, by improved densification methods,
(c) Enhancement of the filling factor,
(d) Mechanical and electrical stability and reliability and
(e) Reduction of filament size for reducing the losses in AC applications.

This requires further studies on cold and hot deformation techniques. The question of mass density in superconducting filaments is becoming more and more important, the filaments in an increasing number of superconducting wires being produced by powder metallurgical techniques. In addition to MgB_2, this is the case for Nb_3Sn wires produced by the PIT technique, for Bi-2212 wires and for 122 pnictide wires. It follows that the improvement of critical current density by application of an external pressure is a subject of high actuality, even for hypothetical future superconducting materials produced by powder metallurgy. In the following chapters, our attention will be concentrated on the effect of high pressure on *in situ* and on *ex situ* MgB_2 wires.

References

1. J. M. Rowell, *Supercond. Sci. Technol.* **16**, p. R17 (2003).
2. E. W. Collings, M. D. Sumption, M. Bhatia, M. A. Susner and S. D. Bohnenstiehl, *Supercond. Sci. Technol.* **21**, p. 103001 (2008).
3. A. V. Pan, S. Zhou, H. K. Liu and S. X. Dou, *Supercond. Sci. Technol.* **16**, p. 639 (2003).
4. A. Serquis, L. Civale, D. L. Hammond, X. Z. Liao, J. Y. Coulter, Y. T. Zhu, M. Jaime, D. E. Peterson and F. M. Mueller, *Appl. Phys. Lett.* **82**, p. 2847 (2003).
5. T. A. Prikhna, W. Gawalek, Y. A. Savchuk, V. E. Moshchil, M. Wendt and S. N. Dub, *J. Phys.: Conf. Ser.* **97**, p. 012022 (2008).
6. H. Yamada, M. Igarashi, Y. Nemoto, Y. Yamada, K. Tachikawa, H. Kitaguchi, A. Matsumoto and H. Kumakura, *Supercond. Sci. Technol.* **23**, p. 045030 (2010).
7. A. Abrahamsen, J.-C. Grivel, N. H. Andersen, M. Herrmann, W. Hässler, B. Birajdar, O. Eibl and K. Saks, *J. Phys.: Conf. Ser.* **97**, p. 012315 (2008).
8. T. Matsushita, M. Kikuchi, A. Yamamoto, J. I. Shimoyama and K. Kishio, *Supercond. Sci. Technol.* **21**, p. 015008 (2008).
9. W. Pachla, P. Kovac, R. Diduszko, A. Mazur, I. Husek, A. Morawski and A. Presz, *Supercond. Sci. Technol.* **16**, p. 7 (2003).
10. P. Kovac, I. Husek, W. Pachla and M. Kulczyk, *Supercond. Sci. Technol.* **20(7)**, pp. 607–610 (2007).
11. W. Pachla, P. Kovac, I. Husek, T. Melisek, M. Muler, V. Strbik, A. Mazur and A. Presz, The effect of hydrostatic extrusion on the J_c(B) characteristics of *ex situ* MgB_2 wires, *Supercond. Sci. Technol.* **18**(4), pp. 552–556 (2005).
12. G. Giunchi, S. Raineri, R. Wesche and P. L. Bruzzone, *Physica C* **401**, 310 (2004).
13. J. M. Hur, K. Togano, A. Matsumoto, H. Kumakura, H. Wada and K. Kimura, *Supercond. Sci. Technol.* **21**, p. 032001 (2008).
14. K. Togano, J. M. Hur, A. Matsumoto and H. Kumakura, *Supercond. Sci. Technol.* **22**, p. 015003 (2009).
15. R. Flükiger, M. S. A. Hossain and C. Senatore, *Supercond. Sci. Technol.* **22**, p. 085002 (2009).

16. M. S. A. Hossain, C. Senatore, R. Flukiger, M. A. Rindfleisch, M. Tomsic, J. H. Kim and S. X. Dou, *Supercond. Sci. Technol.* **22**, p. 09504 (2009).
17. R. Flükiger, M. S. A. Hossain, M. Kulich and C. Senatore, *Adv. Cryo. Engrg.* **58**(1435), pp. 353–362 (2012).
18. C. Senatore, M. S. A. Hossain and R. Flükiger, *IEEE. Trans. Appl. Supercond.* **21**, p. 2680 (2011).
19. R. Flükiger, M. S. A. Hossain, C. Senatore, F. Buta and M. A. Rindfleisch, *IEEE. Trans. Appl. Supercond.* **21**, p. 2649 (2011).
20. The authors thank A. Morawski for sending us the scheme of the high pressure machine (unpublished results).
21. J. B. Zwayer, M. A. Susner, M.A, C. J. M. Tomsic and E. W. Collings, *Supercond. Sci. Technol.* **26**(9), p. 095007 (2013).
22. M. S. A. Hossain, C. Senatore, M. A. Rindfleisch and R. Flükiger, *Supercond. Sci. Technol.* **24**, p. 075013 (2011).
23. A. Matsumoto, K. Togano, T. Ohmura and H. Kumakura, *Supercond. Sci. Technol.* **26**, p. 105027 (2013).
24. M. S. A. Hossain, A. Motaman, S. Barua, D. Patel, M. Mustapic, J. H. Kim, M. Rindfleisch, M. Tomsic, O. Cicek, T. Melisek, L. Kopera, A. Kario, B. Runtsch, S. X. Dou, W. Goldacker and P. Kovac, *Supercond. Sci. Technol.* **27**, p. 095016 (2014).
25. M. Kulich, R. Flükiger, C. Senatore, M. Tropeano and R. Piccardo, *Supercond. Sci. Technol.* **26**, p. 105019 (2013).
26. M. Tomsic, M. Rindfleisch, J. Yue, K. McFadden, D. Doll and J. Phillips, *Physica C* **456**, pp. 203–208 (2007).
27. P. Kovac, T. Melisek and I. Husek, *Supercond. Sci. Technol.* **18**(998), pp. L45–L48 (2005).
28. J. H. Kim, S. Zhou, M. S. A. Hossain, A. V. Pan and S. X. Dou, *Appl. Phys. Lett.* **89**, p. 142505 (2006).
29. W. Hässler, P. Kovac, M. Eisterer, A. B. Abrahamsen, M. Herrmann, C. Rodig, K. Nenkov, B. Holzapfel, T. Melisek, M. Kulich, M. V. Zimmermann, J. Bednarcik and J.-C. Grivel, *Supercond. Sci. Technol.* **22**, p. 065011 (2010).
30. C. Zhou, P. Gao, M. Dhallé, M. D. Sumption, M. Rindfleisch, M. Tomsic, M. Kulich, C. Senatore and A. Nijhuis, *Supercond. Sci. Technol.* **27**, p. 075002 (2014).
31. M. Eisterer, *Supercond. Sci. Technol.* **20**, pp. R47–73 (2007).

Chapter 4e

Fabrication of MgB_2 Wires by Internal Mg Diffusion (IMD)

Hiroaki Kumakura

National Institute for Materials Science
Tsukuba, Ibaraki 305-0047, Japan
Kumakura.Hiroaki@nims.go.jp

Internal Mg diffusion (IMD) is an effective method for improving the filling factor and for enhancing J_c as well as J_e values of MgB_2 wires. IMD processed wires exhibit much higher J_c values than PIT processed, *ex situ* or *in situ* wires. Since the short diffusion range of Mg in compacted B layers limits the formation of MgB_2 in the wire, the filament size has to be decreased in order to obtain a full reaction between Mg and B layer. This condition is fulfilled in multifilament wires. During the heat treatment, coronene ($C_{24}H_{12}$) additions to the starting B powder lead to a uniform $C_{24}H_{12}$ coating of each B particle, to the C substitution on B sites in the MgB_2 structure and thus to enhanced J_c values. MgB_2 wires with $C_{24}H_{12}$ additions exhibit higher J_c values than those with the well-known SiC addition. The addition of Mg powder to the B powder layer in IMD wires partly solves the problem of the short diffusion distance of Mg in B. With increasing the Mg content in the B layer, the wire structure can gradually be changed from an IMD to a powder-in-tube (PIT) type. This intermediate hybrid process between the IMD and PIT processes is effective in increasing the MgB_2 volume fraction in the wires and leads to enhanced J_e values.

1. Introduction

The powder-in-tube (PIT) process, the most commonly used method for the fabrication of MgB_2 wires and tapes, can be extended to scale-up production. However, being a powder metallurgical processes, serious difficulties are encountered when fabricating wires with dense structures and almost no voids. At present, PIT processed MgB_2 wires and tapes produced either by the *ex situ* or the *in situ* technique exhibit relatively low critical current density (J_c) values due to their low MgB_2 mass density and the correlated low connectivity of MgB_2. A possible method for enhancing the MgB_2 mass density and thus the J_c values consists in the application of hot pressing on cold worked wires and tapes.[1] However, it is difficult to fabricate practical long conductors by means of hot pressing techniques. Another method for increasing the MgB_2 density consists in forming MgB_2 layers by a diffusion reaction between separate Mg and B components of a composite, rather than between Mg and B powder mixtures as in PIT wires. Several investigations demonstrated that such a composite diffusion process leads to a higher mass density of MgB_2. Among these diffusion processes, the most promising one is the internal Mg diffusion (IMD) process, based on the *Infiltration Process* originally developed by Giunchi *et al.*[2] which can be easily applied for the fabrication of long MgB_2 wires and tapes.

2. Fabrication and Properties of Internal Mg Diffusion (IMD) Processed MgB_2 Wires

Figure 1 illustrates the fabrication of mono and multifilamentary MgB_2 wires using the IMD process.[3,4] The process starts with a monofilamentary wire: a pure Mg rod is fixed at the center of a metal tube and the hollow space between the inner wall of the metal tube and the Mg rod is filled with B powder. A typical diameter of the Mg rod is 2 mm, while the outer and inner diameters of the metal tube are 6 mm and 3.5 mm, respectively. The use of high quality B powder is essential to obtain wires with high J_c values, similarly to the case of PIT processed wires. Usually we used nanosize B powder of 99.9% purity. Various kinds of metal tubes can be used as a sheath, e.g., Fe, Ta and Nb tubes. Nanosize SiC powder can be added to the B powder in order to improve the J_c values at high magnetic fields,[5] similarly to the case of PIT processed wires. This composite

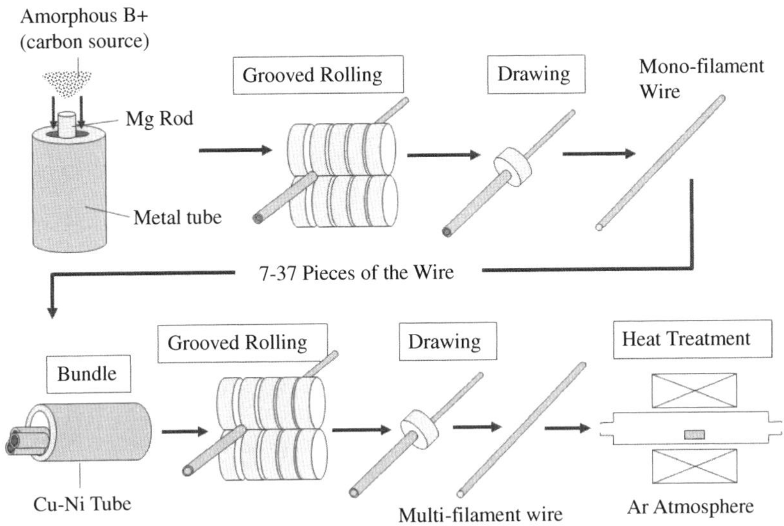

Fig. 1. Schematic illustration of IMD process for the MgB_2 wire fabrication.

is cold worked into a wire at room temperature without intermediate annealing.

For example, the composite rod is first cold worked by groove-rolling into a square rod with a cross-section of $\sim 2.3 \times 2.3\,\text{mm}^2$, followed by drawing into a monofilamentary wire. The initial composite can then be cold worked at room temperature into a wire of $\sim$1 mm diameter without any breakage. Multifilamentary MgB_2 wires can be easily fabricated, as shown schematically in Fig. 1. As an example, either 7, 19 or 37 monofilamentary wires are bundled and then inserted into a Cu–Ni tube with 3.9 mm inner and 6 mm outer diameter.[5] This composite rod is again cold worked into a final wire of 1.0–1.3 mm diameter by both groove-rolling and drawing. Intermediate annealing is carried out at 250°C for 1 h. to achieve good workability of the Mg filaments. Finally the wires are heat treated at 600$\sim$700°C for 0.25–10 h. in an argon gas atmosphere. During the heat treatment, Mg diffuses into the B powder layer and reacts with B to form MgB_2 (diffusion reaction).

C substitution for B sites in the MgB_2 lattice increases electron scattering, thus causing a decrease of the electronic mean free path and an increase of the upper critical magnetic field, B_{c2}. The latter leads to an enhancement of J_c in MgB_2 wires at high magnetic fields. A widely used additive for a C substitution on the B sites is SiC.[6] In a PIT MgB_2 wire, nanosize SiC powder additions to a Mg and B powder mixture result in a C substitution, thus enhancing B_{c2} and J_c at high magnetic fields.[7] In the IMD process nanosize SiC powder is added to the starting B powder to obtain a C substitution on the B sites and thus high J_c values at high magnetic fields. C coated nanosize B powders (supplied by SMI company)[8] can be used for this purpose.[9,10]

Figures 2(a) and 2(b) show optical micrographs of transverse and longitudinal cross-sections of IMD processed monofilamentary wires before and after the heat treatment at 670°C for 2 h. A Fe tube is used as a metal sheath, the final diameter of the wire being 1.2 mm. The wire cross-sections before the heat treatment reflect a uniform deformation of the composite. The poor deformability of Mg at room temperature is considerably improved by the presence of B powder around the Mg rod, thus yielding a uniform Mg filament structure. During the heat treatment, the Mg at the center of the wire diffuses

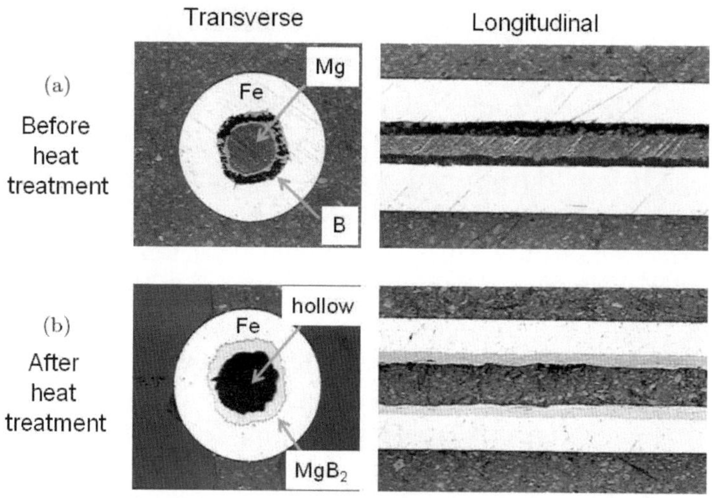

Fig. 2. Optical micrographs of transverse and longitudinal cross-sections of IMD processed monofilament wire, (a) before and (b) after the heat treatment at 670°C. From Togano *et al.*[11] (Reproduced with permission from IOP Publishing.)

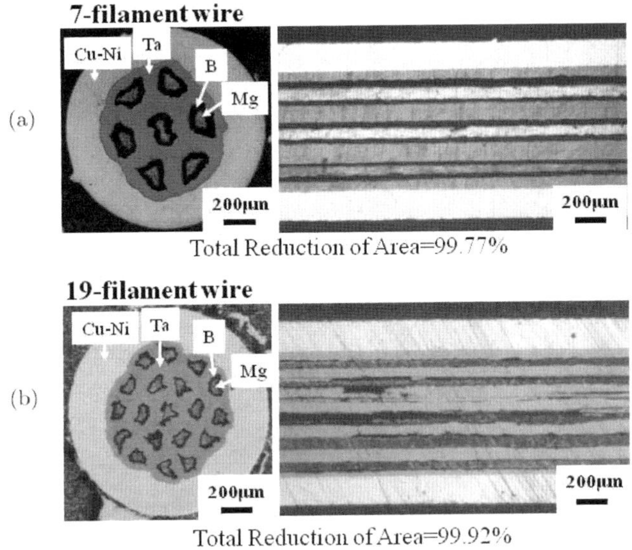

Fig. 3. Optical micrographs of transverse and longitudinal cross-sections of (a) 7- and (b) 19-filament wires before heat treatment. From Togano *et al.*[11] (Reproduced by permission of IOP Publishing.)

into the B powder layer and a MgB_2 layer is uniformly formed along the inner wall of the Fe sheath, as shown in Fig. 2(b). At the end of reaction, the center of the wire is hollow: a large void is generated due to the Mg diffusion into the B layer. A very thin layer of Fe_2B is observed at the interface between Fe sheath and B layer after the heat treatment.

In Figs. 3(a) and 3(b), optical micrographs of transverse and longitudinal cross-sections of 7- and 19-filament wires before heat treatment are shown.[11] Ta and Cu–Ni tubes are used as inner and outer sheath materials of these wires. Despite the large total area reduction, exceeding 99% from the starting billet to the final 7- and 19-filament wires, the cross-sections reflect a fairly uniform deformation. After the heat treatment a reacted layer with a thickness of 10–30 μm extends along the inner wall of the Ta sheath, while holes appear at the center of each filament, as in the case of monofilamentary wires. No reaction is observed between the Ta sheath and the reacted layer. The longitudinal cross-section of the

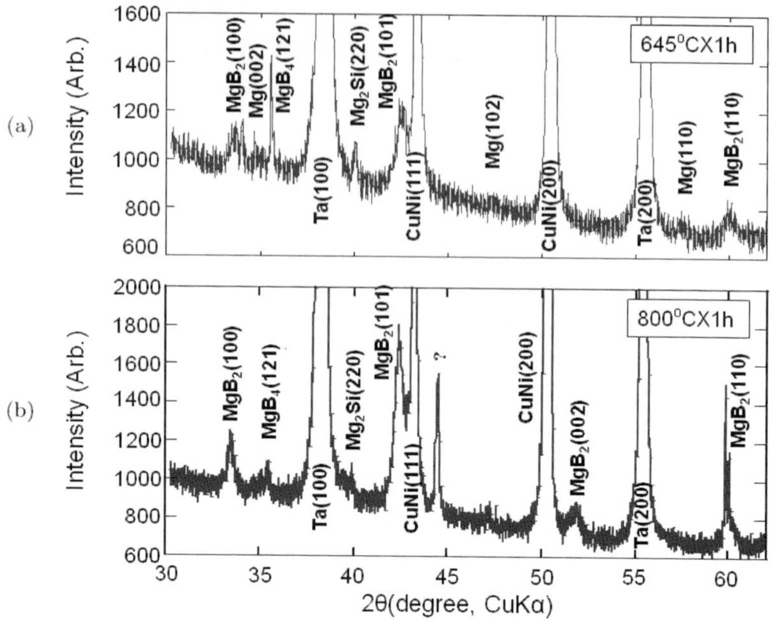

Fig. 4. X-ray diffraction patterns taken from the polished longitudinal cross-sections of the 7-filament MgB_2 wires heat treated at (a) 645°C and (b) 800°C for 1 h. From Ref. 11. (Reproduced with permission from IOP Publishing).

wires after IMD processing shows that the reacted layers form quite uniformly along the length of the wire, too.

The X-ray diffraction analysis of IMD wires shows that the major phase in the reacted layer is MgB_2. The reacted layer contains, however, a certain amount of impurity phases, e.g., MgB_4, Mg and Mg_2Si (the latter in case of SiC additives). Figures 4(a) and 4(b) show X-ray diffraction patterns of the cross-sections of SiC added 7-filament wires heat treated at 645°C and 800°C, respectively.[11] Strong peaks corresponding to the Ta and the Cu–Ni sheaths are observed. The peak intensities corresponding to MgB_4 are relatively high, especially after the treatment at 645°C, thus indicating the presence of substantial amounts of MgB_4 in the reacted layer. After the heat treatment at 800°C, the height of the MgB_4 peaks is lower, larger amount of Mg diffusing into the B layer, thus resulting in an enhanced amount of MgB_2. The a-axis lattice parameter calculated

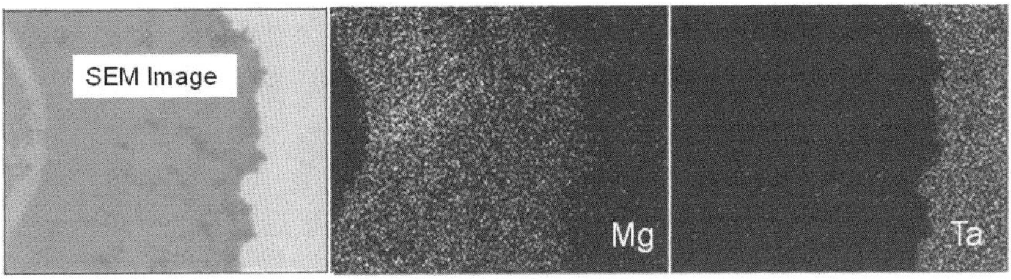

Fig. 5. SEM image and Mg and Ta EDX mappings at the cross-section of a 7-filament wire heat treated at 645°C for 1 h. From Togano *et al.*[3] (Reproduced with permission from IOP Publishing.)

from the (110) peak is 0.308 nm, indicating that the amount of C substitution is about $x = 0.1$ in the formula $Mg(B_{1-x}C_x)_2$. This confirms that during the IMD process C atoms from the nanosize SiC powder can effectively substitute Boron in the MgB_2 structure, in the same manner as in the PIT process.

Figure 5 shows a scanning electron microscope (SEM) image and energy dispersive X-ray (EDX) mappings of the Mg and Ta contents in the MgB_2 layer of a 7-filament wire heat treated at 645°C.[3] The Mg content in the MgB_2 layer clearly decreases from the Mg core towards the Ta sheath. The SEM image shows darker spots in the MgB_2 layer, their amount being enhanced in the vicinity of the Ta sheath. Taking into account the X-ray diffraction pattern in Fig. 4(a), these results suggest that Mg poor compounds such as MgB_4 are present in the MgB_2 layer, especially in the region near the Ta sheath.

The thickness of the reacted MgB_2 layer is sensitive to the heat treatment conditions. It rapidly increases with increasing temperature, while the unreacted B layer decreases. Above 640°C, however, J_c decreases with increasing heat treatment temperature, as will be discussed later. Thus, the initial B layer thickness has an important influence on the final thickness of the unreacted B layer.

Figure 6 shows high magnification optical microscope images on the transverse cross-sections of the mono, 7- and 19-filament Ta-sheathed wires after 640°C for 1 h.[11] In the single filament wire, a thick unreacted B layer remained between the reacted layer and the Ta wall, as a consequence of a too large initial thickness of the B layer. The presence of unreacted material is strongly reduced in the

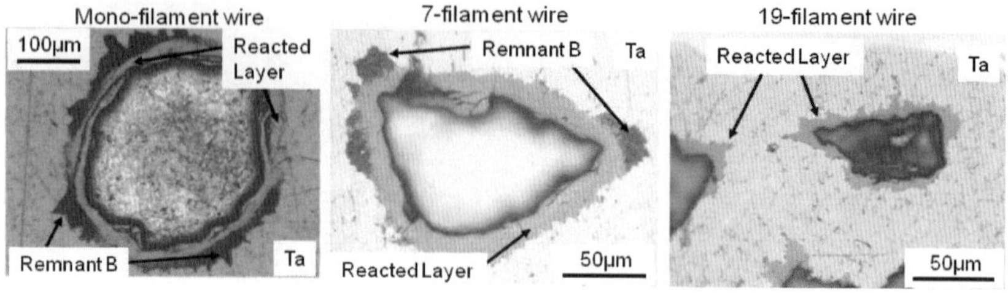

(a) (b) (c)

Fig. 6. Optical microscope images of (a) mono, (b) 7-wire and (c) 19-filament wires after the heat treatment at 640°C for 1 h. From Togano *et al.*[11] (Reproduced with permission from IOP Publishing.)

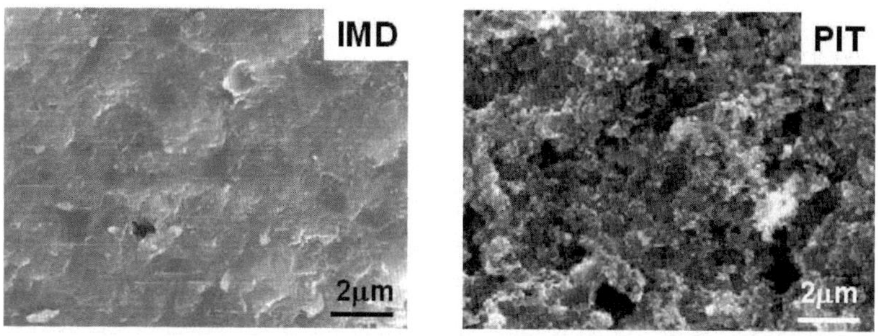

(a) (b)

Fig. 7. SEM images of the fractured MgB_2 cross-sections of (a) IMD and (b) PIT processed wires.

7-filament wire and almost disappears in the 19-filament wire heated at 640°C.[11] From these observations, it appears that the diffusion distance of Mg in the B powder layer is limited to about 20–30 μm at this temperature. Therefore, in order to obtain full reaction, it is necessary to reduce the thickness of the B layer to less than 20 μm, a value which is realized for wires with 19 or more filaments.

Figure 7 compares the SEM images of the fractured cross-sections of MgB_2 layers between IMD and PIT wires. The PIT wire has a very porous MgB_2 microstructure, in contrast to the IMD wire, which has a much denser MgB_2 layer. The Vickers hardness of the reacted MgB_2 layer in the IMD processed wires is around Hv1300, i.e., much higher than that of PIT processed wire and comparable to that of

MgB_2 bulk material synthesized at high pressures.[12] All these data indicate that the density of the MgB_2 layer in IMD wires is much higher than that in PIT wires.

The transport J_c value is sensitive to the heat treatment conditions. The J_c value at 4.2 K of IMD wires tends to increase with increasing temperature, reaching a maximum at around 640°C before decreasing again at higher temperatures.[11] The variation of J_c with the heat treatment temperature of IMD wires is different from that one observed in PIT wires, where J_c tends to decrease for increasing heat treatment temperatures above 600°C. In IMD wires, lower heat treatment temperatures, in the range between 600 and 640°C, result in larger amounts of impurity phases such as MgB_4 and unreacted B due to the lack of Mg supply from the Mg core and thus in lower J_c values. The optimum heat treatment temperature depends on the quality of the B powder used and on the filament diameter. The optimized heat treatment temperature seems to be around 670°C for IMD processed wires with diameters as small as ~0.6 mm fabricated using the C coated fine B powder.[10]

Figure 8 shows the variation of J_c versus B at several temperatures of a monofilamentary IMD MgB_2 wire with 0.6 mm diameter, fabricated using C coated B powder and heat treated at 670°C for 1 h.[10] For comparison, the results for Ta-sheathed, SiC added 7-filament MgB_2 wires[11] with 1.0 mm diameter are shown. The J_c–B curve at 4.2 K of the most commonly used Nb–Ti metallic superconducting wire[13] is also shown in the same figure. Both MgB_2 wires show excellent J_c values, exceeding 10^5 A/cm^2 at 10 T/4.2 K. The wires based on C coated B show excellent J_c values, too. The engineering critical current density J_e of the C coated wire, calculated by dividing I_c by the total area of the wire, is above 10^4 A/cm^2 the thickness of the B powder layer before heat treatment was sufficiently small, resulting in an almost complete reaction of B with Mg in the MgB_2 layer and thus high J_c and J_e values. At 20 K, the wires based on C coated B exhibit higher J_c values at magnetic fields below 6 T. At 20 K and 5 T, the J_c value of the wire based on C coated B is 7.6×10^4 A/cm^2, i.e., higher than that of the SiC added wire. At 20 K and 4 T the J_c values exceed 10^5 A/cm^2. At 10 K, the J_c values of the

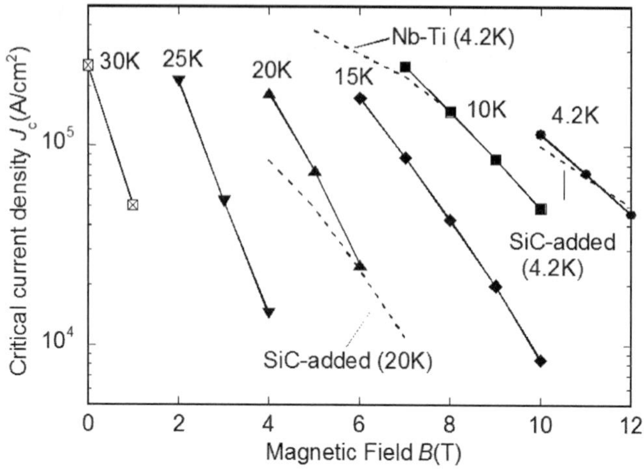

Fig. 8. J_c versus B at several temperatures for a monofilamentary IMD processed MgB_2 wire based on C coated B powder.[10] For comparison, the data of SiC added 7-filament MgB_2 wires[11] as well as a commercial Nb–Ti wire[13] is also shown. (Reproduced with permission from IOP Publishing).

wire based on C coated B are higher than those at 4.2 K of Nb–Ti commercial wires at the magnetic fields. The slope of the J_c versus B curve for the wire with C coated B is steeper than that for the 7-filament wire, both at 4.2 and 20 K. This is due to the fact that the 7-filament wire has a higher C content ($x = 0.1$ in $Mg(B_{1-x}C_x)_2$) for B than the C coated wire ($x = 0.03$).

The J_c values of IMD MgB_2 wires have a superior mechanical tolerance with respect to PIT wires. Figure 9 shows I_c as a function of applied tensile strain (ε) for monofilamentary IMD and PIT MgB_2 wires at 10 T and 4.2 K.[14] Solid markers indicate I_c values measured at fixed strain. Open markers indicate I_c values measured at zero applied stress to check the reversibility of the I_c value. The strain does not go back to zero for both wires after releasing because of the plastic deformation of the wires. The variation of I_c–ε for both wires is similar in shape. I_c increases linearly with increasing tensile strain up to the reversible limit, the irreversible strain, ε_{irr}, at which the I_c reaches its maximum value. At strains $\varepsilon > \varepsilon_{irr} I_c$ deteriorates irreversibly. These $I_c - \varepsilon$ relationships are qualitatively similar to the previously reported ones for MgB_2 wires.[15–18] It follows that IMD

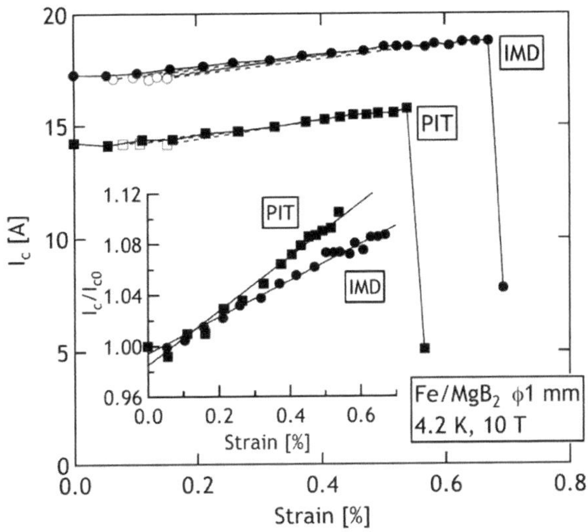

Fig. 9. Tensile strain (ε) dependence of I_c for monofilamentary IMD and PIT MgB_2 wires at 10 T, 4.2 K. From Nishijima *et al.*[14] (Reproduced with permission from IOP Publishing.)

MgB_2 wires exhibit a larger ε_{irr} value than PIT wires, i.e., 0.67 and 0.54%, respectively.

The inset in Fig. 9 shows the behavior of I_c/I_{c0} in the reversible region ($\varepsilon < \varepsilon_{irr}$) of I_c–ε curves, I_{c0} being the value of I_c at zero strain. Each I_c/I_{c0} versus ε plot can be fitted with a linear function, its slope corresponding to the strain sensitivity. The deduced strain sensitivity values are 0.14 and 0.21 for IMD and PIT wires, respectively. In addition, it should be mentioned that the slope for IMD MgB_2 wires increases with increasing magnetic field (not shown in the figure). The tendency is similar to the results of Kitaguchi *et al.*,[16] although it does not agree quantitatively.

Generally, the increase of I_c with increasing applied tensile strain (stress) in the reversible region ($\varepsilon < \varepsilon_{irr}$) is explained by the release of the residual compressive strain in the superconducting filament. Increasing the applied tensile stress, the compressive strain in the filament decreases and eventually changes to the tensile strain. When the applied strain ε exceeds ε_{irr}, the curve becomes irreversible, due to the introduction of cracks in the MgB_2 filament. One would expect that a higher reaction temperature would lead to larger ε_{irr} values, due to the larger compressive strain in the filament. However, in the present result, IMD MgB_2 wires reacted at 670°C show larger ε_{irr} values than PIT wires reacted at 850°C. This different behavior may have its origin in the different mass densities of the MgB_2 filaments in PIT and in IMD wires. As previously mentioned, the MgB_2 filaments in PIT wires are porous, with a reduced grain connectivity: cracks are easily introduced when applying small tensile stresses to the filament. The filaments in IMD wires are much denser and the grain connectivity is stronger, and microcracks are not introduced as easily as in PIT filaments.

3. Hydrocarbon Additions

As mentioned above, the C substitution on the B sites of the MgB_2 lattice enhances B_{c2} and J_c at high magnetic fields. Various C containing materials are used for both PIT and IMD MgB_2 wires. Aromatic hydrocarbons have high C and low H contents and should thus be effective C sources for MgB_2 superconducting wires. Yamada *et al.*[19,20] and Ye *et al.*[21] used liquid aromatic hydrocarbons in both, PIT and IMD MgB_2 wires. They observed a C substitution for B and an enhancement of J_c in their wires. Kim *et al.*[22] and Maeda *et al.*[23] used aromatic hydrocarbon pyrene ($C_{16}H_{10}$) additives in PIT MgB_2 wires and obtained a large increase of the J_c values.

Coronene ($C_{24}H_{12}$) is a useful C source for MgB_2 and is very effective in increasing J_c values of IMD and PIT MgB_2 wires.[24] The addition of $C_{24}H_{12}$ to the B powder is obtained by mixing $C_{24}H_{12}$ and B powders. However, $C_{24}H_{12}$ coating of B powders seems to be more effective than powder mixing, leading to a more uniform C distribution. $C_{24}H_{12}$ coating on B powder can be obtained by heating the mixture of B and $C_{24}H_{12}$ powders at a temperature above the melting point of $C_{24}H_{12}$, i.e., 438°C. During heating $C_{24}H_{12}$ melts and forms a $C_{24}H_{12}$ coating on each B particle.

Figure 10 shows the variation of the C content substituted on the B site (x in $Mg(B_{1-x}C_x)_2$) as a function of the C content in a $C_{24}H_{12}$ added PIT MgB_2 wire. The x value is calculated from the

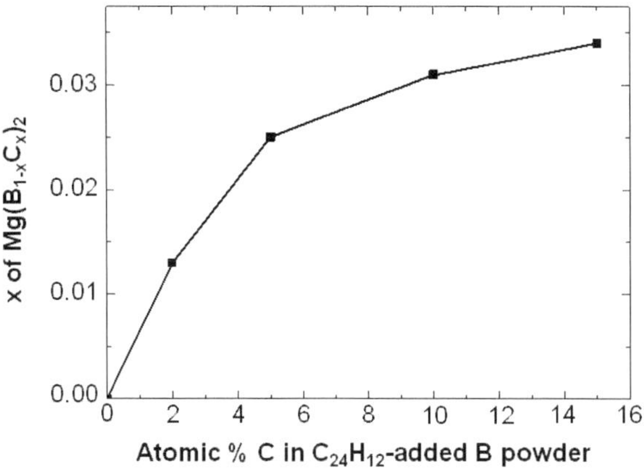

Fig. 10. The amount of C substitution for B site (x in $Mg(B_{1-x}C_x)_2$) as a function of $C_{24}H_{12}$ addition (at.% C) for a PIT MgB_2 wire. From Ye *et al.*[24] (Reproduced with permission from IOP Publishing.)

MgB_2 (110) peak in the X-ray diffraction pattern according to the method of Lee *et al.*[25] In this figure the horizontal axis shows the C content in the powder mixture (in atomic %). A similar result is expected for IMD wires. The C content can be easily changed by changing the amount of $C(C_{24}H_{12})$ addition and is 0.013, 0.025 and 0.031 for 2, 5 and 10 at.%C. At 5 at.%C, half of the added C substitutes B in the MgB_2 lattice. It follows that $C_{24}H_{12}$ is an active C source in MgB_2 superconducting wires. At $C_{24}H_{12}$ content less than 10 at.% C carbon effectively substitutes B, while at $C_{24}H_{12}$ contents exceeding 10 at.% C the factor x shows a saturation.

The variation of T_c with the C content (T_c defined at zero resistivity) of PIT MgB_2 wires is similar to that reported by Lee *et al.*,[25] who determined the T_c value by means of normalized zero-field magnetization and by the onset of the magnetization curve. Both T_c values show a similar trend, confirming the change with the C substitution for B shown in Fig. 10. Furthermore, the amount of the C substitution for B is also consistent with the slope of the J_c–B curve. Thus, the increase of the in-field J_c obtained by $C_{24}H_{12}$ additions can mainly be attributed to the enhancement of B_{c2} induced by the C substitution for B.

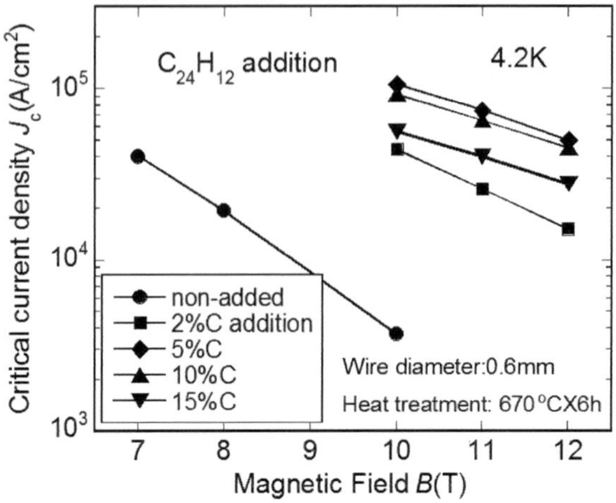

Fig. 11. J_c–B curves at 4.2 K for $C_{24}H_{12}$ added IMD processed MgB_2 wires. From Ye *et al.*[24] (Reproduced with permission from IOP Publishing.)

Figure 11 shows J_c–B curves at 4.2 K for IMD MgB_2 wires for 0, 2, 5, 10 and 15 at.% C($C_{24}H_{12}$) additions. All the wires have a diameter of 0.6 mm.[24] The slope of the J_c–B curve decreases with increasing $C_{24}H_{12}$ addition, which is the same as for the PIT wires. With increasing C($C_{24}H_{12}$) content, the J_c value first increases and then decreases.

A similar behavior is also obtained for PIT processed wires.[24] However, the highest J_c values are obtained for the wire with 5 at.% C($C_{24}H_{12}$) addition, which is different from the case of PIT wires, where the 10 at.% C addition yielded the highest J_c values. Actually, the addition of 10 at.% C($C_{24}H_{12}$) caused the highest C substitution for B, as shown in Fig. 10, but also produced more impurities (such as H_2 gas) compared to a 5 at.% C addition ($C_{24}H_{12}$). These impurities can be absorbed in the PIT wires due to the presence of many voids in the MgB_2 layer. IMD MgB_2 wires, on the other hand, have a very dense MgB_2 layer, which is mainly responsible for the high critical current properties. However, the impurities mentioned above decrease the effective mass density of the MgB_2 layer and thus the J_c values.[10]

Figure 12 shows SEM images of the polished MgB_2 layer along the longitudinal direction for MgB_2 wires[24] with 0 and 10 at.%

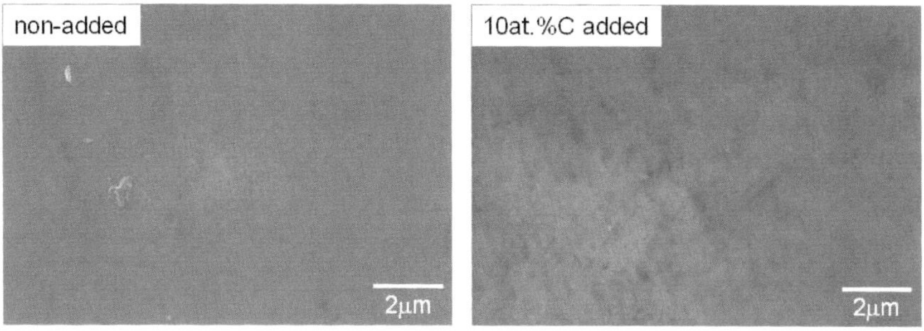

Fig. 12. Comparison between SEM images of polished MgB_2 layers of wires without additives and with 10 at.% $C(C_{24}H_{12})$. From Ye *et al.*[24] (Reproduced with permission from IOP Publishing.)

$C(C_{24}H_{12})$ additions. Both wires have MgB_2 layers with a high mass density. For the 10 at.% $C(C_{24}H_{12})$ added MgB_2 wires, small voids are found to be uniformly dispersed in the MgB_2 layer. These voids are considered to be caused by the H_2 gas produced during the decomposition of the uniformly dispersed $C_{24}H_{12}$. Voids can also be observed in wires without additives, but their number is much smaller and their distribution is random.

Figure 13 shows the (a) J_c–B curves and (b) J_e–B curves at 4.2 K for IMD MgB_2 wires fabricated by different methods[24]: a wire without additions, two wires with 5 at.% $C(C_{24}H_{12})$ and starting Mg rod diameters of 2.0 mm (termed 5% C) and of 1.8 mm (termed 5% C#), a wire with 10 mol.% SiC and a IMD wire based on B powder from SMI, coated with 4 mol.% C (termed SMI (4% C)) shown in Fig. 8.[10] A high J_c value of 1.1×10^5 A/cm^2 at 4.2 K and 10 T is obtained for the $C_{24}H_{12}$ added wire (5% C), while J_e reaches 1.1×10^4 A/cm^2 at 4.2 K and 10 T for 5% C#. These values are similar to the high J_c and J_e values obtained for wires fabricated using SMI B powder[9] and pretreated SMI B powder,[10] though the average particle size in the B powder used in this section is four times larger than that in the SMI B powder. Furthermore, the slopes of the J_c–B curves for the wires with 5% C and 5% C# additions are steeper than those for the SMI (4% C) wires, suggesting that the amount of C substitution for B in the 5% C wires is in reality lower than that in the SMI (4% C) wire. This result also confirms

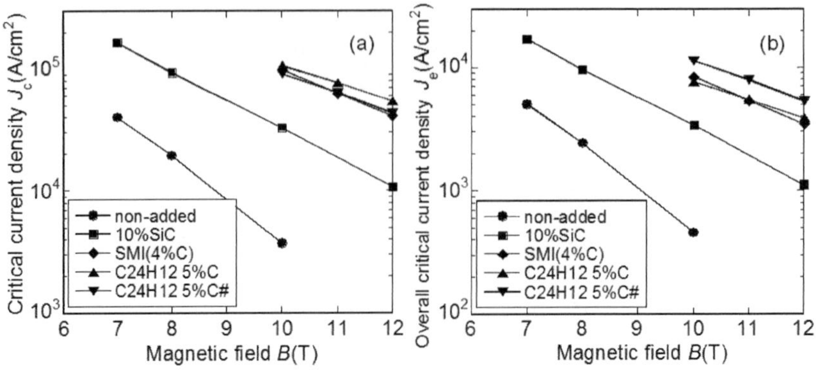

(a) (b)

Fig. 13. (a) J_c–B curves and (b) J_e–B curves at 4.2 K for IMD MgB_2 wires fabricated by different methods:[24] without additives, with 5 at.% C($C_{24}H_{12}$) (2.0 mm Ø Mg rod):(5% C), with 5 at.% C($C_{24}H_{12}$) (Mg rod: 1.8 mm Ø) (5% C#), with 10 mol.% SiC IMD wire and IMD wire based on 4 mol.% C coated SMI B powder (SMI (4% C)).[10] (Reproduced by permission of IOP Publishing.)

that $C_{24}H_{12}$ is an active C source, well suited for being used as an additive in MgB_2 superconducting wires.

At present, SiC is the most commonly used additive for MgB_2 wires. However, the J_c and J_e values for SiC added wires are lower than those for $C_{24}H_{12}$ added wires. One reason is linked to the fact that both B and SiC used in this experiment are nanosize powders: it is very difficult to obtain a uniform mixture for two kinds of nanosize powders, due to agglomeration. A main advantage in using $C_{24}H_{12}$ additives consists in the fact that there is no agglomeration problem. The critical current properties are expected to be further improved by the optimization of fabrication parameters such as the $C_{24}H_{12}$ content, the heat treatment conditions and the B powder particle size.

Figure 14 shows the J_c values of a 10 at.% C($C_{24}H_{12}$) added IMD MgB_2 wire at 4.2 K and magnetic fields up to 28 T.[24] The data for the IMD MgB_2 wire without additives, for the PIT MgB_2 tape with 10 mol.% SiC[26] and for the PIT MgB_2 tape with 20 mol.% SiC[27] are also included in the figure. The slope of the J_c–B curve for the 10 at.% C($C_{24}H_{12}$) added wire remains almost constant up to 23 T and

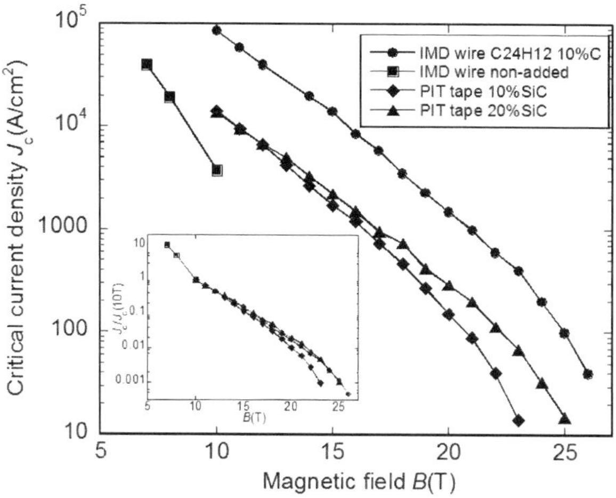

Fig. 14. J_c–B curves of a MgB_2 wires without additives and of a IMD MgB_2 wire with 10 at.% $C(C_{24}H_{12})$ at 4.2 K.[24] The data of PIT MgB_2 tapes with 10 mol.% SiC[26] and with 20 mol.% SiC[27] are also included in the figure.

then increased slightly for higher fields. The IMD wire with 10 at.% $C(C_{24}H_{12})$ shows higher J_c values at high magnetic fields as well as higher B_{irr} values than those for the SiC added MgB_2 tapes. The inset in Fig. 14 shows the ratio $J_c/J_c(10\,T)$ versus B for each wire. The slopes for the 10 at.% $C(C_{24}H_{12})$ added IMD wires are smaller than those for the 10 mol.% SiC added PIT tapes, and are almost the same as those for the 20 mol.% SiC added wires. The slope for the 10 at.% $C(C_{24}H_{12})$ added wire is much smaller than that for the IMD wire without additives, fabricated under the same experimental conditions. These results confirm that $C_{24}H_{12}$ is an effective C source for MgB_2 wires. If the irreversibility field B_{irr} is determined with the criterion of 100 A/cm^2, its value for these wires is 25 T. This value is higher than those for C_6H_{10} added PIT wires (21 T)[28] and for PIT wires based on nanosize SMI B powder doped with 6.48 mol.% C (24.4 T).[29]

4. Hybrid Method of IMD and PIT

As mentioned above the IMD process between bulk Mg and compacted B powder leads to MgB_2 layers with a high mass density.

In this process, a composite billet composed of a metal tube, an intermediate layer of compacted B powder and a Mg rod at the center is cold worked into a fine wire and then heat treated to convert the B layer into an MgB_2 layer. However, the reacted MgB_2 layer in the IMD wire usually contains a certain quantity of unreacted B particles as well as B rich compounds such as MgB_4. This is caused by an insufficient Mg supply to the B layer, due to the short diffusion distance of Mg. Therefore, a reduction of both, unreacted B and the B rich non-superconducting compound MgB_4, corresponds to an enhanced MgB_2 volume fraction and is expected to further enhance the critical current properties of IMD MgB_2 wires. As mentioned in the previous section, the reduction of the filament size in multifilament wires is effective in decreasing the volume fraction of the unreacted B and B rich compounds. Another approach to reduce these impurities in IMD MgB_2 wires consists in the addition of Mg powder to the B layer, which reduces the required Mg diffusion distance for a full reaction between Mg and B.[30,31] By adding Mg powder to the B layer of IMD wires one can change the structure of the wire from the IMD to the PIT type: this can be obtained by sequentially increasing the concentration of the Mg powder in the B layer and correspondingly decreasing the diameter of the Mg rod placed at the center of the wire in order to maintain a Mg:B ratio of 1:2, corresponding to stoichiometric MgB_2.

Figure 15 shows the change of wire structure from a IMD to a PIT wire type.[31] At the stoichiometric ratio of Mg to B and assuming the Mg rod diameter as zero, the wire tends towards a PIT type wire. Table 1 lists the specifications of the wires fabricated in this section.[31] The purity of the B powder used in this section was 99%. The Mg powder concentration in B powder layer was enhanced from 0 to 50 at.%, which corresponds to a reduction of the Mg rod diameter from 2.0 to 0 mm in order to maintain the stoichiometric composition. This means that the B layer thickness increased with increasing Mg powder content.

All wires contained an amount of 10 mol.% SiC nanosize powder, which was added to the compacted B powder for improving the J_c values at high magnetic fields. In Table 1, M0 and P0 define a IMD

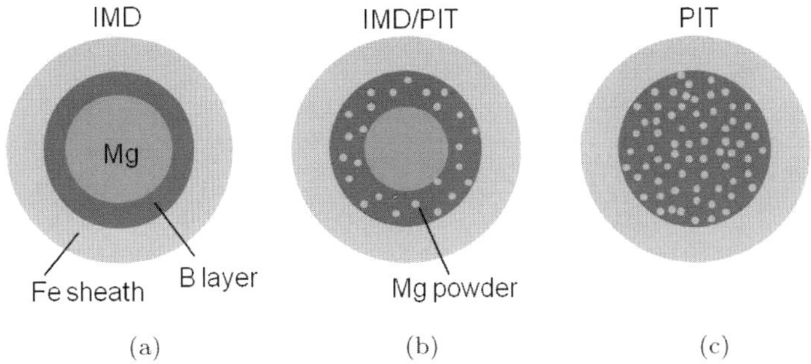

Fig. 15. Schematic illustration of (a) IMD wire type, (b) Wire fabricated with the hybrid IMD/PIT method, (c) PIT wire type.

Table 1: Specifications of the investigated wires.

	M0	M1	M2	M3	P0
Method	IMD	IMD + PIT	IMD + PIT	IMD + PIT	PIT
Wire diameter (mm)	1.16	1.16	1.2	1.26	1
Mg rod (mm)	2	2	1.7	1.4	0
Fe tube Inner diameter (mm)	3.5	3.5	4	4	3.5
Mg powder in B (at.% ratio of Mg to B)	0	6	17	25	50
MgB_2 fraction (%)	8	8.1	21.4	23.8	25.3

and a PIT wire, respectively. Three wires, termed M1, M2 and M3, were fabricated using the hybrid IMD/PIT method. The ratios of added Mg powder to the B powder were 6, 17 and 25 at.% for M1, M2 and M3, respectively. The reaction heat treatment was carried out at 670°C for 2 h. under Ar gas atmosphere for all wires except for PIT wire P0. The reaction temperature was gradually increased from 640–670° C to ensure a sufficient Mg diffusion from the Mg rod to the B layer. The PIT wire was submitted to a heat treatment, of 850°C for 1 h, maintaining this temperature for 1 h, followed by furnace cooling. Although the higher temperature heat treatment causes an unwanted interface reaction between Fe and B,[32,33] the temperature of 850°C was found to yield a higher critical current

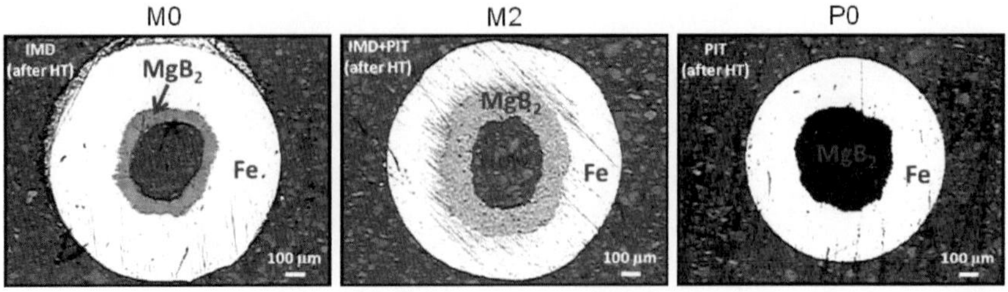

Fig. 16. Optical microscopy images of the various wire types after the heat treatment: (a) IMD wire (M0), (b) hybrid IMD/PIT wire (M2) and (c) PIT wire (P0). From Ye *et al.*[31] (Reproduced by permission of IOP Publishing).

for the same starting materials. For wires prepared by the hybrid method, we have obtained about 30% higher J_c values than in PIT wires heat treated under the same conditions.

Optical micrographs of the transverse cross-sections of the wires, M0, M2 and P0 after the heat treatment, are shown in Fig. 16.[31] In the IMD wire M0 a hole filled with resin (due to the polishing process) was found in the wire center. The MgB_2 layer appears in orange, which is similar to the color of MgB_2 in the single crystals reported by Lee *et al.*[25] The MgB_2 layer for heat treated PIT wire P0 in Fig. 16 shows a dark color, due to the voids in the PIT MgB_2 layer absorbing the light. Thus, the orange color in M0 is due to the reflection of light, suggesting that the MgB_2 layer in IMD wires has a high mass density. In the hybrid IMD/PIT wire M2 the entire MgB_2 layer shows a similar orange color but dark spots are visible, thus suggesting that the B layer is fully reacted and that neither unreacted B nor B rich regions exist. This is confirmed by SEM analysis, showing that the dark spots are in reality voids. It follows that the MgB_2 layer in the hybrid M2 wire has a lower mass density than in the IMD wire M0, but a much higher density than in the PIT wire P0. As indicated in Table 1, the MgB_2 area fractions for M2 (21.4%) and M3 (23.8%) are considerably higher than for M0, and comparable to P0 (25.3%). Thus, the hybrid process yields a higher MgB_2 layer mass density than the PIT process, but a higher MgB_2 area fraction than the IMD process.

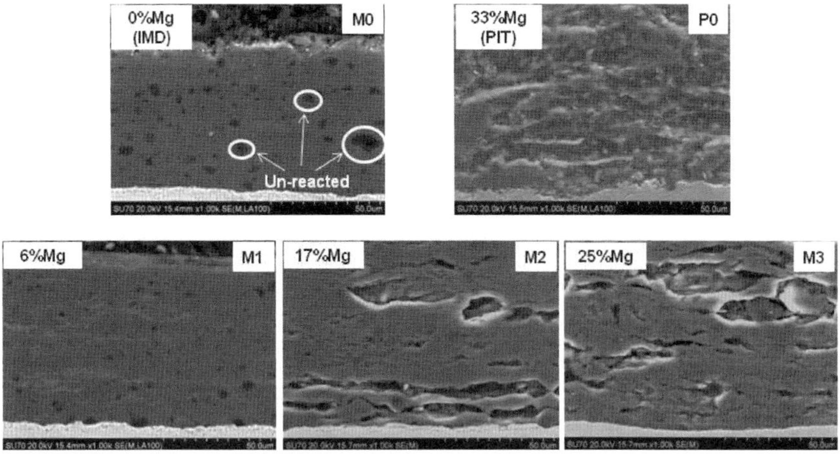

Fig. 17. SEM images of five wires listed in Table 1 after the heat treatment. From Ye *et al.*[31] (Reproduced from IOP Publishing.)

Figure 17 shows SEM images of the longitudinal cross-sections of the wires M0, M1, M2, M3 and P0 after reaction.[31] The IMD wire has a high mass density MgB_2 layer but contains some unreacted B particles (dark spots). With increasing the Mg powder content in the B layer, the number and the size of unreacted B particles decrease while the number of voids increases. This clearly indicates that increasing the Mg powder content in the B layer leads to a gradual transition of the microstructure from the IMD to the PIT type. The small number of voids observed in M1 is due to the small amount of added Mg powder. In M2 and M3, a large number of ellipsoidal voids can be seen, arising from the elongation of the Mg powder particles during the groove-rolling and drawing process. The microstructure of wire M3 is granular, as in the PIT wire P0, due to the large amount of Mg powder addition, 25 at.%, to the B layer, thus half the value of P0 (50 at.%).

The J_c–B curves for all wires in this section have almost the same slope, the SiC addition being always 10 mol.%. This indicates that a similar amount of C substitution for B occurs in all wires. Figure 18 shows the change of J_c and J_e values at 10 T and 4.2 K with increasing Mg powder content in the B layer.[31] The J_e values

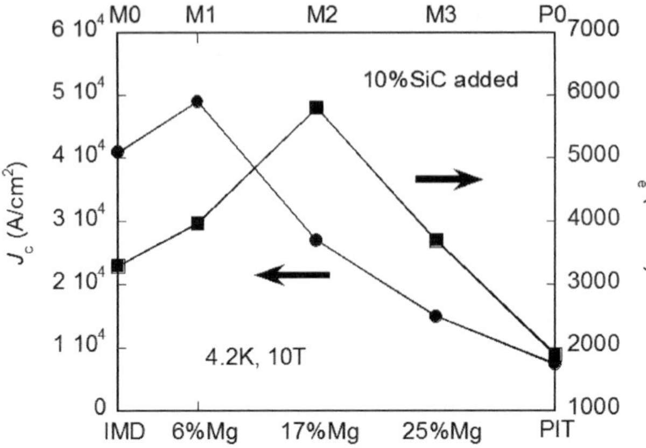

Fig. 18. The change of J_c and J_e values at 10 T and 4.2 K with increasing Mg powder addition in the B layer.[31]

were calculated by dividing I_c through the total cross-sectional area of the wires. As the Mg content increases, J_c first increases to a maximum for M1 (6 at.% Mg) and then decreases. The increase of J_c is due to a reduction in the number of unreacted B particles, while the subsequent decrease is the result of void formation, illustrated in Fig. 17. A similar trend is also seen for the engineering critical current density J_e, but the maximum J_e value in the M2 wire was obtained at higher Mg powder contents. The increase J_e from the IMD wire M0 to the hybrid wire M1 is due to the increase of J_c.

From M1 to M2, J_e continues to increase to a maximum value at M2, although J_c decreases. This J_c decrease is more than compensated by the increase of the MgB_2 area fraction. The J_e value for the hybrid wire M2 is more than twice as high as for the IMD wire M0. Thus, the hybrid method is much effective in enhancing J_e values. At 4.2 K and 10 T, the maximum J_c and J_e values are 4.9×10^4 A/cm^2 and 5.8×10^3 A/cm^2, respectively. This J_c value is not as high as that for the SiC added wires shown in Fig. 8.[11] One reason is that small unreacted B particles and/or B-rich compounds still remain in the MgB_2 layer due to the large particle size of the B powder (325-mesh) used in this section.

5. Conclusions

The IMD process is an effective method for increasing the filling factor and enhancing the J_c values of MgB_2 wires. Much higher J_c values than those of PIT wires are obtained by the IMD process. However, the diffusion distance of Mg in the B layer is not as long and the typical thickness of reacted MgB_2 layer is limited to 20–30 μm. Thus, the MgB_2 filament size must be decreased in order to obtain a full reaction between the Mg central rod and the B layer. This can be realized in a multifilament wire. Mono, 7- and 19-filament MgB_2 wires were fabricated. C or C containing impurity additions to the B layer bring about C substitution for B site in the MgB_2 layer and strongly enhance the J_c values. High J_c values of 1.1×10^5 A/cm^2 (4.2 K, 10 T) and 7.6×10^4 A/cm^2 (20 K, 5 T) are obtained using C coated nanosize B powder. Addition of $C_{24}H_{12}$ to the B layer is also very effective in enhancing J_c values. $C_{24}H_{12}$ is solid at room temperature and melts during the heat treatment, each B particle being uniformly coated. This simple method is suitable to the fabrication of long wires. Another method to solve the problem of short Mg diffusion distance is the addition of Mg powder in the B layer of IMD wire. The addition of Mg powder in the B layer and correspondingly the decrease of the diameter of Mg rod placed at the center of the wire produce a hybrid method of IMD and PIT methods. Much higher J_e than that of the IMD wire is obtained for the wire fabricated with this hybrid method.

References

1. H. Yamada, M. Igarashi, Y. Nemoto, Y. Yamada, K. Tachikawa, H. Kitaguchi, A. Matsumoto and H. Kumakura, Improvement of the critical current properties of *in situ* powder-in-tube-processed MgB_2 tapes by hot pressing, *Supercond. Sci. Technol.* **23**, p. 045030(4pp) (2010).
2. G. Giunchi, S. Ceresara, G. Ripamonti, A. D. Zenobio, S. Rossi, S. Chiarelli, M. Spadoni, R. Wesche and P. L. Bruzzone, High performance new MgB_2 superconducting hollow wires, *Supercond. Sci. Technol.* **16**, pp. 285–291 (2003).

3. K. Togano, J. M. Hur, A. Matsumoto and H. Kumakura, Fabrication of seven-core multi-filamentary MgB_2 wires with high critical current density by an internal Mg diffusion process, *Supercond. Sci. Technol.* **22**(1), p. 015003(5pp) (2009).
4. H. Kumakura, J. Hur, K. Togano, A. Matsumoto, H. Wada and K. Kimura, Superconducting properties of diffusion-processed multifilamentary MgB_2 wires, *IEE Trans. Appl. Supercond.* **21**, pp. 2643–2648 (2011).
5. S.-J. Ye, A. Matsumoto, K. Togano, T. Ohmura and H. Kumakura, The critical current properties of 37-filament internal Mg diffusion-processed MgB_2 wires, *Superconductor Sci. Technol.* **26**, p. 105027 (2013).
6. S.-X. Dou, S. Soltanian, J. Horvat, X.-L. Wang, S.-H. Zhou, M. Ionescu, H.-K. Liu, P. Munroe and M. Tomsic, Enhancement of the critical current density and flux pinning of MgB_2 superconductor by nanoparticle SiC doping, *Appl. Phys. Lett.* **81**, pp. 3419–3421 (2002).
7. H. Kumakura, A. Matsumoto, H. Fujii and K. Togano, Upper critical fields of powder-in-tube-processed MgB_2/Fe tape conductors, *Appl. Phys. Lett.* **84**, pp. 3669–3671 (2004).
8. Specialty Materials, Inc. (SMI), Lavel, MA 01851, USA; info@specma terials.com
9. G.-Z. Li, M. Sumption, M. Susner, Y. Yang, K. Reddy, M. Rindfleisch, M. Tomsic, C. Thong and E. Collings, The critical current density of advanced internal-Mg-diffusion-processed MgB_2 wires, *Supercond. Sci. Technol.* **25**, p. 115023(8pp) (2012).
10. S.-J. Ye, M. Song, A. Matsumoto, K. Togano, M. Takeguchi, T. Ohmura and H. Kumakura, High performance MgB_2 superconducting wires for use under liquid-helium-free conditions fabricated using an internal Mg diffusion process, *Supercond. Sci. Technol.* **26**, 125003 (2013).
11. K. Togano, J. M. Hur, A. Matsumoto and H. Kumakura, Microstructures and critical currents of single- and multi-filamentary MgB_2 superconducting wires fabricated by an internal Mg diffusion process, *Supercond. Sci. Technol.* **23**, p. 085002(7pp) (2010).
12. Y. Takano, H. Takeya, H. Fujii, H. Kumakura, T. Hatano, K. Togano, H. Kito and H. Ihara, Superconducting properties of MgB_2 bulk materials prepared by high-pressure sintering: *Appl. Phys. Lett.* **78**, pp. 2914–2916 (2001).
13. S. Hong, D. Geschwindner, A. Mantone, W. Marancik, S. Zalek and R. Zhou, High current density of Nb-Ti composite, *EEE Trans. Magn.* **25**, pp. 1934–1936 (1989).
14. G. Nishijima, S.-J. Ye, A. Matsumoto, K. Togano, H. Kumakura, H. Kitaguchi and H. Oguro, Mechanical properties of MgB_2 superconducting wires fabricated by internal Mg diffusion process, *Supercond. Sci. Technol.* **25**, p. 054012 (2012).

15. H. Kitaguchi, A. Matsumoto, H. Hatakeyama and H. Kumakura, Critical current of magnesium diboride/stainless steel composite tapes under tensile or compressive strains, *Supercond. Sci. Technol.* **16**, pp. 976–979 (2003).
16. H. Kitaguchi and H. Kumakura, Superconducting and mechanical performance and the strain effects of a multifilamentary MgB_2/Ni tape, *Supercond. Sci. Technol.* **18**, pp. S284–S289 (2005).
17. M. Dhallé, H. van Weeren, S. Wessel, A. den Ouden, H. H. J. ten Kate, I. Hušek, P. Kováč, S. Schlachter and W. Goldacker, Scaling the reversible strain response of MgB_2 conductors, *Supercond. Sci. Technol.* **18**, pp. S253–S260 (2005).
18. P. Kováč, T. Melišek, M. Dhallé, A. den Ouden and I. Hušek, Critical currents of MgB_2 wires prepared *in situ* and *ex situ* subjected to axial stress, *Supercond. Sci. Technol.* **18**, pp. 1374–1379 (2005).
19. H. Yamada, M. Hirakawa, H. Kumakura and H. Kitaguchi, Effect of aromatic hydrocarbon addition on *in situ* powder-in-tube processed MgB_2 tapes, *Supercond. Sci. Technol.* **19**, pp. 175–177 (2006).
20. H. Yamada, N. Uchiyama, A. Matsumoto, H. Kitaguchi and H. Kumakura, The excellent superconducting properties of *in situ* powder-in-tube processed MgB_2 tapes with both ethyltoluene and SiC powder added, *Supercond. Sci. Technol.* **20**(6), pp. L30–L33 (2007).
21. S.-J. Ye, A. Matsumoto, K. Togano and H. Kumakura, Enhancement of the critical current density of internal Mg diffusion processed MgB_2 wires by the addition of both SiC and liquid aromatic hydrocarbon, *Physica C* **471**, pp. 1133–136 (2011).
22. J.-H. Kim, X. Xu, M. Hossain, D. Shi, Y. Zhao, X.-L. Wang, S.-X. Dou, S. Choi and T. Kiyoshi, Influence of disorder on the in-field J_c of MgB_2 wires using highly active pyrene, *Appl. Phys. Lett.* **92**, p. 042506(1–3) (2008).
23. M. Maeda, J.-H. Kim, Y. Zhao, Y.-U. Heo, K. Takase, Y. Kubota, C. Moriyoshi, F. Yoshida, Y. Kuroiwa and S.-X. Dou, In-field J_c improvement by oxygen-free pyrene gas diffusion into highly dense MgB_2 superconductor, *J. Appl. Phys.* **109**, p. 023904 (2011).
24. S.-J. Ye, A. Matsumoto, Y.-C. Zhang and H. Kumakura, Strong enhancement of high-field critical current properties and irreversibility field of MgB_2 superconducting wires by coronene active carbon source addition via the new B powder carbon-coating method, *Supercond. Sci. Technol.* **27**, p. 085012 (2014).
25. S. Lee, T. Masui, A. Yamamoto, H. Uchiyama and Setsuko Tajima, Crystal growth of C-doped MgB_2 superconductors: accidental doping and inhomogeneity, *Physica C* **412–414**, pp. 31–35 (2004).

26. H. Kumakura, A. Matsumoto, H. Fujii and K. Togano, Upper critical fields of powder-in-tube-processed MgB_2/Fe tape conductors, *Appl. Phys. Lett.* **84**, pp. 3669–3671 (2004).
27. A. Matsumoto, H. Kitaguchi and H. Kumakura, Superconducting properties of MgB_2/Fe tape heavily doped with nanosize SiC, *Supercond. Sci. Technol.* **21**, p. 065007 (2008).
28. M. Maeda, J.-H. Kim, Y.-U. Heo, S. Kwon, H. Kumakura, S. Choi, Y. Nakayama, Y. Takano and S.-X. Dou, Superior MgB_2 Superconducting Wire Performance through Oxygen-Free Pyrene Additive, *Appl. Phys. Express* **5**, p. 013101(1–3) (2012).
29. M. Susner, Y. Yang, M. Sumption, E. Collings, M. Rindfleisch, M. Tomsic and J. Marzik, Enhanced critical fields and superconducting properties of pre-doped B powder-type MgB_2 strands, *Supercond. Sci. Technol.* **24**, 012001 (2011).
30. S.-J. Ye, M. Song, A. Matsumoto, K. Togano, Y.-C. Zhang, H. Kumakura, M. Takeguchi, R. Teranishi and T. Kiyoswhi, Enhancing the critical current properties of internal Mg diffusion-processed MgB_2 wires by Mg addition, *Supercond. Sci. Technol.* **25**, p. 125014 (2012).
31. S.-J. Ye, A. Matsumoto, K. Togano, Y.-C. Zhang, T. Ohmura and H. Kumakura, Fabrication of MgB_2 superconducting wires with a hybrid method combining internal-Mg-diffusion and powder-in-tube process, *Supercond. Sci. Technol.* **27**, p. 055017 (2014).
32. C. Grovenor, L. Goodsir, C. Slater, P. Kovac and I. Husek, Interfacial reactions and oxygen distribution in MgB_2 wires in Fe, stainless steel and Nb sheaths, *Supercond. Sci. Technol.* **17**, pp. 479–484 (2004).
33. P. Kovac, I. Husek, T. Melisek and V. Strbik, Basic properties of rectangular MgB_2/FeNiCo and MgB_2/Fe wires made *in situ*, *Supercond. Sci. Technol.* **18**, pp. 856–860 (2005).

Chapter 4f

Development and Properties of Advanced Internal Magnesium Infiltration (AIMI) Processed MgB_2 Wires

E. W. Collings[*,‡], G. Z. Li[*], M. D. Sumption[*] and M. A. Susner[*,†]

[*]*Center for Superconducting and Magnetic Materials, MSE*
The Ohio State University 2041 College Road North
Columbus, OH, 43210, USA

[†]*Materials Science and Technology Division,*
Oak Ridge National Laboratory
1 Bethel Valley Road, Oak Ridge, TN 37831, USA
[‡]*collings.2@osu.eu*

The development, processing, properties and formation mechanisms of Advanced Internal Magnesium Infiltration (AIMI) MgB_2 wires are discussed against a background of the related and original processes, Internal Magnesium Diffusion (IMD) and Magnesium-Reactive-Liquid-Infiltration (Mg-RLI). First reviewed are the formation, properties and applications of Mg-RLI bulks as basis for discussions of Mg-RLI-processed and IMD-processed wires. The transition from Mg-RLI- and IMD- to AIMI wires is explained, and the relative performances of powder-in-tube (PIT), IMD and AIMI wires are summarized in the form of an iso-J_e diagram of $J_{c,nb}$ versus A_{nb}/A_{tot} in which A_{tot}, A_{nb}, $J_{c,nb}$ and J_e are, respectively, the wire's cross-sectional area, the area inside the chemical barrier, the critical current (I_c) normalized to A_{nb}, and I_c normalized to A_{tot}. After introducing the details of AIMI wire fabrication — selection of starting powders, dopants and reaction heat treatments (HT) — the report goes on to describe in detail the development of high performance AIMI wires: layer J_c, fill factors, J_e, and the effects of wire size, multifilamentarization, doping with C, and co-doping with C and Dy_2O_3. The two-stage mechanism of layer formation in AIMI wires is discussed: first the reactive infiltration of liquid Mg into a porous B pack, a process that terminates with the formation of a dense MgB_2 layer; second the slow diffusion of Mg into any remaining B through that MgB_2 layer. The report concludes with a brief general discussion of anisotropy, current percolation and the J_c field dependence of MgB_2 wires.

1. Introduction

Pending in 2001 was a patent applied for by EDISON SpA for Giunchi's invention of the Mg-Reactive-Liquid-Infiltration (Mg-RLI) process. At about the same time, and subsequently, other related techniques for producing MgB_2 bulks and simple wires were being proposed, such as those due to: (i) Canfield *et al.* (2001)[1] who diffused Mg into B fiber, (ii) Dunand (2001)[2] who synthesized Mg/MgB_2 composites by the liquid Mg pressure-assisted infiltration of Mg into compacts of either MgB_2 powder or Mg powder, (iii) Ueda *et al.* (2005)[3] whose powder-in-closed-tube (PICT) method consisted of the reactive infiltration at 800–900°C of Mg from a Mg powder compact into a B powder compact, (iv) Pan *et al.* (2008)[4] in whose proposed "improved Mg-diffusion" method a mixture of B, graphite (as dopant) and 10% Mg powder was reacted for 24 h/900°C in an iron tube along with a sufficient quantity of additional Mg powder. But as evidenced by a copious volume of papers published from 2001 to the present, the Mg-RLI process has emerged not only as the dominant method for producing dense MgB_2 bulks but also as the forerunner of the several wire-processing techniques (IMD, AIMI) to be discussed in other sections of this chapter.

1.1 *Bulks processed by RLI*

1.1.1 *Formation of Mg-RLI bulks*

The Mg-RLI technology consists of the reactive diffusion of liquid Mg into a compacted body of crystalline β-rhombohedral B powder, a process that has been discussed by Giunchi *et al.*[5–9] and others.[10] The ground-and-sieved B powder (<100 μm) is pressed (~1 kbar) into a non-reactive container (Nb, stainless steel) to ~54% theoretical density along with some massive pieces of pure Mg. The container is sealed by welding and heat treated for typically 3 h/900–950°C. After reaction an ingot of poly-crystalline MgB_2 with a density more than 90% of the theoretical value replaces the B powder compact, and holes replace the original pieces of Mg. As a refinement of the process the infiltration depth may be reduced by replacing large pieces of Mg with a suitable number of small (0.2–0.5 mm) Mg spheres[11] and, according to Gonzales *et al.*[12] residual intergranular Mg may be reduced by aggressive dynamic low-vacuum annealing for 23 h/900°C. With regard to the choice of precursor materials, Saglietti *et al.*[13] have investigated the influences of: (i) B types (commercial amorphous, high purity amorphous, crystalline) and purification, (ii) milling processes and (iii) preform packing density, on the behavior of the B and Mg reactants during Mg-RLI processing.[14] The influences of starting B types on mechanical properties,[15] thermal conductivity,[16] and superconducting properties[17,18] have also been investigated. The Mg infiltration process has been considered in detail by Giunchi *et al.*[5,6] with particular reference to the CALPHAD-computed pressure–temperature phase diagram.[19,20] In a reaction-kinetics study based on SEM/EDS and XRD Albisetti *et al.*[21] have presented evidence indicating that Mg_2B_{25} (observed earlier by Giunchi *et al.*)[9,22] is a precursor to MgB_2 formation during Mg-RLI (see Sec. 4). This Mg_2B_{25} phase (which has a crystal structure similar to β-rhombohedral B) is replaced by MgB_2 as the reaction proceeds. Residual Mg_2B_{25} if able to be present at the nanoscale level was considered as a possible flux-pinning inclusion.[21] During the Mg-RLI process, according to Giunchi *et al.*[5–9] the B reacts with liquid Mg to form MgB_2/Mg_2B_{25} particles which, as they grow, release B to react with more Mg to form interstitial MgB_2 that welds the particles together. Thus the bulk product of the process can be regarded as a composite formed by growing MgB_2/Mg_2B_{25} grains welded together by intergranular MgB_2.[23] Based on this mechanism Giunchi *et al.*[5–9] have described several Mg-RLI-based techniques for forming superconducting joints between bulks and between bulks and wires.[23]

1.1.2 *Doping of Mg-RLI bulks*

SiC-doping of Mg-RLI bulk samples was achieved by adding 10 wt.% nano-SiC,[12,24,25] also 20 wt.% nano-SiC,[11] to the starting B powder.

The pinning energy,[24,25] and the critical fields B_{c2} and B_{irr}[12,25] were measured. Also investigated were the effects of Zn-doping achieved by performing Mg-RLI process by medium of a multiply-melted Mg-10 wt.% Zn alloy,[12] also alloys of 10 at.% Zn, 3 and 5 at.% Cu, and 5, 10 and 20 at.% Co,[11] after which various superconductive properties were measured. The doping of MgB_2 with borides of the rare-earth (RE) elements Ce, Sm, and Gd as well as with the RE metals themselves has been investigated by Saglietti *et al.*[26] The samples for measurement were prepared using Giunchi's Mg-RLI process.[6] During heat treatment (HT) (2 h/850–900°C) the Mg attacked the boride particles, transforming hexaborides into the corresponding tetraborides. In most cases the dopants reduced J_c, their large size leading to surface reactions and not to RE diffusion into the MgB_2 matrices. But of particular interest was the crystalline-B-based SmB_6-doped sample whose magnetic $J_c(B)$ increased slightly e.g., at 10 K, 5 T from 3.6×10^3 to 1.0×10^4 A/cm^2. It also showed a complete absence of low-field (0–1 T) flux jumping, an effect that we attribute to the large low temperature heat capacity of the Sm-based inclusions.

1.1.3 *Physical properties of Mg-RLI bulks*

Cavallin *et al.*[16] and Giunchi *et al.*[15] have investigated the influence of B types on the thermal conductivity and mechanical properties, respectively, of Mg-RLI-processed bulk samples. Both properties depended strongly on the choice of starting B namely: (i) crystalline B, ground and sieved (<100 μm), ingot-IA, (ii) amorphous B grade-I (~95%), ingot-IB, (iii) amorphous B grade II (92%), ingot-IC, which yielded differences in grain size and residual impurity content. For example, ingot-IA possessed the highest Mg_2B_{25} content, and at ~100 μm its grain size was greater than those of ingots 1B and IC (1–5 μm).[16] The mechanical properties measured were Vickers microhardness (H_v), Young's modulus (E), ultimate strength and fracture toughness (K_{IC}). The average E-values for IA and IB were, respectively, 125 and 86 Gpa.[15] The thermal conductivities (κ), measured between 4 K and 300 K varied widely from sample to sample, κ_{20K} for example increasing in the sequence IB < IC < IA, the largest value being 50 W/m · K.[16]

1.1.4 *Superconducting properties of Mg-RLI bulks*

With applications in mind (for 15–30 K, up to 2–3 T) Giunchi[8] described the various bulk forms — cylinders, tubes, disks — that can be produced by Mg-RLI processing. Recognizing their potential for low temperature, low-field flux jumping was calculated, for critical-current-carrying bulks, the thickness limits for flux-jump stability as functions of temperature and field. The superconducting properties were found to depend on the choice of starting B powders, whether they be amorphous (~1 μm),[17,18] or crystalline ground and sieved to 40 μm,[17,18] 100 μm or 180 μm.[17] For example, $J_c(B)$ decreased as size increased in the sequence 1, 40, 100, 180 μm. The transport $J_c(B)$ of a piece of MgB_2 bulk was measured at 4.2–30 K in fields of up to 8 T[27] yielding at 4.2 K, 5 T, for example, a J_c of 2.6×10^3 A/cm^2 with an n-value of 17. The J_c values of Mg-RLI-processed bulks are generally low, although those processed from amorphous B can be higher than those originating in crystalline B. For example, $J_{c,\text{magnetic},\text{-mostly } 4K}$ (in kA/cm^2) for the former at 5 T is 4.5[17] and 4.2,[18] and for the latter at 5 T is 1.6[18] and 1.9.[7] Using AC susceptibility measurements over a frequency range of 2 Hz–10 kHz, a DC field range of up to 5 T, a temperature range of up to ~40 K and an AC field of 1.34 T, Ghigo *et al.*[27] investigated the creep properties of Mg-RLI prepared bulks near the irreversibility line and measured the field- and temperature dependence of the pinning energy.

1.1.5 *Applications of Mg-RLI bulks*

(a) *General Applications*

Giunchi and colleagues have suggested numerous applications for Mg-RLI bulks. Frequently cited, and discussed in detail, are their possible uses in: (i) magnetic shielding or screening[8,17,18] of AC or DC fields[6,9]; (ii) magnetic levitation[9] and magnetic bearings[6,18] in particular for energy storage flywheels[8]; (iii) fault current limiters,[8] inductive,[6,18] or resistive[6]; (iv) trapped-field ("permanent") magnets[6,8,17] for motors and physics experiments[18]; (v) conductors or current leads[6]; (vi) linear and rotating electric motors.[6]

The thickness (say L) of a bulk superconducting component is limited by the possibility of adiabatic flux jumping which scales with LJ_c. But favoring the stability of Mg-RLI-processed bulk MgB_2 is its large grain size and hence relatively low J_c ($1{\sim}2\,kA/cm^2$ at 20 K, 3 T)[8] compared to that of undoped powder-in-tube (PIT) wire ($35\,kA/cm^2$).[28] The large grain size stems from the coarse B powder needed to ensure a thick reaction depth, see Sec. 4.2. For a SC of given $J_c(B,T)$ Giunchi has estimated the upper limit of L for a range of L and T values and concluded that in the case considered L_{max} is generally of the order of millimeters.[8] Finally it is useful to note that SmB_6-doped material with its resistance to flux jumping[26] (see above) may enable the fabrication of thicker bulk components.

(b) *Microwave Properties*

The microwave responses of high density bulk samples prepared by Mg-RLI have been investigated by Gallitto *et al.*[14] who concluded that materials processed in this way are particularly promising for microwave resonant cavities.[14]

(c) *Bulk Magnets by Pulsed Field Magnetization (PFM)*

Suggested uses for superconducting bulk magnets have been: sputtering cathodes, magnetic separators, and drug delivery systems. In this context, Fujishiro *et al.*[29] have investigated the pulsed-field magnetization (PFM) of a cryocooled Mg–RLI-prepared disc of bulk MgB_2. Produced by capacitor-discharge though a LN-cooled normal solenoid, it was possible to trap a maximum field of 0.47 T at 23 K.

(d) *Superconductive Inserts in Metallic Substrates (SIMS)*

Using a variant of the Mg–RLI process thick MgB_2 strips are imbedded in the grooves of metallic or ceramic substrates.[30] These stabilized and/or strengthened composites were referred to as superconducting inserts in metallic (ceramic) substrates — SIMS or SICS, respectively. The ends of the MgB_2 spirals or coils may be closed (hence amenable to persistent current flow) or left open. So-called "closed-loop" SIMS has been offered as innovative alternatives to monolithic bulk superconductors for applications such as magnetic shields and superconducting trapped-field ("permanent") magnets. Suggested applications of SIMS in open loop, or current-driven, configurations are resistive fault current limiters and variable-field magnets.

1.2 *Wires processed by RLI and related processes*

1.2.1 *Introduction*

Beginning in 2003 Giunchi *et al.* described adaptations of the Mg–RLI process, initially developed for bulks, to the processing of MgB_2 wires. Then beginning in 2007 researchers at Japan's National Institute for Materials Science (NIMS) and their collaborators at Kyushu and Tokyo Universities adopted their own version of the above process referring to it as "internal magnesium diffusion" (IMD). More recently a collaboration between Ohio State University and Hyper Tech Research Inc. (OSU/HTR) has resulted in the development of an improved version of the earlier wires by the "advanced internal magnesium infiltration (AIMI)" process, the principal subject of this chapter.

1.2.2 *Mg–RLI-processed wires*

Giunchi *et al.*[28] were the first to recognize that the Mg-RLI technique for producing MgB_2 bulks could be developed into a wire-manufacturing technique. The new method depended on the surprisingly high drawing ductility of suitably clad Mg rods. Mg is generally regarded as having very poor ductility, with an elongation-to-fracture of $\sim$7% under static strain. The key to the new manufacturing technique was the discovery by Giunchi *et al.* that Mg could be extensively co-reduced to very high area reduction ratios when embedded in fine B powder. Amorphous B powder was packed around a Mg rod axially mounted in a Nb-lined steel tube. After diameter reductions of some 24:1 the monocore composite wire was heat treated for 30 min/850°C in preparation for study and measurement. In general, the HT temperatures for the Mg-RLI wires ranged from 750°C,[25,31] 850°C,[28] and 900°C,[28,32] to 950°C.[25] The layer-J_cs of the early wires were rather low, e.g., at 4.2 K, 5 T 2.6×10^4 A/cm^2 for the monocore and 2.1×10^4 A/cm^2 for a 7-core multifilament. Subsequent developments included fill factor increase, wire diameter reduction,[33] and the introduction of C doping via colloidal C[32,33] and nano-SiC[25] have led to a recently reported layer $J_{c,4K,5T}$ of 1.2×10^5 A/cm^2.[23] Microstructural studies relating to the packing density and choice of B powder type were undertaken.[31] Looking ahead to future large scale applications of MgB_2 wire (operating in the $\leq$20 K, $\leq$4 T regime) the need for high current conductors has emerged in response to which several designs have been proposed: (i) a 12 mm OD hollow monocore, (ii) a 12 mm OD 7×7-bundled multicore, (iii) a massive rectangular monocore based on a side-by-side arrangement of solid Mg and B powder.[34]

1.2.3 *IMD-processed wires*

The Mg–RLI process with its HT temperatures from 750°C to 900°C, well above the melting point of Mg (650°C) is clearly based on liquid *infiltration.* The NIMS group and colleagues have taken the earlier Mg–RLI process to a new level. The fact that their HT temperatures were sometimes below the melting point of Mg,[35–39] sometimes a little above it, but seldom above 670°C,[40–44] justified their use of the term "internal magnesium *diffusion*" (IMD). In specifying the critical current density (CCD) of MgB_2 wires three metrics are in use: (i) the engineering CCD, J_e, which is the critical current I_c normalized to the wire's overall cross-sectional area, (ii) the non-barrier CCD, $J_{c,nb}$, (analogous to the non-Cu CCD of Nb_3Sn wires) which in the case of a monocore is I_c normalized to the area inside the chemical barrier, (iii) the layer CCD, $J_{c,\text{layer}}$ (or frequently just

J_c) which is I_c normalized to the area of reacted MgB_2. Caution must be used when comparing the CCDs of PIT and IMD strands since a $J_{c,layer}$ cannot be assigned to the former. As evidenced by a long list of publications (2007–current) the NIMS group has extensively developed the IMD process, for descriptions of which the review article by Kumakura[35] and summaries by Ye *et al.*[42,44,45] are recommended.

The basic IMD process starts with a tube of non-reactive metal (Fe, Nb or Ta) containing an axially mounted Mg rod packed in B powder. Such a monocore can be cold-worked into wire or rebundled into a Cu–Ni–alloy tube for drawing into a multifilamentary composite. Many variants of this process have been explored. An initial concern was the drawability of the Mg rod. It is well known that Mg, an HCP metal, has limited ductility with an elongation-to-fracture of ~7% under static strain. In order to obtain a more ductile BCC version, the alloy LA141 (Mg-14 wt.% Li-1 wt.% Al) was chosen as a substitute for pure Mg. In one version of the IMD process this alloy was used for the internal rod,[46] in another version the B was packed into an alloy tube.[47] The inclusion of Li is accompanied by poor superconducting properties: a lowering of T_c and B_{c2} and unusually low J_c values. Subsequent research by the NIMS group quickly led to very large increases in both $J_{c,5T}$ and $J_{c,10T}$. A common ingredient in many of the powder mixtures was 5 mol.% and 10 mol.% nano-SiC (a few tens of nm in size). The 4.2 K J_cs of such SiC-doped Fe-clad monocores were 4.27×10^5 A/cm^2 (5 T) and 4.17×10^4 A/cm^2 (10 T)[48] and those of Ta-clad monocores were 5.6×10^5 A/cm^2 (5 T) and 8.13×10^4 A/cm^2 (10 T).[49] Issues considered during IMD process development included the choices of: (i) primary (starting monocore) sheath material, (ii) dopant, (iii) HT time/temperature, (iv) the properties of multicore wires.

Regarding the influence of primary sheath material measurements of 7-core wires indicated that after HT for 1 h/645°C the 5 T and 10 T 4.2 K J_cs would decrease in the sequence Ta > Nb > Fe. In the units 10^5 A/cm^2 the 5 T J_cs were: 4.17 and 4.53 (Ta),[38,50] 2.57 (Nb)[50] and 1.45 (Fe)[51] and in the units 10^4 A/cm^2 the 10 T

J_cs were: 8.71 and 8.22 (Ta),[38,50] 4.42 (Nb),[50] and 1.74 (Fe).[51] The seven-stack Ta and Nb monocores had been bundled into a Cu-10 wt.% Ni tube and Fe monocores into a Fe tube.

It is well known that property improvements accompany the substitution of C into the B sublattice. The "standard" vector for C-doping was nano-SiC (tens of nanometers in size) at concentrations of 10 wt.% or 10 mol.%. Other variants were (i) SiC co-doped with hydrocarbons yielding $J_{c,4K,10T}$s of 5.00 and 4.73×10^4 A/cm^2 (toluene),[40,52] 1.84×10^4 A/cm^2 (ethyltoluene),[40] and 4.68×10^4 A/cm^2 (dimethylbenzene);[40] (ii) C-doped B from Specialty Materials Inc. (SMI) treated with dimethylbenzene to remove unwanted Cl[45] yielding an NIMS-record $J_{c,4K,10T}$ of 1.2×10^5 A/cm^2; (iii) Cl-free B from Pavezyum doped with hexabenzobenzene (coronene, $C_{24}H_{12}$) a hydrocarbon with 96 wt.% C. Coronene-doped (5% C) wires[42] yielded $J_{c,4K,10T}$s as high as 1.07×10^5 A/cm^2.

One of the incentives for processing multicore wires was the need to reduce the thickness of the B layer to facilitate its complete conversion into MgB_2. Thus clusters of seven monocores,[36–38,49–51] 19 monocores,[49,50] and even 37 monocores[44] or subelements were stacked into drawable tubes usually of Cu-10 wt.% Ni. Reduced "diffusion distance" is also achieved by mechanical size reduction to small diameters, e.g., monocore reduction of from 4 mm OD to 0.6 mm OD.[42] To the same end some PIT-based variants of the standard IMD architecture have been introduced: (i) Ye *et al.*[41] packed around the central Mg rod powder mixtures of B and with 3, 6 and 9 mol.% Mg, and in a later version of this "hybrid IMD/PIT process" they accompanied the increase in Mg powder content by reductions of the central Mg rod diameter so as to preserve the Mg:B ratio and included a B-content-based 10 mol.% of nano-SiC.[43] The $J_{c,4K,10T}$ values achieved were, respectively, 1.15×10^4 and 4.89×10^4 A/cm^2. Maeda *et al.* replaced the central Mg rod by an array of Mg stringers formed by the elongation during drawing of coarse Mg particles in the (Mg + B) powder mixture.[53] Assisted by cold high pressure densification (CHPD)[54,55] the resulting wire achieved a $J_{c,4K,10T}$ of 5.32×10^3 A/cm^2.

Among the many IMD wires processed by the NIMS group and colleagues some could be characterized by:

(a) *Exceptionally high joint 5 T and 10 T 4.2 K* J_c *values:*
   (i) Ta-sheathed, 10 mol.% nano-SiC, 7-core in Cu-10 wt.% Ni tube, HT 1 h/640°C.[50]
   (ii) Ta-sheathed, 10 mol.% nano-SiC, monocore in Cu-10 wt.% Ni tube, HT 1 h/640°C.[49]

(b) *Exceptionally high 10 T, 4.2 K* J_c *values:*
   (i) Ta-sheathed, 10 mol.% nano-SiC, 7-core in Cu-10wt.% Ni tube, HT 1 h/645°C.[50]
   (ii) Ta-sheathed, 10 mol.% nano-SiC, 7-core in Cu-10wt.% Ni tube, HT 1 h/640°C.[49]
   (iii) Fe-sheathed monocore, C-doped with coronene (2, 5, 10 and 15% C) heavy area reduction, HT 6 h/670°C.[42]
   (iv) Ta-sheathed monocore, SMI-B treated with dimethylbenzene, HT 8 h/675°C.[45]

The J_c values referred to are listed in Table 1.

1.2.4 *Fill factor*

From an engineering standpoint the important transport property is not the layer J_c as such but the engineering CCD, J_e, which is J_c

Table 1: List of layer J_c of IMD wires at 5 T and 10 T, 4.2 K.

Strand	Reference	Layer J_c (5 T), 10^5 A/cm^2	Layer J_c (10 T), 10^5 A/cm^2
	Exceptionally high joint 5 T and 10 T 4.2 K J_c *values*		
(a)-(i)	50	5.23	1.00
(a)-(ii)	49	5.60	0.81
	Exceptionally high 10 T, 4.2 K J_c *values*		
(b)-(i)	50	—	1.00
(b)-(ii)	49	—	1.00
(b)-(iii)	42	—	1.07
(b)-(iv)	45	—	1.19

multiplied by the fill factor, $FF = A_{SC}/A_{total}$, where A_{SC} is the area of the SC layer. The *FF*s of a number of monocores range from 6.9%[40] to 11.9%[43] with an average value $\langle FF \rangle_7 = 8.0\%$. In forming multicore wires by bundling the IMD monocores into thick Cu–Ni tubes the *FF*s become smaller. For example, the 7-core wire of Togano *et al.*[38] had an *FF* of 3.0%, while that of Ye *et al.*'s 37-core conductor[44] was only 1.8%. In the light of these numbers an important goal of advanced processing is an increase in the *FF*.

2. Advanced Internal Magnesium Infiltration (AIMI) Processed Wires

2.1 *The transition from Mg-RLI and IMD to AIMI*

As mentioned above in specifying the CCD of MgB_2 wires three metrics are in use: (i) the engineering CCD, J_e, which is the critical current I_c normalized to the wire's overall cross-sectional area, A_{tot}, (ii) the non-barrier CCD, $J_{c,nb}$, (analogous to the non-Cu CCD of Nb_3Sn wires) which in the case of a monocore is I_c normalized to the area inside the chemical barrier, A_{nb}, this is just the "J_c" of a PIT wire, (iii) the layer CCD, $J_{c,layer}$, which is I_c normalized to the area of a reacted MgB_2 layer, A_{layer}, hence useful for Mg-RLI/IMD. The core (non-barrier) of a PIT wire is a porous mixture of MgB_2 and macroscopic vacancies in place of the original Mg particles; that of an Mg-RLI/IMD wire is a cylinder of dense MgB_2 surrounding an axial hole in place of the original Mg rod. For this reason when evaluating the properties of PIT and Mg-RLI/IMD wires, apart from comparing their J_es, the only other valid comparisons that can be made are between J_c (PIT) and $J_{c,nb}$ (Mg-RLI/IMD) since the area of normalization in each case takes into account both MgB_2 and the absent Mg. In considering the evolution of PIT through Mg-RLI/IMD to AIMI it is useful to compare the CCDs of PIT and early OSU/HTR-made IMD wires.[56] In doing so we note that at 4.2 K, 10 T, a J_c of 1.43×10^4 A/cm^2 for 2% C-doped PIT is accompanied by a $J_{c,nb}$ of 4.0×10^4 A/cm^2 for 2% C-doped IMD. Although the cores of both wires consist of MgB_2 plus "Mg vacancies" the difference in CCD occurs because the PIT core's MgB_2

is porous with reduced connectivity while the IMD core contains a dense well-connected MgB_2 layer. This result has prompted recent MgB_2 conductor research to be directed towards improving the properties of IMD-type wires, special attention being given to the selection and properties of the B powder, increasing the MgB_2-layer thickness (diffusion distance and hence *FF*), and to the processing of longer and longer lengths of multifilamentary wire. Recognizing that the critical current, I_c, is given by $J_e \times A_{tot} = J_{c,nb} \times A_{nb}$ it follows that $J_e = J_{c,nb} \times A_{nb}/A_{tot}$. Thus plots of $J_{c,nb}$ versus A_{nb}/A_{tot}, which are rectangular hyperbolae corresponding to selected values of J_e as in Fig. 1, enable fair comparisons to be made of the performances of various types of MgB_2 wire.

OSU/HTR's variant of the Mg–RLI/IMD approach includes the use of fine-grain B, selection of the best sheath material and filament sizes, optimization of C-doping level and HT condition, and increasing SC area fraction, all directed towards increasing J_e. OSU/HTR has named the results of this process "advanced internal Mg infiltration" or "AIMI".[62] The improved properties called for the new name furthermore the term "infiltration" describes the

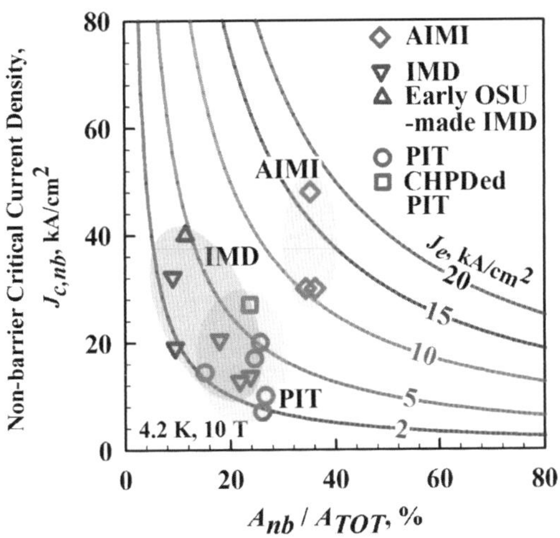

Fig. 1. Plots of $J_{c,nb}$ versus A_{nb}/A_{tot} for PIT,[55–58] early IMD,[38,40,48,51,59,60] and AIMI[61] type MgB_2 wires at 4.2 K, 10 T. Rectangular hyperbolae correspond to selected values of J_e from 2–20 kA/cm^2.

primary physical process that accompanies HT — infiltration of liquid Mg.

As indicated above the 4.2 K, 10 T, $J_{c,nb}$s of equal-sized PIT and early OSU/HTR-made IMD wires were 1.5×10^4 and 4.0×10^4 A/cm^2, respectively, leading to J_es of 2.30×10^3 and 4.23×10^3 A/cm^2, respectively. The superior performance of the early HTR processed IMD wire stimulated research into the development of even better properties. An obvious step towards J_e improvement was to increase the thickness of the MgB_2 layer. Initially the layer growth stopped shortly after the onset of HT, leaving a maximum MgB_2 layer thickness of at most 20–30 μm. To overcome this limitation, the OSU/HTR group initiated a detailed study of the layer growth mechanism. It was found that Mg infiltration into the surrounding B powder dominated the process and the infiltration depth determined the MgB_2 layer thickness. This discovery led to modifications of wire geometry/design, powder choice and preparation, B/Mg ratios, doping species, HT conditions, and hence to higher performance wires. Although the 4.2 K, 10 T, $J_{c,\text{layer}}$ remained more-or-less unchanged at 1.0×10^5 A/cm^2, the 4.2 K, 10 T J_e increased by a factor of 4 to 1.7×10^4 A/cm^2. The next step was to develop an increased $J_{c,\text{layer}}$ in the pursuit which OSU/HTR turned to C-doping and measured $J_{c,\text{layer}}$s and *FF*s in response to the variation of C concentration within the range of 0–4 mol.%. A series of studies showed[61,62] that although the 4.2 K, 10 T $J_{c,\text{layer}}$ maximized at 1.5×10^5 A/cm^2 with 3 mol.% C an accompanying suppression of MgB_2 layer formation severely reduced *FF* and with it J_e. Co-doping with nano-Dy_2O_3 tended to remedy this situation.[63]

2.2 *Fabrication and powder selection*

2.2.1 *Wire formation*

IMD wires are batch-processed at NIMS by first selecting the chemical-barrier, a tube of Ti, Nb, Ta or Fe, and packing B powder into it around an axially mounted Mg rod. The composite is then groove-rolled to rod and cold-drawn to wire in preparation for HT. Applied at some stage is a Cu–10Ni outer tube to serve

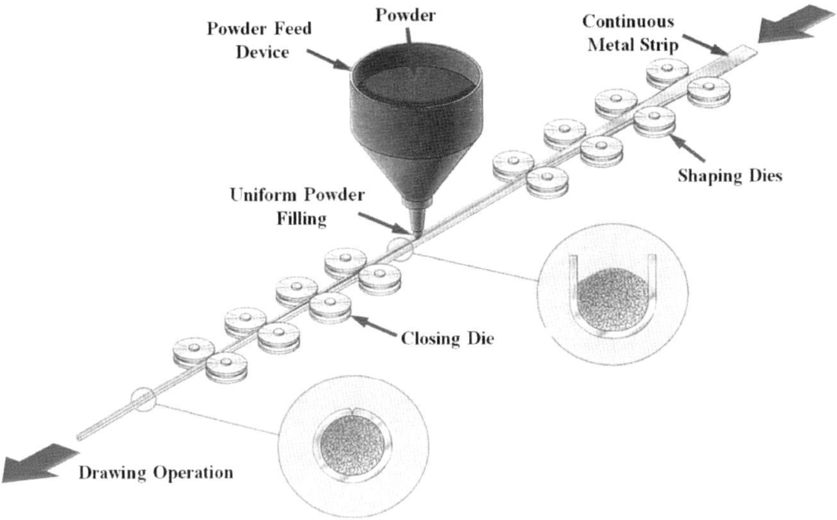

Fig. 2. Schematic picture of HTR's "continuous tube forming-and-filling, CTFF" process.

as a drawing aid. Following this process NIMS has been able to produce wire lengths of up to 100 m.[64] HTR is able to continuously process PIT and AIMI wires by its patent "continuous tube forming-and-filling" (CTFF) process (Fig. 2). Feeding into the machine is the chemical barrier in the form of a flat strip soon to be formed into a U-shaped channel. An Mg wire and B powder are fed into the channel which is then closed by an overlap seam or (more recently) by laser-beam butt-welding.[65,66] The composite, enclosed at some stage in a Monel® outer tube, continues on through a series of drawing dies. Using a recently installed CTFF machine HTR is able to make 1–3 km of AIMI monocore wire.

2.2.2 *Boron (B) powder*

Along with strand design and the C-doping method, the selection of B powder is a key point for the optimization of MgB_2's superconducting properties.[58,59,67]

All HTR AIMI starting wires are filled with B produced by Specialty Materials Inc. from $BCl_3 + H_2$ and enough CH_4 to enable

C-doping in the range of 2–4 mol.% (see subsequent paragraph). The SMI B, produced in an RF plasma from a mixture of BCl_3, H_2, and CH_4 (the source of C) gases, remains contaminated with BCl_3 gas. No attempt is made to remove BCl_3 although a toluene wash has been applied from time to time to adjust the density of the B powder.[68]

For their pioneering Mg-RLI studies the group of Giunchi *et al.* used crystalline and amorphous B powders chiefly from H.S. Starck in various grades, e.g., crystalline, grade-I (95%)[31]; amorphous grade II (90%),[31] amorphous grade III (88%).[23] Mostly they were "micrometric" or $<100\,\mu m$ in size. The NIMS IMD starting wires were generally packed with amorphous B, 99.9% or 99.99% pure, around 300 mesh in size. More recently, however, some much finer B powders were introduced: (i) nanosized (10–100 nm) C-coated amorphous B powder from SMI,[45] (ii) amorphous B powder (PVZ B, average size, 250 nm)[42] from the Pavezyum chemical company.[69] To remove Cl, the NIMS group treated the B powder with *p*-dimethylbenzene (C_8H_{10}).[45] On the other hand PVZ B,[42] produced by the decomposition of gaseous B_2H_6, has no associated BCl_3.

The size of the B particles and their packing density are two of the factors that influence the thickness of the MgB_2 reaction layer. In IMD samples formed from 300-mesh (40–50 μm) B powder the NIMS group noted that diffusion of the Mg was limited to about 20–30 μm.[35] On the other hand the fine ($\sim 40\,\mu m$) B powder from SMI turns out to be "fluffy" and of low density. As a result layer thicknesses of up to 90 μm have been achieved in 0.55 mm OD HTR-fabricated monocores.[61]

2.2.3 *Dopants*

By now the doping of MgB_2 with nitrides, borides and silicides, metallic elements, metal oxides, carbides, organic compounds, and C[70] has been widely studied. In selecting dopants, reaction chemistry is an important consideration. Particular attention has been paid to C, the only element known to substitute into the B sublattice.[70] Following its initial recognition by Dou *et al.*[71] as a dopant for C-substitution, SiC powder has been frequently mixed with starting Mg and B powders. The NIMS group, for example, has included typically 10 mol.% nano-SiC (tens of nm) in many of its wires.[40,41,46,52] The success of SiC[71] as a dopant for C relies on the fact that, during *in situ* processing, the Mg decomposes the SiC and releases atomic C before the onset of the $Mg + 2B \rightarrow MgB_2$ reaction. A disadvantage of SiC is that a byproduct of the reaction, Mg_2Si, may show up in grain boundaries and impair connectivity. For this reason researchers have begun to turn to doping with hydrocarbons (e.g., C-rich hexabenzobenzene, $C_{24}H_{12}$, "coronene")[42] or C itself. In what follows we discuss the properties of AIMI wires C-doped using C-coated SMI B powders.

2.2.4 *Reaction HT*

An extensive study of CTFF processed PIT wires showed little variation in J_c in response to HTs at 600–700°C for suitable times, viz, at least 71 h at 600°C but as short at 30 min, at 700°C. As a result HTs at 650–675°C for 2–8 h were selected for most of the AIMI wires, in particular 4 h/675°C[61] and 1 or 2 h/675°C.[62] The HTs were preceded by an 80 min ramp to soak temperature and followed by a furnace cool to room temperature.

2.3 ***Development of high performance AIMI wires***

To enable its use in the windings of practical superconducting devices the MgB_2 wire needs a high engineering J_e, for the attainment of which several essential quantities need to be maximized — among them layer J_c and *FF*. In a case study of these issues the properties of a pair of AIMI monocores of differing diameters were investigated as function of HT time at fixed temperature.[61] The initial CTFF processed monocore consisted of a Mg rod positioned along the axis of a B-filled double tube of Nb (the chemical barrier) and Monel® (the hard outer sheath). The B powder, from SMI, had been 2 mol.% C-doped by mixing the appropriate volume fraction of CH_4 with the process gases. Having been drawn down to final ODs (0.83 mm for wire "D" and 0.55 mm for wire "d") the wires were HT at 675°C for 0.5–8 h (D) and 0.5–4 h (d). The reaction layers and the J_cs of these two series of wires were studied in detail.

2.3.1 *The 0.83 mm OD wires, HT for 30 min (D1), 1 h (D2), and 8 h (D3)*

For these wires the B layer remained only partially reacted even after 8 h and consisted of two parts: (i) inner regions of thicknesses 0–30 μm (D1), 10–50 μm (D2), and 15–60 μm (D3) (the variation in thicknesses being due to the Mg rods' being slightly off axis), (ii) outer B-rich regions. As a result the MgB_2 *FF*s were 4–8% instead of 19–20%. The 4.2 K, 10 T layer J_cs increased strongly with HT time, ranging from 3.5×10^4 A/cm^2 (D1) to 1.06×10^5 A/cm^2 (D3). D1's small layer J_c was due to the presence of unreacted B within the MgB_2 layer; high values exhibited by both D2 and D3 indicated the presence of dense fully reacted layers (albeit of small volume fraction), Table 2.

2.3.2 *The 0.55 mm OD Wires, HT for 30 min (d1), 1 h (d2), and 4 h (d3)*

For the d-series wires the fraction of B converted increased from 52% through 57% to 100% and the *FF* increased from 10.1 through 11.2 to 18.8 as the HT time increased from 30 min to 4 h, Table 2. The MgB_2 layers of all the d-series wires were of good quality, the layer J_c of d1 being 1.07×10^5 A/cm^2 with that of d3 being only 13% lower, possibly as a result of grain growth. The high J_c of d1 comes at the expense of a significantly lower *FF*. So from a practical

Table 2: Fill factors and 4.2 K, 10 T CCDs of 2 mol.% C-doped AIMI wires.

Strand	Wire OD, mm	HT time at 675°C, h	Starting powder *FF*, %	MgB_2 *FF*, %	Layer J_c, 10^5 A/cm^2	Eng. J_e, 10^4 A/cm^2
D1	0.83	0.5	21.1	4.4	0.35	0.15
D2	0.83	1	19.7	6.3	1.00[a]	0.64
D3	0.83	8	19.1	7.5	1.06	0.80
d1	0.55	0.5	19.6	10.1	1.07	1.08
d2	0.55	1	19.7	11.2	1.02	1.14
d3	0.55	4	18.8	18.8	0.93	1.75

[a]Flux jumped at 10 T, therefore J_c extrapolated back from higher fields.

standpoint the highest performing wire of the set is d3 with a 4.2 K, 10 T $J_e = J_c \times FF$ of 1.75×10^4 A/cm^2 which is more than $4\frac{1}{2}$ times higher than that previously reported for IMD wires.[40]

2.3.3 *Engineering J_e, wire architecture, and critical current*

To be considered here are the layer J_c, the fill factor of the starting powder, the ability of the powders to be fully reacted, and the current carrying capacity of the final conductor. (i) Although the Mg/B ratio within the chemical barrier is the same in both cases IMD/AIMI has an advantage over conventional PIT. During the HT of PIT wire the reacting Mg particles leave behind pores and create a partially connected MgB_2 body. Since the only "porosity" present in the IMD/AIMI wire is the axial hole, the site of the prior Mg rod, the existence of a dense well-connected MgB_2 layer results in a much larger non-barrier CCD, $J_{c,nb}$. Attention can then be directed to increasing the $J_{c,\text{layer}}$ and hence the engineering J_e. (ii) The next step towards increasing J_e is to increase the fill factor of the elemental ingredients; the above d-series wires have powder fill factors of about 20%. (iii) It is then necessary to ensure that the powders are fully reacted. In this regard both HT and strand architecture come into play, Table 2. In the larger diameter D-series wires significant fractions of the B layers remained unreacted. After HT at 675°C for 0.5 h 79% of the starting B remained unreacted, some residual B within the MgB_2 layer itself inducing a relatively small $J_{c,\text{layer}}$. But although longer HT times produced dense well-connected MgB_2 layers with high $J_{c,\text{layer}}$s ($\sim 1.0 \times 10^5$ A/cm^2) much of the B still remained unreacted; 68% after 1 h, and 61% after 8 h. To ensure complete reaction, the starting B layer must be made suitably thin by a scaling down of the size of the composite. For example, a reduction of the wire diameter to 0.55 mm enabled full conversion of the B to MgB_2 during HT at 675°C for 4 h.

The maximization of J_e is not the end of the story. Practical applications will require not only high J_e but also a particular transport current. Thus the wire's diameter having been reduced to ensure complete B conversion, steps will need to be taken to restore its current-carrying capacity. Given that $J_e = J_c \times FF$, it follows that $I_c = J_e \times A_{tot}$. In other words the optimized J_e needs to be associated with a suitably large wire c/s area, A_{tot}. This implies either a multiwire cable or a multicore (multifilamentary) wire. Earlier 7-core, 19-core and 37-core wires formed by bundling IMD monocores into thick Cu–Ni tubes with *FF*s of only 3.0%,[38] 3.2%[49] and 1.8%,[44] respectively, may not lead to practical conductors. Not so for the more recent 18-core AIMI wires to be considered next, with their *FF*s of 10–14%.[62]

3. Doped AIMI Wires

3.1 *Studies of C concentration, HT time, and filament number*

The substitution of 2 mol.% C into the B sublattice increases the 4.2 K, 10 T, $J_{c,\text{layer}}$ from 1.5–100 kA/cm^2 — nearly two orders of magnitude. It is therefore important to find out whether further increases in $J_{c,\text{layer}}$ would accompany the introduction of higher concentrations of dopant. It was noted above (e.g., Table 2) that a scaling down of the wire cross-section increases the ability of the Mg to fully react the B layer, and that achieving this while preserving the conductor's current-carrying capacity requires a multicore strand architecture. The results of increasing C concentration and increasing filament number from 1–18 are summarized in Table 3 and Fig. 3.

3.1.1 *C-doped AIMI monocores*

A series of monocore AIMI wires were CTFF-processed by HTR using mostly amorphous 10–100 nm size SMI-B powder *in situ* doped with 2, 3 and 4 mol.% C. The starting billet, a Nb-lined Monel® tube containing the B powder and the axially mounted Mg rod, was drawn down to an OD of 0.55 mm. After HT for 1–4 h at 675°C the short-sample (~50 mm) J_cs of the wires were measured at 4.2 K, see Table 3(a) and Fig. 3. After an initial rapid increase with C-doping level $J_{c,\text{layer}}$ maximizes at 3 mol.% C. The subsequent decrease at 4 mol.%, where at some regions of B-rich and C-rich impurity phases have been observed,[72] is likely a result of reduced connectivity.

Table 3: Fill factors and 4.2 K, 10 T CCDs of 0.55 mm diameter monocore AIMI wires.[62]

Strand	Length[a]	mol.% C	HT time at 675°C, h	MgB_2 *FF*, %	Layer J_c, 10^5 A/cm^2	Eng. J_e, 10^4 A/cm^2
		(a) 0.55 mm diameter monocore AIMI wires				
d2	S	2	1	11.2	10.2	11.4
d3	S	2	2	18.8	9.3	17.5
m1	S	3	1	4.2	13.3	5.6
m2	S	3	2	5.1	14.5	7.4
m3	S	3	4	5.2	15.2	7.9
m4	S	4	1	3.7	10.1	3.7
m5	S	4	2	5.0	5.7	2.9
m6	S	4	4	5.0	8.8	4.4
		(b) 0.83 mm diameter 18-core AIMI wires				
M1	L	2	1	13.2	7.13	9.4
M2	L	2	2	14.4	4.73	6.8
M3	S	3	1	13.1	7.15	9.4
M4	S	3	2	13.0	6.71	8.7
M5	S	4	1	9.8	6.48	6.3
M6	S	4	2	12.0	4.48	5.4

[a]S: short sample (~50 mm in length); L: long sample (ITER barrel, ~1 m in length).

Comparing Figs. 3(a) and 3(b) (see also Table 3(a)) we note that J_e maximizes for 2 mol.%, the subsequent decreases being due to reductions in *FF* from ~18% (d3 with 2% C) to 4–5% (m1–m6 with 3 and 4% C). Microscopy performed on the 2% C-doped wires d1–d3 revealed layer thicknesses of 40–90 μm, wire d3 being fully reacted. On the other hand in 3% C-doped m3 the reaction layer, although dense, was only 8–25 μm thick.[62] The suppression of MgB_2 layer growth in response to SiC (hence C) doping has also been noted by Kumakura *et al.*[50]

Taken together it can be seen that increases in C concentration beyond 3 mol.% decrease $J_{c,\text{layer}}$ itself while increases beyond just 2 mol.% decrease J_e as the *FF* of the layer decreases. What is needed is a dopant (or co-dopant) that increases J_c but at the same time maximizes the *FF* by enabling more complete conversion of the B layer. It turns out that Dy_2O_3 is such a dopant, see below.

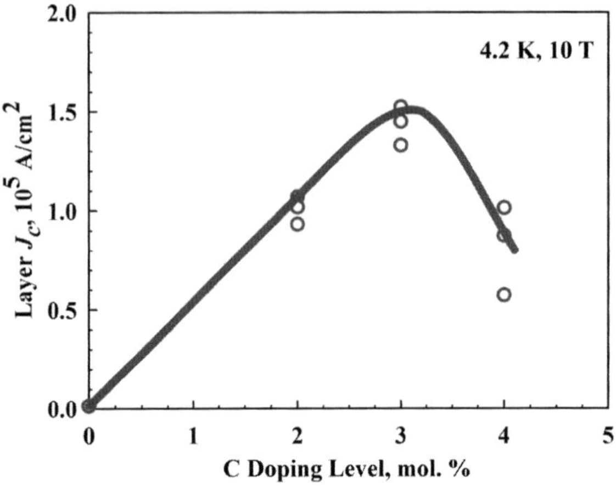

(a)

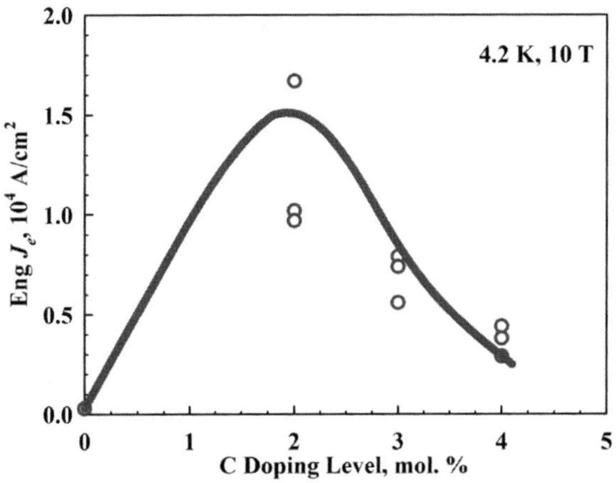

(b)

Fig. 3. Effect of C-doping level on (a) the 4.2 K, 10 T, $J_{c,\mathrm{layer}}$s and (b) the 4.2 K, 10 T, J_es of monocore AIMI wires.[62]

3.1.2 *C-doped AIMI multicore (multifilamentary) wires*

In fabricating the multicore wires, Table 3(b), Nb-clad B/Mg monocores were drawn down to 1.5 mm OD. After which 18 of them and one central Nb-clad Cu–Ni alloy rod were stacked into a Monel® tube (9.5 mm OD and 8.6 mm ID) and drawn down to 0.83 mm OD in preparation for of 1 and 2 h at 675°C. At final size the filaments were about 100 μm in diameter and the average B precursor layer thickness turned out to be less than 20 μm.[62]

The best performing wires at 4.2 K, 10 T, are M1 (long sample, 2 mol.% C) and M3 (short sample, 3 mol.% C) with $J_{c,layer}$s of 7.13 and 7.15 $\times$ 10^4 A/cm^2, respectively, and J_es of 9.4 kA/cm^2 after HT at 675°C for only 1 h. Increasing the HT time to 2 h reduced $J_{c,layer}$ significantly, e.g., from 7.13 $\times$ 10^4 A/cm^2 (M1, 1 h) to 4.73 $\times$ 10^4 A/cm^2 (M2, 2 h). The 1 h HT time is preferable for multicore wires since for them the average B-precursor layer thickness of $<$20 μm enables its complete transformation to MgB_2 in that time. Extending the HT time, unnecessary for layer growth, can cause grain coarsening, oxidation, or Cu contamination.[62] The $J_{c,layer}$s of the multicore wires responded only weakly to C-doping, e.g., 7.13, 7.15, and 6.48 $\times$ 10^4 A/cm^2 for 1 h heat treated M1 (2% C), M3 (3% C) and M5 (4% C).

The ideal *FF* of a multicore wire must be the same as that of its constituent (and unsheathed) monocores. The *FF* of fully reacted d3, Table 3(a), is 18.8%. Given that the 18-core multi contains a central CuNi alloy filament its ideal *FF* should be 17.8%. The observed departures from this value, Table 3(b), are attributed to incomplete reaction and/or filament defects of various kinds. Multicore AIMI wires with their finer filaments suffer more easily than monocores from degradation during wire fabrication. Most of the wires evince some unreacted B. Occasional Nb barrier breakage observed in M1, and Cu leakage detected in M3 and M5, has degraded their effective $J_{c,layer}$s below those of their monocore counterparts, Tables 3(a) and 3(b).

3.1.3 *Conclusion*

Multicore wires should be more amenable to high $J_{c,layer}$ attainment than the monocores. This is because the smaller filaments should enable full reaction during short HT times. In other words since the maximum reaction-layer thickness is severely limited (e.g., to $\sim$25 μm for 3 mol.% C-doping) the thickness of the B precursor layer should be less than that value. This can be realized in the multicore.

The starting ingredients having been optimized (optimal Mg core diameter plus 2–3 mol.% C-doped SMI B) a record 4.2 K, 10 T, J_e for multicore wire of 9.4 kA/cm^2 had been achieved. Improved wire processing techniques that eliminate breakages and barrier leakages and by so doing increase $J_{c,layer}$ and FF offer the promise of still further increases in J_e to much more than 10 kA/cm^2.

3.2 *Advanced doping of AIMI wires*

3.2.1 *Introduction*

The importance of doping MgB_2 with C is illustrated in Fig. 3(a) which shows the 4.2 K, 10 T, $J_{c,layer}$ rising from 1.5–100 kA/cm^2 with the addition of 2 mol.% C. As illustrated in Fig. 4 this high-field increase is in response to a dopant-induced increase in the 4.2 K B_{irr}. The beneficial influence of C is two-fold: the increase of $J_{c,layer}$ at high fields is accompanied by a decrease at low fields, a property important for stability in that region. But many applications of MgB_2 will require operation at much higher temperatures than 4.2 K. In this regime (e.g., at 20 K) it turns out that C-doping produces little change in B_{irr}, Fig. 5. What was needed was a dopant or co-dopant with the potential for increasing the higher temperature performance. Since RE dopants had been shown to do this for typical PIT wires,[73,74] it seemed likely that they would do the same for AIMI wires. The particular RE oxide chosen was Dy_2O_3 since its lower cost made it a suitable ingredient of commercially produced wire.

3.2.2 *Processing and properties*

A series of monocore AIMI wires were CTFF-processed by HTR using mostly amorphous 10–100 nm size SMI-B powder *in situ* doped with 2 mol.% C mixed with 1 or 2 wt.% of Dy_2O_3 powder (99.9%, 100 nm in size).[63] The starting billet, a Nb-lined Monel® tube containing the powder mixture and the usual axially mounted Mg rod, was drawn down to an OD of 0.52 mm. After HT at 650°C and 675°C for 1–8 h, optical- and electron microstructures were studied and CCDs measured by V–I and magnetic techniques,

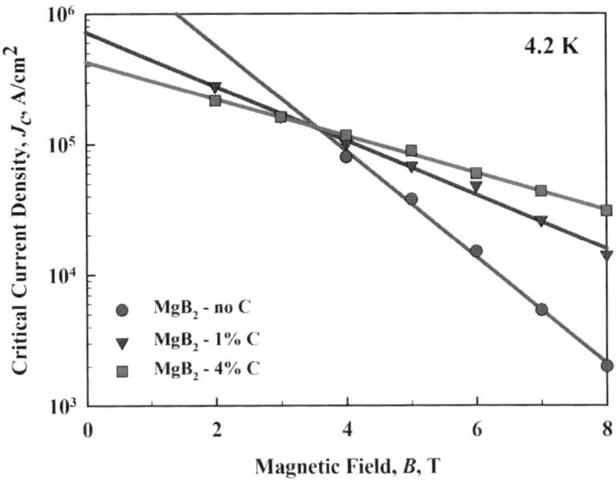

(a)

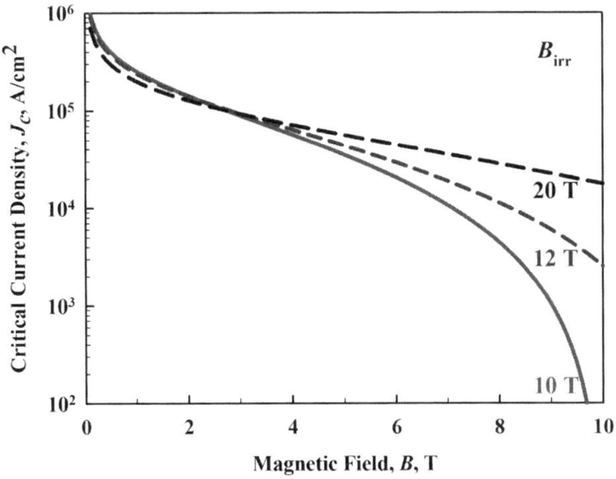

(b)

Fig. 4. (a) $J_c(B)$ at 4.2 K for HTR-manufactured MgB_2 monocore wires for dopant concentrations of 0, 1.29, and 4.01 mol.% C,[75] (b) Model calculation of $J_c(B)$ for three values of B_{irr} based on the Dew–Hughes grain boundary pinning function.[75,76] Reprinted with permission from Ref. 75. ©2013 Elsevier.

generally in the temperature- and field ranges of 4.2–30 K and 0–14 T, respectively.

As for reaction speed and completeness, the MgB_2 layer growth rate appeared to increase with either an increase in the Dy_2O_3

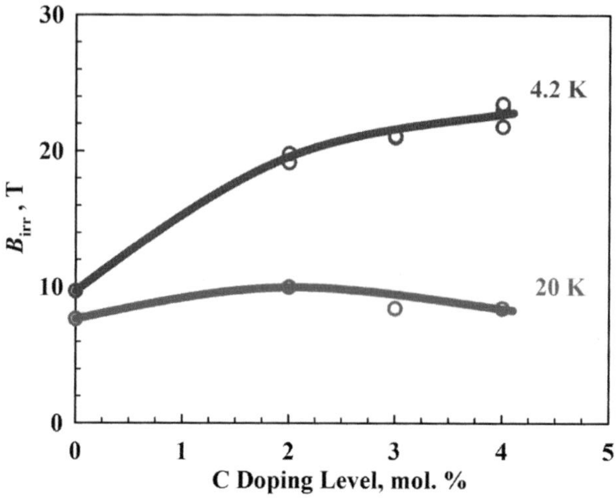

Fig. 5. B_{irr} versus C-doping level at low and high temperatures. $B_{\mathrm{irr}}(4.2\,\mathrm{K})$ is based on transport J_c measurement with criterion $J_{c,nb} = 1000\,\mathrm{A/cm^2}$ for the fully reacted undoped and 2% C-doped wires, and $J_{c,\mathrm{layer}} \sim 2000\,\mathrm{A/cm^2}$ for the partially reacted 3% and 4% C-doped wires. $B_{\mathrm{irr}}(20\,\mathrm{K})$ is based on resistivity measurement (10% criterion) in the PPMS.

co-doping level from 1 wt.% to 2 wt.% or an increase of the HT temperature from 650°C to 675°C. Full reaction within 8 h was achieved by the 1 wt.% Dy_2O_3 wires at 675°C and the 2 wt.% Dy_2O_3 wires at 650°C.

The superconducting properties of 0.52 mm OD 2 wt.% Dy_2O_3/C co-doped wires F2 and F3, a selected two of the many wires tested, are presented in Tables 4–6, and Fig. 6 for comparison with those of other 2% C-doped IMD/AIMI wires.

Doping with Dy_2O_3 does not significantly improve the 4.2 K, 10 T $J_{c,\mathrm{layer}}$s of 2% C-doped wires the average values of which are $10.8 \times 10^4\,\mathrm{A/cm^2}$ (F2, F3) and $9.8 \times 10^4\,\mathrm{A/cm^2}$ (d2, d3, I2), Table 4. On the other hand in the higher temperature range, frequently of practical interest, Dy_2O_3 addition has a very strong effect. This can be seen in the $J_c(B)$ curves of Fig. 6 as well as in Table 5 (a reformulation of the Fig. 6 data) which show for example that at 30 K, 1.4 T, F3 has a $J_{c,\mathrm{layer}}$ of $10^4\,\mathrm{A/cm^2}$, whereas I2 can achieve that value only in self-field. The reason for the difference lies in D3's superior higher

Table 4: Critical current densities of Dy_2O_3/2%-C co-doped wires F2 and F3, and 2% C-doped wires d2, d3 and I2.

Strand	HT, h/°C	MgB_2 *FF*, %	Layer J_c, 10^4 A/cm^2	Eng. J_e, 10^3 A/cm^2
F2	2/675	7.1	13.5	9.6
F3	4/675	12.8	8.1	12.8
d2[a]	1/675	11.2	10.2	11.2
d3[a]	4/675	18.8	9.3	18.8
I2[b]	0.5/675	—	10.0	—

[a]data from Table 3(a).
[b]I2: an early OSU/HTR-made, 2%C-doped, IMD wire.[56]

Table 5: Fields at which $J_{c,\text{layer}} = 10^4$ A/cm^2 at temperatures of 10–30 K for wires F3 and I2. (unit: T)

Strand	10 K	15 K	20 K	25 K	30 K
F3	12.2	9.6	6.8	4.0	1.4
I2[a]	12.0	9.1	5.9	2.8	0
Field increase, T	0.2	0.5	0.9	1.2	1.4

[a]I2: an early OSU/HTR-made, 2%C doped, IMD wire.[56]

Table 6: Irreversibility fields, [a]B_{irr}(T), of 2 mol.% C-doped wire and the same co-doped with 2 wt.% Dy_2O_3 wire. Reprinted with permission from Ref. 63. ©2013 AIP Publishing LLC.

Strand	10 K	15 K	20 K	25 K	30 K
Dy_2O_3/C co-doped, F3	14.7	11.4	8.2	4.9	2.0
C-doped, I2	14.4	10.9	7.0	3.4	0.4

[a]based on $J_{c,nb}$ extrapolated to 1000 A/cm^2.

temperature B_{irr} which, as Table 4 shows, increases rapidly beyond that of I2 at temperatures above 25 K.

The advanced doping of AIMI wires is a work in progress involving further studies of reaction rate, reaction depth and completeness, the influence of dopants on these issues, and the effectiveness of the dopants.

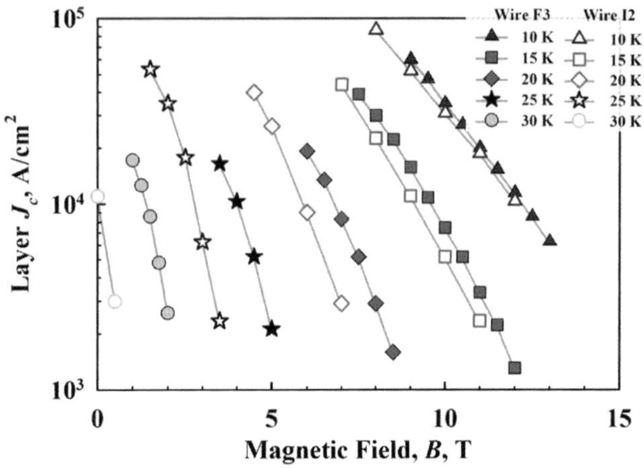

Fig. 6. $J_{c,\text{layer}}$ versus B for wires F3 and I2 at temperatures of 10–30 K. Reprinted with permission from Ref. 63. ©2013 AIP Publishing LLC.

4. Mechanisms of MgB_2 Layer Formation in AIMI Wires

4.1 *Introduction*

The formation of MgB_2 from elemental ingredients as a result of the reaction $Mg + 2B \rightarrow MgB_2$ takes place within the B matrix according to $B \rightarrow MgB_7 \rightarrow MgB_4 \rightarrow MgB_2$ each stage of which is accompanied by expansion. The overall expansion resulting from the conversion of 2 moles of B (4.59 cm^3/mol) to 1 mole of MgB_2 (17.46 cm^3/mol) is $17.46/(2 \times 4.59) = 1.9$ which is more than sufficient to collapse the interstices between the B powder particles. Accordingly the HT of Mg-RLI/IMD/AIMI precursor wire leaves a Mg-depleted core surrounded by a thin cylinder of almost fully dense MgB_2 produced by the initial rapid infiltration of liquid Mg (Stage I). Reaction of the B remaining outside this cylinder requires the diffusion of Mg through this dense inner layer, a much slower process (Stage II). In the fairly short HT times required to produce reaction while restricting grain growth and contamination the layer thickness is, to a first approximation, that of Stage I. For example: (i) in a 2 mol.% C-doped monocore heat treated at 675°C the reacted area increased by 26% between 0.5 h and 1 h then by only 1.3% after another 7 h, and after 0.5, 1 and 8 h the maximum MgB_2 reaction layer thicknesses were 30, 50, and 60 μm[61]; (ii) in a 3 mol.% C-doped monocore heat treated at 675°C the reacted area increased by 26% between 1 and 2 h then again by only 1.3% after another 2 h.[62] Microscopy shows that the dense MgB_2 layer is surrounded first by a $MgB_2 + B$ "transition region" and then by an "unreacted B layer" which, taken together, are regarded as a "B-rich layer". In what follows we discuss the layer formation mechanisms in detail.

4.2 *Stage-I, dense MgB_2 layer formation*

Stage I is the reactive infiltration of liquid Mg into a porous B compact. Many studies have been made of liquid infiltration into porous bodies, all of them assuming driving forces of various kinds.[77,78] In the present case the driving force is the reactive diffusion of Mg into the B nanoparticles. During reactive infiltration the particles expand and eventually block the paths to further infiltration, resulting after time t_0 in a dense MgB_2 layer of thickness x_0. The analysis proceeds as follows:

If A_{SC} is the area of the Mg/B interface then the volume of the future reaction layer is $A_{SC}x_0$. The volume of precursor B powder occupying that volume is $A_{SC}x_0X$, where X is the powder packing fraction (e.g., ~0.65 for close-packed spheres). The corresponding pore volume within the future reaction layer is $A_{SC}x_0(1 - X)$. Due to volume expansion only a fraction of the B powder is needed to fill the porosity, thus

$$(1.9 - 1)V_B = A_{SC}x_0(1 - X). \qquad (\mathrm{m}^3) \tag{1}$$

Reactive infiltration is completed in time t_0 at an MgB_2 formation rate.

$$J_{\mathrm{form}} = k_{\mathrm{MB}}C_{\mathrm{S}}, \qquad (\mathrm{mol/m^2 \cdot s}) \tag{2}$$

where C_S is the concentration (mol/m^3) of the Mg at the reaction interface and k_{MB} is the 2-D rate constant (m/s) for the chemical reaction Mg + 2B $\rightarrow$ MgB_2. According to chemical kinetics, k_{MB} only depends on temperature and reaction activation energy such that $k_{\mathrm{MB}} = k^{\circ}_{\mathrm{MB}} \exp(Q_{\mathrm{MB}}/RT)$ in which k°_{MB} is the 2D reaction rate constant at standard conditions and Q_{MB} is the activation energy for MgB_2 formation. The amount of MgB_2 (molar volume, M_{MB}) formed in time t_0 is therefore

$$J_{\mathrm{form}} A_{\mathrm{SC}} t_0 = \frac{V_B}{M_{\mathrm{MB}}}. \qquad \text{(mol)} \tag{3}$$

Upon combining Eqs. (1), (2) and (3) we find

$$x_0 = (2.6 k_{\mathrm{MB}} C_{\mathrm{S}} M_{\mathrm{MB}}) t_0, \qquad \text{(m)} \tag{4}$$

wherein $2.6 k_{\mathrm{MB}} C_{\mathrm{S}} M_{\mathrm{MB}}$ is the infiltration velocity (m/s). The reactive infiltration terminates after time t_0 which is the time for Mg to diffuse fully into a B particle of size x_p according to

$$x_{\mathrm{p}} = \sqrt{D_{\mathrm{p}} t_0}, \qquad \text{(m)} \tag{5}$$

in which D_p is the Mg $\rightarrow$ B diffusion coefficient.

Combining Eqs. (4) and (5) we find

$$x_0 = \frac{2.6 k_{\mathrm{MB}} C_{\mathrm{S}} M_{\mathrm{MB}}}{D_{\mathrm{p}}} x_{\mathrm{p}}^2. \qquad \text{(m)} \tag{6}$$

Equation (6) shows x_0 increasing with the square of the B-particle size. For this reason Giunchi *et al.* were able to produce large dense MgB_2 objects ("manufacts") from crystalline B ground to sizes of 40–180 μm (Sec. 1.1.4), albeit at the expense of $J_c(B)$ which decreased as x_p increased from 1 μm $\rightarrow$ 40 μm $\rightarrow$ 100 μm $\rightarrow$ 180 μm.

4.3 *Stage II, prolonged HT*

After the completion of the first stage, further reaction requires Mg to diffuse through a dense MgB_2 layer. With a Mg concentration at the initial Mg/B interface of C_S dropping to C_i at the MgB_2/B interface the diffusion flux through the dense layer of thickness x_0 is

$$J_{\mathrm{diff}} = \frac{D_{\mathrm{II}}(C_{\mathrm{S}} - C_{\mathrm{i}})}{x_0}, \qquad (\mathrm{mol/m^2 \cdot s}) \tag{7}$$

where D_{II} is the Mg $\rightarrow$ MgB_2 diffusion coefficient.

This is equal to the rate of MgB_2 formation starting at the MgB_2/B interface (cf. Eq. (2)) viz:

$$J_{\text{form}} = k_{\text{MB}} C_{\text{i}}. \qquad (\text{mol/m}^2 \cdot \text{s}) \tag{8}$$

Since J_{diff} must equal J_{form}, we can eliminate C_i between Eqs. (7) and (8) to find

$$J_{\text{diff}} = \frac{D_{\text{II}} k_{\text{MB}} C_{\text{S}}}{k_{\text{MB}} x_0 + D_{\text{II}}}. \qquad (\text{mol/m}^2 \cdot \text{s}) \tag{9}$$

After inserting N_v (mol/m^3), the amount of Mg needed to produce unit volume of MgB_2, the layer growth rate can be written

$$\frac{dx_0}{dt} = \frac{J_{\text{diff}}}{N_{\text{v}}} = \frac{D_{\text{II}} k_{\text{MB}}}{k_{\text{MB}} x_0 + D_{\text{II}}} \cdot \frac{C_{\text{S}}}{N_{\text{v}}}. \qquad (\text{m/s}) \tag{10}$$

Then since Stage-II is a diffusion-controlled process, such that $k_{\text{MB}} x_0 \gg D_{\text{II}}$, Eq. (10) reduces to

$$\frac{dx_0}{dt} = \frac{D_{\text{II}} C_{\text{S}}}{N_{\text{v}}} \cdot \frac{1}{x_0}, \qquad (\text{m/s}) \tag{11}$$

the solution to which is

$$x_0^2 = \frac{2D_{\text{II}} C_{\text{S}}}{N_{\text{v}}} \cdot t + const. \qquad (\text{m}^2) \tag{12}$$

The constant in Eq. (12) can be evaluated by considering as a starting point for Stage II the terminal conditions for Stage I, viz x_0 and t_0 which for the present purpose we now write as x_0^* and t_0^*. After doing so we find

$$x_0^2 = \frac{2D_{\text{II}} C_{\text{S}}}{N_{\text{v}}} \cdot (t - t_0^*) + (x_0^*)^2. \qquad (\text{m}^2) \tag{13}$$

4.4 *Conclusion*

In order to compare the Stage I and Stage II layer thicknesses we need to insert numerical values into Eq. (13): (i) the diffusion coefficient $D_{\text{II}} = 10^{-17}\ \text{m}^2/\text{s}$,[79] (ii) the concentration of liquid Mg,

$C_S = 6.52 \times 10^4$ mol/m^3, and (iii) the amount of Mg per m^3 of MgB_2, $N_v = 5.73 \times 10^4$ mol/m^3. Equation (13) then becomes

$$x_0^2 = 2.28 \times 10^{-5}(t - t_0^*) + (x_0^*)^2. \qquad (\mu\text{m}^2) \tag{14}$$

As pointed out in Sec. 2.3.1, wire D2 acquired a Stage I reaction layer thickness (x_0^*) of 50 μm after 60 min at 675°C. Suppose we extend the HT into Stage II for a further time, $t - t_0^* = 60$ min. After substituting these numbers into Eq. (15) we find the new layer thickness, x_0, to be given by

$$x_0^2 = 0.082 + 2500, \qquad (\mu\text{m}^2) \tag{15}$$

which represents a 0.002% increase in x_0^* beyond the Stage I value, in accord with Sec. 4.1. In other words, the very small diffusivity of Mg through the dense Stage I layer prevents a large increase in layer thickness. Only grain-boundary-induced increases in D_{II}[80] would lead to significant amounts of Stage II MgB_2. A detail discussion of MgB_2 reactive infiltration, taken into account a wide range of B particle size and the influence of C-doping, is presented elsewhere.[81]

5. The Effects of Porosity, Connectivity and Anisotropy on the $J_c(B)$ of MgB_2 Wires

5.1 *Eisterer's percolation model*

How far can MgB_2 performance be pushed with the AIMI approach, and which constraints are now the limiting ones for J_c and J_e? In order to answer these questions, we now evaluate the $J_c(B)$ of AIMI wires in terms of a percolation model with particular reference to connectivity, porosity, and upper critical field anisotropy. The MgB_2 layer in AIMI wires is almost fully dense and well connected, not so for typical PIT-type MgB_2 wires. A comparison of the properties of these two classes of wire provides insight into the ways that connectivity and anisotropy affect the in-field CCD.

The percolation model was developed by Eisterer *et al.*[82] for describing the current-flow characteristics in MgB_2 superconductors.

By adjusting a few variables, the percolation model precisely fits the $J_c(B)$ curves of MgB_2 at all temperatures and fields.[83,84] The model, based on percolation theory,[85,86] assumes an orientation-dependent upper critical field, $B_{c2}(\theta)$, and grain-boundary pinning mechanism. Accordingly, the field dependence of J_c for a single MgB_2 grain can be expressed as

$$J_c(B) = F_m \frac{\left(1 - \frac{B}{B_{c2}(\theta)}\right)^2}{\sqrt{B_{c2}(\theta)B}}, \tag{16}$$

where F_m is the maximum pinning strength.

If the J_c distribution for all grains is known, then resulting transport J_c can be estimated as follows:

$$J_c(B) = \int_0^{J_{c,\max}(B)} A_p(p(J), p_c)dJ, \tag{17}$$

where the percolation threshold p_c is the minimum fraction of superconducting grains necessary for a continuous current path, $p(J)$ is the fraction of MgB_2 grains whose J_cs are higher than J, the limit $J_{c,\max}(B)$ is defined by $p[J_{c,\max}(B)] = p_c$, and the percolation cross-section A_p, on which the current effectively flows, is given by

$$A_p(p(J), p_c) = \left(\frac{p(J) - p_c}{1 - p_c}\right)^{1.76}. \tag{18}$$

It is noted that Eq. (18) is valid for fully dense well-connected samples. However, in most PIT-type MgB_2 wires, non-superconducting particles and voids reduce the superconducting cross-section and thus $p(J)$ is reduced to $p_S \cdot p(J)$. The term p_S denotes the fraction of superconducting MgB_2 grains in the specimen. Hence, A_p can be rewritten as

$$A_p(p(J), p_c) = \left(\frac{p_S \cdot p(J) - p_c}{1 - p_c}\right)^{1.76}. \tag{19}$$

The percolation threshold p_c should only rely on the arrangement of MgB_2 grains and must be a constant with respect to temperature or field changes. According to van der Marck,[87] a reasonable p_c for closed packed three dimensional objects is 0.2. For detailed information about the percolation model papers by Eisterer *et al.*[82,88–90] are suggested.

To apply the percolation model, a system with over 10,000 units is built. Each unit is assigned a value of $\sin(\theta)$ randomly distributed between 0 and 1 to simulate the randomly oriented grains observed in MgB_2 wires. Four parameters are taken into account in fitting the $J_c(B)$ performance, viz: the maximum pinning strength F_m, the upper critical field B_{c2}, the anisotropy parameter γ and the fraction of MgB_2 grains p_s. The quantities F_m and B_{c2} have been well studied by researchers in terms of their influences on $J_c(B)$.[70,91] They can be obtained by way of $J_c(B)$ and $\rho(T)$ measurements. Both of them depend on the choices of B powder and C-doping level, but they are nearly independent of the wire fabrication route (i.e., PIT or AIMI). On the other hand the information on γ and p_s, and their impact on J_c is very limited.[88,92,93] Those quantities are difficult to measure and to control during wire fabrication. However, they have a strong influence on J_c, particular the high-field J_c. Figure 7 illustrates the individual effects of γ and p_s on J_c. The J_cs are calculated using the percolation model, assuming that $B_{c2} = 20\,\text{T}$, $F_m = 5 \times 10^6\,\text{A}\cdot\text{T/cm}^2$, and $p_c = 0.2$. As shown in Fig. 7(a), γ increases from one for an ideal isotropic case to as high as six for the highly anisotropic MgB_2 samples.[93] The figure shows, the high-field J_c degrading drastically with γ increases. The 10 T J_c falls by nearly two orders of magnitude from $8.8 \times 10^4\,\text{A/cm}^2$ to $1 \times 10^3\,\text{A/cm}^2$, as γ increases from one (representing isotropic MgB_2) to 3 (typical of many undoped MgB_2 wires).[72] The effect of changing p_s is displayed in Fig. 7(b). By setting other parameters constant and only decreasing p_s from 1.0 to 0.6, the 10 T J_c drops more than an order of magnitude from $2.7 \times 10^3\,\text{A/cm}^2$ to $141\,\text{A/cm}^2$. Since p_s is primarily determined by porosity and connectivity, the J_c drop in Fig. 7(b) indicates the benefits of dense well-connected MgB_2 over typical PIT MgB_2 wires.

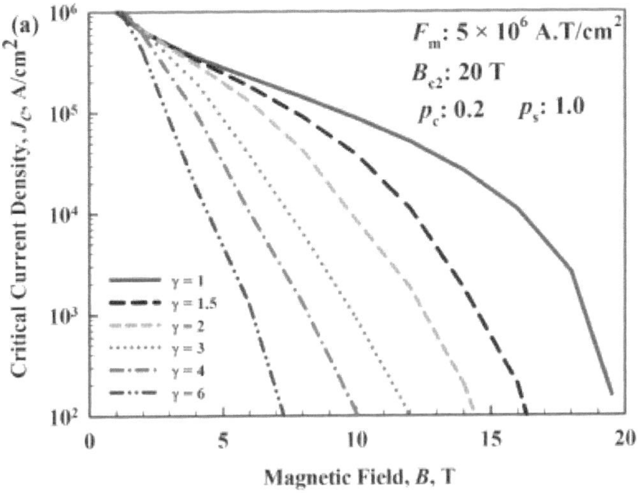

(a)

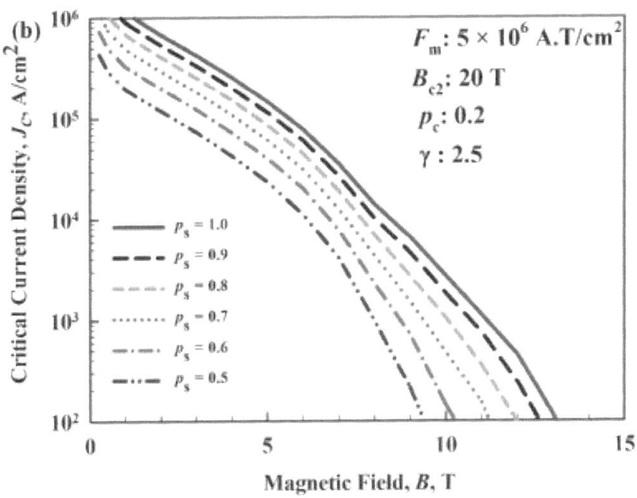

(b)

Fig. 7. Calculated $J_c(B)$ curves based on Eisterer's percolation model by (a) increasing anisotropy γ from 1 to 6, and (b) changing p_s from 1.0–0.5. Other parameters are constant, with $B_{c2} = 20\,\text{T}$, $F_m = 5 \times 10^6\,\text{A} \cdot \text{T/cm}^2$, and $p_c = 0.2$.

Table 7: Fitting parameters to MgB_2-AIMI and MgB_2-PIT.

Fitting parameters	MgB_2-AIMI	MgB_2-PIT
F_m, $A \cdot T/cm^2$	1.1×10^7	1.1×10^7
B_{c2}, T	31.8	33
γ	2.6	2.6
p_c	0.2	0.2
p_s	1.00	0.67

5.2 *Fitting of the percolation model to experimental data*

An AIMI-fabricated MgB_2 wire (hereafter MgB_2-AIMI) and a PIT-type wire (hereafter MgB_2-PIT) were studied. Both samples used the same C-doped SMI B powders and underwent the same HT scheme. So, for both wires, the superconducting properties F_m, B_{c2} and γ of the MgB_2 grains should be the same and the major difference is p_s.

The $J_c(B)$ curves are fitted by the percolation model. The fitting parameters are listed in Table 7, and the fitting plots are presented in Fig. 8. The B_{c2}s are 32 T and 33 T for MgB_2-AIMI and MgB_2-PIT, respectively, according to the results of $\rho(T)$ measurements. Table 7 demonstrates that all variables are the same for both MgB_2-AIMI and MgB_2-PIT except the porosity and connectivity term "p_s". The value of p_s yielding the best fit in the case of MgB_2-PIT is found to be 0.67, indicating 33% total porosity. The value of p_s achieves one for MgB_2-AIMI, indicating a fully dense structure. In other words p_s values indicate a porous structure in PIT wires and a dense well-connected structure in AIMI wires. Therefore, it is reasonable to conclude that the significant increase in J_c of AIMI wires is attributable to the enhanced p_s.

The J_cs of both MgB_2-AIMI and MgB_2-PIT wires exhibit the effects of anisotropy in addition to which the J_c of PIT wire is degraded by porosity and weak connectivity. To illustrate these effects, by way of reference the $J_c(B)$ curve of a SC with well-connected isotropic grains is constructed based on the above MgB_2-AIMI fitting parameters but with γ set equal to 1.0. The $J_c(B)$s

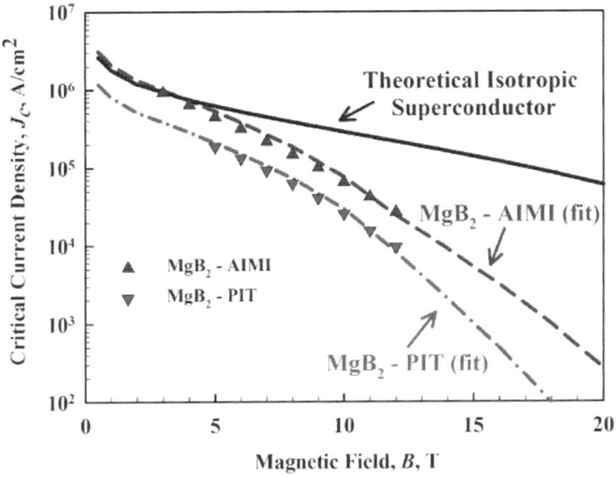

Fig. 8. Percolation model fittings for samples MgB_2-AIMI and MgB_2-PIT, together with plot of an isotropic grain boundary SC with the same F_m and B_{c2}.

Table 8: Contributions of anisotropy and connectivity on J_c loss with respect to "ideal" isotropic, well-connected MgB_2.

	2 T (%)	6 T (%)	10 T (%)	14 T (%)	18 T (%)
Porosity and Connectivity	75	65	40	19	11
Upper Critical Field Anisotropy	25	35	60	81	89

of both classes of wire are then calculated based on the Table 7 data. The results, presented in Fig. 8, show the 4.2 K, 10 T J_c of MgB_2-AIMI dropping by a factor 3.8 from 2.9×10^5 A/cm^2 to 7.7×10^4 A/cm^2, while afterwards that of MgB_2-PIT drops by a factor 2.3 from 7.7×10^4 A/cm^2 to 3.0×10^4 A/cm^2. For the typical PIT wire the drop in 4.2 K, 10T J_c from that of an ideal isotropic SC comes from two parts: 60% due to anisotropy and 40% due to porosity and weak connectivity. For both classes of wire the deleterious effect of the anisotropy increases as the applied magnetic field increases. Table 8 lists the contributions of anisotropy and connectivity for J_c loss at fields of 2–18 T.

5.3 *Conclusion*

A percolation model is applied to MgB_2 wires so as to enable the effects of anisotropy and porosity/connectivity to be deconvoluted. The $J_c(B)$s of both MgB_2-AIMI and MgB_2-PIT wires are degraded by B_{c2} anisotropy especially at high fields. The existence of porosity and weak connectivity reduces the $J_c(B)$ of MgB_2-PIT still further. Referred to the 4.2 K, 10 T J_c of an isotropic fully connected model SC 60% of that reduction is due to anisotropy and 40% to porosity/connectivity. The AIMI process provides a route towards enhanced J_c at all fields.

6. Conclusions

The development, processing, properties, and formation mechanisms of AIMI MgB_2 wires were discussed against a background of the related and original processes, IMD and Mg–RLI. The transition from Mg–RLI- and IMD- to AIMI wires has been explained, and the relative performances of PIT, IMD and AIMI wires were summarized in the form of an iso-J_e diagram of $J_{c,nb}$ versus A_{nb}/A_{tot}, Fig. 1, in which A_{tot}, A_{nb}, $J_{c,nb}$ and J_e are, respectively, the wire's cross-sectional area, the area inside the chemical barrier, the critical current (I_c) normalized to A_{nb}, and I_c normalized to A_{tot}. The AIMI wires clearly exhibited superior $J_{c,nb}$s as compared to traditional PIT wires and superior J_es as compared to IMD wires. The property enhancements in AIMI wires results from both improved connectivity (with respect to PIT) as well as increased MgB_2 fill factor (compared to IMD). A two-stage mechanism of layer formation in AIMI wires was described. During the first stage a dense MgB_2 layer is formed by the reactive infiltration of liquid Mg into a porous B-powder-packed region. In the second stage the initially formed dense MgB_2 layer inhibits Mg infiltration, further MgB_2 layer growth being generated by the slow diffusion of Mg through that layer into any remaining B.

Our analysis of AIMI wires using Eisterer's percolation model in conjunction with a carefully chosen set of related samples (PIT and AIMI made with the same starting powders) revealed that even for fields in the 5–10 T regime, the J_c suppression due to porosity/connectivity in PIT conductors is comparable to that due to anisotropy. It seems that previous researchers had sometimes overstated the anisotropy contribution to J_c suppression (inadvertently including the effects of porosity/connectivity). It was shown that the successful fabrication of fully dense AIMI wires solved the porosity/connectivity issue and overcame about 40% of the previously noted reduction of the 4.2 K, 10 T, J_c. Non-barrier 4 K, 10 T, $J_{c,nb}$s as high as 4×10^4 A/cm^2 (or J_e values of 10^4 A/cm^2) have revived once more the high-field possibilities for MgB_2. New results for Dy_2O_3-doped AIMI wires have shown Dy_2O_3 to be the most potent dopant since C. But in contrast to C, which is effective at 4.2 K, Dy_2O_3 was shown to be effective in the 20–30 K regime, e.g., at 30 K it increased B_{c2} by a factor of five. Early PIT-produced MgB_2 wires exhibited adequate flux pinning but left porosity/connectivity and anisotropy open for improvement.[70] The described developments in AIMI wires in which the porosity/connectivity problem has been overcome have led to useful conductors with substantially increased operating ranges of B and T. Based on such an advanced conductor, further improvements in B_{c2} can be expected to result in concomitant increases in the regimes of application.

Acknowledgments

This work is funded by the Department of Energy, High Energy Physics division under Grant No. DE-FG02-95ER40900, a DOE SBIR, and a program from the Ohio Department of Development. For MAS partial funding for writing of the chapter was sponsored by US Department of Energy, Office of Science, Basic Energy Sciences, Materials Science and Engineering Division and supported in part by DOE Grant No. DE-FG02-09R46554.

References

1. P. C. Canfield, D. K. Finnemore, S. L. Bud'ko, J. E. Ostenson, G. Lapertot, C. E. Cunningham and C. Petrovic, Superconductivity in dense MgB_2 wires, *Phys. Rev. Lett.* **86**, pp. 2423–2426 (2001).

2. D. C. Dunand, Synthesis of superconducting Mg/MgB_2 composites, *Appl. Phys. Lett.* **79**, pp. 4186–4188 (2001).
3. S. Ueda, J. Shimoyama, I. Iwayama, A. Yamamoto, Y. Katsura, S. Horii and K. Kishio, High critical current properties of MgB_2 bulks prepared by a diffusion method, *Appl. Phys. Lett.* **86**, p. 222502 (2005).
4. X. F. Pan, Y. Zhao, Y. Feng, Y. Yang and C. H. Cheng, Effects of graphite doping on critical current density and microstructure of MgB_2 bulks by an improved Mg-diffusion method, *Physica C* **468** (15–20), pp. 1169–1173 (2008).
5. G. Giunchi, S. Ceresara, G. Ripamonti, S. Chiarelli and M. Spadoni, MgB_2 reactive sintering from the elements, *IEEE T. Appl. Supercon-duct.* **13**, pp. 3060–3063 (2003).
6. G. Giunchi, High density MgB_2 obtained by reactive liquid Mg infiltration, *Int. J. Mod. Phys. B.* **17**, pp. 453–460 (2003).
7. G. Giunchi, S. Raineri, R. Wesche and P. L. Bruzzone, The voltage-current relations for MgB_2 obtained by reactive liquid infiltration, *Physica C* **401**, pp. 310–315 (2004).
8. G. Giunchi, The superconducting properties of high density bulk MgB_2 obtained by reactive liquid Mg infiltration, *AIP Conf. Proc.* **52**, pp. 813–819 (2006).
9. G. Giunchi, G. Ripamonti, T. Cavallin and E. Bassani, The reactive liquid Mg infiltration process to produce large superconducting bulk MgB_2 manufacts, *Cryogenics* **46**, pp. 237–242 (2006).
10. J.-H. Ahn, S. Oh, X. L. Wang and S. X. Dou, Infiltration of magnesium in porous boron skeletons, *Int. J. Mod. Phys. B* **2009**, pp. 3503–3508 (2009).
11. G. Giunchi, E. Perini, C. Orecchia, G. Ripamonti, E. Bassani, G. Carcano, G. DeMarzi, R. Viola and S. Turtu, Analysis of various dopants on the MgB_2 superconducting properties, *IEEE T. Appl. Superconduct.* **19**, p. 2802 (2009).
12. L. Gozzelino, R. Gerbaldo, G. Ghigo, F. Laviano, G. Lopardo, G. A. Ummarino, G. Giunchi, E. Perini, E. Bassani, A. Agostino and B. Minetti, Effects of annealing and nanoparticle doping on electrical properties of MgB_2 bulks, *IEEE T. Appl. Superconduct.* **19**, pp. 3524–3228 (2009).
13. L. Saglietti, E. Perini, G. Ripamonti, E. Bassani, G. Carcano and G. Giunchi, Boron purity effects on structural properties of the MgB_2 obtained by Mg reactive liquid infiltration, *IEEE T. Appl. Superconduct.* **19**, pp. 2739–2744 (2009).
14. A. A. Gallitto, G. Bonsignore, G. Giunchi, M. L. Vigni and Y. A. Nefyodov, Microwave response of bulk MgB_2 samples of different granularity, *J. Phys. Conf. Ser.* **43**, pp. 480–483 (2006).

15. G. Giunchi, T. Cavallin, E. Bassani and S. Guicciardi, The mechanical properties of the MgB_2 bulk materials obtained by reactive liquid Mg infiltration, *AIP Conf. Proc.* **54**, p. 396 (2008).
16. T. Cavallin, E. A. Young, C. Beduz, Y. Yang and G. Giunchi, Thermal conductivity of bulk MgB_2 produced by infiltration of different boron powders, *IEEE T. Appl. Superconduct.* **17**, p. 2770 (2007).
17. G. Giunchi, S. Ginocchio, S. Raineri, D. Botta, R. Gerbaldo, B. Minetti, R. Quarantiello and A. Matrone, High density MgB_2 bulk materials of different grains size Supercurrents instability and losses in variable magnetic fields, *IEEE T. Appl. Superconduct.* **15**, p. 3230 (2005).
18. G. Giunchi, G. Ripamonti, S. Raineri, D. Botta, R. Gerbaldo and R. Quarantiello, Grain size effects on the superconducting properties of high density bulk MgB_2, *Supercond. Sci. Technol.* **17**, pp. S583–S588 (2004).
19. Z. K. Liu, D. G. Schlom, Q. Li and X. X. Xi, Thermodynamics of the Mg-B system: implications for the deposition of MgB_2 thin films, *Appl. Phys. Lett.* **78**, p. 3678 (2001).
20. Z. K. Liu, Y. Zhong, D. G. Schlom, X. X. Xi and Q. Li, Computational thermodynamic modeling of the Mg-B system, *Calphad.* **25**, pp. 299–303 (2001).
21. A. F. Albisetti, L. Saglietti, E. Perini, C. Schiavone, G. Ripamonti and G. Giunchi, The Mg_2B_{25} formation and its role in the preparation of bulk MgB_2 superconductors, *Solid State Sci.* **14**, pp. 1632–1635 (2012).
22. G. Giunchi, C. Orecchia, L. Malpezzi and N. Masciocchi, Analysis of the minority crystalline phases in bulk superconducting MgB_2 obtained by reactive liquid Mg infiltration, *Physica C* **433**(3–4), pp. 182–188 (2006).
23. G. Giunchi, L. Saglietti, A. F. Albisetti, E. Perini, C. Schiavone, G. Ripamonti, D. Uglietti and P. L. Bruzzone, MgB_2 hollow wires and cables embedded in a Mg alloy matrix, *IEEE T. Appl. Superconduct.* **23**, p. 6200605 (2013).
24. L. Gozzelino, B. Minetti, R. Gerbaldo, G. Ghigo, F. Laviano, G. Lopardo, G. Giunchi, E. Perini, T. Cavallin and E. Mezzetti, Pinning properties in pure and SiC doped MgB_2 bulk obtained by reactive Mg liquid infiltration technique, *Physica C* **460–462**, pp. 604–605 (2007).
25. L. Gozzelino, B. Minetti, R. Gerbaldo, G. Ghigo, F. Laviano, G. Lopardo, G. Giunchi, E. Perini, T. Cavallin, A. Agostino, M. Panetta and E. Mezzetti, Effects of nanoparticle doping on electrical properties of MgB_2 bulks and wires obtained by reactive Mg liquid infiltration technique, *IEEE T. Appl. Superconduct.* **17**, p. 2726 (2007).
26. L. Saglietti, E. Perini, A. F. Albisetti, G. Ripamonti, C. Schiavone and G. Giunchi, MgB_2 doping by rare earth metals and borides, *IEEE T. Appl. Superconduct.* **22**, p. 6200504 (2012).

27. G. Ghigo, D. Botta, A. Chiodoni, R. Gerbaldo, L. Gozzelino, E. Mezzetti, B. Minetti, S. Ceresara, G. Giunchi and G. Ripamonti, Creep properties of MgB_2 bulk near the irreversibility line, *Supercond. Sci. Technol.* **14**, p. 722 (2001).
28. G. Giunchi, S. Ceresara, G. Ripamonti, A. Di Zenobio, S. Rossi, S. Chiarelli, M. Spadoni, R. Wesche and P. L. Bruzzone, High performance new MgB_2 superconducting hollow wires, *Supercond. Sci. Technol.* **16**, pp. 285–291 (2003).
29. H. Fujishiro, T. Tamura, T. Arayashiki, M. Oyama, T. Sasaki, T. Naito, G. Giunchi and A. F. Albisetti, Pulsed field magnetization of large MgB_2 bulk fabricated by reactive liquid Mg infiltration, *Jpn. J. Appl. Phys.* **51**, p. 103005 (2012).
30. G. Giunchi, MgB_2 superconductive inserts products between bulks and wires, *IEEE T. Appl. Superconduct.* **21**, p. 1564 (2011).
31. L. Saglietti, C. Orecchia, A. F. Albisetti, E. Perini and G. Giunchi, Microstructure of the MgB_2 wires resulting by the infiltration process, *IEEE T. Appl. Superconduct.* **21**, p. 2655 (2011).
32. G. Giunchi, G. Ripamonti, E. Perini, T. Cavallin, E. Bassani, U. Gambardella, Y. Yang, E. A. Young, M. Bianchetti and C. Beduz, Superconducting characteristics of MgB_2 wires obtained by reactive liquid Mg infiltration, *Mater. Res. Soc. Symp. Proc.* **946**, pp. 9–18 (2006).
33. G. Giunchi, G. Ripamonti, E. Perini, T. Cavallin and E. Bassani, Advancements in the reactive liquid Mg infiltration technique to produce long superconducting MgB_2 tubular wires, *IEEE T. Appl. Superconduct.* **17**, p. 2761 (2007).
34. G. Giunchi. The reactive Mg-liquid infiltration to obtain long superconducting MgB_2 cables. In *CERN Proceedings Workshop WAMSDO*, pp. 88–93, Geneve, Switzerland (2008).
35. H. Kumakura, Development of high performance MgB_2 tapes and wires, *J. Phys. Soc. Jpn.* **81**, p. 011010 (2012).
36. Y. Shimada, Y. Kubota, S. Hata, K. Ikeda, H. Nakashima, A. Matsumoto, K. Togano and H. Kumakura, Electron microscopy observations of MgB_2 wire prepared by an internal Mg diffusion method, *Physica C* **471**(21–22), pp. 1137–1141 (2011).
37. Y. Shimada, Y. Kubota, S. Hata, H. Ikeda, H. Nakashima, A. Matsumoto, K. Togano, J. M. Hur and H. Kumakura, Microstructure in High density MgB_2 wires prepared by an internal Mg diffusion method, *IEEE T. Appl. Superconduct.* **21**(3), p. 2668 (2011).
38. K. Togano, J. M. Hur, A. Matsumoto and H. Kumakura, Fabrication of seven-core multi-filamentary MgB_2 wires with high critical current

density by an internal Mg diffusion process, *Supercond. News Forum.* **22**(1), p. 015003 (2009).
39. P. Kováč, I. Hušek, T. Melišek, L. Kopera and J. Kováč, Critical currents, Ic-anisotropy and stress tolerance of MgB_2 wires made by internal magnesium diffusion, *Supercond. Sci. Technol.* **27**(6), p. 065003 (2014).
40. S. J. Ye, A. Matsumoto, K. Togano and H. Kumakura, Enhancement of the critical current density of internal Mg diffusion processed MgB_2 wires by the addition of both SiC and liquid aromatic hydrocarbon, *Physica C* **471**(21–22), pp. 1133–1136 (2011).
41. S. J. Ye, M. Song, A. Matsumoto, K. Togano, Y. Zhang, H. Kumakura, M. Takeguchi, R. Teranishi and T. Kiyoshi, Enhancing the critical current properties of internal Mg diffusion-processed MgB_2 wires by Mg addition, *Supercond. Sci. Technol.* **25**(12), p. 125014 (2012).
42. S. J. Ye, A. Matsumoto, Y. C. Zhang and H. Kumakura, Strong enhancement of high-field critical current properties and irreversibility field of MgB_2 superconducting wires by coronene active carbon source addition via the new B powder carbon-coating method, *Supercond. Sci. Technol.* **27**, p. 085012 (2014).
43. S. Ye, A. Matsumoto, K. Togano, Y. Zhang, T. Ohmura and H. Kumakura, Fabrication of MgB_2 superconducting wires with a hybrid method combining internal-Mg-diffusion and powder-in-tube processes, *Supercond. Sci. Technol.* **27**(5), p. 055017 (2014).
44. S. J. Ye, A. Matsumoto, K. Togano, T. Ohmura and H. Kumakura, The critical current properties of 37-filament internal Mg diffusion-processed MgB_2 wires, *Supercond. Sci. Technol.* **26**(10), p. 105027 (2013).
45. S. Ye, M. Song, A. Matsumoto, K. Togano, M. Takeguchi, T. Ohmura and H. Kumakura, High-performance MgB_2 superconducting wires for use under liquid-helium-free conditions fabricated using an internal Mg diffusion process, *Supercond. Sci. Technol.* **26**(12), p. 125003 (2013).
46. K. Togano, H. Fujii, T. Takeuchi and H. Kumakura, An internal Mg–Li alloy composite process for the fabrication of MgB_2 wire, *Supercond. Sci. Technol.* **20**(7), pp. 629–633 (2007).
47. K. Togano, H. Fujii, T. Takeuchi and H. Kumakura, Composite process for the fabrication of MgB_2 wire using ductile Mg–Li alloy, *Supercond. Sci. Technol.* **20**(3), pp. 239–243 (2007).
48. J. M. Hur, K. Togano, A. Matsumoto, H. Kumakura, H. Wada and K. Kimura, Fabrication of high-performance MgB_2 wires by an internal Mg diffusion process, *Supercond. Sci. Technol.* **21**, p. 032001 (2008).
49. K. Togano, J. Hur, A. Matsumoto and H. Kumakura, Microstructures and critical currents of single- and multi-filamentary MgB_2

superconducting wires fabricated by an internal Mg diffusion process, *Supercond. Sci. Technol.* **23**(8), p. 085002 (2010).
50. H. Kumakura, J. M. Hur, K. Togano, A. Matsumoto, H. Wada and K. Kimura, Superconducting properties of diffusion-processed multifilamentary MgB_2 wires, *IEEE T. Appl. Superconduct.* **21**, p. 2643 (2011).
51. J. M. Hur, K. Togano, A. Matsumoto, H. Kumakura, H. Wada and K. Kimura, High critical current density MgB_2/Fe multicore wires fabricated by an internal Mg diffusion process, *IEEE T. Appl. Superconduct.* **19**, p. 2735 (2009).
52. S. J. Ye, M. Song, A. Matsumoto, K. Togano, M. Takeguchi, H. Kumakura, R. Teranishi and T. Kiyoshi, Comparison of SiC and/or toluene additives to the critical current density of internal Mg diffusion-processed MgB_2 wires, *Physica C* **484**, pp. 167–170 (2013).
53. M. Maeda, M. S. Hossain, A. Motaman, J. H. Kim, A. Kario, M. Rindfleisch, M. Tomsic and S. X. Dou, Synergetic combination of LIMD with CHPD for the production of economical and high performance MgB2 wires, *IEEE T. Appl. Superconduct.* **23**(3), p. 6200704 (2013).
54. R. Flükiger, M. S. Hossain and C. Senatore, Strong enhancement of J_c and B_{irr} in binary in situ MgB_2 wires after cold high pressure densification, *Supercond. Sci. Technol.* **22**, p. 085002 (2009).
55. M. S. Hossain, C. Senatore, R. Flükiger, M. Rindfleisch, M. Tomsic, J. H. Kim and S. X. Dou, The enhanced J_c and B_{irr} of in situ MgB_2 wires and tapes alloyed with $C_4H_6O_5$ (malic acid) after cold high pressure densification, *Supercond. Sci. Technol.* **22**, p. 095004 (2009).
56. G. Z. Li, Y. Yang, M. A. Susner, M. D. Sumption and E. W. Collings, Critical current densities and n-values of MgB_2 strands over a wide range of temperatures and fields, *Supercond. Sci. Technol.* **25**, p. 025001 (2012).
57. C. Wang, D. Wang, X. Zhang, C. Yao, C. Wang, Y. Ma, H. Oguro, S. Awaji and K. Watanabe, Improved J_c–B properties of MgB_2 multifilamentary wires and tapes, *Supercond. Sci. Technol.* **25**(12), p. 125001 (2012).
58. M. A. A. Mahmud, M. A. Susner, M. D. Sumption, M. A. Rindfleisch, M. Tomsic, J. J. Yue and E. W. Collings, Comparison of critical current density in MgB_2 with different B sources and nano-particle dopant additions, *IEEE T. Appl. Superconduct.* **19**, pp. 2756–2759 (2009).
59. G. Z. Li, J. B. Zwayer, C. J. Kovacs, M. A. Susner, M. D. Sumption, M. Rindfleisch, C. J. Thong, M. Tomsic and E. W. Collings, Transport critical current densities and n-values of multifilamentary MgB_2 wires at various temperatures and magnetic fields, *IEEE T. Appl. Superconduct.* **24**, p. 6200105 (2014).

60. G. Nishijima, S. J. Ye, A. Matsumoto, K. Togano, H. Kumakura, H. Kitaguchi and H. Oguro, Mechanical properties of MgB_2 superconducting wires fabricated by internal Mg diffusion process, *Supercond. News Forum.* **25**(5), p. 054012 (2012).
61. G. Z. Li, M. D. Sumption, M. A. Susner, Y. Yang, K. M. Reddy, M. Rindfleisch, M. Tomsic, C. J. Thong and E. W. Collings, The critical current density of advanced internal-Mg-diffusion MgB_2 wires, *Supercond. Sci. Technol.* **25**, p. 115023 (2012).
62. G. Z. Li, M. D. Sumption, J. B. Zwayer, M. A. Susner, M. A. Rindfleisch, C. J. Thong, M. J. Tomsic and E. W. Collings, Effects of carbon concentration and filament number on advanced internal Mg infiltration-processed MgB_2 strands, *Supercond. Sci. Technol.* **26**(9), p. 095007 (2013).
63. G. Z. Li, M. D. Sumption, M. Rindfleisch, C. J. Thong, M. Tomsic and E. W. Collings, Enhanced higher temperature (20–30 K) transport properties and irreversibility field in nano-Dy_2O_3 doped advanced internal Mg infiltration processed MgB_2 composites, *Appl. Phys. Lett.* **105**, p. 112603 (2014).
64. S. J. Ye. private communication. Oct. 2014.
65. M. Tomsic, M. Rindfleisch, J. J. Yue, K. McFadden, J. Phillips, M. D. Sumption, M. Bhatia, S. D. Bohnenstiehl and E. W. Collings, Overview of MgB_2 superconductor applications, *Int. J. Appl. Ceram. Technol.* **4**, p. 250 (2007).
66. M. Tomsic, M. Rindfleisch, J. J. Yue, K. McFadden, D. Doll, J. Phillips, M. D. Sumption, M. Bhatia, S. D. Bohnenstiehl and E. W. Collings, Development of magnesium diboride (MgB_2) wires and magnets using in situ strand fabrication method, *Physica C* **456**, pp. 203–208 (2007).
67. Q. Cai, Y. C. Liu, Z. Q. Ma and L. M. Yu, Superconducting properties and growth mechanism of layered structure in MgB_2 bulks with Cu/Y_2O_3 co-doping, *J. Mater. Sci. Mater. Electron.* **24**, pp. 1451–1457 (2013).
68. M. Rindfleisch. private communication. 2014.
69. Pavezyum chemical company. Available from: http://pavezyum.com.tr.
70. E. W. Collings, M. D. Sumption, M. Bhatia, M. A. Susner and S. D. Bohnenstiehl, Prospects for improving the intrinsic and extrinsic properties of magnesium diboride superconducting strands, *Supercond. Sci. Technol.* **21**, p. 103001 (2008).
71. S. X. Dou, V. Braccini, S. Soltanian, R. Klie, Y. Zhu, S. Li, X. L. Wang and D. Larbalestier, Nanoscale-SiC doping for enhancing J_c and H_{c2} in superconducting MgB_2, *J. Appl. Phys.* **96**(12), p. 7549 (2004).
72. M. A. Susner, Y. Yang, M. D. Sumption, E. W. Collings, M. A. Rindfleisch, M. Tomsic and J. V. Marzik, Enhanced critical fields and

superconducting properties of pre-doped B powder-type MgB_2 strands, *Supercond. Sci. Technol.* **24**, p. 012001 (2011).
73. P. Mikheenko, S. K. Chen and J. L. MacManus-Driscoll, Minute pinning and doping additions for strong 20K in-field critical current improvement in MgB_2, *Appl. Phys. Lett.* **91**, p. 202508 (2007).
74. S. K. Chen, M. Wei and J. L. MacManus-Driscoll, Strong pinning enhancement in MgB_2 using very small Dy_2O_3 additions, *Appl. Phys. Lett.* **88**, p. 192512 (2006).
75. M. A. Susner, M. D. Sumption, M. A. Rindfleisch and E. W. Collings, Critical current densities of doped MgB_2 strands in low and high applied field ranges: The J_c(B) crossover effect, *Physica C* **490**, pp. 20–25 (2013).
76. D. Dew-Hughes, Flux pinning mechanisms in type II superconductors, *Philos. Mag.* **30**, pp. 293–305 (1974).
77. G. Rajesh and R. B. Bhagat, Infiltration of liquid metals in porous compacts: modelling of permeabilities during reactive melt infiltration, *Transp. Porous Media* **36**, pp. 43–68 (1999).
78. T. Bárczy and G. Kaptay, Modeling the infiltration of liquid metals into porous structures, *Mater. Sci. Forum* **473**, pp. 297–302 (2005).
79. J. D. DeFouw, J. P. Quintana and D. C. Dunand, In situ X-ray synchrotron study of MgB_2 synthesis from elemental powders, *Acta Mater* **56**, pp. 1680–1688 (2008).
80. Y. Mishin and C. Herzig, Grain boundary diffusion: recent progress and future research, *Mater. Sci. Eng* **260**, pp. 55–71 (1999).
81. G. Z. Li, M. D. Sumption and E. W. Collings, Microstructural investigation and kinetic analysis of infiltration-processed MgB_2 wires, *in preparation.* (2015).
82. M. Eisterer, M. Zehetmayer and H. W. Weber, Current percolation and anisotropy in polycrystalline MgB_2, *Phys. Rev. Lett.* **90**, p. 247002 (2003).
83. J. H. Kim, S. Oh, Y.-U. Keo, S. Hata, H. Kumakura, A. Matsumoto, M. Mitsuhara, S. Choi, Y. Shimada, M. Maeda, J. L. MacManus-Driscoll and S. X. Dou, Microscopic role of carbon on MgB_2 wire for critical current density comparable to NbTi, *NPG Asia Mater.* **4**, pp. 1–7 (2012).
84. G. Z. Li, K. M. Reddy, J. B. Zwayer, M. A. Kuldell, M. A. Susner, Y. Yang, M. D. Sumption, J. J. Yue, M. Rindfleisch, M. Tomsic, C. J. Thong and E. W. Collings, Critical current density and current transfer length of multifilamentary MgB_2 strands of various design, *IEEE T. Appl. Superconduct.* **23**, p. 6200204 (2013).
85. S. W. Kenkel and J. P. Straley, Percolation theory of nonlinear circuit elements, *Phys. Rev. Lett.* **49**, pp. 767–770 (1982).

86. J. P. Straley and S. W. Kenkel, Percolation theory for nonlinear conductors, *Phys. Rev. B.* **29**, pp. 6299–6305 (1984).
87. S. C. van der Marck, Percolation thresholds and universal formulas, *Phys. Rev. E.* **55**, p. 1514 (1997).
88. M. Eisterer, C. Krutzler and H. W. Weber, Influence of the upper critical field anisotropy on the transport properties of polycrystalline MgB_2, *J. Appl. Phys.* **98**, p. 033906 (2005).
89. M. Eisterer, Magnetic properties and critical currents of MgB_2, *Supercond. Sci. Technol.* **20**, pp. R47–R73 (2007).
90. M. Eisterer and H. W. Weber, The influence of weak texture on the critical currents in polycrystalline MgB_2, *Supercond. Sci. Technol.* **23**, p. 034006 (2010).
91. M. Putti and G. Grasso, MgB_2, a two-gap superconductor for practical applications, *MRS Bulletin.* **36**, pp. 608–613 (2011).
92. M. Angst, R. Puzniak, A. Wisniewski, J. Jun, S. M. Kazakov, J. Karpinski, J. Roos and H. Keller, Temperature and field dependence of the anisotropy of MgB_2, *Phys. Rev. Lett.* **88**, p. 167002 (2002).
93. C. Buzea and T. Yamashita, Review of the superconducting properties of MgB_2, *Supercond. Sci. Technol.* **14**, R115–R146 (2001).

Chapter 5a

Material and Conductor Properties Relevant for Applications: A Fundamental Study

Carmine Senatore* and Marco Bonura

Department of Quantum Matter Physics (DQMP) and
Department of Applied Physics (GAP)
University of Geneva, 24 Quai Ernest Ansermet
1211 Geneva, Switzerland
**carmine.senatore@unige.ch*

This chapter focuses on selected aspects of the material properties and of the conductor design that are being developed to satisfy the requirements of the envisaged applications of MgB_2 (cryogen-free MRI systems, fault current limiters, large current DC cables, etc.). Scope of the chapter is also to show how experimental techniques generally used for fundamental investigations, such as measurements of specific heat, magnetic relaxation and thermal conductivity, become operative tools to obtain the practical understanding necessary to advance in the material and conductor processing. We present here three case studies based on recent results obtained at the University of Geneva: (i) the cold high pressure densification (CHPD) was developed for improving the critical current density of MgB_2 wires and tapes. Here we investigate the influence of CHPD on the current percolation, based on a combined analysis of specific heat and resistivity measurements; (ii) we report a study of the magnetic relaxation on MgB_2 bulks and wires to evaluate the possibility to operate MgB_2 magnets in the persistent mode; (iii) we analyze by thermal conductivity measurements the influence of the MgB_2 conductor layout and architecture on the parameters governing the thermal stability of a magnet.

1. Introduction

Over the last 20 years, the portfolio of available technical superconductors has expanded. Apart from the conventional metallic superconductors NbTi and Nb_3Sn, Bi-based first generation high temperature superconductors (1G HTS) are commercially available in long lengths, whilst second generation ceramic REBCO conductors (2G HTS) are in the early stage of commercialization. After 15 years from its discovery, MgB_2 is also in this early commercialization stage, in view of cryogen-free devices operating between 20 K and 30 K. Nevertheless, the commercially available wires are still not able to satisfy all requirements of the envisaged applications for MgB_2 (cryogen-free MRI systems, fault current limiters, large current DC cables, etc.), the performance being still below the inherent potential of the material.[1,2] Further developments of MgB_2 focus on the improvement of the current capability at increasing temperatures and magnetic fields as well as on the engineering of the thermal and mechanical properties of practical conductors to ensure the stability of the superconducting devices. In this paper, we propose three case studies concerning both the material properties and the conductor design:

(i) Recently, cold high pressure densification (CHPD) was introduced at the University of Geneva as a new route for enhancing the critical current density in binary and doped MgB_2.[3,4] This process consists of submitting MgB_2 wires to high pressures ($p \sim 1$ GPa) at room temperature after the final drawing and before the heat treatment (see Chapter 4d). In order to understand the fundamental mechanism behind the strong improvement of J_C, the properties of binary and alloyed MgB_2 wires have been investigated without and with CHPD, using specific heat and resistivity measurements. The results are discussed in Sec. 2.

(ii) One stringent condition for the application of MgB_2 wires in NMR and MRI magnets is the ability to operate in persistent mode, which requires a low relaxation rate of the superconducting current with time. This depends in turn on intrinsic properties of the material — the activation energy for the motion of thermally activated vortices — and on the development of a superconducting joint technique for MgB_2 conductors. In Sec. 3, we present magnetic relaxation studies performed on MgB_2 bulks and tapes. The results for Nb_3Sn and REBCO are also shown for comparison.

(iii) To improve the design of MgB_2 wires suitable for operation in cryogen-free devices, the main unresolved issue concerns the behavior of the normal zones induced by local thermal disturbances. In Sec. 4, we report a study on thermal conductivity of MgB_2 tapes with different thermal stabilization layouts, composite materials and architecture. We also discuss the influence of the conductor design on minimum quench energy and normal zone propagation velocity.

2. Effects of the CHPD Process Probed by Resistivity and Specific Heat Measurements

The CHPD was recently developed in Geneva for improving the in-field critical current density J_C of *in situ* binary and alloyed MgB_2 wires and tapes.[3,4] A high pressure step is performed on the wires at the end of the deformation process, in order to densify the precursors mixture before the reaction heat treatment. After reaction, this corresponds to an enhancement of the MgB_2 filament mass density and, as a consequence, to a considerable increase of the critical current density J_C. After applying 1.5 GPa to malic acid ($C_4H_6O_5$) added wires, the highest J_C values so far reported for *in situ* MgB_2 wires were obtained, as high as J_C (4.2 K) $= 10^4$ A/mm^2 at $B \approx 14$ T.[4]

In order to understand the fundamental mechanism behind this strong improvement of J_C, we investigated the properties of binary and alloyed MgB_2 wires without and with CHPD, using resistivity and specific heat measurements.[5] In particular, a deconvolution of the specific heat data was used to determine the distribution of T_C in the samples. This analysis, in combination with the results of the resistivity measurements, allowed us to estimate the minimum superconducting volume fraction needed for percolation with zero resistivity and the influence of CHPD on its value.

Here we report on the analysis performed on four samples, two binary MgB_2 wires from the same batch, non-densified (Bin0) and densified at $p = 2\,GPa$ (Bin2), and two 10 wt.% $C_4H_6O_5$ doped wires from the same batch, non-densified (Mal0) and densified at $p = 2\,GPa$ (Mal2). Both binary and malic acid doped MgB_2 wires were fabricated at the University of Geneva. Wires are monofilamentary and have a Fe matrix. The cross-section before densification is a $2\,mm \times 2\,mm$ square. All the measurements were performed on the superconducting filaments after extraction from the matrix.

2.1 *Electrical resistivity and connectivity analysis*

The application of 2 GPa on the original square binary MgB_2 wire caused an enhancement of J_C at 4.2 K and 10 T by ~50% (see Table 1). The increase of J_C with the applied pressure is accompanied by a reduction of the normal state resistivity. Figure 1 shows the comparison of the $\rho(B, T = 25\,K)$ curves for the binary MgB_2 samples reacted in the as-drawn state and after cold densification.

The resistivity of the MgB_2 core at the superconducting transition decreases from $72\,\mu\Omega\,cm$ for the non-densified sample to $53\,\mu\Omega\,cm$ for the wire densified at 2 GPa. The decrease of ρ with pressure represents a strong evidence of the improvement of the grain connectivity. The correlation between electrical resistivity and grain connectivity is discussed in a seminal article by Rowell.[6] Rowell defines the connectivity, i.e., the fractional area of the sample that

Table 1: Summary of the sample properties: critical current density, resistivity at T_C, RRR, onset T_C and $T(\rho = 0)$.

Wire sample	J_c(4.2 K,10 T) [A/cm^2]	J_c(4.2 K,15 T) [A/cm^2]	ρ_0 [$\mu\Omega$ cm]	RRR	Onset T_c [K]	$T(\rho = 0)$ [K]
Bin0	910	n/a	72	1.67	37.4	36.2
Bin2	1330	n/a	53	1.73	37.5	36.5
Mal0	n/a	980	154	1.40	34.6	33.0
Mal2	n/a	2100	146	1.33	33.8	32.6

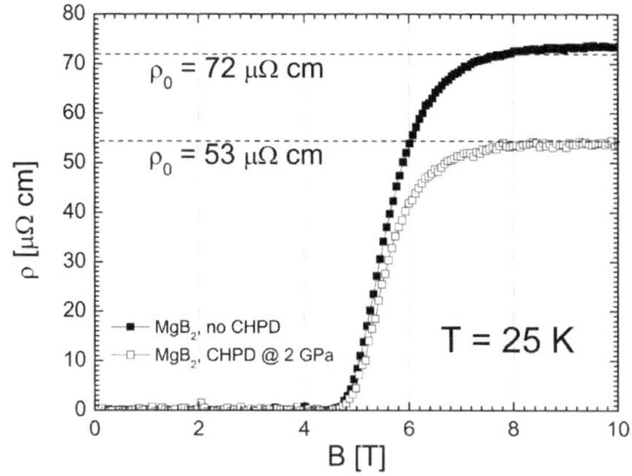

Fig. 1. Plot of the resistivity at $T = 25\,\mathrm{K}$ versus magnetic field for the non-densified binary MgB_2 sample (solid symbols) and for the binary MgB_2 sample submitted to CHPD at p = 2 GPa (open symbols). The normal state resistivity is reduced for the densified sample.

carries current, as $K = \Delta\rho_{sc}/\Delta\rho$, where $\Delta\rho = \rho(300\,\mathrm{K}) - \rho(40\,\mathrm{K}) = \rho(40\,\mathrm{K})(RRR - 1)$ is the phononic term of the electric resistivity, $RRR = \rho(300\,\mathrm{K})/\rho(40\,\mathrm{K})$, and $\Delta\rho_{sc}$ is the corresponding value for a fully connected single crystal. Rowell assumes $\Delta\rho_{sc} = 4.482$, using the single crystal results from Eltsev *et al.*[7] According to this model, the connectivity increases by ~25% in the sample densified at $p = 2\,\mathrm{GPa}$, the values of K are $K(p = 0) = 9.3\%$ and $K(p = 2\,\mathrm{GPa}) = 11.5\%$.

After densification at $p = 2\,\mathrm{GPa}$, the malic acid added MgB_2 wire exhibits an enhancement of J_C at 4.2 K and 15 T by >100% (see Table 1). In the case of doped MgB_2, densification produces two simultaneous effects on the resistivity versus magnetic field, as shown in Fig. 2:

- the resistivity at the superconducting transition is reduced, but the variation of ρ_0 is less pronounced compared to the effect on the binary sample. ρ_0 after CHPD decreases by 7% for the doped sample and by 35% for the binary MgB_2;
- the superconducting transition is shifted to higher magnetic fields, i.e., the upper critical field is increased.

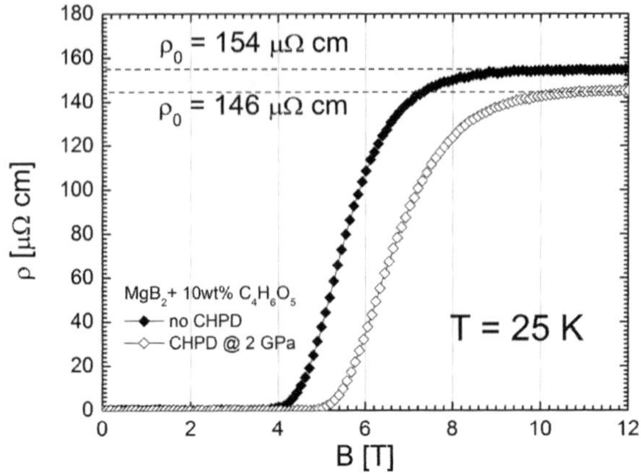

Fig. 2. Plot of the resistivity at $T = 25\,\mathrm{K}$ versus magnetic field for the malic acid doped samples, non-densified (solid symbols) and densified at $p = 2\,\mathrm{GPa}$ (open symbols). CHPD causes the reduction of the normal state resistivity and a shift of the transition towards higher magnetic fields.

These two phenomena are the consequence of the interplay of two competing mechanisms caused by densification:

- connectivity is improved, leading to the reduction of ρ_0;
- the precursor powders in the densified filaments are better packed and thus the reaction paths become shorter. In doped wires this makes the C substitution more effective, leading to a reduction of T_C, an enhancement of B_{C2} at low temperatures, accompanied by an increase of ρ_0.

Unfortunately, the enhanced C content after cold densification does not allow estimating the variation in K using the Rowell method for the malic acid doped wires.

2.2 *Specific heat and T_C distribution analysis*

As a complement to the connectivity studies, we analyzed the temperature dependence of the specific heat for the samples in Table 1 in order to get insights on the effects of CHPD on the intrinsic superconducting properties of MgB_2. Measurements were performed using the long relaxation technique.[8] The superconducting contributions to specific heat were then analyzed by means of a particular deconvolution method in order to determine the distribution of T_C.[9]

In the case of the binary samples, the $c(T,B)$ curves measured on the filaments extracted from the non-densified wire and from the wire densified at 2 GPa exhibit the same transition widths. In particular, from the data at $B = 0\,\mathrm{T}$, we determined the T_c distributions shown in Fig. 3 and it is concluded that CHPD does not influence the width of the T_C distribution in binary MgB_2, which in the present samples is >8 K.

Figure 4 shows the calorimetric distributions of the critical temperature for the samples doped with 10 wt.% malic acid. After densification, the onset T_C is reduced by 1 K. As already pointed out in Sec. 2.1, CHPD causes the substitution of a higher content of C on the B sites. This explains the lower onset T_C and the slightly broader T_C distribution for the densified sample. It is interesting to point out that the conventional methods to increase the substitution level of C on the B sites involve a higher annealing temperature and/or a longer annealing time. In both cases, the higher C content is accompanied by an increase of the grain size, higher annealing temperature and/or a longer annealing time. In both cases, the higher C content is accompanied by an increase of the grain size and thus by a reduction of the grain boundary density. On the other hand, CHPD enhances the amount of substituted C, thanks to a reduced reaction path under the same reaction conditions.

2.3 *Influence of CHPD on the percolation threshold*

The novelty of the present study resides in the combination of the T_C distribution analysis with the results of the $\rho(T)$ measurements to obtain insights on the influence of CHPD on the superconducting current percolation.

In Table 1, we report the onset T_C and the temperature corresponding to $\rho = 0$, $T(\rho = 0)$, for the four examined samples, all measured with a probing current density $J = 0.6\,\mathrm{A/cm^2}$. By integrating the T_C distributions in Figs. 3 and 4 between the onset T_C

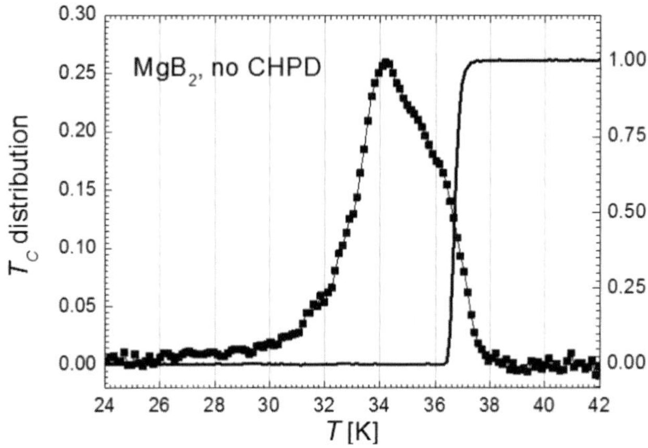

(a)

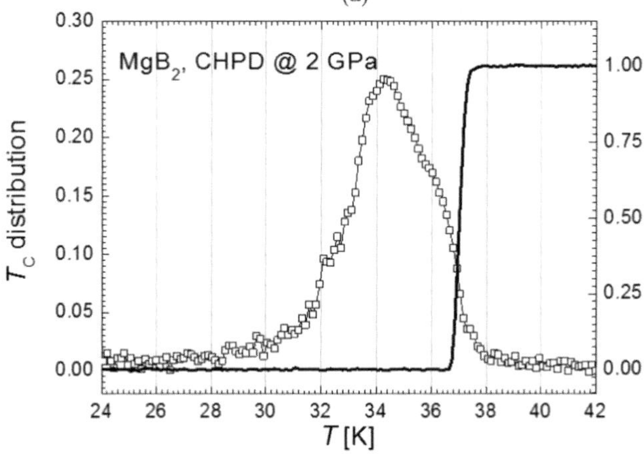

(b)

Fig. 3. Distribution of T_C obtained by the deconvolution of the calorimetric data and normalized resistivity for the non-densified binary MgB_2 sample (a) and for the binary MgB_2 sample after densification at $p = 2\,\text{GPa}$ (b). It is seen that the T_C distribution is not influenced by CHPD in the case of binary MgB_2.

and $T(\rho = 0)$ we determined the minimum superconducting volume fraction needed for a full percolation path with zero resistivity. For the non-densified binary sample, 14.9% of the filament volume needs to become superconducting in order to get percolation with zero resistivity. This fraction decreases to 8.4% after CHPD. By analogy, percolation with zero resistivity in the malic acid doped wires is

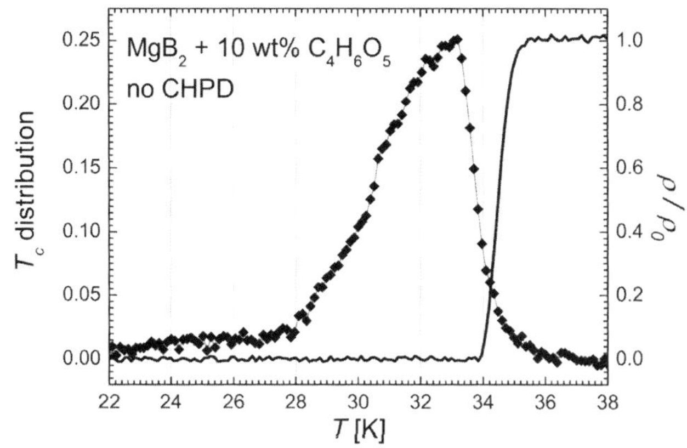

(a)

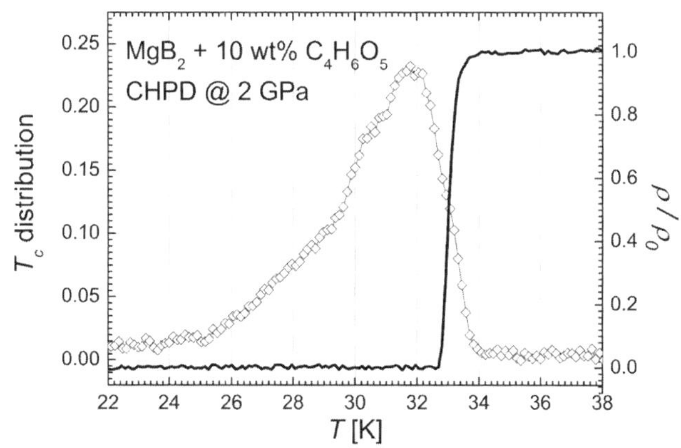

(b)

Fig. 4. Distribution of T_c and normalized resistivity for the malic doped samples, non-densified (solid symbols) and densified at $p = 2\,\text{GPa}$ (open symbols).

obtained for a superconducting filament volume of 25.3% in the non-densified sample. This threshold is lowered to 12.5% for the sample densified at $p = 2\,\text{GPa}$.

In conclusion, the positive effect of the CHPD process on the electrical transport properties of *in situ* MgB_2 wires is related to the improved grain connectivity and percolation. In both binary and doped wires the minimum superconducting volume needed for the percolation of the superconducting current is reduced by ~50% in samples treated by CHPD. As a side effect of the cold densification process, we observed also that the average C content in doped wires is enhanced, thanks to the reduced reaction path of the precursor powders in the densified filaments.

3. Study of the Magnetic Relaxation Rates in View of Persistent Mode Operation

The operation at 4.2 K of NbTi magnets is one of the reasons that make the Magnetic Resonance Imaging (MRI) units currently installed in the hospitals very complex and expensive. MgB_2 magnets would enable the development of low cost, easy-to-use MRI units able to operate above 4.2 K in a cryogen-free environment. Because most MRI magnets have to run in persistent mode, the main requirement for the application of MgB_2 is a very low decay of the superconducting current in the SC as well as across the joints.

The condition for persistent mode operation is a long-term drift of the magnetic field below 0.1 ppm/h. This translates into a total magnet resistance to be kept below $10^{-9}\,\Omega$. The largest portion of this resistance is typically allocated to the joints. An individual joint should not exceed a resistance value of $10^{-10} - 10^{-11}\,\Omega$, depending on the wire piece lengths used for the winding. Recently, the group of Iwasa at Massachusetts Institute of Technology (MIT) reported on a novel splicing technique for monofilamentary MgB_2 round wires.[10,11] The technique applies to unreacted wires: the filaments are exposed removing the matrix by chemical etching and the wires are joined using an Mg + 2B powder mixture. Finally, the superconducting joint fabrication requires a reaction step with a combination of temperature and duration that is compatible with the reaction schedule of the winding itself. By measuring the field decay rate of a test coil wound with 50 m monofilamentary wire and operating at 100 A, 15 K, the MIT team estimated a joint resistance of less than $1.3 \times 10^{-10}\,\Omega$.[10] The present challenge consists in transferring the technique to multifilamentary wires.

The other mechanism behind the decay of "persistent" currents has a microscopic origin related to the pinning properties of the superconducting material and to the motion of the flux lines due to thermal activation. The experimental n-value of the voltage–current (V–I) transition is commonly used as a quality index to this respect. A high n-value is the consequence of a high pinning energy, thus corresponding to a slow decay of the persistent current. The practical criterion is to require an n value above 30 for persistent designs. However, it is not straightforward to associate weak pinning to a low n value. Even in the presence of strong pinning, an inhomogeneous distribution of the current across the superconducting layer, i.e., a distribution of J_C would lead to the broadening of the V–I transitions and, thus, to reduced n values. We will show in the following that the relaxation rate extracted from the time decay of the irreversible magnetization represents a more reliable quality index of the pinning properties.

The irreversible magnetization M in a type II SC is proportional to the current density J when the sample is fully penetrated by the magnetic field. Anderson and Kim[12] predicted the logarithmic time dependence of the magnetization M for the thermally activated motion of vortex bundles in the presence of correlated defects:

$$M(t) = M_0 \left[1 - \frac{k_B T}{U(B,T)} \ln\left(\frac{t}{t_0}\right)\right], \tag{1}$$

where M_0 is the unrelaxed magnetization, $U(B,T)$ is the volume pinning energy and t_0 is a characteristic time scale. Therefore the normalized relaxation rate $S(B,T)$ is inversely proportional to the volume pinning energy $U(B,T)$:

$$S(B,T) = \frac{1}{M_0}\frac{dM}{d\ln(t)} = -\frac{k_B T}{U(B,T)}. \tag{2}$$

If the superconducting properties of the material are not uniform, the derivative of the magnetization M with respect to $\ln(t)$ represents the average value $\langle S\rangle$ of the relaxation rate S over the sample volume. The time dependence of the magnetization for a SC is given by the expression:

$$\begin{aligned}\mathbf{M}(t) &= \frac{1}{2V}\int d^3r\mathbf{r}\times\mathbf{J}(\mathbf{r},t)\\ &= \frac{1}{2V}\int d^3r\mathbf{r}\times\mathbf{J}(\mathbf{r},t=0)\left[1+S(\mathbf{r})\ln\left(\frac{t}{t_0}\right)\right].\end{aligned} \tag{3}$$

It follows that

$$\begin{aligned}\frac{1}{M_0}\frac{dM}{d\ln(t)} &= \frac{1}{\displaystyle\int d^3r\mathbf{r}\times\mathbf{J}(\mathbf{r},t=0)}\int d^3r\mathbf{r}\\ &\quad\times\mathbf{J}(\mathbf{r},t=0)S(\mathbf{r}) = \langle S\rangle\,.\end{aligned} \tag{4}$$

Indeed, in the presence of inhomogeneities the relaxation rate still provides information on the average pinning properties of the sample, as follows from Eq. (4). On the other hand, the inhomogeneities lead to the smearing of the V–I curves and thus to an underestimation of the pinning energy.

Here we present the results of the relaxation rate measurements performed on various MgB_2 samples, differing in composition and preparation. We measured the time relaxation of the irreversible magnetization for

- binary and C-doped polycrystalline bulks; C-doped bulks were prepared using either SiC or C nanopowders as a C source.[13]
- the MgB_2 core extracted from an *ex situ* tape; due to the partial texturing achieved in *ex situ* tapes, we performed the measurements on the MgB_2 core for two different orientations of the magnetic field, parallel and perpendicular to the crystallographic c-axis of MgB_2, respectively.[14]
- the MgB_2 core extracted from an *in situ* wire doped with malic acid, without and with CHPD; the effect of CHPD on the relaxation rates was tested.

In our magnetic relaxation measurements, the external field is ramped up to a fixed value B_0 with a sweep rate of $2\,\mathrm{T/min}$ and is then decreased to the setpoint value B_{SP} with a sweep rate of $0.1\,\mathrm{T/min}$ in order to reduce the effects due to the magnetic field overshoot. Magnetization is then measured as a function of time for 5000 s, keeping the external field constant. The values for B_0 and B_0–B_{SP} are chosen higher than the full penetration field in order to obtain an initial magnetic state in the sample with fully entrapped flux. Measurements have been performed at different dc fields from 0 to 8 T in the temperature range 5–20 K.

The standard logarithmic relaxation law $M \propto \ln(t)$ is recovered in our samples for $t > 10$ s, as shown in Fig. 5. The non-logarithmic behavior at short times is due to the magnetic field overshoot occurring when the external field ramp is stopped (at the nominal stop of the ramp, the field exceeds the target value B_{SP}, reaching it usually after few seconds). This overshoot produces a shielded flux zone in proximity of the surface of the sample, which affects the relaxation process in the first few seconds.[15]

Figure 5 shows the comparison of the magnetic relaxation measurements performed at $T = 20$ K and $B = 1$, 2 and 4 T on the cores of MgB_2 extracted from the wires with malic acid additive, reacted in the as-drawn state and after cold densification at $p = 2$ GPa. In the previous section, we have identified the positive influence of CHPD

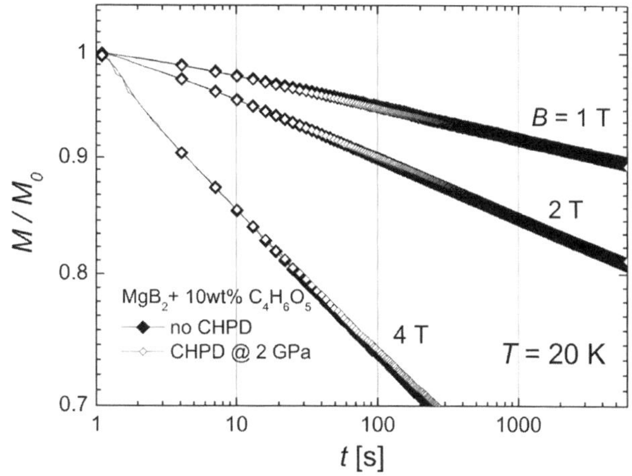

Fig. 5. Magnetic relaxation measurements at $T = 20$ K and $B = 1$, 2 and 4 T for the malic acid-doped MgB_2, non-densified (solid symbols) and densified at $p = 2$ GPa (open symbols). CHPD does not affect the relaxation rate.

on the grain connectivity and the current percolation mechanism. Here we do not observe any sizeable effect of the cold densification process on the relaxation rate of magnetization and, thus, on the pinning energy.

The relaxation rates at $T = 5$ and $20\,\mathrm{K}$ are summarized for all the examined MgB_2 samples in Fig. 6, where the values are also

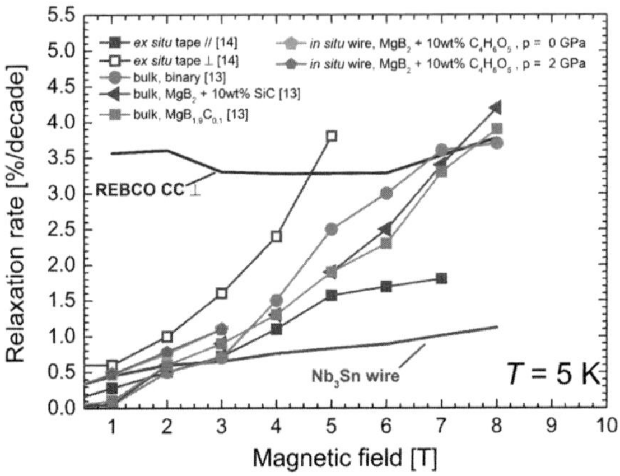

(a)

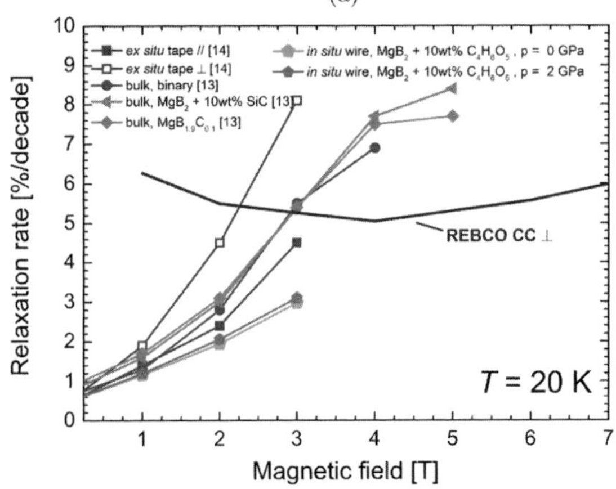

(b)

Fig. 6. Relaxation rates of various MgB_2 samples, (a) at $T = 5\,\mathrm{K}$ and (b) at $T = 20\,\mathrm{K}$. Binary and C-doped polycrystalline bulk samples, filaments (extracted from an *in situ* wire-doped with malic acid, with and without CHPD) and from an *ex situ* tape. Relaxation rates of Nb_3Sn and REBCO in perpendicular field are also reported for comparison.

compared to the relaxation rates for a high performance Nb_3Sn wire (only at $T = 5\,K$) and a REBCO coated conductor.

At $T = 5\,K$, the cap and floor of the relaxation rates among the examined MgB_2 samples were found for the MgB_2 core of the *ex situ* tape in perpendicular and parallel field, respectively. The relaxation rates for the untextured polycrystalline bulks and for the filaments extracted from the *in situ* wires are below 1.5% per decade up to $B = 4\,T$. These values are certainly higher than those of Nb_3Sn, but definitely lower than the relaxation rates measured for the REBCO coated conductor in perpendicular field. At 20 K, the relaxation rates increase steadily with the magnetic field but the values are still significantly lower than those of REBCO up to $B \approx 3\,T$ (Fig. 6(b)).

A low relaxation rate in the SC is a stringent condition for magnets operating in persistent mode. Even if it is not straightforward to establish a direct correlation between the results of our measurements, performed on short samples, and the behavior of a coil, the comparison between the relaxation rates of MgB_2, Nb_3Sn and REBCO together with the recent developments of the superconducting joint technology suggest that MgB_2 could represent a viable option for persistent magnets operating in cryogen-free environment and at magnetic fields up to 3 T.

4. Thermal Conductivity and Stability of Industrial MgB_2 Wires

Beyond the optimization of the electrical performances, the thermal stability of technical superconductors remains one of the most relevant issues in the development of superconducting devices. Due to different kinds of disturbances, the temperature of a superconducting wire in a magnet may locally rise up to values at which a sharing of the current between the SC and the metal matrix occurs. In this case, heat is generated by Joule effect and a quench of the whole system occurs if the perturbation is not reabsorbed. The thermal conductivity, κ, of the conductor is a key parameter in the evaluation of the stability margins of superconducting devices and more generally in the description of the physics behind a quench. κ is anisotropic in technical superconductors, and the heat transfer along the conductor is the main channel for dissipating heat. In the following, reference to the thermal conductivity of technical SC will mean its longitudinal component.

Technical superconductors are usually stabilized by adding copper to the composite, as the thermal conductivity of bare superconductors is much smaller than that of normal metals.[16,17] Thermal conductivity of copper is determined by the purity of the material, defined by the residual resistivity ratio (*RRR*).[18] The use of copper with a high *RRR* ratio improves the thermal conductivity of the conductor. However, this gain in thermal conduction is less pronounced when the conductor is subjected to magnetic fields. Indeed, the magnetic field lowers κ, and the reduction is more intense the higher the *RRR* of the metal.[18] The control of the *RRR* of the stabilizer as well as the optimization of the Cu to non-Cu ratio are fundamental elements for ensuring a reliable quench protection without reducing excessively the engineering critical current density. Certainly, this optimization relies on an adequate understanding of thermal conduction properties of composite conductors.

In this section, we present a study of thermal properties of MgB_2 conductors produced by Columbus Superconductors differing in architecture, stabilization and composite materials. We demonstrate that the thermal transport through the stabilizer rules the overall thermal conduction of the composite conductor at cryogenic temperatures. We study the effect of magnetic field on the thermal conduction and evaluate the accuracy of the analytical procedures normally used to estimate the in-field κ from an electrical characterization of the stabilizer. From the experimental investigation of thermal conduction properties, we also determine the effect of the operating conditions on the minimum quench energy and normal zone propagation velocity — fundamental parameters in the definition of the stability margins of SC-based devices. The analysis of correlations between thermal properties and conductor layout allows us to discuss solutions to make the conductor design more suitable for specific applications.

4.1 *Experimental technique and samples*

We have investigated the temperature and field dependence of the thermal conductivity by means of a setup specifically designed for measurements in technical superconductors. Heat Q supplied to the sample creates a temperature gradient ΔT along it.

The thermal conductivity is defined as:

$$\kappa = \frac{Q}{\Delta T} \cdot \frac{d}{S_{\text{tot}}}, \tag{5}$$

where d is the distance between the temperature taps and S_{tot} is the surface of the conductor cross-section. Figure 7 shows a schematic drawing of the sample holder. Heat power is supplied to one end of the sample, ensuring a good thermal contact with the heater over the entire conductor cross-section. Thus, all the materials present in the tape contribute to the thermal transport. Once the steady-state heat flow is established, the temperature gradient, ΔT, is measured by two Cernox bare chips directly glued on the sample surface using GE varnish. The two bare chips are calibrated during each measurement, allowing precise temperature measurements. Two calibrated Cernox thermometers are also present in the setup, one on the heat sink and one on the heater holder. The thermometers have been calibrated in magnetic fields up to 21 T. The measurement uncertainty $\approx$5% is mostly due to inaccuracy in assessing the distance between the thermometers. Further details on the experimental setup can be found in Refs. 19 and 20.

In this chapter, we present results obtained in three MgB_2 tapes with different layout produced by Columbus Superconductors. The

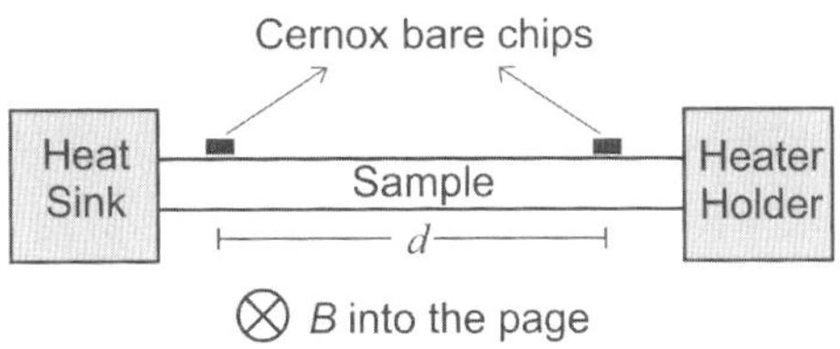

Fig. 7. Sketch of the sample holder for longitudinal thermal conductivity measurements on technical superconductors. In case of in-field measurements, B is applied perpendicularly to the thermal current direction.

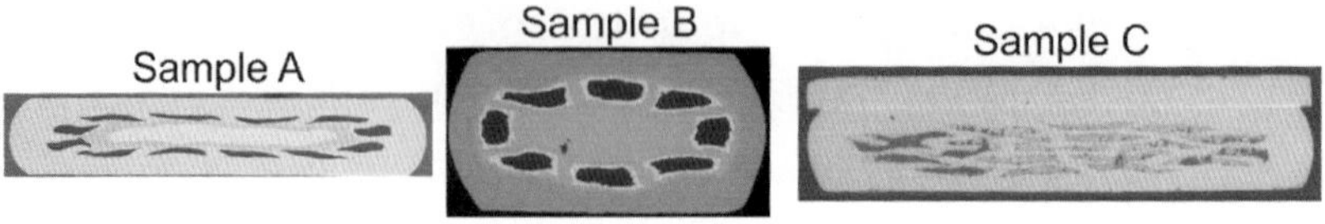

Fig. 8. Pictures of the cross-sections of the investigated MgB_2 conductors.

Table 2: Geometrical characteristic of MgB_2 conductors.

	Sample A	Sample B	Sample C
Width (mm)	3.65	2.25	3.00
Thickness (mm)	0.65	1.10	0.65
Number of filaments	12	19	8
Cu stabilization	Internal	Internal	External

conductor cross-sections are shown in Fig. 8, while the maximum width and thickness of the tapes are listed in Table 2 along with other structural parameters, e.g., the number of filaments and the type of stabilization. Sample A is composed of 12-filaments of MgB_2 and has a Ni matrix. It is stabilized by a Cu core coated with pure Fe in order to prevent chemical poisoning during the heat treatment. Sample B is a 8-filament conductor with an inner Cu core. The stabilizer is shielded with CuNi/Nb. The matrix consists of stainless steel (SS). Ni is used as material for the matrix in Sample C. This conductor presents 19 partly bridged MgB_2 filaments. A Cu strip of thickness $\approx$0.2 mm is soldered on one side of the tape for thermal stabilization. No barrier for the chemical protection of the stabilizer is needed with this architecture. The volume fraction occupied by each composite material is listed in Table 3.

4.2 *Thermal conductivity: temperature and field dependence*

We have investigated κ in the range of temperatures from 5–40 K at $B = 0$ and $B = 7\,\mathrm{T}$, this field value being an upper limit for MgB_2 conductors in typical applications. Furthermore, we have measured an extra data point at $T \approx 78\,\mathrm{K}$ and $B = 0$ inserting the

Table 3: Composition of the investigated MgB_2 conductors.

	Sample A	Sample B	Sample C
MgB_2	12%	13%	16%
Cu	15%	14%	27%
Ni	63%	—	57%
Fe	10%	—	—
CuNi/Nb	—	26%	—
Stainless Steel	—	47%	—

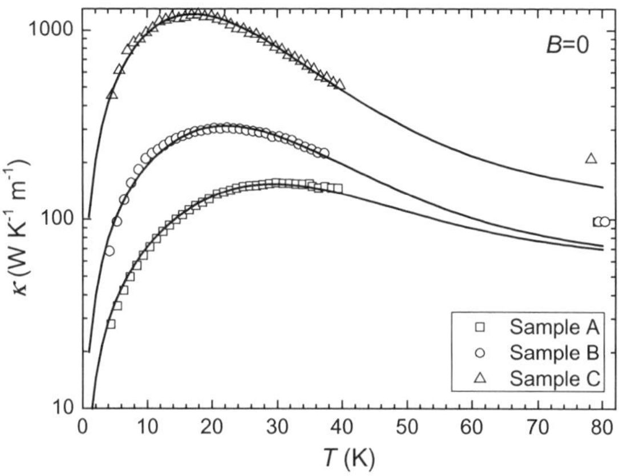

Fig. 9. Temperature dependence at $B = 0$ of thermal conductivity of MgB_2 conductors with different layouts. Continuous lines are the best-fit curves obtained as described in the text.

Source: Owb data.

probe in liquid nitrogen. In Fig. 9, we have plotted the experimental $\kappa(T)$ curves obtained at $B = 0$. The conductor layout has a major influence on the thermal response. The best thermal conduction properties have been observed in Sample C, which presents an external stabilization. The $\kappa(T)$ curves in Fig. 9 exhibit a maximum at $T_M \approx 30\,\text{K}$, $\approx 22\,\text{K}$ and $\approx 17\,\text{K}$ for Sample A, B and C, respectively. A correlation exists between peak position and κ magnitude, the best thermal conduction properties being realized in the sample with the lowest T_M.

Longitudinal thermal transport in composite conductors can be analyzed with a formalism analogous to the case of electrical resistances connected in parallel. The overall thermal conductivity is the weighted sum of the thermal conductivity of each component, κ_i, with weights $s_i \equiv S_i/S_{tot}$: $\kappa = \sum \kappa_i s_i$. S_i and S_{tot} are the surface of the cross-section occupied by the ith component and the total cross-section area of the conductor, respectively. In metals, thermal conduction occurs via the electronic and lattice channels, the former being the most relevant one.[21] The electronic contribution is in turn determined by electron-defect and electron-phonon scattering processes. Thus, the temperature dependence of the thermal conductivity of metals is modeled as:

$$\kappa = (W_0 + W_i + W_{i0})^{-1}, \tag{6}$$

where $W_0 = \frac{\beta}{T}$, $W_i = \frac{P_1 T^{P_2}}{1+P_1 P_3 T^{(P_2+P_4)} e^{-(P_5/T)^{P_6}}}$, $W_{i0} = P_7 \frac{W_i W_0}{W_i + W_0}$, W_0 and W_i represent the electron-defect and electron–phonon contribution, respectively. W_{i0} is an interaction term between W_0 and W_i.[18] β is a function of the residual resistivity ratio, $RRR \equiv \rho(273\,\mathrm{K})/\rho_{res}$, where ρ_{res} is the residual electrical resistivity measured at low temperature. The parameters β and P_i differ from one metal to another. They have been determined by least squares fit of experimental data taken at $B = 0$. For Cu, the values are: $\beta \approx 0.634/RRR$, $P_1 = 1.754 \times 10^{-8}$, $P_2 = 2.763$, $P_3 = 1102$, $P_4 = -0.165$, $P_5 = 70$, $P_6 = 1.756$, $P_7 = 0.235 \cdot RRR^{0.1661}$ in SI units.[18] At temperatures lower than about 50 K the electron-defect term, W_0, is predominant and κ is mainly determined by the RRR of the metal. In particular, samples with high RRR are particularly efficient in conducting heat.

We expect that the main contribution to κ in the MgB_2 conductors investigated in this work comes from Cu. This is a consequence of the thermal conduction properties of the composite materials[17,21–23] and of the fact that only the stabilizer has been optimized for heat conduction, whilst the other metals of the tape have RRR lower than $\approx$10.[24] Thus, we have analyzed the experimental results obtained at $B = 0$ in the approximation $\kappa \approx \kappa_{Cu} \cdot s_{Cu}$ using Eq. (6), the values of s_{Cu} reported in Table 3, and keeping the RRR of Cu as the only free fitting parameter. The best fit curves have been reported in Fig. 9 as

Table 4: Electrical and thermal-stability parameters of the investigated MgB_2 conductors. *MQE* and *NZPV* have been evaluated at $T_{Op} = 20\,K$, $I_{Op} = 100\,A$, $B = 1\,T$. Critical current, I_C, measured at $T = 20\,K$, $B = 1\,T$.

	Sample A	Sample B	Sample C
RRR_{Cu} from $\kappa(T)$	30	90	238
RRR_{Cu} from $\rho(T)$	37	86	255
MQE [mJ]	6	21	34
NZPV [cm/s]	10	6	12
I_c [A]	345	240	300

continuous lines. A very good agreement between experimental and calculated curves has been obtained in the whole temperature range. Data measured at liquid nitrogen temperature are up to ≈30% higher than calculated values. This could be due to non-Cu contributions to the overall thermal conductivity observable at high temperatures, when κ_{Cu} becomes smaller.

We have listed the *RRR* values resulting from the best-fit procedure in Table 4. The smallest value, $RRR \approx 30$, has been obtained in Sample A; in samples B and C we have found $RRR \approx 90$ and $RRR \approx 238$, respectively. A measurement of the temperature dependence of the electrical resistivity performed on the superconducting tape does not allow extracting the *RRR* value of the stabilizer, the superconducting transition preventing the measurement of the value ρ_{res} of Cu. In order to circumvent this problem and to check the consistency of the best fit parameter, we have measured the electrical resistivity with a standard four probe technique on Cu specimens extracted from the MgB_2 tapes. For samples A and B, we have extracted the stabilizer from the interior of the conductor, removing the other layers by a combination of chemical etching and mechanical abrasion. For Sample C, we have unsoldered the Cu strip using a heating plate at 250°C. The process lasts about 1 minute, thus limiting possible annealing effects. The procedure has been repeated on several Cu samples taken from the same batch for each MgB_2 conductor, finding *RRR* variations within the 15%. The average *RRR* values obtained by electrical measurements are also shown in Table 4.

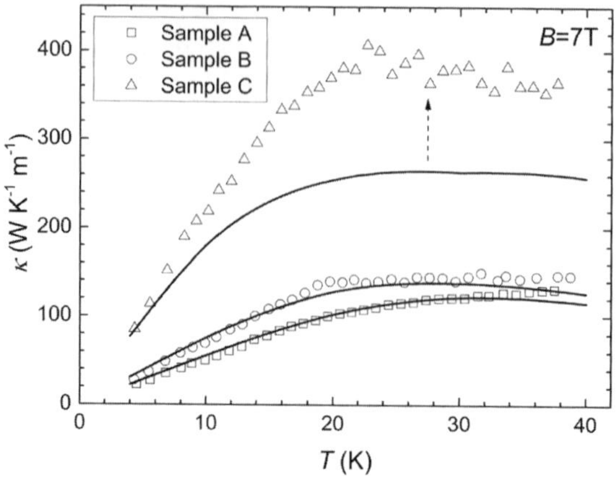

Fig. 10. Temperature dependence of the thermal conductivity of MgB_2 conductors with different layouts at $B = 7\,T$. Continuous lines are calculated from curves obtained as described in the text.

Source: Owb data.

A comparison of measured and best fit *RRR* indicates differences <20%, and this is within the accuracy of Eq. (6)[21] This result confirms the validity of the approximation $\kappa \approx \kappa_{Cu} s_{Cu}$ for the investigated MgB_2 conductors.

We report in Fig. 10, the experimental $\kappa(T)$ curves obtained at $B = 7$ T. The field has been applied perpendicularly to the thermal current, as indicated in Fig. 7, since this configuration is the most relevant one for coils. A comparison with the results obtained at $B = 0$ shown in Fig. 9 reveals the field induced effects on the thermal transport: κ is reduced after the application of field, and the variation is higher with larger *RRR* values. It follows that the differences in κ measured at $B = 0$ in the three samples are reduced after the application of magnetic field. The influence of field on the thermal conductivity can be qualitatively understood considering that κ in metals is proportional to the electron mean free path, l, which is in turn reduced by the Lorenz force. In metals, not only κ but also the electrical resistivity, ρ, is related to l. The Wiedemann–Franz law establishes that their product is proportional to the temperature, the constant being the Lorenz number, $L : \kappa\rho = LT$. The Lorenz number is expected to be field independent if the electron scattering events are elastic.[25,26] In this case, in-field κ can be deduced from the magnetoresistance of the metal:[19]

$$\kappa(T, B) = \frac{\rho(T, 0)}{\rho(T, B)} \kappa(T, 0), \tag{7}$$

The magnetoresistance of Cu has been widely reported in the literature for fields applied perpendicularly to the electrical current. Simon *et al.*[18] have collected in-field data from different references and summarized them in a unique Kohler plot where the fractional increase in resistance $\Delta\rho(T, B) \equiv [\rho(T, B) - \rho(T, 0)]/\rho(T, 0)$ is plotted as a function of the product $B \times [\rho(273\,\text{K}, B = 0)/\rho(T, B = 0)]$. The authors have derived an average Kohler curve to evaluate the in-field electrical resistivity starting from ρ values measured at $B = 0$.[18]

Due to practical difficulties associated with in-field thermal conductivity measurements, κ at $B \neq 0$ is usually evaluated by Eq. 7, starting from the *RRR* value determined at $B = 0$. In fact, $\kappa(T, 0)$ can be calculated from the *RRR* value of Cu by means of Eq. 6. Furthermore, *RRR* allows calculating the theoretical $\rho(T,\ B = 0)$ curve of Cu and, consequently, the ratio $\rho(273\,\text{K}, B = 0)/\rho(T, B = 0)$ necessary to evaluate the magnetoresistance from the Kohler plot.[18] In Fig. 10, are plotted the theoretical lines obtained by this procedure using the *RRR* values deduced by electrical resistivity measurements performed on the stabilizer (data listed in Table 4). The comparison between experimental and calculated results shows a very good agreement in the case of Samples A and B, whilst calculated values for sample C are up to 30% smaller than the measured data, the largest variation having been observed at $T \approx 25\,\text{K}$. This relatively large discrepancy is within the accuracy associated with the use of Eqs. (6) and (7). Indeed, one finds from the Kohler plot that $\rho(25\,\text{K}, 0\,\text{T})/\rho(25\,\text{K}, 7\,\text{T}) = 0.27 \pm 0.1$ for $RRR = 255$, which is the value measured for Sample C. This corresponds to an uncertainty in assessing the magnetoresistance of about $\pm 30\%$, that has to be added to the one associated with Eq. (6).

Therefore, the total uncertainty is about $\pm 50\%$. It is noteworthy that an evident discrepancy between calculated and experimental curves has been obtained in the sample with the highest *RRR* of Cu. This is related to the fact that field induced effects on κ are more pronounced in pure metals. We would also like to underline that we have achieved a very good agreement between calculated and experimental curves, especially in the case of Samples A and B, thanks to the precise measurement of *RRR* of the stabilizer. The *RRR* value of raw materials is often used for evaluating the thermal properties of the composite. However, this procedure excludes all modifications resulting from the fabrication process and can lead to much larger errors in the estimation of κ. Our study has allowed assessing the accuracy of the procedure we used for calculating in-field κ from an electrical characterization of stabilizer specimens extracted from the reacted tapes. In-field measurements remain necessary when an accurate evaluation of the thermal stability is needed, especially for stabilizers with high *RRR*.

4.3 *Conductor thermal stability*

The study of thermal conduction in technical superconductors can provide important insights into the thermal stability of superconducting devices. Different kinds of disturbances can trigger a quench of the whole system if the heat generated in the perturbed area is not dissipated by thermal conduction. Indeed, when the perturbed zone is smaller, the so-called minimum propagation zone, *MPZ*, it will eventually shrink, whilst if it is larger, it will grow indefinitely, originating a quench. In adiabatic conditions, the length of the minimum propagation zone can be estimated from: $l_{MPZ} \approx \sqrt{2\kappa(T_{\mathrm{CS}} - T_{\mathrm{Op}})/J_{\mathrm{Cu}}^2 \rho_{\mathrm{Cu}}}$, where T_{Op} is the operating temperature, J_{Cu} and ρ_{Cu} the current density and the electrical resistivity of the stabilizer.[16] The minimum quench energy, *MQE*, i.e., the smallest energy able to trigger a quench, is a key parameter in the evaluation of the stability of superconducting devices. In case of superconducting wires or tapes, $MQE \approx l_{MPZ} S_{\mathrm{tot}} \int_{T_{\mathrm{Op}}}^{T_{\mathrm{CS}}} c(T)dT$, where S_{tot} and c are the cross-section area and the volumetric specific heat of the conductor, respectively.[16,27] Previously, we have demonstrated that $\kappa \approx \kappa_{\mathrm{Cu}} s_{\mathrm{Cu}}$; using the Wiedemann–Franz law we deduce:

$$MQE \approx \frac{\kappa S_{\text{tot}}^2}{I_{op}} \sqrt{\frac{2 s_{\text{Cu}}(T_{\text{CS}} - T_{\text{Op}})}{L T_{\text{CS}}}} \int_{T_{\text{Op}}}^{T_{\text{CS}}} c(T) dT. \tag{8}$$

The $c(T)$ values can be calculated from data reported in the literature for raw materials[22,23,28] considering that $c(T) = \sum_i v_i c_i$, where v_i is the volume fraction occupied by the ith material shown in Table 3. *MQE* evaluated from Eq. (8) at $T_{\text{Op}} = 20\,\text{K}$, $B = 1\,\text{T}$, and $I_{\text{Op}} = 100\,\text{A}$ is reported in Table 4. These parameters define typical operating conditions envisaged in cryogen-free MRI magnets. In the calculation, we have supposed $\kappa(1\,\text{T}) \approx \kappa(0\,\text{T})$, as this approximation has been shown to be valid for Cu over a wide range of temperatures and *RRRs*.[19] T_{CS} has been extracted from critical current data supplied by Columbus Superconductors.[24] The calculated *MQEs* span from $\approx$6 mJ to $\approx$34 mJ, the values associated with samples A and C, respectively.

The investigated samples differ not only by their thermal properties, but also by the electric ones. At $T = 20\,\text{K}$, $B = 1\,\text{T}$ the critical current is $I_{\text{C}} \approx 345$ A, ≈ 240 A and ≈ 300 A in samples A, B and C, respectively (see Table 4). In order to evaluate the thermal stability of the conductors for the same current margin, we have reported in Fig. 11(a) the *MQE* calculated for different values of the reduced current $I_{\text{Op}}/I_{\text{C}}$. The curves associated with Samples B and C are found to be very close each other. On the contrary, *MQE* assumes markedly smaller values in Sample A. The origin of this difference has to be sought in the high κ values of Sample C and in the use of stainless steel as matrix material in Sample B. Indeed, the specific heat of stainless steel is larger than that of nickel, leading to an increase of the *MQE*.[22,23] The *MQE* values obtained in this work are of the same order of magnitude as the values determined experimentally in various MgB_2 technical conductors.[28–30] Higher *MQEs* of the order of 1 J (at $T \approx 0K$, $B \approx 1T$, $I_{\text{Op}} \approx 100\,\text{A}$) have been measured in MgB_2 conductors values of low T_{C} superconductors.[27,28] Furthermore, large Cu/non-Cu ratios lead to a reduction of the engineering current density with much larger Cu/non-Cu ratios ($s_{\text{Cu}} \approx 45\%$) with respect to the tapes

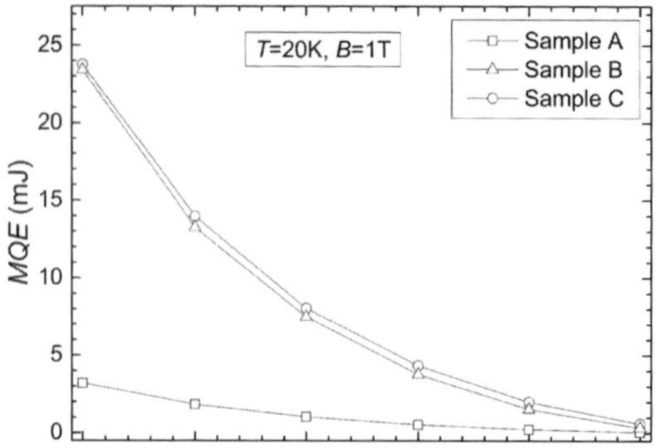

(a)

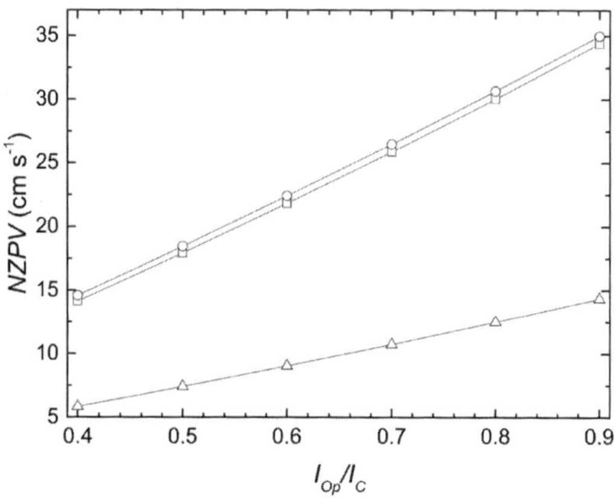

(b)

Fig. 11. Minimum quench energy (a) and normal zone propagation velocities, (b) calculated as described in the text for different values of $I_{\mathrm{Op}}/I_{\mathrm{C}}$ at $T = 20\,\mathrm{K}$ and $B = 1\,\mathrm{T}$. $I_{\mathrm{C}} \approx 345\,\mathrm{A}$ for Sample A, $I_{\mathrm{C}} \approx 240\,\mathrm{A}$ for Sample B, and $I_{\mathrm{C}} \approx 300\,\mathrm{A}$ for Sample C.

investigated in this work.[31] However, it is worth noting that *MQE* of the order of 10 mJ is at least two order of magnitudes higher than typical.

When designing a superconducting device, one wants to keep the *MQE* as large as possible. Nevertheless, this is not the only parameter that has to be kept under control. If a quench occurs, the temperature rise has to be limited, otherwise serious damages could occur in the system. This requires a rapid propagation of the perturbation, i.e., large values for the normal zone propagation velocity, *NZPV*. High *NZPV* values are also important to quickly detect the quench and activate the protection system.[32] In adiabatic conditions, we can use the equation derived by Wilson for estimating the longitudinal component of *NZPV*:

$$NZPV \approx \frac{I_{Op}}{c_{Av}S_{\text{tot}}}\sqrt{\frac{\rho\kappa}{T_S - T_{Op}}} \approx \frac{I_{Op}}{c_{Av}S_{\text{tot}}}\sqrt{\frac{LT_S}{T_S - T_{Op}}}, \tag{9}$$

where $c_{Av} = \sqrt{c_s c_n}$ is the geometric average of the volumetric specific heat at T_{Op} (c_s) and at T_C (c_n), and $T_S = (T_{CS} + T_C)/2$.[16] We would like to emphasize that, since *NZPV* depends on the product $\rho\kappa$, one obtains from the Wiedemann–Franz law a field-independent expression. The *NZPV* value calculated by Eq. 9 at $T_{Op} \approx 20\,\text{K}$ and $I_{Op} = 100\,\text{A}$ has been listed in Table 4. In Fig. 11(b) *NZPV* has been plotted for different values of the reduced current. Samples A and C exhibit comparable propagation velocities being approximatively twice the value calculated for Sample B. This difference is due to the use of stainless steel instead of nickel in Sample B, since this leads to larger c_{Av}. *NZPV* values obtained in this work are of the same order of magnitude than those measured experimentally or determined numerically for other MgB_2 technical conductors.[28–31,33]

From the study of the thermal properties, one can deduce valuable information about the correlation between conductor layout and thermal stability. The use of Cu with high *RRR* is very effective in increasing *MQE*, especially in case of applications at relatively low magnetic fields. This originates from the fact that $MQE \propto \kappa$. Even in case of applications in intense magnetic field, which should not be the case of MgB_2, the use of high purity Cu contributes to enhance *MQE*. The purity of the stabilizer is not important in determining the propagation velocity of the perturbation because *NZPV* is proportional to the product $\kappa\rho$, i.e., to the Lorenz number, rather than to κ. This is why we have found comparable *NZPV*s in Samples A and C, which have similar composition but very different *RRR* values of the stabilizer. With regard to the conductor matrix, the use of SS in place of Ni has a double effect on thermal stability: on one hand, it increases *MQE* and thus the stability of the conductor; on the other, it lowers *NZPV*. The two effects are due to the fact that the specific heat of stainless steel is larger than that of nickel. From our study, we also deduce that the procedure used for stabilizing the conductor is very important since it determines the final thermal properties of the conductor. The choice of stabilizing the conductor via a Cu strip soldered on the reacted tape was found to be very effective: the stabilizer being not subjected to deformation processes that would affect its *RRR* value.

References

1. V. Braccini *et al.*, High-field superconductivity in alloyed MgB_2 thin films, *Phys. Rev. B* **71**, p. 012504 (2005).
2. A. Matsumoto, Y. Kobayashi, K. Takahashi, H. Kumakura and H. Kitaguchi, MgB_2 thin films fabricated by a precursor and post-annealing method have a high J_c in high magnetic fields, *Appl. Phys. Expr.* **1**, p. 021702 (2008).
3. R. Flükiger, M. S. A. Hossain, C. Senatore, Strong enhancement of J_c and B_{irr} in binary *in situ* MgB2 wires after cold high pressure densification, *Supercond. Sci. Technol.* **22**, p. 085002 (2009).
4. M. S. A. Hossain, C. Senatore, M. A. Rindfleisch, M. J. Tomsic, J. H. Kim, S. X. Dou and R. Flükiger, The enhanced J_c and B_{irr} of *in situ* MgB_2 wires and tapes alloyed with $C_4H_6O_5$ (malic acid) after cold high pressure densification, *Supercond. Sci. Technol.* **22**, p. 095004 (2009).
5. C. Senatore, M. S. A. Hossain and R. Flükiger, Enhanced connectivity and percolation in binary and doped in situ MgB_2 wires after Cold High Pressure Densification, *IEEE Trans. Appl. Supercond.* **21**(3), pp. 2680–2685 (2011).
6. J. M. Rowell, The widely variable resistivity of MgB_2 samples, *Supercond. Sci. Technol.* **16**, pp. R17–R27 (2003).
7. Y. Eltsev, S. Lee, K. Nakao, N. Chikumoto, S. Tajima, N. Koshizuka and M. Murakami, Anisotropic superconducting properties of MgB_2 single crystals probed by in-plane electrical transport measurements, *Phys. Rev. B.* **6**, p. 140501 (2002).
8. D. Sanchez, A. Junod, J.-Y. Genoud, T. Graf and J. Muller, Low-temperature specific heat of the 123, 124 and 247 phases of Y-Ba-Cu-O in magnetic fields up to 14 Tesla, *Physica C* **20**, pp. 1–11 (1992).

9. C. Senatore, D. Uglietti, V. Abächerli, A. Junod and R. Flükiger, Specific heat, a method to determine the T_c distribution in industrial Nb_3Sn wires prepared by various techniques, *IEEE Trans. Appl. Supercond.* **17**(2), pp. 2611–2614 (2007).
10. J. Ling, J. Voccio, Y. Kim, S. Hahn, J. Bascuñán, D. K. Park and Y. Iwasa, monofilament MgB_2 wire for a whole-body MRI magnet: superconducting joints and test coils, *IEEE Trans. Appl. Supercond.* **23**(3), p. 6200304 (2013).
11. W. Yao, J. Bascuñán, S. Hahn and Y. Iwasa, A superconducting joint technique for MgB_2 Round Wires, *IEEE Trans. Appl. Supercond.* **19**(3), pp. 2261–2264 (2009).
12. P. W. Anderson and Y. B. Kim, Hard superconductivity: theory of the motion of abrikosov flux lines, *Rev. Mod. Phys.* **36**, p. 39 (1964).
13. C. Senatore, P. Lezza, R. Lortz, O. Shcherbakova, W. K. Yeoh, S. X. Dou and R. Flükiger, Specific heat and magnetic relaxation analysis of MgB_2 bulk samples with and without additives, *IEEE Trans. Appl. Supercond.* **17**(2), pp. 2941–2944 (2007).
14. C. Senatore, P. Lezza and R. Flükiger, Critical current anisotropy, pinning properties and relaxation rate of *ex situ* MgB_2/Fe tapes, *Adv. Cryo. Eng.* **53**, pp. 654–661 (2006).
15. D. Zola, M. Polichetti, C. Senatore and S. Pace, Magnetic relaxation of type-II superconductors in a mixed state of entrapped and shielded flux, *Phys. Rev. B* **70**, p. 224504 (2004).
16. M. N. Wilson, *Superconducting Magnets.* Oxford, Clarendon (1983).
17. M. Putti *et al.*, Thermal conductivity of MgB_2 in the superconducting state, *Phys. Rev. B* **67**, p. 064505 (2003).
18. N. J. Simon, E. S. Drexler and R. P. Reed, *Properties of Copper and Copper Alloys at Cryogenic Temperatures.* Boulder, CO: U.S. Dept. of Commerce, Technology Administration, National Institute of Standards and Technology (1992).
19. M. Bonura and C. Senatore, High-field thermal transport properties of REBCO coated conductors, *Supercond. Sci. Technol.* **28**, p. 025001 (2015).
20. M. Bonura and C. Senatore, Transverse thermal conductivity of REBCO coated conductors, *IEEE Trans. Appl. Supercond.* **25**(3), p. 6601304 (2015).
21. J. G. Hust and A. B. Lankford, *Thermal conductivity of Aluminium, Copper, Iron, and Tungsten for Temperatures from 1 K to the Melting Point.* National Bureau of Standards, Boulder, CO, NBSIR 84–3007 (1984).
22. P. Bauer, H. Rajainmaki and E. Salpietro, *EFDA Material Data Compilation for Superconductor Simulation*, CSU, Garching, Germany (2007).

23. W. R. G. Kemp, P. G. Klemens and G. K. White, Thermal and electrical conductivities of iron, nickel, titanium, and zirconium at low temperatures, *Australian Journal of Physics* **9**(2) pp. 180–188 (1956).
24. M. Tropeano, *Columbus Superconductors Private Communication.* (2015).
25. R. W. Arenz, C. F. Clark and W. N. Lawless, Thermal conductivity and electrical resistivity of copper in intense magnetic fields at low temperatures, *Phys. Rev. B* **26**, p. 2727 (1982).
26. G. K. White and R. J. Tainsh, Lorenz number for high-purity copper, *Phys. Rev.* **119**, p. 1869 (1960).
27. Y. Iwasa, *Case Studies in Superconducting Magnets* — Second Edition. Springer, ISBN: 978-0-387-09799-2 (2009).
28. H. van Weeren, *Magnesium Diboride Superconductors for Magnet Applications.* Ph.D. thesis, ISBN 978-90-365-2519-0 (2007).
29. E. Martínez *et al.*, Quench development and propagation in metal/MgB_2 conductors, *Supercond. Sci. Technol.* **19**, pp. 143–150 (2006).
30. J. Pelegrín *et al.*, Experimental and numerical analysis of quench propagation on MgB_2 tapes and pancake coils, *Supercond. Sci. Technol.* **26**, p. 045002 (2013).
31. Liyang Ye *et al.*, Magnetic field dependent stability and quench behavior and degradation limits in conduction-cooled MgB_2 wires and coils, *Supercond. Sci. Technol.* **28**, p. 035015 (2015).
32. D. K. Park, S. Hahn, J. Bascunán and Y. Iwasa, Active protection of an MgB_2 test coil, *IEEE Trans. Appl. Supercond.* **21**(3), p. 2402 (2011).
33. U. Gambardella *et al.*, Stability Measurements in Multifilamentary MgB_2 Tapes, *IEEE Trans. Appl. Supercond.* **17**(2), p. 2937 (2007).

Chapter 5b

AC Losses in MgB_2 Wires

Ján Kováč

Institute of Electrical Engineering of Slovak Academy of Sciences
Dúbravska cesta 9, Bratislava 84104
Jan.Kovac@savba.sk

Experimental AC loss studies in MgB_2 superconductors are reported. Single core as well as filamentary wires of different compositions, sheath materials and filament structures were exposed to external AC magnetic field perpendicular to the wire axis. The magnetization measurements of energy dissipation were performed at various field amplitudes, frequencies and temperatures. It was shown that small dimensions of superconducting filaments are needed to minimize the hysteresis loss. The additional effect caused by metallic parts was studied in wires with different sheath materials. The reduction of coupling current losses in multifilamentary wires has been provided by twisting and/or by resistive barrier between filaments. The degradation of transport currents due to twisting was considered. MgB_2 cables were also measured to show the impact of cabling on AC losses.

1. Introduction

The increasing interest for the use of MgB_2 wires is due to their production at low cost in kilometer lengths, which for selected applications constitutes an advantage with respect to High Temperature Superconductors. The critical temperature T_c of MgB_2, close to 40 K, is high enough for keeping this SC at operating conditions around 20 K using conventional cryocoolers, thus rendering this compound attractive with respect to Low Temperature Superconductors. In case of possible AC applications, e.g., power transmission cables, transformers, fault current limiters, etc., involving dissipative processes generating heat, AC losses play an important role. Therefore, it is very important to minimize AC losses, due to their direct impact on the cooling performance and the associated operating costs. A measurement system allowing experimental studies of AC losses by magnetization at temperatures around 20 K and applied external fields of various amplitudes and frequencies has been developed. The apparatus consists of two parts: the cooling system and the apparatus for the measurement of magnetization losses (see Fig. 1). A two-stage Sumitomo cryocooler and a heater with computer controlled power supply were used for keeping the sample at constant temperatures in the range between 15.5–40 K, with an accuracy better then 10^{-2} K.[1] A special fiberglass vacuum vessel was designed.

The sample holder consists of a high thermal conductivity ceramic and is connected with the second stage of cryocooler by a

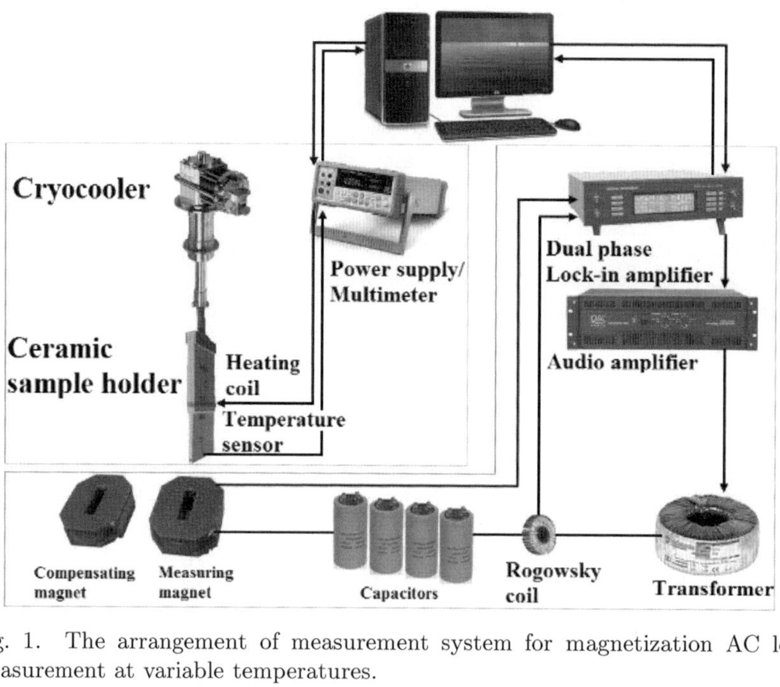

Fig. 1. The arrangement of measurement system for magnetization AC loss measurement at variable temperatures.

flexible Cu strand. A Cu plate and super insulation were fixed to the first stage of the cryocooler in order to eliminate thermal radiation from the vessel walls.

The second part of our apparatus comprises the system for magnetization AC loss measurement, based on the original calibration free method).[2] Two magnets (for measurement and compensation) were designed to allow high current loads without unwanted heating and mechanical vibrations. A massive fiberglass magnet skeleton with system of flow channels for cooling by liquid nitrogen was used.

To increase the accuracy and sensitivity of AC measurements, a set of pick-up and compensation coils was successfully designed. The present apparatus is able to measure magnetization AC losses with high precision and sensitivity in external fields ranging from 10^{-4} T up to 10^{-1} T at frequencies between 72 Hz and 144 Hz. At lower field amplitudes (up to 18 mT) AC loss measurements can be performed in a wide frequency range, from 2.3 Hz up to 1152 Hz.

2. AC Losses in Superconducting Composites

Composite superconducting wires consist of superconducting filaments and are of normal metallic materials. Consequently, the energy dissipation can have an origin in three separate processes: hysteresis loss in the superconducting filaments, losses due to coupling currents (coupling losses) and losses in the normal metallic parts (eddy currents and magnetic hysteresis).[3,4]

The AC loss mechanism is often described in terms of magnetic fields entering the SC in the form of vortices. Losses be understood in terms of viscous motion of magnetic vortices due to pinning forces exerted by the external field.[5–7] Another concept for explaining these effects is that alternating magnetic fields generate electromagnetic forces in the SC, thus giving rise to an electric field E which drives the screening currents in the material (see Fig. 2(a)). Screening currents generate a magnetic moment in the superconducting sample, depending on the magnetic field. These changes in magnetic moment are irreversible and result in AC losses. The screening-current pattern depends on the actual magnetic field as well as on previous field

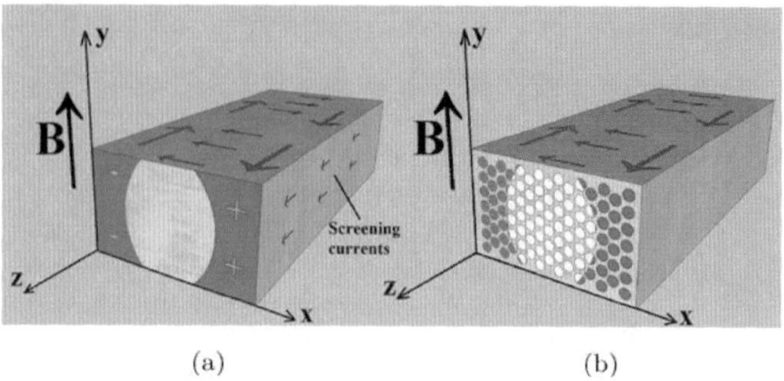

Fig. 2. (a) Illustration of screening currents in a superconducting slab shielding the interior (grey region), (b) illustration of coupling currents in multifilamentary superconducting composite wires.

changes, i.e., on the magnetic history of the sample. The magnetic moment of the sample $\boldsymbol{m}$ can be calculated from the screening current distribution. Then, the magnetization loss is found as the area of the magnetization curve:

$$Q_m = \oint_{\substack{Field\\cycle}} \boldsymbol{B} d\boldsymbol{M} = \oint_{\substack{Field\\cycle}} \boldsymbol{M} d\boldsymbol{B}\boldsymbol{M} \tag{1}$$

where Q_m is the magnetic loss per volume of the SC (in Joule per field cycle), $\boldsymbol{M}$ is magnetization and $\boldsymbol{B}$ is the external magnetic field. The loss mechanism in superconductors has a highly nonlinear character and is invariable with respect to the length of the period of field change. This means that losses in one cycle are not frequency dependent and are therefore often called hysteresis losses. The hysteresis loss can be reduced by decreasing the SC dimension, which can be achieved effectively by filamentary wires.

Another dissipative process plays a role in composite conductors with many thin superconducting filaments embedded in a normal metal matrix. Figure 2(b) shows a multifilamentary composite in a magnetic field. The magnetic field induces an electric field which drives currents around any loop formed by two filaments embedded into the normal metal matrix. Inside the filaments the currents (indicated by arrows) flow practically without resistance. They encounter a resistance only at the ends of the composite where they cross the normal conducting matrix. The currents are therefore much higher than eddy currents, which encounter a resistance along their entire path. The currents are called coupling currents because they couple the filaments together into a single large magnetic system.

The system has thus a magnetic moment higher than the sum of the magnetic moments of the individual filaments. The additional AC loss due to the coupling currents is called coupling-current loss. According to Faraday's law the coupling currents increase with increasing dB/dt and also with the "effective loop area" depending on the sample length (twist pitch) and on the internal geometry. The coupling currents are well described in Refs. 3 and 4. The coupling losses Q_C per unit volume for partly coupled filaments and sinusoidal applied magnetic field with angular frequency ω are described by the relationship

$$Q_C = \frac{B_a^2}{2\mu_0}\left[\frac{2\pi n_S \omega\tau}{1+\omega^2\tau^2}\right], \tag{2}$$

where n_S is a shape factor of the multifilamentary core, while the time constant τ characterizing the shielding ability of sample is

$$\tau = \frac{\mu_0^2}{2\rho_{\text{eff}}}\left(\frac{L_t}{2\pi}\right)^2, \tag{3}$$

where L_t is the twist pitch and ρ_{eff} is the effective transverse resistivity. The coupling currents can be decreased by twisting of the filaments, by reducing the dimensions of the composite and by increasing the resistivity of metallic matrix.[3,4,8–11]

The *Eddy current losses* represent the dissipation of the energy in the ohmic parts of the wire due to eddy currents in the normal conducting sheath. This type of loss depends significantly on the conductivity of the used metals as well as on the orientation and frequency of applied field.[12]

3. AC Losses in Single-core MgB_2 Wires and the Impact of Sheath Material to the Overall Loss

As mentioned above, AC losses of filamentary composites have usually two main components: hysteresis losses and coupling losses between the filaments. It is well known that hysteresis losses depend on the critical current density and can efficiently be reduced by reducing the filament size. Additional losses (e.g., eddy current and hysteresis losses in magnetic material) can be generated in the metallic material used for composite wire. Therefore, not only the properties of MgB_2 filaments are important, but low loss conductors must also have an adequate sheath material. To demonstrate the effect of applied metallic sheaths on AC losses in single core MgB_2 wires with no coupling currents, the set of four single core MgB_2 wires with different outer sheaths was selected for the magnetization AC loss measurements. Figure 3 shows the cross-sections of these wires named 1Gl (a), 1SS (b) 1Cu (c) and 1Mo (d). The basic properties of these wires are described by Table 1. Wire 1Cu was made by the IMD process,[13] while 1Gl and 1SS were fabricated using the *in situ* powder-in-tube (PIT) technique[14,15] and 1Mo by the so-called continuous tube filling and forming (CTFF) technique.[16]

Wire 1Gl consists of a Nb barrier, GlidCop AL-60 material (copper reinforced by Aluminum Oxide inclusions) being used for the outer sheath. Sample 1SS has a Nb barrier and a SS 316L outer

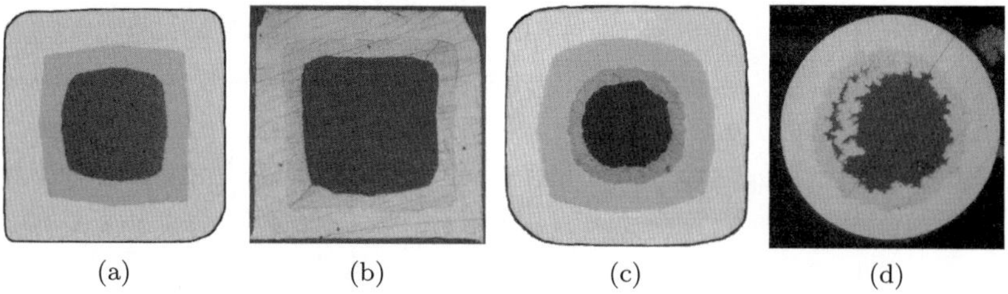

Fig. 3. Cross-sections of the analyzed single core MgB_2 wires: 1Gl (a), 1SS (b), 1Cu (c) and 1Mo (d).

Table 1: Basic properties of single-core MgB_2 wires.

Sample	1Gl	1SS	1Cu	1Mo
J_c (6T, 4.2K) [Acm^{-2}]	19896	6861	53097	19285
Core. size [μm]	542	970	582	417
Sheath material	GlidCop	SS	Cu	Monel

sheath. Pure copper with a high conductivity was used as an outer sheath of wire 1Cu, while 1Mo was sheathed by a standard Monel 400 alloy, containing mainly Ni (~63.0%), Cu (28–34%) and also Mn (≤2.0%).

Figure 4(a) shows the temperature dependence of magnetization AC losses per volume in one cycle for the samples 1Gl and 1SS at a field of 37 mT and frequencies of 72 Hz and 144 Hz. In both cases, the AC losses have very low values at 18–20 K due to high J_c and low-field penetration into the sample. The AC losses increase with temperature, which is attributed to the decreased current density and the subsequent deeper penetration of the magnetic flux into the MgB_2 core at a given AC magnetic field. The AC loss maximum represents the full penetration of the magnetic flux (B_{fp}) at given temperature. A further increase of temperature causes a lowering of the AC losses, due to the lowering of J_c with temperature. AC losses are not detected anymore in the normal state, when the critical temperature is reached. In sample 1SS the full penetration is reached at lower temperature, due to its lower J_c value (see Table 1). The present results show a hysteretic character of only AC losses, in agreement with the theoretical assumption of a single-core SC. This

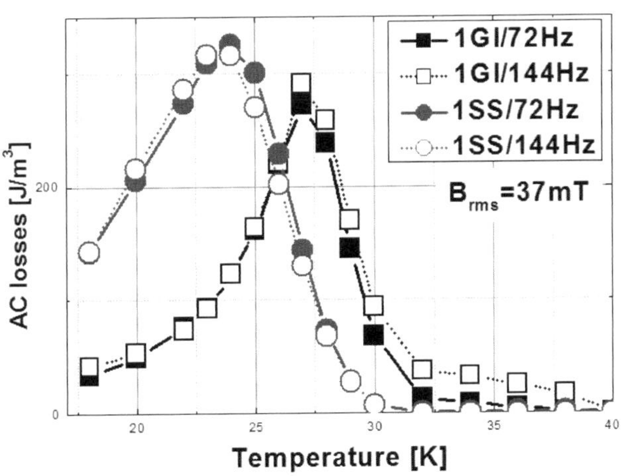

Fig. 4. (a) Temperature dependence of magnetization AC losses measured in sample 1Gl and 1SS at constant fields.

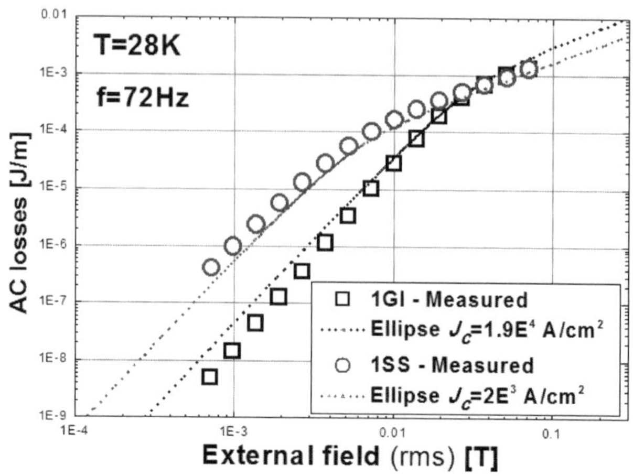

Fig. 4. (b) Measured field dependence of AC losses of wires 1Gl and 1SS at constant temperature compared with the calculation using the ellipse model.[17]

is confirmed by the fact that measurements at frequencies of 72 Hz and 144 Hz show the same result in both samples.

Figure 4(b) shows the magnetic field dependences of AC losses in samples 1Gl and 1SS at 28 K compared with the theoretical calculation using an ellipse model with J_c as fitting parameter.[17] The slope of presented $Q(B)$ at low fields corresponds approximately to B^3. The shape of $Q(B)$ measured for sample 1Gl at 28 K changes the tendency from B^3 to B^1 at about 25 mT, which represents the point of full penetration B_{fp}. This means that above this field, sample 1Gl is fully penetrated by magnetic flux and the AC loss is proportional to the applied field B. A similar change of $Q(B)$ is observed in sample 1SS, at lower B_{fp}~5 mT due to its lower J_c. The difference between measurement and calculation can be explained by the real cross-section of the SC, which does not has a perfect elliptical shape. The dependence on the temperature and field presented here confirm the necessity of small superconducting core size to minimize the hysteresis loss, thus justifying the choice of a multifilamentary configuration. In both cases, sample 1SS exhibits higher overall losses despite its lower J_c. Figure 5(a) shows the temperature dependence of magnetization AC loss for 1Cu wire, which is different from the previous wires 1Gl and 1SS. A strong frequency dependence is well

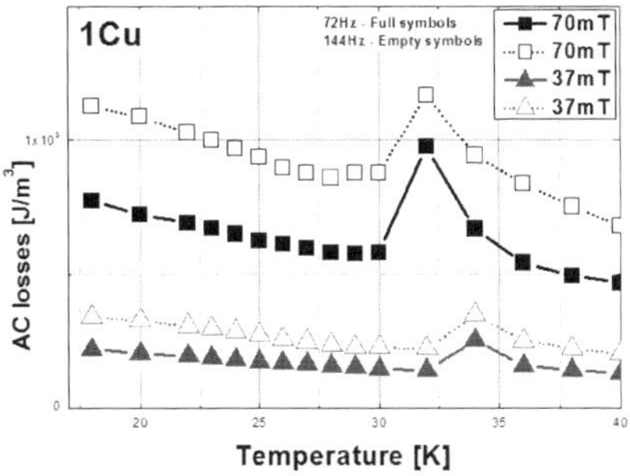

(a)

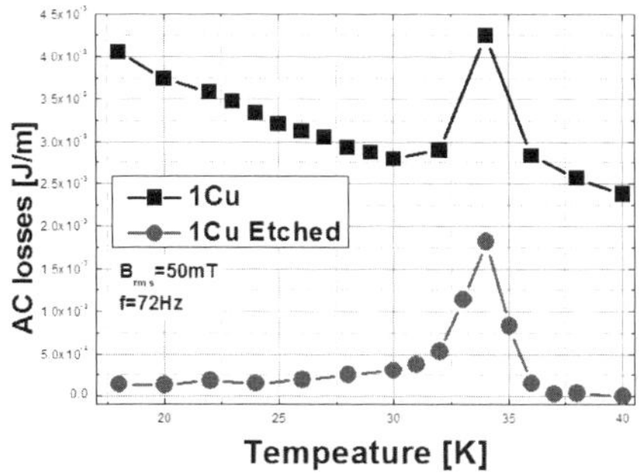

(b)

Fig. 5. (a) Temperature dependence of the magnetization AC loss measured in sample 1Cu at 37 and 70 mT and frequencies of 72 and 144 Hz, (b) AC magnetization loss of samples 1Cu, before and after etching away the Cu sheath (1Cu Etched).

visible for both applied fields. The measured losses remain relatively high at low temperatures (18–25 K) and surprisingly decrease with increasing temperature, which does not correspond to the expected dominant hysteresis losses in MgB_2 core.

The presented behavior can be attributed to eddy currents in the high conductive copper sheath. Then, the total losses Q are composed of eddy current loss in Cu and hysteretic loss of the superconducting MgB_2 core. The with the maximum loss occurs at a higher temperature (33 K) in comparison to 1Gl and 1SS: this is caused by the considerably higher current density of the MgB_2 layer made by IMD (see Table 1).[18] The measured $Q(B)$ with a slope of the AC loss being proportional to B^2 as well as the dependence on frequency correspond well to the expected eddy current loss, where the slope is proportional to B^2. In order to demonstrate the impact of eddy currents in sample 1Cu, the Cu sheath was etched away and the AC losses of MgB_2 sheathed by Ti were only measured. Figure 5(b) shows the apparent loss reduction after removing the Cu sheath. While the AC losses remain relatively high in 1Cu ($4.0 \times 10^{-3}\,\mathrm{Jm}^{-1}$) near 18 K and then decrease again at higher temperatures, the AC losses showed an opposite tendency for the 1Cu etched sample, which showed a typical behavior of hysteretic losses only. The AC loss of the etched wire at 20 K is lowered by a factor of ~27, which emphasizes the importance of appropriate sheath materials for obtaining low loss MgB_2 conductors. These measurements confirm that in the presence of well conducting metals the overall loss of a composite SC can be considerably increased.

Monel is frequently used as a shielding material for MgB_2 wires due to its good ductility and acceptable mechanical properties. In contrast to the case of DC applications, where the magnetic properties of monel are not of primary importance, they can play a crucial role in the AC regime. Therefore, the AC losses of the single core wire 1Mo supplied by HyperTech[19] have been studied. The temperature dependence of AC losses of 1Mo in Fig. 6(a) shows that the measured $Q(T)$ practically does not depend on temperature, which does not agree with dominant hysteresis losses in the MgB_2 core. Only small peaks can be observed at the highest fields (50 mT and 70 mT) at $T = 30$ K. Due to frequency independent $Q(T)$ characteristics, eddy current losses are excluded for the highest fields (50 mT and 70 mT) at $T = 30$ K. Due to frequency independent $Q(T)$characteristics, eddy current losses are excluded for this case. AC losses in wire 1Mo indicate dominant losses in the magnetic sheath material, the MgB_2 core being completely shielded by monel at low applied fields. The magnetic field penetrates partially into MgB_2 core, but only at high-field amplitudes. The contribution of the SC to the total loss is visible only by the small peaks above 30 K. Figure 6(b) confirms the magnetic nature of measured losses, where the slope of $Q(B)$ is exactly proportional to B^2 and is identical for all temperatures and both frequencies.

4. AC Losses of Filamentary MgB_2 Wires and "Decoupling Effect" by Twisting

The influence of coupling current losses and the decoupling effect by twisting have been examined in several different MgB_2 wires. The untwisted filamentary wire 30Gl (30 MgB_2 filaments, Ti barriers and GlidCop outer sheath, see Fig. 7(a))[20] and twisted ones with variable twist pitch: 10 mm (30Gl-10), 5mm (30Gl-5) and 2.5 mm (30Gl-2.5) made by standard PIT technology[21] were studied. Similarly, four-core

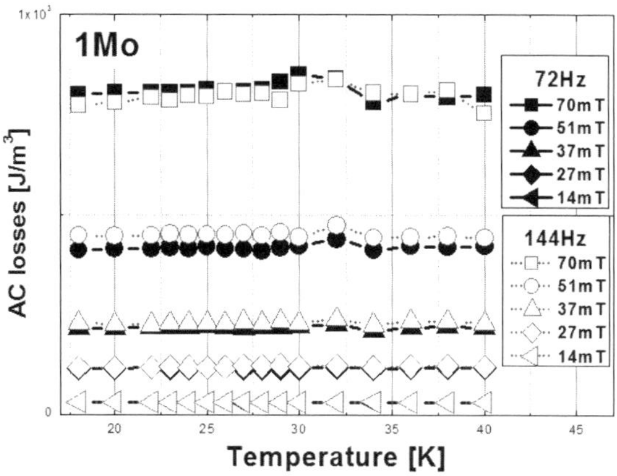

(a)

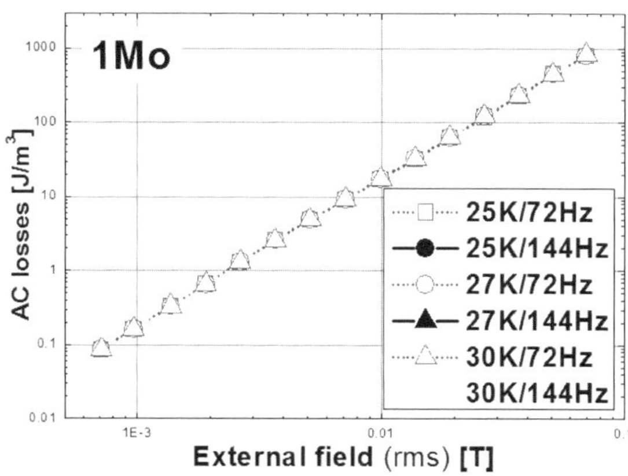

(b)

Fig. 6. (a) Temperature dependence of the AC losses for sample 1Mo measured at various magnetic fields and at the frequencies of 72 Hz and 144 Hz, (b) field dependence of the AC losses.

wires 4Gl untwisted and twisted with variable twist pitches (L_p) down to 3 mm (4Gl-17, 4Gl-10, 4Gl-5 and 4Gl-3)[22] were measured. Cross-sections of these wires are shown in Fig. 7(b). The AC losses of MgB_2 wires made by PIT[23,24] were compared with IMD wires.

Figure 8(a) shows the temperature dependence of the AC magnetization loss (per cycle and length) for 30Gl compared with 1Gl at an external magnetic field $B_{\text{ext}} = 19\,\text{mT}$ and frequencies of

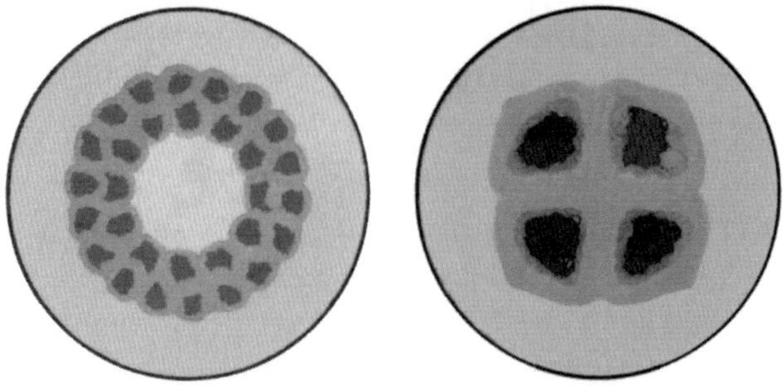

(a) (b)

Fig. 7. Cross-sections of multicore MgB_2 wires: (a) 30Gl and (b) 4Gl.

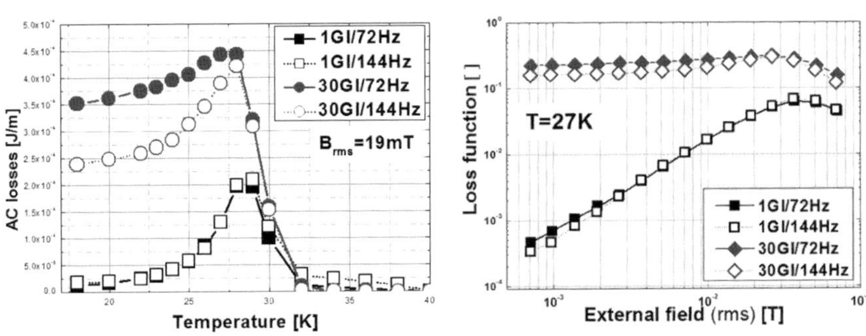

(a) (b)

Fig. 8. (a) Temperature dependence of AC magnetization loss of single-core and 30-filamentary untwisted sample in a constant external field of 19 mT and two frequencies, (b) loss function of examined wires measured at temperature 27 K.

$f = 72$ Hz (full line with full symbols) and 144 Hz (dashed line with empty symbols).

The losses of the 30Gl wire remains relatively high at low temperatures (18–25 K) in comparison to single-core 1Gl wires, which indicates the presence of coupling losses. The loss per cycle is frequency dependent, which indicates the presence of coupling currents. By increasing the temperature the hysteretic loss is added to the coupling losses reaching the maximum at the full flux penetration (similarly as for 1Gl). This explains the higher loss maxima for wire 30Gl in comparison to 1Gl. Figure 8(b) compares the dimensionless loss function[25] $\Gamma = \mu_0 Q/2B_a^2$ versus B_{ext}, where

μ_0 is the permeability of vacuum, Q is the AC loss in J per field cycle and per m^3 of the SC and B_a is the amplitude B_{ext} of the external field. In this representation a plateau of the loss function of sample 30Gl in low external magnetic field reflects the influence of coupling current losses. The position of the maximum loss function reflects a full penetration of magnetic field. One of the possibilities to decrease this loss is to shorten the loop area in order to reduce the induced electrical field, which can be effectively performed by filament twisting. Figure 9(a) shows the effect of filaments decoupling by twisting for different pitches (10, 5 and 2.5 mm) of wires 30Gl, 30Gl-10, 30Gl-5 and 30Gl-2.5. Figure 9(b) shows the same effect for the 4-filament IMD samples 4Gl, 4Gl-10 and 30Gl-5. As expected, the total AC losses are reduced for shorter twist pitches.

Figure 10(a) shows the effect of twisting on the loss function of wire 30Gl. As expected, the total value of the loss function is reduced when the twist pitch is shortened, the frequency dependence being influenced, too.

Considering the Eqs. (2) and (3), one can assume that the maximum of Q as a function of frequency is shifted to higher frequencies for shorter twist pitches. Figure 10(b) shows this effect for 4-filament IMD wires compared with analytical solutions. Our results show that filament twisting significantly reduced the AC

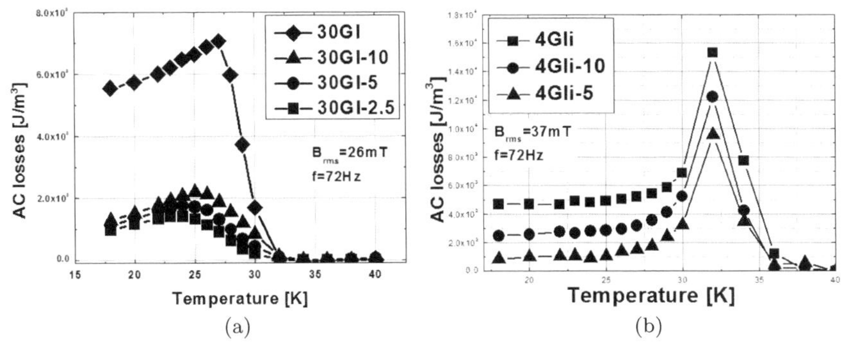

Fig. 9. (a) Comparison of the temperature dependence of AC magnetization loss for 30-filament untwisted and twisted samples with different twist pitch lengths, (b) the same comparison for 4-filament IMD wires.

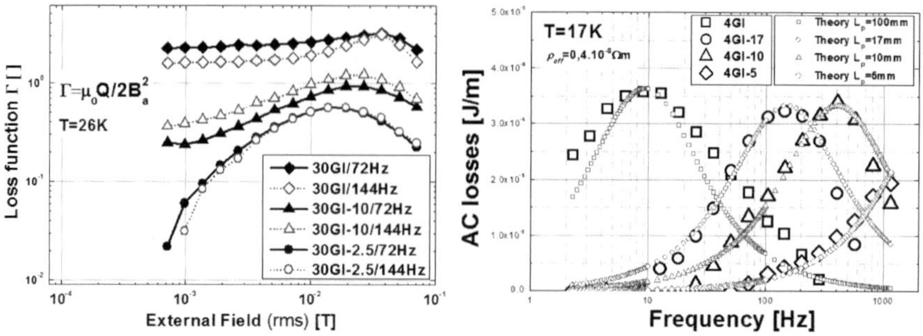

(a) (b)

Fig. 10. (a) Loss functions of 30-filament untwisted and twisted samples with different twist pitch lengths, (b) frequency dependence of measured AC losses in 4-filamentary IMD wires compared with analytical solutions.

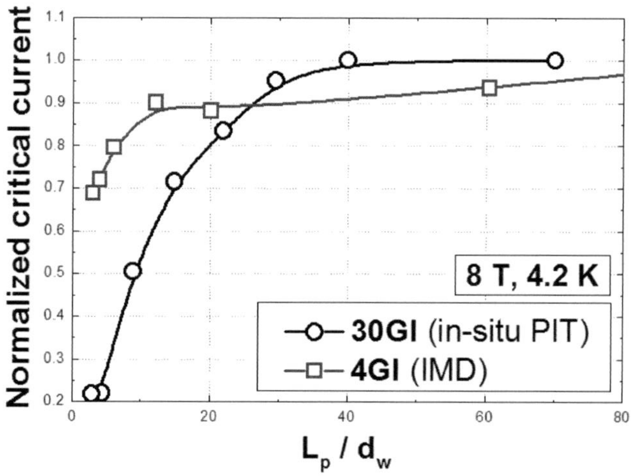

Fig. 11. Comparison of normalized critical currents versus the twist pitch for sample 30-filament SiC-doped wire **30Gl** made by *in situ* PIT[21] and **4Gl** sample made by IMD.[22]

losses in the low frequency range. On the other hand, twisting of as-deformed wires with powdered filaments reduces the filament density: consequently, the transport currents are lowered with decreased twist pitch.[23] These two effects are contradictory, decoupling being requested but the I_c reduction being unwanted. Figure 11 shows the normalized critical currents versus the ratio of twist pitch L_p

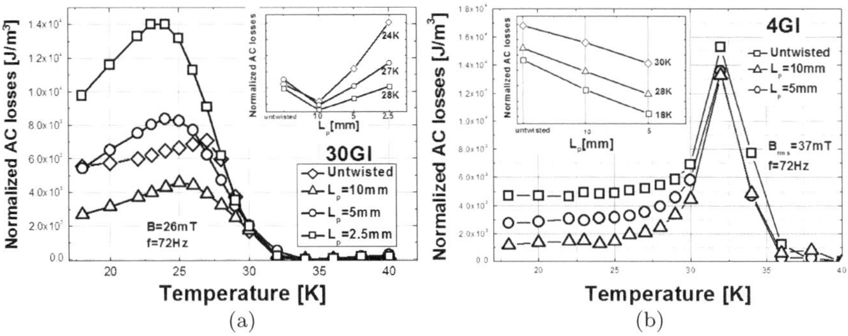

Fig. 12. (a) Normalized AC losses of 30-filament wire produced by the PIT process,[26] (b) and 4-core wire made by IMD.[24]

to wire diameter d_w for 30-filament PIT wire and 4-core IMD one. An apparent degradation of critical current is observed for 30Gl wire at $L_p/d_w < 25$. In the case 4Gl wire, the radical current degradation starts only for $L_p/d_w < 10$.

Taking into account the degradation of critical current density by twisting, the normalized plot of $Q_{\rm norm} = Q(I_{cN}/I_{cT})$ has been made, where Q is the measured AC loss, I_{cT} the critical current of the twisted wire and I_{cN} the current of the untwisted wire. The correction factor I_{cN}/I_{cT} was determined from the measured I_c data.

Figure 12 shows the normalized AC losses for different twist pitch while $Q_{\rm norm}$ versus L_p is shown in the insets. One can see that $Q_{\rm norm}$ has a minimum at $L_p = 10$ mm for PIT wires (see Fig. 12(a)) but decreases for all applied twist pitches in IMD wires (see Fig. 12(b)). The normalized losses of PIT wires with $L_t < 10$ mm are even higher than for untwisted one. Therefore, MgB_2 wires produced by the IMD process seem to be more promising for possible AC applications than PIT.

5. AC Losses of Untwisted Multifilamentary MgB_2 Wires and the "Decoupling Effect" due to a Resistive Barrier

The second possibility for the reduction of coupling losses consists in increasing the resistance between filaments. This effect was

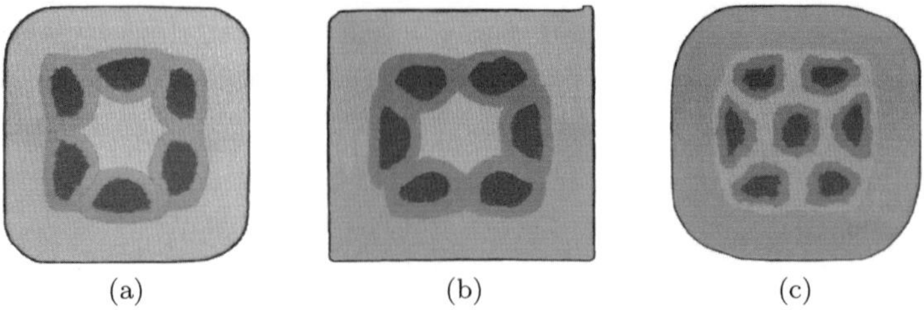

Fig. 13. Cross-sections of un-twisted multicore wires: 6/Nb (a), 6/Ti (b) and 7/NbTi (c).

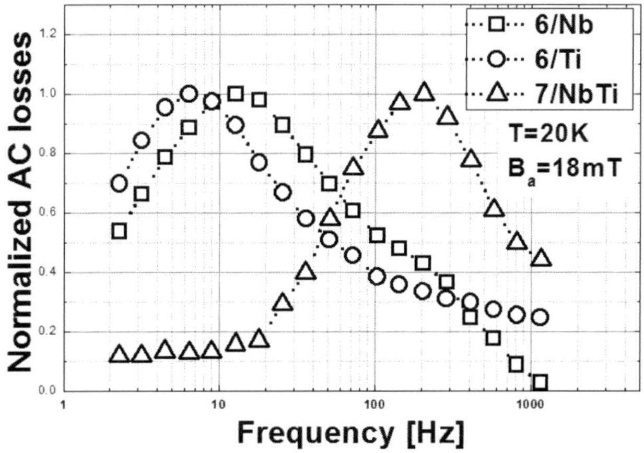

Fig. 14. Frequency dependence of normalized AC losses in 6- and 7-filamentary untwisted wires with Nb, Ti and NbTi barriers.

experimentally studied in untwisted 6- and 7-filament wires with various barriers. Sample 6/Nb consists of a Nb barrier, a GlidCop sheath and a Cu inner core. Sample 6/Ti has a Ti barrier, a GlidCop sheath and a Cu inner core. Wire 7/NbTi contains a NbTi barrier, a Cu barrier and a stainless steel outer sheath. Cross-sections of the examined samples are shown in Fig. 13. The best way to show the impact of the barrier resistance is to study the frequency dependence. Figure 14 shows the comparison between the frequency dependence of normalized AC losses (to be in the range) measured at 20 K with frequencies ranging from 2.4–1152 Hz.

The AC losses presented here have their maximum at a frequency of 16 Hz for wire 6/Nb, 6.8 Hz for wire 6/Ti and approximately 200 Hz for wire 7/NbTi. Using the time constants and Eq. (3), the calculated effective barrier resistivity is then obtained: $0.105 \times 10^{-8}\Omega$m for wire 6/Nb, $0.062 \times 10^{-8}\Omega$m for wire 6/Ti and $1.91 \times 10^{-8}\Omega$m for wire 7/NbTi. The difference between the resistivity of Nb and Ti is approximately twice, which agrees with the data taken from literature.[27] The assumed normal state resistivity of NbTi is about one order of magnitude higher than the resistivity of Nb. It is important to note that the resistivity depends on temperature (in the examined temperature range).

The present results prove that an increasing barrier resistance causes a similar decoupling effect as twisting of the wires. It follows that the most suitable barrier material for MgB_2 wires intended to work at temperatures about 20 K with low AC losses is the ductile NbTi, which exhibits a much higher resistivity than Nb or Ti.

6. AC Losses in MgB_2 Cables

Finally, the AC losses of MgB_2 cables as a function of temperature has been measured. The cross-sections of the cables 7 × 30Gl and 7 × 1Cu are shown in Fig. 15. Both cables consist of six wires surrounding a central wire, with 15 mm transposition length. Cable 7 × 30Gl contains 30-filament untwisted wires with the same configuration as previously discussed for wire 30Gl, but with a lower diameter. Cable

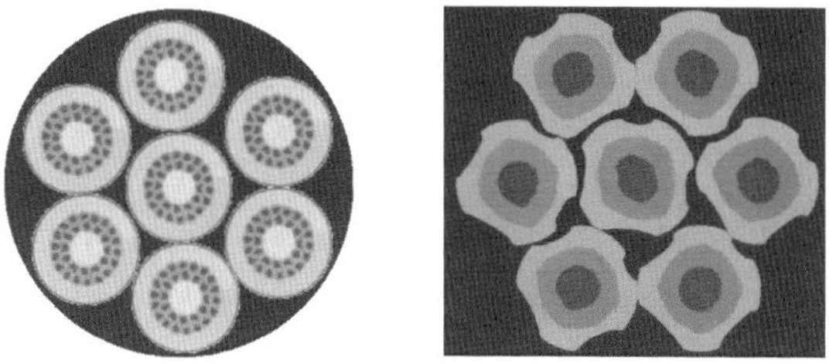

(a) (b)

Fig. 15. Cross-sections of measured wires and cables (a) 7 × 30Gl and (b) 7 × 1Cu.

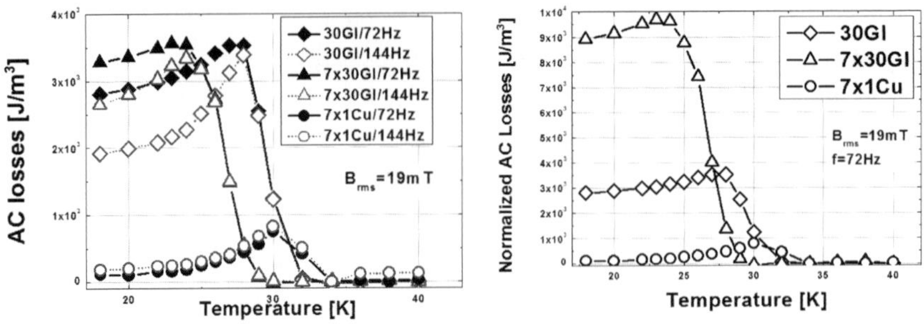

(a) (b)

Fig. 16. (a) Temperature dependence of AC magnetization loss of 7-strands MgB_2 cables 7 × 30Gl, 7 × 1Cu and comparison to non-twisted 30Gl and (b) the same dependences normalized by factor of critical current degradation.

7 × 1Cu contains seven single core wires with a Ti barrier and Cu sheath.

The results shown in Fig. 16(a) are compared to the AC loss of the untwisted wire 30Gl. It appears that cabling leads to enhanced coupling losses compared to untwisted multifilamentary wires (triangles) in the cable. The peak of losses is shifted to lower temperatures in comparison to the untwisted wire 30Gl (diamonds). This can be caused by two factors, both reducing the penetration field: the filament diameter are significantly smaller and the critical current has been lowered as well. The 7 × 30Gl cable behaves like a wire with thinner filaments, but without any effect on the coupling loss reduction, which is clearly visible from the frequency dependence of AC loss and from the behavior at low temperatures. The normalized loss Q_{norm} (Fig. 16(b)) shows that after taking into account the transport properties, the 7 × 30Gl cable seems to exhibit considerably higher losses than the other ones. Different results have been obtained for the 7 × 1Cu cable made of monocore wires. The shape of loss dependence reflects dominantly a hysteretic character. The small frequency dependence can be explained by eddy current losses in the copper matrix. Moreover, the full penetration at constant field 19 mT is reached at higher temperature. It reflects the larger size of superconducting filaments in the cable made of single core wires and a negligible degradation of critical current by the cabling operation. The lowest normalized magnetization AC losses Q_{norm} has been measured in the $7 \times 1Cu$ cable that also exhibit no coupling loss. Consequently, this cable structure appears to be particularly promising for future AC applications.

7. Conclusions

The AC losses of MgB_2 wires and cables with various wire configurations, sheaths and barrier materials, number of filaments and twist pitch lengths were studied experimentally. It has been found that hysteresis loss depends on the size of SC as well as on its critical current density. The applied sheath material can affect significantly the overall loss of composite. Therefore, high conductive and magnetic sheath materials are not suitable for AC applications. The decoupling effect by twisting was demonstrated for wires made by different technologies. Wires made by the IMD process appear to be the better choice than PIT processed wires. Another possibility to eliminate the coupling losses is to use resistive barriers. The most effective barrier material at temperatures around 20 K appears to be NbTi. It was found that losses of individual wires in cables have a dominant effect. Consequently, cable made of thin single-core wires appear as suitable solutions for required low AC losses, in analogy to Nb_3Sn and to NbTi multifilamentary wires.

References

1. J. Šouc, F. Gömöry and M. Vojenčiak, *Supercond. Sci. Technol.* **18**, pp. 592–595 (2005).
2. J. Kováč, J. Šouc and P. Kováč, *Physica C* **475**, pp. 1–4 (2012).
3. M. N. Wilson, *Superconducting Magnets*, Oxford University Press, (1983).
4. P. M. Oomen, *AC loss in superconducting tapes and cables*, Gegevens Koninklijke Bibliotheke, Den Haag, (2000).
5. H. London, *Physi. Lett.* **6**, pp. 162–165 (1963).
6. C. P. Bean, *Rev. Mod. Phys.* **36**, pp. 31–39 (1964).
7. M. Magnusson, *Supercond. Sci Technol.* **27**, p. 105003 (2014).
8. W. J. Carr, Jr., *AC loss and macroscopic theory of superconductors*, Gordon and Breach, New York, USA (1983).

9. B. Seeber, *Handbook of Applied Superconductivity*, Institute of Physics Publishing, Bristol, UK (1998).
10. A. M. Campbell, *A general treatment of losses in multifilamentary superconductors, Cryogenics* **22**, p. 3 (1982).
11. S. Takacs, *Supercond. Sci. Technol.* **23**, p. 065023 (4pp) (2010).
12. K. V. Namjoshi and P. P. Biringer, *Low-frequency eddy current loss estimation in long conductors by using the moment of inertia of cross-sections, IEEE Trans.* Magnetics **24**, p. 2181 (1988).
13. I. Hušek, P. Kováč, A. Rosová, W. Pachla, T. Melišek and M. Hain, *J of All and Compounds* **588**, p. 366 (2014).
14. J. Viljamaa *et al.*, *J. Physics: Conference Series* **234**, p. 022041 (2010).
15. P. Kováč, M. Reissner, T. Melišek, I. Hušek and S. Mohammad, *J. Appl. Phys.* **106** p. 013910 (2009).
16. M. D. Sumption *et al.*, *Sup. Sci. and Technol.* **18**, p. 730 (2005).
17. F. Gömöry *et al.*, *Sup. Sci. and Technol.* **15**, p. 1311 (2002).
18. P. Kováč, I. Hušek, T. Melišek, L. Kopera and J. Kováč, *Sup. Sci. and Technology* **27**, p. 065003 (2014).
19. Hyper Tech Research, Inc., 539 Industrial Mile Road, Columbus OH, USA, www.hypertechresearch.com.
20. P. Kováč *et al.*, *J. Alloys Compounds* **509**, p. 8783 (2011).
21. P. Kováč, I. Husek, T. Melisek and L. Kopera, *Supercond. Sci. Technol.* **24**, p. 115006 (5pp) (2011).
22. P. Kováč, I. Hušek, T. Melišek, L. Kopera and J. Kováč, *Supercond. Sci. Technol.* **27**, p. 065003 (2014).
23. P. Kováč, T. Melišek, L. Kopera, J. Kováč and I. Hušek, *Sup. Sci. Technol.* **25**, p. 095008 (2012).
24. J. Kováč, J. Šouc, P. Kováč and I. Hušek, *Supercond. Sci. Technol.* **28**, p. 015013 (7pp) (2015).
25. M. N. Wilson, *Superconducting Magnets* ISBN-13-978-0198548102 (1983).
26. J. Kováč, J. Šouc, P. Kováč, I. Hušek and F. Gömöry, *Physica C* **495**, p. 182 (2012).
27. J. W. Ekin, *Experimental Techniques For Low-Temperature Measurements*, Oxford University Press, (2011), ISBN 978-0-19-857054-7.

Chapter 5c

Effect of Mechanical Load on J_c of MgB_2 Wires

Pavol Kováč

Institute of Electrical Engineering of Slovak Academy of Sciences
Dúbravska cesta 9, Bratislava 84104
Pavol.Kovac@savba.sk

The effect of tensile stress on the critical current of MgB_2 wires has been studied experimentally. Filamentary wires of different compositions, sheath materials and filament structures subjected to mechanical load allowed estimating the stress and strain tolerances of MgB_2 composites. It was shown that the outer sheath plays a dominant role on these properties, but that the filament quality (additions, density and twisting) has also a considerable effect on the electro–mechanical behavior of MgB_2 wires.

1. Introduction

MgB_2 wires are composites consisting of brittle (ceramic) superconducting MgB_2 filaments placed inside one or more ductile metallic sheaths. Consequently, the performance of MgB_2 wires is not only limited by the electrical properties (J_c) of the SC, but also by the resulting mechanical stresses during operation. In a similar way to continuous fiber reinforced metal matrix composites, the mechanical behavior of multifilamentary MgB_2 wires is influenced by fibre characteristics, by the strength and nature of the interface and by matrix hardening and strength characteristics.[1] Therefore, the effects of mechanical load on J_c of MgB_2 composites have been studied very early after the discovery of superconductivity in MgB_2.[2–4] The first measurements under uniaxial tension have shown a linear increase of the critical current with tension up to the irreversible strain ε_{irr}, where a radical degradation occurs due to cracking of the MgB_2 filaments.[4] It was shown that the behavior of MgB_2/metal composite wires is similar to that of A15 wires[5] for a compressive internal strain state of the filaments and similar to that of ceramic high temperature superconductors (HTS)[6] when this internal strain becomes tensile. Just like in ceramic HTS, the MgB_2 filaments will not be damaged by an applied uniaxial tensile stress as long as this stress does not fully compensate the thermally induced compressive stress (i.e., as long as the internal strain state of the filaments remains compressive).[4]

2. Electro-mechanical Properties of MgB_2 Wires

Several investigations have been carried out with devices exploiting basic methods for the mechanical characterization of MgB_2 composites soldered to metallic spring (e.g., U-shape, Pacman or Walter spiral).[2,4,7] The advantage of these methods lies in the possibility to use long samples with a significant distance between potential taps as well as in the possibility of measuring the effect of pressure. On the other hand, the sample is mechanically and electrically bonded to the holder (by soldering or by frictional force), and therefore the measured data are smoothed. Rabara *et al.*[8] have shown that under uniform tensile load, the probability of the brittle filament fracture follows a Weibull distribution. It turned out that the various experimental set up systems influence the real load acting on the wire: consequently, the results of critical current degradation with strain $I_c(\varepsilon)$ are different.[9] The electromechanical characteristics, e.g., stress versus strain or I_c versus strain of various free standing samples under variable tensions, magnetic fields and temperatures were measured.[10] The results obtained for the free-standing sample (without any parallel metallic spring) give the most realistic picture of the critical current degradation inside composite wires subjected to uniaxial tension.[3,9,10]

Figure 1 shows the tension test rig used in our laboratory for the electro-mechanical characterization of MgB_2 wires. Samples of

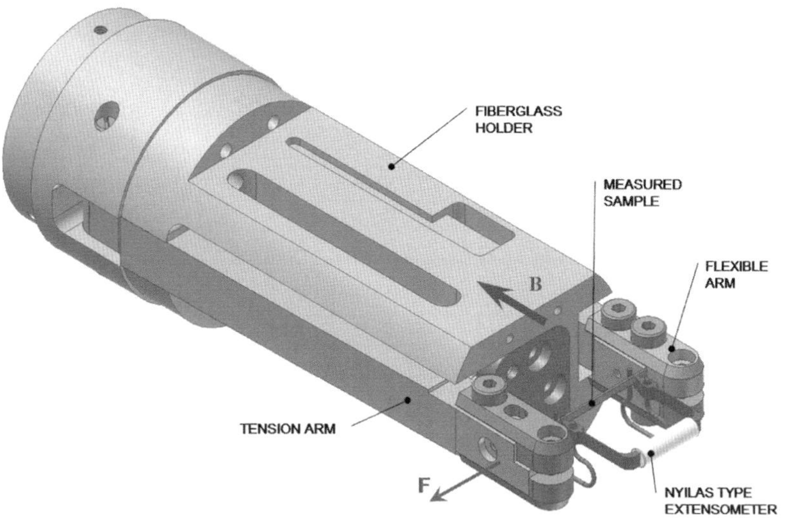

Fig. 1. Drawing of the tension test rig: (a) fiberglass holder, (b) tension arm, (c) flexible arm (load cell), (d) Nyilas type extensometer, (e) measured sample.

50 mm length are joined to flexible current leads and fixed by the ends into the holding grips between the tension and flexible arms. A tensile load applied to the specimen is transferred from sliding motion of the pusher along the bevel edge of the tension arm. Four strain gauges connected to a Wheatstone bridge circuit are glued to the flexible arm and serve as sensitive load sensors. The sample elongation is measured by a Nyilas type (NT) extensometer with a gauge length of 25 mm clamped to the wire by springs.[10]

The presented system allows obtaining the stress $\sigma(\varepsilon)$ and the corresponding $I_c(\sigma)$ and $I_c(\varepsilon)$ characteristics for the tested MgB_2/metal samples. Figure 2(a) shows the typical $\sigma(\varepsilon)$ and $I_c(\varepsilon)$ behavior obtained for a MgB_2/Ti/Cu/Monel tape (see the inserted cross-section), which enables to estimate the values of irreversible stress (σ_{irr}) and irreversible strain (ε_{irr}), above which the critical current (I_c) is irreversibly degraded due to filament cracking. An interesting feature of $I_c(\varepsilon)$ is the linear increase of the critical current with axial strain (K_I) and Young modulus (E). K_I can be obtained from the equation $I_c(\varepsilon) \equiv (1 + K_I\varepsilon)I_c(0)$, while E is obtained from the elastic part of $\sigma(\varepsilon)$. The Young's modulus E was estimated from

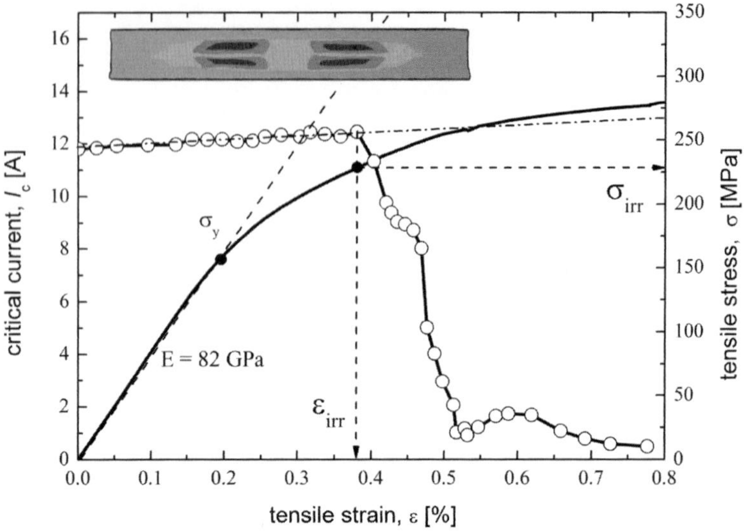

Fig. 2(a). Example of $I_c(\varepsilon)$ and $\sigma(\varepsilon)$ characteristics of a MgB_2/Ti/Cu/Monel tape of 2.96 mm × 0.24 mm cross-section.

the linear part of the $\sigma(\varepsilon)$ curve determined by the points between the value of $\varepsilon \sim 0.02\%$ and σ_y, the point of yielding. The point of yielding (σ_y) was determined as the 0.005% deviation from linearity of the $\sigma(\varepsilon)$ curve. For the sample presented by Fig. 2(a) the Young modulus of the MgB_2/Ti/Cu/Monel tape was estimated to $E = 82$ GPa. Figure 2(b) compares the $I_c(\varepsilon)$ characteristics of the same 4-core MgB_2/Fe wire measured by the U-shape spring system[4] and by the free standing sample.[10]

A comparison between the results shows nearly identical irreversible strain values $\varepsilon_{irr} = 0.34\%$, but different $I_c(\varepsilon)$ variations. Brittle behavior above ε_{irr} cannot be observed by means of U-spring. The I_c increase upon strain is due to the release of the compressive stress in MgB_2 filaments caused by the higher thermal contraction of metallic sheath material down to 4.2 K. The different slopes K_I for the same wire measured by different systems are clearly visible, thus reflecting that MgB_2 is under different stress conditions. This correlates well with the results reported for Bi-2223/Ag tapes, where the U-shape spring yielded a higher ε_{irr} and less steeped critical

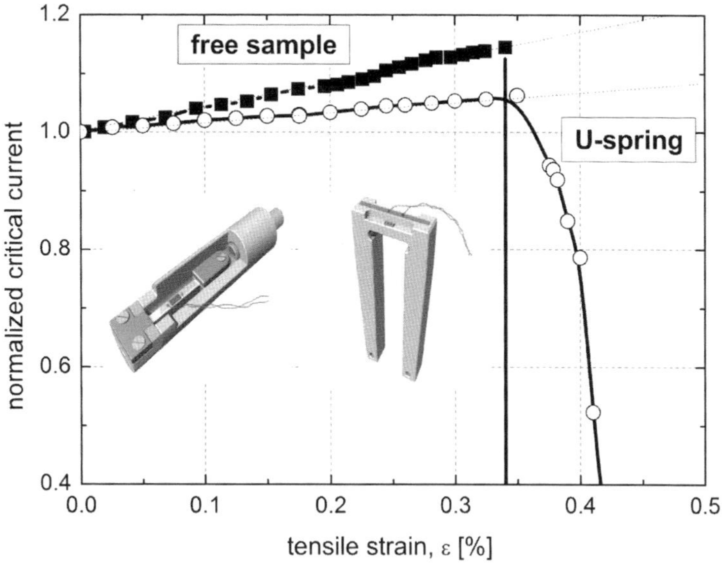

Fig. 2(b). Comparison between the $I_c(\varepsilon)$ characteristics measured by the U-shape spring system and the free standing MgB_2/Fe wire.

current degradation with respect to the free sample method due to higher pre-compression and mechanical shunting of the sample by the spring material (brass or Ti-alloy).[9]

Generally, $\sigma(\varepsilon)$ of composite wires can be divided into three basic parts: (i) metallic matrix and filaments are elastic, (ii) matrix is plastic and filaments elastic, and (iii) matrix and filaments are elastic. For a two-component composite in part (i) the Young modulus of the whole composite can be accurately described by the rule of mixtures[11]:

$$E = f_f E_f + f_m E_m, \tag{1}$$

where f_f and f_m are the volume fractions of filaments and metallic matrix. The largest portion of $\sigma(\varepsilon)$ belongs usually to the stage, where the curve is not linear and the composite modulus may be determined at the different stress–strain levels by:

$$E = f_f E_f + \left(\frac{d\sigma_m}{d\varepsilon_m}\right)_{\varepsilon f} \cdot f_m, \tag{2}$$

where $(d\sigma_m/d\varepsilon_m)_{\varepsilon f}$ is the slope of the matrix stress–strain curve at different strains ε_f of the filaments.[10] In stage (iii) the modulus of composites with ductile filaments must be modified more by a similar way to the reduction of matrix modulus:

$$E = \left(\frac{d\sigma_f}{d\varepsilon_f}\right)_{\varepsilon f} \cdot f_f + \left(\frac{d\sigma_m}{d\varepsilon_m}\right)_{\varepsilon f} \cdot f_m, \tag{3}$$

in case of real multicomponent composite wires, it is not simple to precisely subdivide the individual $\sigma(\varepsilon)$ into the parts described above.

The stress–strain curves (Fig. 2(a)) show a typical behavior of composites during monotonic loading with a very short elastic part. Microplasticity then takes place, at a fairly low stress, which corresponds to the original deviation from linearity in the stress–strain curve. Microplasticity in the composites is attributed to stress concentrations in the matrix. With increasing strain, microplasticity increases in magnitude to global plasticity in the metallic matrix.

As one can see, the variation of $I_c(\varepsilon)$ for strain values lower than ε_{irr} is linear and reversible. This effect was earlier reported in Ref. 4 and is further elaborated in Ref. 12. Concentrating on the reversible behavior at different temperatures and magnetic fields (i.e., $I_c(T, B, \varepsilon)$ for $\varepsilon < \varepsilon_{irr}$), the critical current varies linearly with applied strain:

$$I_c(T, B, \varepsilon) = I_c(T, B, 0)[1 + K_I(T, B)\varepsilon], \tag{4}$$

where $K_I(T, B)$ is the (normalized) slope of the reversible strain response:

$$K_I(T, B) \equiv \frac{1}{I_c(T, B, 0)} \left(\frac{\partial I_c(T, B, \varepsilon)}{\partial \varepsilon}\right). \tag{5}$$

Using the approach presented in Ref. 12, the K_I relation for the constant temperature (4.2 K) and magnetic field (≈6 T) will be:

$$K_I = \frac{1}{I_c}\frac{dI_c}{d\varepsilon} = \frac{1}{I_c}\frac{dI_c}{dp}\frac{dp}{d\varepsilon} = \frac{E}{I_c}\frac{dI_c}{dp}, \tag{6}$$

where p is external pressure acting on the MgB_2 compound.[13] We can conclude that the $I_c(\varepsilon)$ behavior below the irreversible strain limit is compatible with published results on the pressure dependence of the critical temperature $T_c(p)$.[13] Thus, it could well be caused by *intrinsic* variations of the superconducting energy gap, due to changes in the electronic structure as the crystallographic unit cell is compressed. Therefore, K_I should be directly related to the electronic properties of the material, and thus presumably be a material constant.

2.1 *Effect of applied sheath material*

As mentioned in the introduction, the mechanical behavior of MgB_2 composite wires may be influenced by the matrix (sheath) strength.[1] MgB_2 wires with different metallic sheaths have been tested under uniaxial tension at 4.2 K.[14–19] Figure 3(a) shows the stress–strain characteristics of three 4-filament MgB_2 wires sheathed with AgMg, Monel and Fe. As one can see, $\sigma(\varepsilon)$ of these wires are strongly affected by the used sheath. The softest $\sigma(\varepsilon)$ characteristic is measured for the wire with highly ductile AgMg alloy ($E_{AgMg} = 46$ GPa) and the strongest one for pure iron with $E_{Fe} = 174$ GPa. On the other side, the Fe sheathed wire shows very brittle fracture at $\varepsilon = 0.34\%$, while the wire with Nb/AgMg sheath has a large plastic part and can be stressed well above $\varepsilon = 1\%$ without any disruption. The $\sigma(\varepsilon)$ characteristics of the Monel-sheathed wire shows an elastic behavior for $\varepsilon > 0.15\%$ with $E_{Monel} = 101$ GPa and an apparent plastic deformation region for $\varepsilon > 0.2\%$. The dashed lines in Fig. 3(a) mark the point of irreversible critical current degradation (σ_{irr} and ε_{irr}), which occurs at different parts of $\sigma(\varepsilon)$. While the current degradation of AgMg-sheathed wire starts inside the elastic part of $\sigma(\varepsilon)$ at $\varepsilon_{irr} = 0.23\%$ and $\sigma_{irr} = 106$ MPa, it occurs above it for the Fe sheath ($\varepsilon_{irr} = 0.33\%$ and $\sigma_{irr} = 555$ MPa) and far in the plastic part for the Monel sheathed wire ($\varepsilon_{irr} = 0.456\%$ and $\sigma_{irr} = 178$ MPa).

The corresponding $I_c(\varepsilon)$ behavior is illustrated in Fig. 3(b). As one can see, the wire with a strong and brittle sheath (75% Fe) shows the largest current increase (by more than 14% at ε_{irr}), while the wire with soft AgMg exhibits the smallest increase of I_c (only by

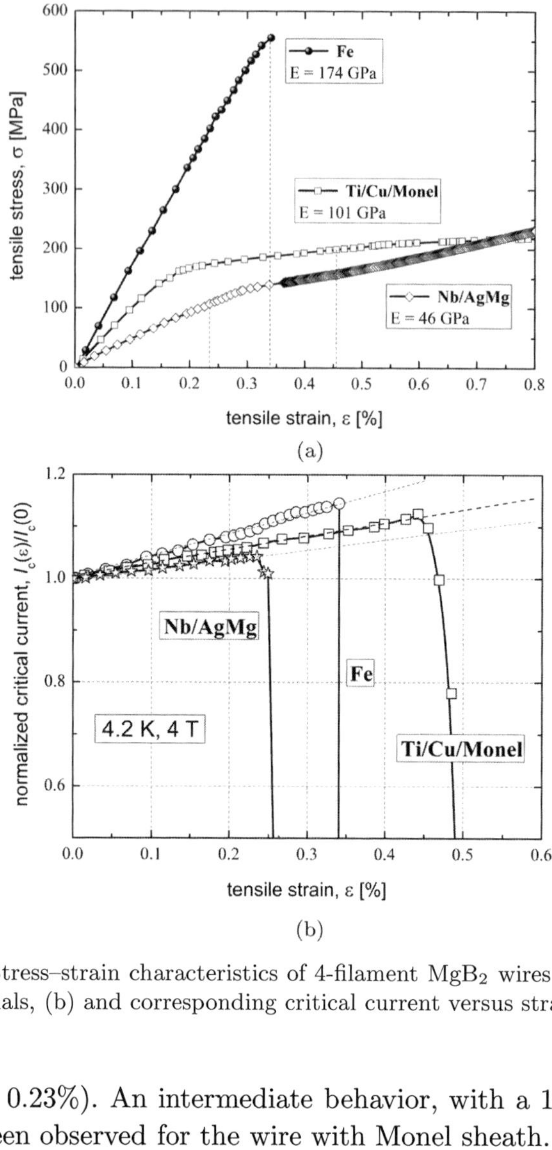

Fig. 3. (a) Stress–strain characteristics of 4-filament MgB_2 wires with different sheath materials, (b) and corresponding critical current versus strain.

4.3% $\varepsilon_{irr} = 0.23\%$). An intermediate behavior, with a 12% increase of I_c has been observed for the wire with Monel sheath.

Figure 4 presents $I_c(\sigma)$ characteristics for the set of MgB_2 wires with variable sheath materials. It appears that the mechanical strength of the outer metallic sheath has a decisive role on the stress tolerance of MgB_2 wires. While for mechanically soft sheath materials (AgMg, Cu and Ni) the MgB_2 wires can only be used up to axial

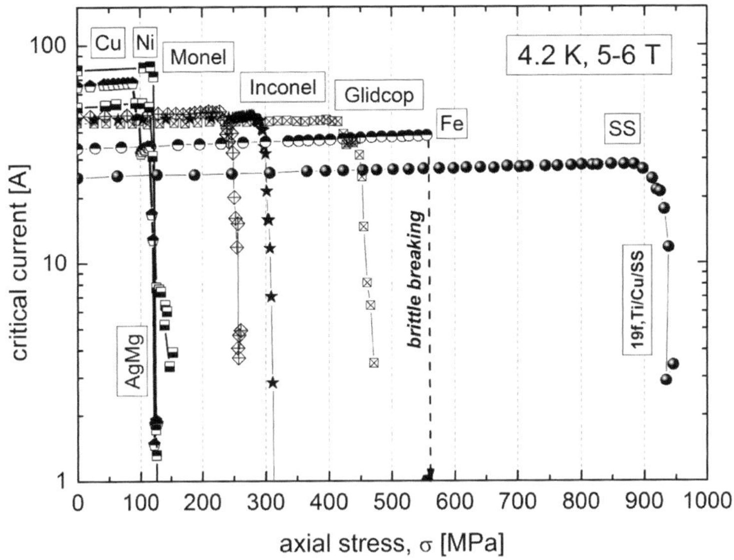

Fig. 4. Stress tolerances of MgB_2 wires with different sheath materials.

stresses of the order of 100–120 MPa, the strongest sheaths (Fe and SS) allow applying stresses as high as 500–900 MPa. However, the brittle behavior of pure iron at low temperatures makes its practical application problematic, while work hardening of the SS sheath during the wire deformation also causes difficulties. It appears that monel, and GlidCop are more practical sheath materials for MgB_2. Finally, it should be mentioned that the applied heat treatment has also some effects due to sheath softening and lowered σ_{irr}.[20]

With axial stress values in the range of 250–400 MPa, monel and GlidCop are acceptable for the most perspective perspectives of this SC.

2.2 *Additions into filaments*

The characteristics of 19-filament MgB_2 wires with Nb barriers made by *in situ* processing without and with 15 wt.% W addition and GlidCop® sheath have been measured.[21] Figure 5(a) compares the variation of the critical current versus tensile stress: both tungsten added wires show an apparent improvement of stress tolerance with

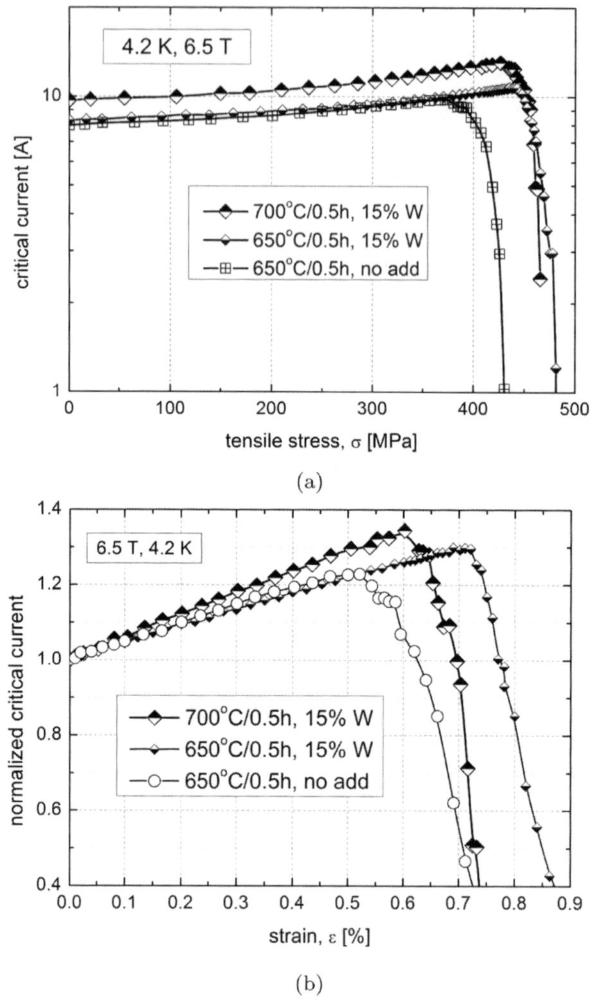

Fig. 5. (a) Critical current versus tensile stress and (b) normalized critical currents versus strain for MgB2 wires annealed at different conditions.

$\sigma_{irr} = 426$–440 MPa with respect to 385 MPa. Figure 5(b) presents the effect of tensile strain on the critical current for the compared wires. Surprisingly, a large pre-compression effect is observed after the heat treatment at 700°C, showing an increase of I_c at ε_{irr} exceeding 30%. Up to now, the highest increase of I_c was observed around 20% for tensile stressed MgB_2 wires at ε_{irr}.[16–19] The wire heat treated at 650°C shows a smaller gradual increase of I_c, but an even higher $\varepsilon_{irr} > 0.7\%$ than the wire with no addition, the slope $I_c(\varepsilon)$ being identical. Following these results, it can be concluded that the addition of tungsten particles into filaments leads to a considerable improvement of the stress and strain tolerance of *in situ* processed MgB_2 wires. Similar improvement has been observed for filaments in GlidCop with different mass density due to applied deformation.[18]

2.3 *Strain tolerances of twisted wires*

Twisted and untwisted 19-filamentary MgB_2 wires manufactured by the *in situ* PIT process have also been mechanically tested. The wire with an outer diameter $d_w = 0.86$ mm with 53% SS-reinforcement, 19.5% Cu stabilizer, 19% Ti barrier and 8.5% MgB_2 filaments. Short wire pieces (≈70 mm) were twisted before (BHT twisting) and also after heat treatment (AHT twisting) with variable twist pitch lengths L_t.[22]

Figure 6(a) compares the critical current degradation for BHT and AHT twisted wires based on the normalized I_c(8T) versus the

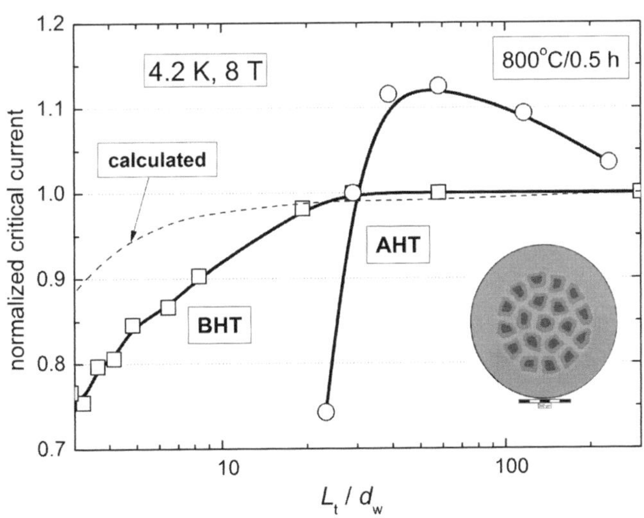

Fig. 6(a). The degradation of J_c after twisting of a 19-filament MgB_2/Ti/Cu/SS wire.

ratio L_t/d_w between twist pitch and wire diameter. No degradation is observed for $L_t/d_w > 40$, but I_c decreases rapidly at $L_t/d_w < 30$. This figure shows $I_c(L_t/d_w)$ the variation for AHT twisting at 800°C/0.5 h in comparison to standard BHT twisting and theoretical I_c degradation for variable twist pitch.

The effective filament cross-section and the corresponding critical current have been calculated for variable twist pitch lengths L_t. The calculated reduction of I_c takes only into account the changed filament area (solid line), which presents an ideal state, assuming no degradation of filament quality. Each filament ring (1 + 6 + 12 filaments) has a different diameter r_f at the same L_t, which results in variable angles θ_f estimated as $\tan\theta_f = 2\pi rf/L_t$. The reduction of the effective area for twisted filaments was obtained by multiplying the cross-sections of individual filament rings by $\cos\theta_f$ and by a subsequent summation of ring areas. The wire with BHT twisting shows slightly increased I_c values for $L_t/d_w > 25$ and considerably reduced ones for $L_t/d_w < 20$. The difference between calculated and measured values can be attributed to reduced filament density and possibly also to microcracks caused by the strong torsion stress.[22]

AHT twisted wire exhibits significant enhancement of the I_c by 12.6% for a heat treatment at 800°C/0.5 h. The highest critical current was measured at $L_t/d_w = 58$ while a stronger reduction was observed for BHT twisting at $L_t/d_w < 30$. A more pronounced current degradation after AHT twisting ($L_t/d_w < 30$) is caused by the elongation of brittle MgB_2 filaments in comparison to (unreacted) Mg–B ones after BHT twisting. These results indicate that twisting may influence the residual strain inside the MgB_2 filaments of the examined wires. To verify the effect of AHT twisting, $I_c(\varepsilon)$ and $\sigma(\varepsilon)$, characteristics of this wire were measured.

Figure 6(b) shows the behavior of $I_c(\varepsilon)$ for wires heat treated at 600°C/2.5 h, differing only by the applied AHT twisting. All samples show typically a linear increase of critical current up to the irreversible strain ε_{irr}. One can see that the slope K_I is not influenced by twisting, while $I_c(\varepsilon = 0)$ and ε_{irr} values are different. Figure 6(b) clearly shows the systematic reduction of ε_{irr} with the shortened L_t.

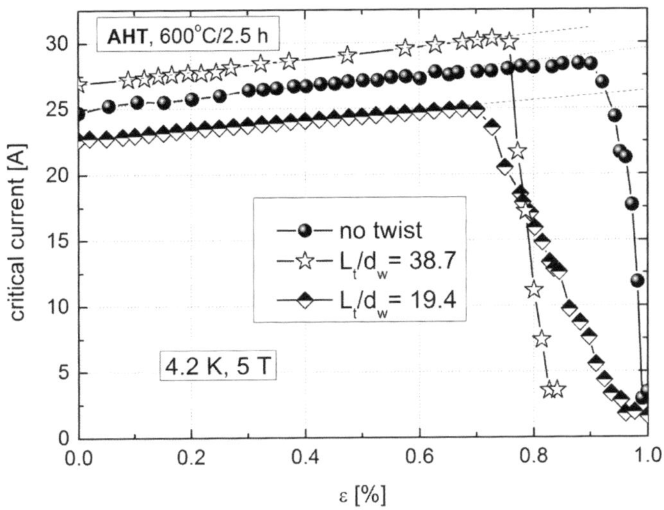

Fig. 6(b). The critical current of a 19-filament MgB_2/Ti/Cu/SS wire versus the uniaxial strain for different twisting modes applied after the final heat treatment.

The value of ε_{irr} of the untwisted wire is reduced from 0.9% to 0.7% for the AHT twisted wire with $L_t/d_w = 19.4$ ($L_t = 17$ mm). On the other hand, $I_c(\varepsilon)$ characteristic with $L_t/d_w = 38.7$ behaves as shifted to lower strains and increased $I_c(\varepsilon = 0)$, which confirms that the residual strain inside the wire is affected by the applied torsion stress. The wire twisted with small a L_t/d_w ratio (19.4) shows lowered $I_c(\varepsilon = 0)$ and also a different slope of $I_c(\varepsilon)$ above ε_{irr}, which is not yet understood.

Figure 6(c) shows how BHT and AHT twisting influences the stress–strain characteristics of the tested 19-filament wire. Generally, all twisted wires show 'stronger' behavior (higher Young modulus $E = 151$–176 MPa) in comparison to untwisted ones ($E = 124$ MPa). The application of twisting deflects the orientation of the filaments from the wire axis, which changes the direction of the stress acting on the filaments. The effect of twisting on the residual strain and $\sigma(\varepsilon)$ is a rather complicated phenomena and its poor understanding needs further systematic analysis.

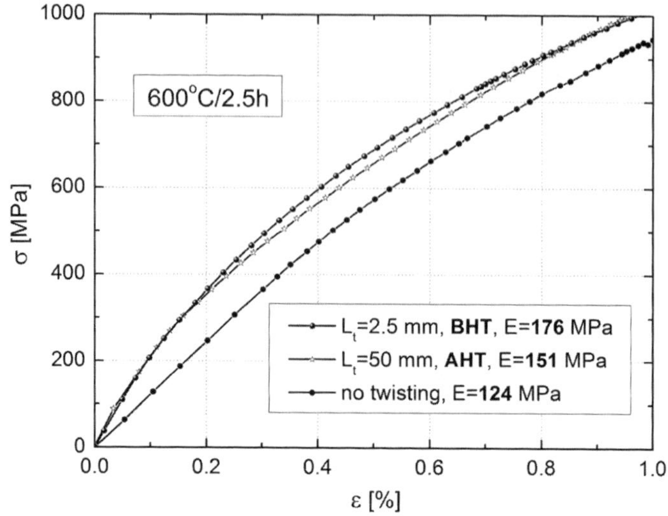

Fig. 6(c). Stress–strain characteristics of MgB_2/Ti/Cu/SS wires after twisting before (BHT) and after (AHT) the final heat treatment.

3. Conclusions

Electro-mechanical characterization has been performed for MgB_2 conductors with different filaments, wire compositions and varied sheath materials. It was shown that the mechanical strength of the outer sheath has a dominant effect on the stress tolerance of MgB_2 composite wire. The stress tolerance can also be affected by an improved filament structure (W-addition, dense filaments). It was shown that twisting of MgB_2 filaments is changing the residual stress inside the wire. Consequently, the stress–strain characteristics of twisted wires is improved, while the value of irreversible strain ε_{irr} is decreased.

References

1. N. Chawla and K. K. Chawla, *Metal Matrix Composites*, Springer, New York (2006).
2. H. Kitaguchi, H. Kumakura and K. Togano, Strain effect in MgB_2/stainless steel superconducting tape, *Physica C* **363**, pp. 198–201 (2001).

3. W. Goldacker and S. I. Schlachter, Influence of mechanical reinforcement of MgB_2 wires on the superconducting properties, *Physica C* **378–381**, pp. 889–893 (2002).
4. P. Kováč M. Dhallé, T. Melišek, H. van Eck, W. A.J Wessel, B. ten Haken and I. Hušek, Dependence of the critical current in *ex-situ* multi- and monofilamentary MgB_2/Fe wires on axial tension and compression, *Supercond. Sci. Technol.* **16**, pp. 600–607 (2003).
5. J. W. Ekin, Strain scaling law for flux pinning in practical superconductors. Basic relationship and application to Nb_3Sn conductors, *Cryogenics* **20**, pp. 611–624 (1980).
6. P. E. Richens, H. Jones, M. van Cleemput and D. P. Hampshire, Strain dependence of critical currents in commercial high temperature superconductors, *IEEE Trans. Appl. Supercond.* **7**, pp. 1315–1317 (1997).
7. N. Cheggour, L. F. Goodsir and J. W. Ekin, *ICSM 2008 Conf. (Side-Antalya, Turkey, Aug.)* (2008).
8. M. Rabara, N. Sekimura, H. Kitaguchi, P. Kováč, K. Demachi and K. Miya, Tensile properties and probability of filament fracture in Bi-2223 superconducting tapes, *Supercond. Sci. Technol.* **12**, pp. 1129–1133 (1999).
9. P. Kováč and P. Bukva, Differences in applied axial strain and I_c degradation of Bi(2223)/Ag tapes, *Supercond. Sci. Technol.* **14**, pp. L8–L11 (2001).
10. L. Kopera, P. Kováč and T. Melišek, Electromechanical characterization of selected superconductors, *Supercond. Sci. Technol.* **21**, p. 115001 (2008).
11. G. Frommeyer, Physical Metallurgy, Part 2 Ed R. W. Cahn and P. Haasen, Amsterdam, p. 1854 (1983).
12. M. Dhalle, H. van Weeren, S. Wessel, A. den Ouden, H. ten Kate, I. Hušek, P. Kováč, S. Schlachter and W. Goldacker, Scaling of the reversible strain response of MgB_2 conductors, *Supercond. Sci. Technol.* **18**, pp. S253–S260 (2005).
13. C. Buzea and T. Yamashita, *Supercond. Sci. Technol.* **14**, p. R115 (2001).
14. P. Kováč, T. Melišek, L. Kopera, I. Hušek, M. Polák and M. Kulich, Progress in electrical and mechanical properties of MgB_2 rectangular wires with different powders and metallic sheaths, *Supercond. Sci. Technol.* **22**, p. 075026 (2009).
15. P. Kováč, T. Melišek, M. Dhallé A. den Ouden and I. Hušek, Critical currents of axially stressed *in-situ* and *ex-situ* MgB_2 wires, *Supercond. Sci. Technol.* **18**, pp. 1374–1379 (2005).

16. P. Kováč, T. Melišek, L. Kopera, J. Kováč and I. Hušek, Selected properties of GlidCop® sheathed MgB_2 wires, *Sup. Sci. and Technology* **25**, p. 095008 (2012).
17. P. Kováč and L. Kopera, Electromechanical properties of filamentary MgB_2 wires, *Special Issue of IEEE Trans. on Appl. Superconductivity* **22**(1), p. 8400106 (2012).
18. P. Kováč, L. Kopera, T. Melišek, M. Rindfleisch, W. Haessler and I. Hušek, Behaviour of filamentary MgB_2 wires subjected to tensile stress at 4.2 K, *Sup. Sci. and Technology* **26**, p. 105028 (2013).
19. P. Kováč, I. Hušek, T. Melišek, L. Kopera and J. Kováč, Critical currents, I_c-anisotropy and resistance to stress of MgB_2 wires made by internal magnesium diffusion, *Sup. Sci. and Technology* **27**, p. 065003 (2014).
20. I. Hušek and P. Kováč, Mechanical properties, interface reactions and transport current densities of multi-core MgB_2/Ti/Cu/SS wire, *Sup. Sci. and Technology* **23**, p. 075012 (2010).
21. P. Kováč, L. Kopera, T. Melišek and I. Hušek, Electro-mechanical behaviour of in-situ W added MgB_2 wire, *Cryogenics* **60**, pp. 5–8 (2014).
22. P. Kováč, I. Hušek, L. Kopera and T. Melišek, Filamentary MgB_2 wires twisted before and after heat treatment, *Sup. Sci. and Technology* **24**, p. 115006 (2011).

Chapter 5d

Properties of Irradiated MgB_2, Bulk and Wires

Ilaria Pallecchi* and Marina Putti

CNR–SPIN and University of Genova, Dipartimento di Fisica
Via Dodecaneso 33, 16146 Genova, Italy
**ilaria.pallecchi@spin.cnr.it*

We review the effects of irradiation on the superconducting and normal state properties of MgB_2 polycrystals, which can be understood by taking into account the peculiar two-band, two-gap nature of superconductivity in this compound. In particular, we focus on fast neutron irradiation which is the most effective and controlled technique to induce defects in bulk materials. Both point defects (vacancies and interstitials) and nanometric size defects are produced, affecting progressively both reversible and irreversible superconducting properties. These effects can be evidenced at best if irradiation is carried out with increasing fluences on very clean pristine samples. It turns out that at low irradiation dose T_c is only weakly suppressed by disorder, while upper critical field and critical current density are significantly improved. This favorable low-disorder regime is relevant to the case of irradiation of superconducting magnets employed in fusion reactors and accelerators, which suggests limited detrimental effect of irradiation on this equipment.

1. Introduction

The effects of irradiation in superconductors have been explored since early times in view of identifying the mechanisms of improvement of in-field performances of superconductors induced by disorder. Indeed, point and extended defects may yield enhanced upper critical fields and pinning properties, thus improving both the value and the field dependence of the critical current density. Fast neutron irradiation under appropriate conditions appears to be the ideal tool to introduce controlled amounts of homogeneous disorder and to study its effects. In general, such investigation is not only targeted at the properties relevant for applications, but also to the understanding of fundamental mechanisms of superconductivity, closely related to the presence of disorder.

Remarkable signatures of two-band two-gap, superconductivity are peculiar characteristics of MgB_2, which can be observed thanks to the negligible interband scattering between σ band and π band carriers in clean samples. In this scenario, the effect of disorder on superconducting properties is multifold, in particular in terms of interband and intraband scattering rates.[1] It is theoretically predicted that interband scattering by non-magnetic impurities has a pair-breaking effect and cause a suppression of the critical temperature T_c down to its isotropic saturation value of 19–22 K,[2–4] corresponding to the point where the two gaps $\Delta_\sigma \sim 7\,\text{meV}$ and $\Delta_\pi \sim 2\,\text{meV}$, associated to σ and π carriers, respectively, merge to the BCS value. This prediction is at odds with single-gap superconductors in the strong scattering limit, for which Anderson's theorem predicts no pair-breaking effect of non-magnetic scattering, to the first order. On the other hand, intraband scattering affects the electrical resistivity, causes smearing of the density of states[5] and, in a two-band superconductor, is expected to cause a change in magnitude and shape of the upper critical field $H_{c2}(\text{T})$.[6,7] The latter can exhibit a very different behavior as expected from the BCS model, depending on the relative diffusivities of the two bands.

Thereby, the theoretical and experimental exploration of the effect of disorder on the reversible properties of MgB_2 is paramount, both for fundamental understanding of multiband superconductivity and for the achievement of the full application potential of this material. Indeed, atomic scale disorder is purposely introduced in superconductors to enhance H_{c2}, by decreasing the mean free path and consequently decreasing the coherence length. However, the concomitant decrease of T_c may counterbalance and eventually neutralize this H_{c2} enhancement. In parallel, disorder is introduced in order to tailor irreversible properties of MgB_2, such as magnetization, critical current density and irreversibility field. Nanometer scale disorder in the form of pinning centers, whose size match the coherence length, allows to improve the in-field behavior of the critical current density $J_c(H)$. This beneficial effect of disorder may be neutralized if the decrease of the condensation energy prevails, thus suppressing H_{c2} and T_c.

Atomic disorder can be introduced in controlled increasing amounts by chemical substitutions. In particular, substitution in either the Mg or B sites may selectively affect both the π and σ band scattering rates, thus allowing to tailor different properties. The most successful substitutions are carbon (C) (on the B site) and aluminium (Al) (on the magnesium (Mg) site), the former being particularly effective in enhancing H_{c2} and reducing the anisotropy.[8] However, chemical substitutions inevitably yield charge-doping, changes in the density of states at the Fermi level as well as possible changes in the Fermi surface topology, as in the case of Al-doping.[9] Moreover, structural distortions associated to chemical-doping may cooperate with increased scattering in affecting superconducting properties. In this respect, particle irradiation appears as a unique method to introduce controlled amounts of disorder and single out the role of disorder itself and associated interband and intraband scattering rates. In addition, the size of the damages induced by irradiation may span over different spatial scales, typically either atomic scale defects or nanometer scale defects, thus affecting both reversible and irreversible properties, respectively. Finally, it is noteworthy that such irradiation induced disorder can be largely recovered, as proper annealing protocols allow to heal irradiation induced damage and virtually restore T_c, resistivity and H_{c2} of pristine unirradiated samples, thus allowing deeper understanding of the irradiation mechanisms into play.[10]

Various experiments of MgB_2 irradiation with protons, neutrons, electrons, α-particles and heavy ions can be found in literature. In the present application oriented review, we mainly present and discuss data obtained in irradiated bulk and wire samples, but the existing data obtained in irradiated thin films and single crystals indicate fully consistent results, giving firm support to the overall picture. We briefly mention the peculiarities and the effects of different kinds of particle irradiation, but devote the main focus to the experimental results of neutron irradiation, which has been most extensively studied in view of the use of superconductors in magnets for fusion reactors and for accelerators. We first recall the theoretical predictions about the effect of disorder in suppressing superconducting properties, evidencing the peculiar case of two-band superconductors. Successively, we consider the morphology of irradiation induced defects and their effect on reversible and irreversible superconducting properties of MgB_2. Finally, we discuss and summarize the results and perspectives of irradiation in MgB_2, assessing positive and detrimental effects and limits, also in comparison with the effect of C-doping.

2. Theoretical Predictions for the Degradation of Superconducting Properties in a Two-band Superconductor

The dominant effect of scattering with impurities in MgB_2 is related to its two-band character. Four scattering rates come into play, namely two intraband scattering rates, $\Gamma_{\sigma\sigma}$ and $\Gamma_{\pi\pi}$, and two interband scattering rates, $\Gamma_{\sigma\pi}$ and $\Gamma_{\pi\sigma}$. Due to the different parity of σ and π orbitals, interband impurity scattering is exceptionally small in clean MgB_2, but disorder enhanced interband scattering has a pair-breaking effect, akin to magnetic impurity scattering in single band superconductors.[1] In the strong interband scattering limit, a complete isotropization of all Fermi surface sheets takes place. This limit is achieved when the interband scattering rate becomes larger than the relevant phonon frequency ($\omega_0 \sim 75\,\mathrm{meV}$ in the case of the E_{2g} mode). Within this limit the critical temperature is predicted to saturate to the isotropic coupling value around 19–22 K,[2–4] and correspondingly it is expected that the smaller gap Δ_π increases while the larger gap Δ_σ decreases, thus eventually merging at the BCS gap value.[11] Details of the models and different assumed values for coupling and pseudopotential matrices as well as for phonon frequencies yield a certain variability about the interband scattering rate Γ at which $T_c(\Gamma)$ saturates, typically several tens meV.[12]

It must be mentioned that the most celebrated model considering electron lifetime effects on reversible superconducting and normal-state properties of materials is the one proposed by Testardi and Mattheiss.[13] This model assumes that the defects just broaden the density of states (DOS) by mixing of electron states in an energy range comparable to the electron relaxation rate Γ. This mechanism well accounts for the T_c degradation disordered in such superconductors as A15 compounds, whose DOS is sharply peaked at the Fermi level. On the contrary, in MgB_2 the DOS is quite flat and featureless at the Fermi level, indicating that this mechanism is certainly present, but less prominent. Yet, experimental data about nuclear magnetic resonance[14] and Sommerfeld constant[15] indicate a conspicuous reduction of the DOS in strongly irradiated MgB_2. In Ref. 5 the effect of broadening is considered separately on the partial DOS of σ and π bands and used to scale the various properties.

3. Mechanisms for the Formation of Defects and Defect Morphology

3.1 *Neutron irradiation*

The interaction reactions of neutron with the matter and the corresponding cross-sections are strongly dependent on the neutron energy, hence fast and thermal neutrons induce different types of defects in MgB_2. Fast neutrons have a sufficient energy to undergo direct inelastic collisions with nuclei and displace them from their lattice positions, thus producing point defects. However, in MgB_2 the cross-section for such events is smaller than in the case of thermal neutrons.[16] Indeed, thermal neutrons, having lower energy, introduce disorder through neutron capture reaction by ^{10}B, which has an extraordinarily large cross-section. This process causes the isotropic emission of a α particle and a ^7Li nucleus, with energy 1.47 MeV and 0.84 MeV, respectively. Along the range of recoil (4.8 μm and 2.1 μm, respectively) these particles lose energy by interacting with electrons and create atom displacements, mainly at the end of the range. These considerations indicate that both sample size and ^{10}B content are crucial in determining the effectiveness and homogeneity of irradiation. Indeed, the large cross-section of the capture reaction of thermal neutrons in MgB_2 prepared with natural B (with 20% ^{10}B content) corresponds to a penetration depth of 200 μm. This causes a sort of shielding effect at the surface, which results in a much disordered region close to the surface and thus to a gradient in defects density throughout the sample. To overcome this hindrance, different strategies are adopted: (i) choice of sample thickness smaller than the penetration depth,[10,17] (ii) use of a cadmium shield during exposure to neutron radiation to absorb the lowest energy (<0.5 eV) neutrons,[16,18] (iii) use of isotopically enriched ^{11}B to prepare MgB_2, as first suggested in Ref. 19. In this latter case, with only 0.5% ^{10}B content, the penetration depth grows up to $\sim$1 cm, becoming larger than typical MgB_2 bulk sample and wire thickness.

It is believed that defects of nm size are created by the nuclear reaction products at the end of their recoil range, whereas point defects are created along the range of the reaction products. Transmission electron microscopy (TEM) investigations reveal indeed large radiation-induced defects. The first image of a typical defect is reported in the dark-field image of an irradiated MgB_2 single crystal (thermal neutron fluence 5×10^{21} m^{-2}),[20] where a defect structure is identified as a contrast due to the strain field associated with locally bent lattice planes. The size of the contrast is about 5 nm, which is indicative of the size of the defect. Another low-magnification TEM image of a MgB_2 polycrystal irradiated at a much higher fluence (thermal neutron fluence 1.4×10^{24} m^{-2}),[21] displayed in Fig. 1, shows homogeneously distributed light zones of size $\sim$3–4 nm. The density of such light zones scales with the neutron dose in different samples. Moreover defocusing observed around these zones results from nanometric amorphous regions, where diffracting electrons are not coherently scattered by lattice planes. These observations clearly indicate that such features are originated by irradiation induced defects.

It is worth noting that, being a ^7Li nucleus a product of the neutron capture reaction by ^{10}B, Li-doping should be in principle taken into account in irradiated samples. However, it can be reckoned

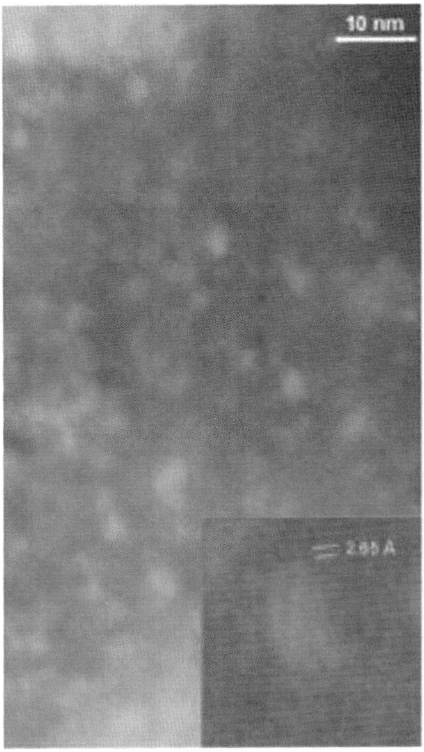

Fig. 1. TEM image of an irradiated MgB_2 polycrystal characterized by the homogeneous distribution of the nanometric defects. The inset is an enlarged view showing one defective zone where lattice fringes lose coherence due to the presence of an amorphous region. (From Ref. 21, © 2008 by Institute of Physics).

that Li-doping may be at most a few percent at the highest irradiation levels, hence its effect is expected to be negligible.[10]

3.2 *Proton irradiation*

Ion irradiation is a reproducible process inducing crystalline disorder. Energetic ions displace atoms from their equilibrium lattice sites, creating a variety of defects, including vacancies and interstitials. Such defects tend to depress the superconducting order parameter locally, and thereby create pinning sites for the vortices. 2 MeV protons can penetrate 40–50 μm into MgB_2.

The probability of displacement per atom (d.p.a.) can be simulated[22] and the proton irradiation parameters can be tailored so as to generate the desired defect density and as uniform a profile as possible through the sample depth. In particular, a monoenergetic proton beam causes an almost flat distribution of damage for most of its range, and then a sharp peak (the Bragg peak) towards the end of the track. Thereby, in order to create fairly uniform damage throughout the depth of a sample, multiple consecutive implants with increasing values of beam energy can be carried out.[22]

3.3 *Ion irradiation*

Since the earliest days of research ion irradiation has been performed and studied, using helium,[23,24] oxygen[25] or heavy ion beams such as silver,[26,27] silicon,[28] gold,[27–29] lead[30] and uranium.[31]

Helium ion (α-particle) irradiation generates predominantly point defects. In Refs. 23 and 24, α-particle irradiation of MgB_2 films is carried out at room temperature using a 2 MeV $^4He^{++}$ ion beam from a Tandem accelerator. Interestingly, this 2 MeV energy is close to the 1.7 MeV energy of the α-particle emitted in the neutron capture reaction by ^{10}B after thermal neutron irradiation, and indeed, considering equivalent particle fluences, α-particle and neutron irradiations have very similar effects on T_c and resistivity of MgB_2.[32] Due to the small penetration depth of such 2 MeV $^4He^{++}$ ions, this irradiation method can be applied only to samples, typically films, whose thickness is thin enough (tens of nm), thus allowing the ions to induce a uniform damage through the entire film cross-section. A total dose of $1.3 \times 10^{21}\ m^{-2}$ was achieved in Refs. 23, 24 after 11 irradiation steps, taking a total time of $\sim$35 h.

The small penetration depth is also an issue in case of oxygen irradiation. With 300 keV O_2^+ ion irradiation,[25] the effect on structural and physical properties such as c-axis, T_c and resistivity of MgB_2 films is very similar to the case of α-particle irradiation, suggesting similar morphology of defects. In this case, of relatively low energy oxygen irradiation, however, oxygen can be doped into MgB_2 grains and form MgO_x or other non-superconducting precipitates, adding further effects.

Heavy ion irradiation is carried out in thin samples as well, due to the limited penetration depth. Heavy ions irradiations are performed with energies of the order of 0.1–1 GeV[26,27,29,31] and create amorphous columnar tracks which in principle could behave as effective pinning centers.[29] However, extended defects are also generated and often result in a degradation of intergrain connectivity in MgB_2, causing only little changes in the superconducting properties.

3.4 *Other irradiation techniques*

In contrast to the nanometer scale defects induced by neutron irradiation, low temperature (20 K) electron irradiation induces only isolated point defects through the formation of Frenkel pairs, as the cryogenic irradiation temperature prevents defect clustering and recombination. Such defects are effective for single vortex pinning mechanism.[33] However, in bulk samples, grain boundaries tend to be heavily damaged by electron irradiation, resulting in degraded global properties.[34]

Finally, γ-rays are also used for MgB_2 irradiation. These methods drives migration of impurity phases and segregation of impurities at the grain boundaries,[35] however intrinsic properties are also likely affected.[36]

4. Effect of Neutron Irradiation on Lattice Parameters, T_c, ΔT_c, Resistivity

In Ref. 37, a series of ^{11}B enriched polycrystalline MgB_2 samples were irradiated with neutrons under different conditions. It is found that the main damages are caused by thermal neutrons, whereas fast neutrons play a minor role. The variation of the lattice parameters, the transition temperature T_c, the transition width ΔT_c and of resistivity curves are studied as a function of the increasing thermal neutron fluence from $1.0 \cdot 10^{21}\,m^{-2}$ up to $1.4 \cdot 10^{24}\,m^{-2}$. The highest fluence corresponds to an exposition time of nearly four months. Another extensive and systematic study of the effects of irradiation with fluences up to $1.9 \cdot 10^{23}\,m^{-2}$ is carried out in Ref. 10 on 140 μm diameter MgB_2 wire segments, obtained by reacting boron (B) filaments with Mg vapor at 950°C.

In the polycrystals of Ref. 37, the X-rays diffraction peaks remain sharp even at the largest fluences and the lattice parameters show a gradual change with increasing irradiation. At the largest fluence $1.4 \cdot 10^{24}\,\mathrm{m}^{-2}$ the a-axis expands by less than 0.4%, whereas the c-axis increases by more than 1%, corresponding to a unit cell volume change of 1.7%. These results are consistent with the anisotropic expansion measured in the irradiated polycrystalline samples Ref. 17, if the same fluences are compared. Somewhat differently, in Ref. 10 a volume expansion as large as 2.6% is already measured at a fluence of $9.5 \cdot 10^{22}\,\mathrm{m}^{-2}$ and a significant (002) peak broadening is observed for the highest fluence ($>10^{23}\,\mathrm{m}^{-2}$), thus indicating strong and anisotropic structural deterioration.

In Fig. 2, the critical temperature T_c (Fig. 2(a)), defined at 50% of the normal state resistivity transition and the residual resistivity ρ at 40 K (Fig. 2(b)) of a series of samples are plotted as a function of thermal neutron fluence, for polycrystals (Refs. 37 and 38) and single crystals (Ref. 16). The critical temperature decreases

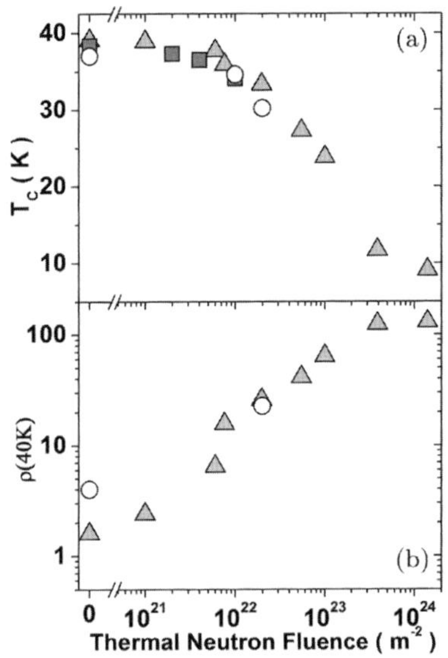

Fig. 2. (a) Critical temperature T_c and (b) resistivity at 40 K, $\rho(40\,\mathrm{K})$, measured in irradiated polycrystals (from data of Tarantini *et al.*[37] and Wang *et al.*[38]) and single crystals (from data of Eisterer[16]) as a function of thermal neutron fluence.

monotonically with neutron fluence; it is suppressed by 2 K only up to $10^{22}\,m^{-2}$ and then reaches 9.1 K at the highest fluence. At the same time resistivity increases by more than two orders of magnitude, from ~1 to 130 $\mu\Omega\cdot$cm. In order to evaluate the intrinsic increase of resistivity, which must be distinguished from the possible loss of intergrain connectivity in highly irradiated polycrystalline samples, the as-measured ρ data must be rescaled according to Rowell criterion, which adjusts $\Delta\rho = \rho(300\,K) - \rho(40\,K)$ to the intrinsic value determined by the π band electron-phonon coupling.[12,39] This scaling procedure is based on the assumption that the temperature dependent transport is always dominated by the π band, which exhibits a stronger phonon coupling. In the sample series of Ref. 37, no systematic increase with irradiation of the temperature dependent part of the resistivity curves $\Delta\rho = \rho(300\,K) - \rho(40\,K)$ is observed, indicating that the produced damage does not severely affect the connectivity of the samples.

The transition widths of these samples remain rather sharp even for the highest fluences, reflecting the homogeneity of defect distribution. As for the wires of Ref. 10, a steady T_c decrease to values below 5 K is measured, that is considerably larger than reported in Ref. 37 at equal fluence. The transition widths remain rather sharp upon irradiation. However, in contrast with Ref. 37, the temperature dependent part of the resistivity curves $\Delta\rho = \rho(300\,K) - \rho(40\,K)$ increases monotonically with increasing irradiation. The simplest explanation for this trend is the increasing deterioration of intergrain connectivity, however other possibilities are (i) switch from π to σ band dominated transport as the system is driven into the dirty limit and (ii) anisotropic irradiation damage, which affects predominantly the *c*-axis direction, yielding and increasing the number of grains which present a high resistivity path to electron transport.

In order to extract information from the above lattice parameters, T_c and resistivity data, the correlations among these properties must be evidenced. In Fig. 3(a), T_c is plotted as a function of the *c*-axis lattice parameter, using the data of the irradiated polycrystals of Ref. 37, showing a decreasing T_c with increasing

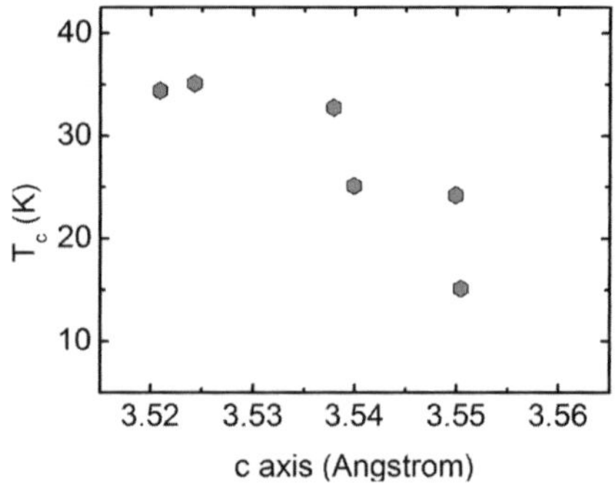

Fig. 3(a). T_c as a function of c-axis parameter, with data from polycrystals of Ref. 37, one polycrystal of Ref. 17 and wires of Ref. 10.

c-axis. Experiments under hydrostatic pressure[40] and on thin films with tensile strain[41] demonstrate that an increase of T_c with the cell expansion is expected, as a consequence of the stiffening of the phonon modes. Hence, the observed trend indicates that disorder (increase of scattering) is more effective in affecting T_c than the changes of the cell parameters (changes in the phonon modes).

The data of T_c as a function of residuals resistivity $\rho_0 \approx \rho(40\,\mathrm{K})$, shown in Fig. 3(b)), are rich in physical information, too. The clear relationship between T_c and ρ_0 indicates that a strong correlation exists between the mechanisms leading to an increase of resistivity (scattering with atomic-scale defects) and the ones reducing T_c. Two sets of data are shown, comparing the as-measured ρ data with the ρ data rescaled by the Rowell criterion.[12,39] The rescaled resistivity is more reliable, T_c versus ρ data of different series of samples merging in a common curve, as shown in Refs. 12 and 37, finally yielding the value $\rho_{\mathrm{rescaled}} \sim 80\,\mu\Omega \cdot \mathrm{cm}$ as the resistivity at which T_c is mostly suppressed. Moreover, Rowell rescaling allows a direct comparison between the dependence of T_c on the scattering rate Γ and theoretical predictions. A linear decrease of T_c versus ρ_{rescaled} is seen, the slope of T_c versus resistivity being weaker in the low- T_c high-/ ρ regime.

This linear correlation between T_c and the rescaled resistivity is universally common to MgB_2 samples of different kinds and

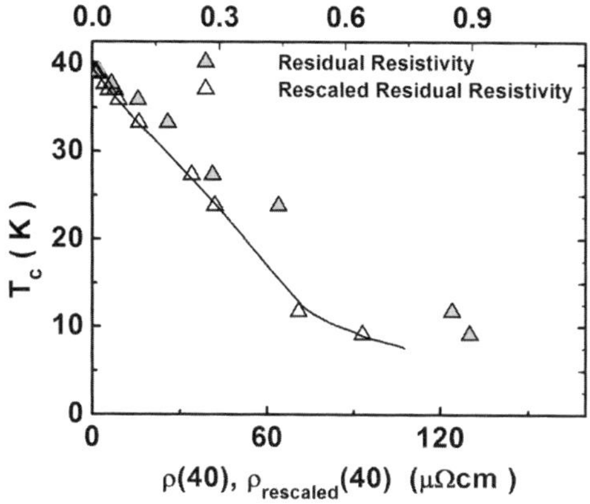

Fig. 3(b). T_c as a function of residual resistivity (filled symbols) and versus residual resistivity corrected by Rowell's criterion (open symbols), with data from Ref. 37 (triangles) and Ref. 38 (circles). In the top axis, the resistivity is converted into scattering rate Γ using the relationship $\Gamma = \rho_0 \varepsilon_0 q \Omega_p^2/\hbar$, where $\Omega_p \approx 7\,\text{eV}$ is the average plasma frequency of MgB_2 (Ref. 2), ε_0 the vacuum dielectric constant, q the electron charge and $\hbar = h/2\pi$ with h the Plank constant. The T_c versus Γ dependence calculated by taking into account both interband and intraband scattering rates (Refs. 5, 12) is shown as a continuous line.

levels of disorder,[12,37] although being in contrast with the two-gap theory of superconductivity which predicts a saturation of T_c around 20 K.[2–4] Indeed, the observed T_c versus ρ_0 dependence can be rationalised by assuming that whereas in the low scattering regime the linear behavior is described by two-band models predicting a pair-breaking effect of the interband scattering, at high scattering rates, the smearing of the DOS due to intraband scattering comes into play and causes the suppression of T_c. Indeed, at high levels of disorder it can be assumed that the effect due to interband scattering is saturated ($\Gamma_{\sigma\pi}$, $\Gamma_{\pi\sigma} > \omega_0$), intraband events becoming the dominating scattering mechanisms. Using the calculated DOS $N_\sigma(\Gamma)$ and $N_\pi(\Gamma)$, smeared by disorder, the critical temperature of irradiated samples can be evaluated by taking into account both interband and intraband electron lifetime effects in different scattering regimes.[5] This framework rationalizes the trend of T_c with ρ_{rescaled} in the whole T_c range, as seen by the agreement of experimental and calculated trends (Fig. 3(b)).

5. Effect of Neutron Irradiation on the Superconducting Gaps

The behavior of the two superconducting gaps is strictly related to the issue of the lowering of T_c by irradiation. Direct measurements of the superconducting gaps in neutron irradiated samples are obtained by specific heat,[15,38] by point contact spectroscopy[42] in polycrystalline bulk samples and by scanning tunnelling spectroscopy[43] in thin films and single crystals.

The electronic specific heat in the superconducting state c_{sc}, evaluated by subtracting the phonon contribution from the experimental data, exhibits evident gap signatures. In particular, the normalized superconducting specific heat $c_{sc}/\gamma T$ versus the reduced temperature $t = T/T_c$ of a two-gap MgB_2 sample presents an excess of specific heat at $t \sim 0.2$, which is a signature of the smaller gap $\Delta\pi$, while the specific heat jump at $t \sim 1$ is a signature of the larger gap Δ_σ. In highly irradiated samples, where T_c is lowered to 11 K, the excess specific heat at $t \sim 0.2$ is not visible anymore and the data can be fitted in a single gap framework, with a lower number of parameters.[15,38] This is seen as a clear indication of the merging of the gaps at high levels of disorder. Indeed, literature data of superconducting gaps in disordered samples, obtained by different techniques and collected in Refs. 12, 15 and 44, are all consistent with a scenario where the larger gap Δ_σ decreases monotonically from 6–7 meV for $T_c \approx 39$ K, to ~1 meV for $T_c \approx 10$ K, while the smaller gap Δ_π varies from ~2 meV for $T_c \approx 39$ K, to ~1 meV for $T_c \approx 10$ K, with a slightly non monotonic behavior, due to the competition between the increase of Δ_π driven by enhanced interband scattering and the decrease of Δ_π driven by the suppression of the DOS and the consequent suppression of the electron–phonon scattering by disorder. This phenomenology slightly departs from theoretical predictions that the smaller gap should increase and merge with the larger gap at the isotropic BCS value.[11] However, most superconductors exhibit smaller gaps than predicted by BCS at high irradiation levels.[12]

6. Effect of Neutron Irradiation on H_{c2}

The upper critical field H_{c2} is a thermodynamic property which depends on the details of electronic structure and on the coupling strength. In the dirty limit, it is strongly affected by defects, which reduce the electron mean free path and, consequently, the coherence length, yielding an increase in H_{c2}. A theoretical model based on Usadel equations for a dirty two-gap anisotropic SC was developed to account for $H_{c2}(T)$ in different scattering regimes.[6] According to this model, interband scattering should be the main cause for the decrease of T_c, whereas intraband scattering, and thus the ratio of the two-band diffusivities $\eta = D_\pi/D_\sigma$, affects the shape of H_{c2} versus T curves in different temperature ranges. If $\eta > 1$, the critical field shows a slight upward curvature near T_c, while at 0 K it exceeds the predicted BCS value; conversely, if $\eta < 1$, H_{c2} remains nearly linear in the high temperature region and presents a remarkable upward curvature at low temperature.

In Fig. 4, the $H_{c2}(T)$ curves of the irradiated polycrystals of Ref. 37 are reported. It must be remarked that in polycrystalline

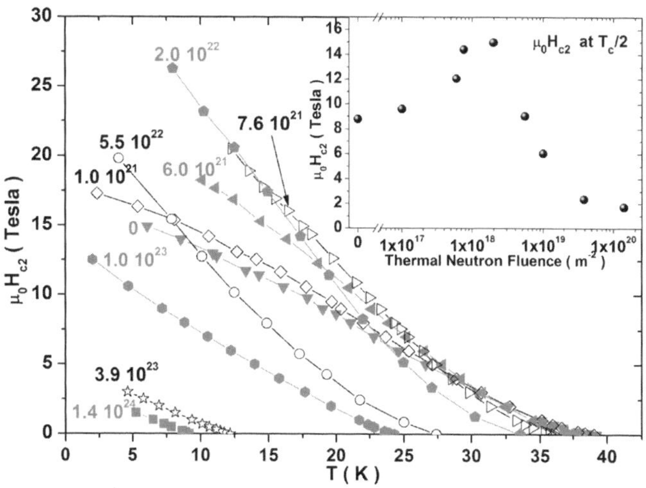

Fig. 4. Upper critical field as a function of temperature estimated at 90% of the resistive transition, with data from Ref. 37. The labels indicate irradiation fluence in m^{-2}. The inset shows critical field behavior versus fluence at $T_c/2$.

samples, with randomly oriented grains, the measured H_{c2} coincides with the largest H_{c2} value, i.e., the critical field parallel to the ab planes. The curves are gradually modified by increasing neutron irradiation and eventually assume a completely different shape. Initially, at low temperature, a remarkable downward curvature is observed. Upon irradiation, the slope increases and the behavior becomes almost linear in the sample irradiated at fluence $2.0 \cdot 10^{22}\,\text{m}^{-2}$, while $H_{c2}(0)$ attains a value larger than 30 T, twice as much as that of the pristine sample. With further irradiation, the curves show an upward curvature at low temperature, particularly evident in the $5.5 \cdot 10^{22}$ and $1.0 \cdot 10^{23}\,\text{m}^{-2}$ samples. The $H_{c2}(\text{T})$ curves of $3.9 \cdot 10^{23}$ and $1.4 \cdot 10^{24}\,\text{m}^{-2}$ samples are linear in the measured temperature range. In the inset, the critical field at $T_c/2$ versus fluence is plotted. H_{c2} increases up to a fluence level of $2.0 \cdot 10^{22}\,\text{m}^{-2}$, while the critical temperature decreases by only a few degrees; instead, in the most irradiated samples, H_{c2} is suddenly reduced both as a consequence of a strong T_c decrease and of the anisotropy reduction (anisotropy reduction is observed in irradiated single crystals[16] and is also extracted from the analysis of the critical current density in the same samples, as described in the next section).

It is remarkable that this is a general trend encountered on all kinds of irradiated samples whose upper critical field presents a maximum, larger than 35 T, when the critical temperature is in the range 28–35 K[12,37] (this is the case for neutron irradiated MgB_2 bulk samples and films, α particle irradiated MgB_2 films, as well as irradiated low T_c superconductors, provided that the temperature is rescaled by the respective T_c). The behavior of H_{c2} as a function of T_c in irradiated samples shown in the inset of Fig. 4 is well reproduced by a generalization of the Testardi–Mattheiss model[13] to the multiband case, where both interband and intraband scattering with impurities are taken into account.[45] On the other hand, the model by Gurevich[6] well accounts for the magnitude and shape of all the $H_{c2}(\text{T})$ curves in the main panel of Fig. 4, except for the highest fluences. It turns out that $\eta = D_\pi/D_\sigma^{ab}$ (D_σ^{ab} diffusivity of the σ band along the ab planes) is larger than unity and monotonically decreases with increasing irradiation, indicating that the σ band is always dirtier than the π band, even if the π band gets dirty faster than the σ band. In the case of the most irradiated samples, the failure of the model may be due to the decrease of the DOS and the electron–phonon coupling.[15] In this case, the critical temperature is low and no upper curvature is observed near T_c, as expected from a WHH[46] single gap behavior. Indeed, the presence of a single gap is supported by specific heat results in the same samples,[15] whose T_c is below 21 K.

7. Effect of Neutron Irradiation on J_c

We now address the effect of neutron irradiation on irreversible properties, in particular on the critical current density J_c. J_c and its field dependence in disordered samples can be improved either via the enhancement of H_{c2} as a consequence of the reduction of the mean free path, or via the introduction of pinning centers. In the former case, atomic scale disorder plays a role, whereas in the latter case nanometer size defects — matching the coherence length — are mostly effective. As neutron irradiation produces both atomic scale and nanometer scale defects, its effect on the $J_c(H)$ curves is particularly interesting.

Figure 5 shows the effect of neutron irradiation on J_c of a polycrystal from Ref. 37, another polycrystal form Ref. 16 and two wires sheathed with Fe and Cu, respectively, from Ref. 16. In Ref. 16, it is concluded that irradiation enhances J_c of all the three samples, especially at high-fields, however inspection of the $H_{c2}(T)$ dependence and fitting of the $J_c(H)$ curves of these samples reveals that irradiation induced defects enhance H_{c2} and reduce anisotropy, but that their contribution to flux pinning is small compared to that of the as-grown defect structure, e.g., grain boundary pinning. On the other hand, it is undeniable that the $J_c(H)$ dependence of the polycrystal from Ref. 37 in Fig. 5 is significantly affected by irradiation. Indeed, in order to get a more quantitative exploration of the irradiation mechanisms, it is necessary to start from samples with low J_c and H_{c2}, namely pristine bulks that are very pure and have very clean grain boundaries, as shown in the following.

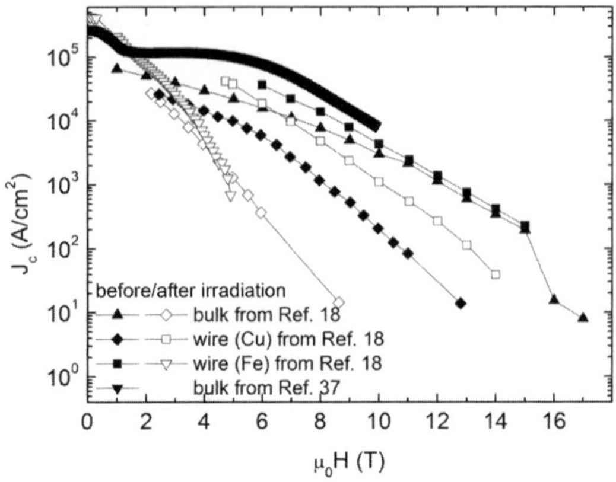

Fig. 5. Critical current density as a function of magnetic field, measured before (open symbols) and after (filled symbols) neutron irradiation under different conditions. Data are extracted from magnetic measurements on a polycrystal from Ref. 37, from magnetic measurements on a polycrystal from Ref. 16 and from transport measurements on Cu- and Fe-sheathed wires from Ref. 16.

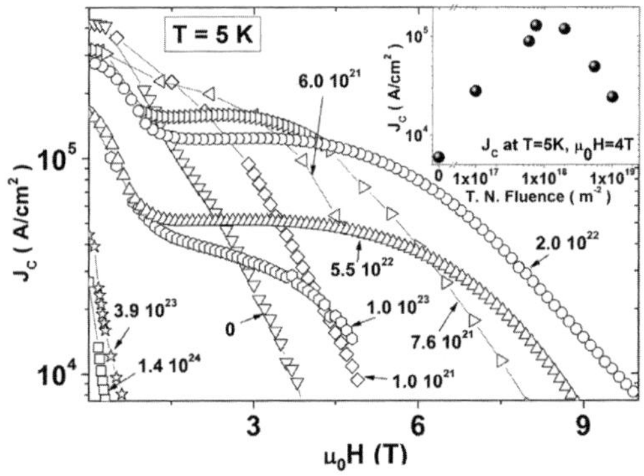

Fig. 6(a). Critical current density J_c at 5 K as a function of applied magnetic field. The labels indicate irradiation fluence in m^{-2}. Inset: J_c as a function of fluence at 4 T.

In Fig. 6, top panel, the $J_c(H)$ curves of the irradiated polycrystals of Ref. 37, extracted from magnetic hysteresis loops using a critical state model, are presented. Clearly, irradiation, despite suppressing T_c, significantly improves the magnetic field dependence of the critical current density. In the inset, J_c is plotted as a function

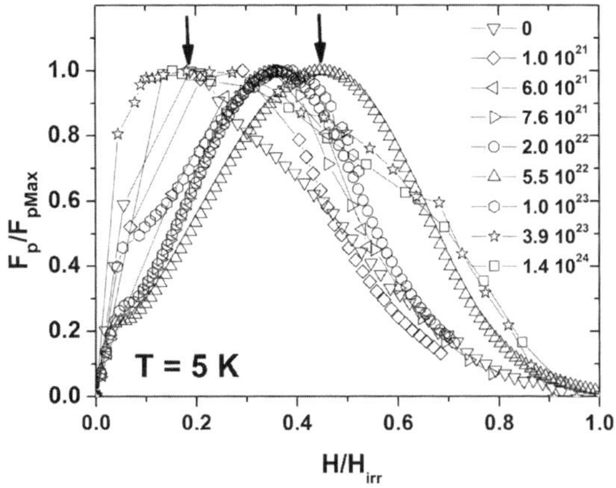

Fig. 6(b). Normalized pinning force $F_p/F_{p\mathrm{Max}}$ as a function of reduced field H/H_{irr} at 5 K. The arrows indicate the position of the maximum for the unirradiated sample and for the sample irradiated with a fluence of $5.5 \cdot 10^{22}\ \mathrm{m}^{-2}$, showing a shift towards higher H/H_{irr}.

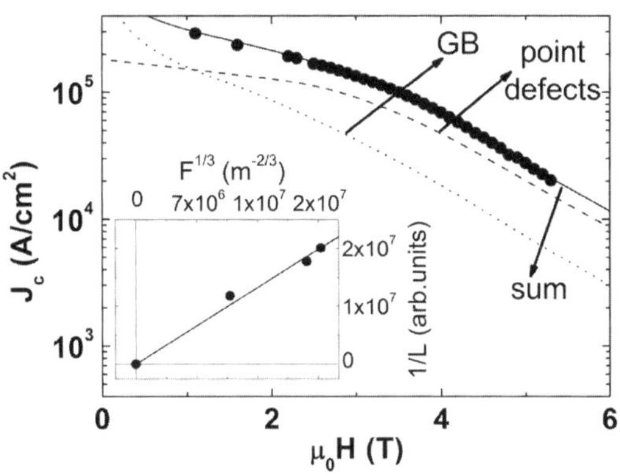

Fig. 6(c). J_c values of the sample irradiated at $6.0 \cdot 10^{21}\ \mathrm{m}^{-2}$ and separate contributions to the fitting curves by grain boundary and by point defect pinning. Inset: inverse average distance between point pinning centers in arbitrary units obtained by the point pinning contribution in the fits of the critical current density, plotted as a function of the 1/3 power of the fluence. The proportionality is evidenced. All data are taken from Refs. 37 and 49.

of the neutron fluence at 5 K and 4 T. Similarly to H_{c2}, the optimal in-field behavior is obtained for the sample irradiated with fluence around $2.0 \cdot 10^{22}\ \mathrm{m}^{-2}$, while J_c reaches $10^4\ \mathrm{A/cm}^2$ at 10 T. However, the J_c improvement cannot be entirely attributed to the change in H_{c2}. Indeed, at low irradiation levels ($10^{21}\,\mathrm{m}^{-2}$), H_{c2} does not change appreciably, whereas the field dependence of J_c is improved progressively by irradiation. This evidence that irradiation induces new pinning mechanisms is even more apparent by plotting the normalized pinning forces, i.e., the product $J_c \cdot H$, as a function of the reduced field H/H_{irr}, where H_{irr} is the irreversibility field defined as the field at which the linear extrapolation of Kramer plots of $J_c^{1/2}H^{1/4}$ versus H crosses the horizontal axis.[47] Such pinning force plot obtained from J_c data at $T = 5\,\mathrm{K}$ is shown in the middle panel of Fig. 6. It can be seen that for the unirradiated sample the maximum of the pinning force lies at $H/H_{irr} \sim 0.2$, the characteristic value of the grain boundary pinning, while with increasing irradiation it shifts to higher values of H/H_{irr}. For a fluence of $5.5 \cdot 10^{22}\,\mathrm{m}^{-2}$ the pinning force maximum occurs at $H/H_{irr} \sim 0.45$, an even higher H/H_{irr} value than expected for point pinning at $H/H_{irr} \sim 0.33$.

Clearly, new pinning mechanisms are introduced by irradiation. With further irradiation beyond $5.5 \cdot 10^{22}\,\mathrm{m}^{-2}$, the pinning force maximum shifts back to the grain boundary value. It is possible to investigate the pinning mechanisms more quantitatively by fitting the $J_c(H)$ curves using the Eisterer percolative model,[48] as done in Ref. 49. Therein, it is reasonably assumed that in the pristine sample the dominant pinning mechanism is by grain boundaries and that this contribution is unaffected by irradiation, so that it should rescale with the condensation energy of the sample $E_c = \mu_0 B_c^2(T) \approx \mu_0 B_c^2(0)(1-(T/T_c)^2)^2 \propto T_c^2(1-(T/T_c)^2)^2$, where $T = 5\,\mathrm{K}$ for the curves of Fig. 6 and T_c varies from sample to sample, decreasing with the irradiation fluence. It turns out that the rescaled grain boundary pinning contribution lies well below the experimental value of J_c for the samples irradiated with fluence lower than $10^{22}\,\mathrm{m}^{-2}$, that is, when the critical temperature is not lowered by more than 10%. The total current can thus be calculated as the sum of the rescaled grain boundary contribution plus a point defect contribution, as shown in the bottom panel of Fig. 6 for the sample irradiated at $6.0 \cdot 10^{21}\,\mathrm{m}^{-2}$. In this analysis, only two free parameters are varied: the point defects pinning multiplicative coefficient A_p and the anisotropy γ. It is found that γ decreases with increasing fluence, as seen on irradiated single crystals,[16] with values around 4–4.5 for samples irradiated at fluences up to $10^{22}\,\mathrm{m}^{-2}$ and around 1 for larger fluences. Even more interestingly, as shown in the inset of the bottom panel of Fig. 6, the coefficient A_p increases proportionally to the 1/3 power of the fluence. This is a check of the validity of this description, as A_p is indeed linearly related to the inverse of the average distance between point defect pinning centers induced by irradiation, which in turns is proportional to the 1/3 power of the fluence. This analysis fails to work for samples irradiated at fluences larger than $10^{22}\,\mathrm{m}^{-2}$, where the rescaled grain boundary contribution is even larger than the experimental J_c data points. Possibly, the crossover from two-band to single band character, as well as the suppression of the DOS play a role. It must be noted that for the samples irradiated at fluences in the range $7.6 \cdot 10^{21}$–$5.5 \cdot 10^{22}\,\mathrm{m}^{-2}$, a smooth plateau/upraise is well visible in $J_c(H)$ curves, rather than a monotonic decrease of J_c versus H. A similar, but more evident, upraise is observed on irradiated single crystals and described as fishtail effect.[20] The peak position is shifted to lower reduced magnetic fields with increasing disorder and it is thus attributed to an order-disorder transition of the flux line lattice.

It is worth mentioning here also some J_c results obtained in neutron irradiated MgB_2 thin films,[50] which confirm the phenomenology observed in bulk samples. The $J_c(H = 0) \sim 10^7\,\mathrm{A/cm^2}$ of unirradiated films is slightly reduced by neutron irradiation, however the in-field behavior exhibits a sizeable improvement after irradiation. In particular, after an intermediate fluence of $\sim 7.6 \cdot 10^{21}\,\mathrm{m}^{-2}$, J_c reaches $1.8 \cdot 10^5\,\mathrm{A/cm^2}$ at $\mu_0 H = 4\,\mathrm{T}$ and $1.1 \cdot 10^4\,\mathrm{A/cm^2}$ at $\mu_0 H = 9\,\mathrm{T}$. A fitting of the pinning force curves evidences an excess critical current at intermediate fields in irradiated films, which indicates the introduction of point-like pinning centers.

8. Effect of Annealing in Neutron Irradiated Samples

In Ref. 10, the effect of annealing at different temperatures up to 500°C and for different times up to 1000 h is systematically studied in neutron irradiated wires. Such study allows to get information on the nature of irradiated induced defects and their characteristic activation energy. It is found that at 500°C, the highest annealing temperature, the samples recover their pristine values of T_c, normal state resistivity and lattice parameters. The upper critical field is also recovered and even slightly improved, growing from 16 T of the undamaged sample to about 18 T. The $H_{c2}(T_c)$ shape slightly changes after annealing at 500°C, showing a more marked upward curvature near T_c and a more linear trend at low temperature. On the contrary, if annealing is carried out in isochronal (24 h) steps at increasingly larger temperature values from 150 to 500°C, corresponding to progressing recovering of T_c from 5 K to the pristine value 37.5 K, much smaller $H_{c2}(0)$ values than those shown in Fig. 4 are found. This suggests that annealing at lower temperature does not heal all irradiation damages. Indeed, irradiation induces different types of defects, such as B and Mg displacements and antisite defects. Annealing performed at a given temperature removes only certain defect types with relatively small activation energy, e.g., the healing of point defects is expected to have a lower activation energy than the dissolving of the defect complexes.

Hence, annealing may remove certain defects that are responsible for the increase of H_{c2}, but not for the suppression of T_c. As for the critical current density, upon 500°C anneal, $J_c(H = 0)$ and $J_c(H)$ exceed those of the unirradiated sample. A characteristic fishtail effect suggests that the remaining defects after post-exposure annealing act as effective pinning values with increasing irradiation. sites. Given the complete recovery of reversible properties, it is plausible that annealing heals atomic scale defects but not nanometer scale defects that act as pinning centers.

Considering T_c as a measure for defect concentration, the activation energy of defects can be extracted by annealing experiments at different times. Activation energies ranging from 1.07 to 2.15 eV are found, possibly corresponding to point defects and defect complexes. Interestingly, the recovery of reversible properties is progressive and monotonic as a function of increasing annealing temperature, with the exception of low temperature annealing at 200°C, which yields an enhancement rather than a decrease of the normal state resistivity. This may result from a change in the defect structure. For example, with increasing annealing temperature, vacancies do not necessarily immediately recombine with interstitials to eliminate defects. Vacancies can initially cluster together to form higher order defects, which can be relatively stable. Thus, the increase in resistivity after annealing at 200°C may result from a reorganization of defects into higher order clusters, which may enhance scattering.

9. Comparison Between the Effects of Neutron Irradiation and of Other Irradiation Techniques

The effect of α particle irradiation is very similar to that of thermal neutrons, which is plausible by considering that α particles are a product of the neutron capture reaction by ^{10}B. In an experiment carried out on MgB_2 thin films,[23] T_c is gradually suppressed down to 7 K and the resistivity increased with increasing fluence, with no effect on intergrain connectivity. Also the behavior of H_{c2}, reaching its maximum for $T_c \approx 35$ K, is very similar to the case of neutron irradiation.

Oxygen ion irradiation of MgB_2 thin films yields very similar gradual T_c decrease and an increase of ρ to α particle irradiation.[25] At moderate radiation levels of $10^{17}\,m^{-2}$, J_c is enhanced and its performance in magnetic fields is significantly improved, as a consequence of both the increase of the upper critical field[51] and the introduction of point-like pinning centers.

At the Bragg peak region (see d.p.a. versus penetration depth in Ref. 18), protons produce inhomogeneous damage. Uniform damage in a proton irradiated sample is only obtained in the steady loss region, i.e., when the proton energies are large enough that the Bragg peak at the end of proton trajectory occurs far outside the sample. This is certainly not the case in Bugoslavsky's experiment on MgB_2 powders having sizes of 50 μm.[22] Indeed, as the damage increases, the onset T_c in Ref. 22 decreases and the $\Delta(T_c)$ broadens, reflecting the inhomogeneity of disorder. However, modest levels of atomic disorder (~1% d.p.a.) induced by proton irradiation already enhance the pinning of vortices, thereby significantly increasing J_c at high field strengths and irreversibility fields. At higher doses, the improvement of the field dependence of J_c saturates, but $J_c(H = 0)$ is lowered. Therefore, there is an optimal level of disorder for enhancement of J_c at high fields.

Heavy ion irradiation is in general by far less effective than the above mentioned irradiation techniques. As it creates extended defects, it often yields degradation of intergrain connectivity,[28] but minor changes in reversible superconducting properties, e.g., T_c. On the other hand, J_c is enhanced by around one order of magnitude after irradiation with 200 MeV Ag and Au at $\sim 10^{16}$ ions/m^2 (Refs. 26, 27 and 29). Among various heavy ions, the best results are obtained with 200 MeV Au particles.[27,29]

10. Comparison of the Effects of Neutron Irradiation and C-Doping on the Superconducting Properties

Controlled disorder in the MgB_2 lattice can also be introduced by chemical substitutions. In general, chemical substitutions, besides disorder, yield charge doping and changes in the density of states and Fermi surface topology, thus making the overall scenario much more complicated than simple introduction of disorder. In particular, C substitution in the B site $Mg(B_{1-x}C_x)_2$ is believed to introduce disorder predominantly in the σ band, thus doping electrons in the system, raising the Fermi level and depleting the σ band of holes. Both effects are expected to lower T_c, as indeed observed experimentally.[8] The theoretical prediction that C induced disorder in the σ band yields an enhancement of H_{c2} at low temperature[6] is confirmed by experiments.[8] A comparison between these effects of C-doping with the corresponding effects of irradiation shows striking similarities in the modification of the reversible parameters, with particular similarities in the decrease of T_c, the enhancement of H_{c2}, the reduction of the H_{c2} anisotropy[52] and the reduction of the larger gap Δ_σ down to the merging with the smaller gap Δ_π (Ref. 53). This suggests on one hand that impurity scattering effects dominate over changes of the Fermi surface in the case of C-doping and on the other hand that point defect disorder introduced by irradiation is mostly effective on σ carriers, as for the case of C-doping. Indeed, this is exactly the conclusion that is drawn by fitting H_{c2} curves of irradiated samples.[45]

Regarding irreversible properties affected by larger scale defects, remarkable results are obtained by adding C in elemental form (microspheres, nanotubes, nanodiamond) or by means of C based compounds such as SiC, TiC, B_4C or organic compounds. The enhancement of J_c obtained by these C inclusions overcome by far all the achievements obtained by any other route.

11. Assessment of Neutron Irradiation Effects in MgB_2 for Applications

Most known neutron irradiation studies on bulk and filamentary MgB_2 samples have been carried out using either samples of thickness smaller than the thermal neutron penetration depth or samples prepared with isotopically enriched ^{11}B. These choices allow to obtain a significant cross-section of the interaction reaction of neutrons with the matter and at the same time a homogeneous damage throughout the sample. Systematic studies over a wide range of fluences have been carried out, up to the complete suppression of superconductivity. On the whole, it turns out that the improvements of superconducting properties in MgB_2 bulk and wires obtained by irradiation are comparable with the effects observed in other superconductors such as A15 type compounds,[54] among which the best results on J_c are found for Nb_3Sn.[55,56] However, if irradiation is carried out on MgB_2 samples having already quite high J_c values, irradiation yields rather limited improvements and in any case smaller improvements than those obtained with C particle additions. We consider two reasons which may explain such outcome.

(i) In bulk polycrystals, pinning by grain boundaries is by far the dominant pinning mechanism, as demonstrated by the shape of the pinning force curves plotted versus the reduced field H/H_{irr}. This pinning mechanism can be enhanced by proper tailoring of the microstructure and by additions. Any further improvement in terms of new pinning mechanisms is hard to obtain.

(ii) As illustrated in this brief review, the optimal irradiation dose in terms of improvement of H_{c2} and J_c is obtained for low fluences, corresponding to a decrease of T_c to values close to 35 K. For higher fluences, not only T_c starts to be severely depressed, but also the DOS and the electron–phonon coupling. Moreover, at the highest fluences the overlapping of the irradiation induced defects, with associated deterioration of all superconducting properties as well as deterioration of grain connectivity, becomes an issue. As a consequence of such trade-off between beneficial and detrimental mechanisms, the effectiveness of irradiation cannot be pushed too far.

It can be concluded that neutron irradiation provides a powerful tool to investigate the physical mechanisms related to disorder and pinning in MgB_2. In particular, these experiments are important in assessing the effect of irradiation on superconducting magnets and current leads in fusion reactors and accelerators. At the relatively low neutron fluence regimes occurring in fusion magnets as well as in accelerators, sizeable improvements of H_{c2} and J_c are obtained in contrast to a weak decrease of T_c, thus resulting in a beneficial rather than a detrimental effect.

References

1. I. I. Mazin, O. K. Andersen, O. Jepsen, O. V. Dolgov, J. Kortus, A. A. Golubov, A. B. Kuz'menko and D. van der Marel, *Phys. Rev. Lett.* **89**, p. 107002 (2002).
2. A. Y. Liu, I. I. Mazin and J. Kortus, *Phys. Rev. Lett.* **87**, p. 087005 (2001).
3. H. J. Choi, D. Roundy, H. Sun, M. L. Cohen and S. G. Louie, *Phys. Rev. B* **66**, p. 020513 (2002).
4. A. Floris, G. Profeta, N. N. Lathiotakis, M. Lüders, M. A. L. Marques, C. Franchini, E. K. U. Gross, A. Continenza and S. Massidda, *Phys. Rev. Lett.* **94**, p. 037004 (2005).
5. M. Putti, P. Brotto, M. Monni, E. Galleani d'Agliano, A. Sanna and S. Massidda, *Europhys. Lett.* **77**, p. 57005 (2007).

6. A. Gurevich, *Phys. Rev. B* **67**, p. 184515 (2003).
7. A. A. Golubov and A. E. Koshelev, *Phys. Rev. B* **68**, p. 104503 (2003).
8. R. H. T. Wilke, S. L. Bud'ko, P. C. Canfield, D. K. Finnemore, R. J. Suplinskas and S. T. Hannahs, *Phys. Rev. Lett.* **92**, p. 217003 (2004).
9. M. Putti, V. Braccini, C. Ferdeghini, I. Pallecchi, A. S. Siri, F. Gatti, P. Manfrinetti and A. Palenzona, *Phys. Rev. B* **70**, p. 052509 (2004).
10. R. H. T. Wilke, S. L. Bud'ko, P. C. Canfield, J. Farmer and S. T. Hannahs, *Phys. Rev. B* **73**, p. 134512 (2006).
11. O. V. Dolgov, R. K. Kremer, J. Kortus, A. A. Golubov and S. V. Shulga, *Phys. Rev. B* **72**, p. 024504 (2005).
12. M. Putti, R. Vaglio and J. M. Rowell, *Supercond. Sci. Technol.* **21**, p. 043001 (2008).
13. L. R. Testardi and L. F. Mattheiss, *Phys. Rev. Lett.* **41**, p. 1612 (1978); L. F. Mattheiss and L. R. Testardi, *Phys. Rev. B* **20**, p. 2196 (1979).
14. A. P. Gerashenko, K. N. Mikhalev, S. V. Verkhovskii, A. E. Karkin and B. N. Goshchitskii, *Phys. Rev. B* **65**, p. 132506 (2002).
15. M. Putti, M. Affronte, C. Ferdeghini, P. Manfrinetti, C. Tarantini and E. Lehmann, *Phys. Rev. Lett.* **96**, p. 077003 (2006).
16. M. Eisterer, *Phys. Status Solidi C* **5**, p. 1606 (2005).
17. A. E. Kar'kin, V. I. Voronin, T. V. D'yachkova, N. I. Kadyrova, A. P. Tyutyunik, V. G. Zubkov, Yu.G. Zainulin, M. V. Sadovski and B. N. Goshchitskii, *JETP Lett.* **73**, p. 570 (2001).
18. M. Eisterer, M. Zehetmayer, S. Tonies, H. W. Weber, M. Kambara, N. H. Babu, D. A. Cardwell and L. R. Greenwood, *Supercond. Sci. Technol.* **15**, p. L9 (2002).
19. M. Putti, V. Braccini, C. Ferdeghini, F. Gatti, P. Manfrinetti, D. Marré, M. Monni, A. Palenzona, I. Pallecchi, C. Tarantini, I. Sheikin, H. U. Aebersold and E. Lehmann, *Appl. Phys. Lett.* **86**, p. 112503 (2005).
20. M. Zehetmayer, M. Eisterer, J. Jun, S. M. Kazakov, J. Karpinski, B. Birajdar, O. Eibl and H. W. Weber, *Phys. Rev. B* **69**, p. 054510 (2004).
21. A. Martinelli, C. Tarantini, E. Lehmann, P. Manfrinetti, A. Palenzona, I. Pallecchi, M. Putti and C. Ferdeghini, *Supercond. Sci. Technol.* **21**, p. 012001 (2008).
22. Y. Bugoslavsky, L. F. Cohen, G. K. Perkins, M. Polichetti, T. J. Tate, R. Gwilliam and A. D. Caplin, *Nature* **411**, p. 561 (2001).
23. R. Gandikota, R. K. Singh, J. Kim, B. Wilkens, N. Newman, J. M. Rowell, A. V. Pogrebnyakov, X. X. Xi, J. M. Redwing, S. Y. Xu and Q. Li, *Appl. Phys. Lett.* **86**, p. 012508 (2005).
24. R. Gandikota, R. K. Singh, J. Kim, B. Wilkens, N. Newman, J. M. Rowell, A. V. Pogrebnyakov, X. X. Xi, J. M. Redwing, S. Y. Xu, Q. Li and B. H. Moeckly, *Appl. Phys. Lett.* **87**, p. 072507 (2005).

25. J. Wang, C. G. Zhuang, J. Li, Z. W. Wu, S. Li, H. F. Chu, Q. R. Feng and D. N. Zheng, *Supercond. Sci. Technol.* **22**, p. 045020 (2009).
26. S. R. Shinde, S. B. Ogale, J. Higgins, R. J. Choudhary, V. N. Kulkarni, T. Venkatesan, H. Zheng, R. Ramesh, A. V. Pogrebnyakov, S. Y. Xu, Qi Li, X. X. Xi, J. M. Redwing and D. Kanjilal, *Appl. Phys. Lett.* **84**, p. 2352 (2004).
27. H. Narayan, R. K. Bhatt, H. M. Agrawal, R. P. S. Kushwaha and H. Kishan, *J. Phys.: Condens. Matter* **19**, p. 136209 (2007).
28. S. D. Kaushik, R. Kumar, P. K. Mishra, J. E. Giencke, D. M. Kim, C. B. Eom and S. Patnaik, *Physica C* **442**, p. 73 (2006).
29. R. Kumar, H. M. Agrawal, R. P. S. Kushwaha and D. Kanjilal, *Supercond. Sci. Technol.* **19**, p. 1047 (2006).
30. N. Chikumoto, A. Yamamoto, M. Konczykowski and M. Murakami, *Physica C* **388/389**, p. 167 (2003).
31. R. J. Olsson, W. K. Kwok, G. Karapetrov, M. Iavarone, H. Claus, C. Peterson and G. W. Crabtree, *arXiv:0201022* (2003).
32. V. Ferrando, I. Pallecchi, C. Tarantini, D. Marré, M. Putti, C. Ferdeghini, F. Gatti, H. U. Aebersold, E. Lehmann, E. Haanappel, I. Sheikin, P. Orgiani and X. X. Xi, *J. Appl. Phys.* **101**, p. 043903 (2007).
33. T. Klein, R. Marlaud, C. Marcenat, H. Cercellier, M. Konczykowski, C. J. van der Beek, V. Mosser, H. S. Lee and S. I. Lee, *Phys. Rev. Lett.* **105**, p. 047001 (2010).
34. S. Okayasu, H. Ikeda and R. Yoshizaki, *Physica C* **378–381**, p. 462 (2002).
35. M. M. A. Sekkina and K. M. Elsabawy, *Physica C* **377**, p. 411 (2002).
36. H. Sözeri, L. Dorosinskii, U. Topal and I. Ercan, *Physica C* **408–410**, p. 109 (2004).
37. C. Tarantini, H. U. Aebersold, V. Braccini, G. Celentano, C. Ferdeghini, V. Ferrando, U. Gambardella, F. Gatti, E. Lehmann, P. Manfrinetti, D. Marré, A. Palenzona, I. Pallecchi, I. Sheikin, A. S. Siri and M. Putti, *Phys. Rev. B* **73**, p. 134518 (2006).
38. Y. Wang, F. Bouquet, I. Sheikin, P. Toulemonde, B. Revaz, M. Eisterer, H. W. Weber, J. Hinderer and A. Junod, *J. Phys.: Condens. Matter* **15**, p. 883 (2003).
39. J. M. Rowell, *Supercond. Sci. Technol.* **16**, p. R17 (2003).
40. M. Monteverde, N. Nunez-Regueiro Rogado, K. A. Regan, M. A. Hayward, T. He, S. M. Loureiro and R. J. Cava, *Science* **292**, p. 75 (2001).
41. A. V. Pogrebnyakov, J. M. Redwing, S. Raghavan, V. Vaithyanathan, D. G. Schlom, S. Y. Xu, Qi Li, D. A. Tenne, A. Soukiassian, X. X. Xi, M. D. Johannes, D. Kasinathan, W. E. Pickett, J. S. Wu and J. C. H. Spence, *Phys. Rev. Lett.* **93**, p. 147006 (2004).

42. D. Daghero, A. Calzolari, G. A. Ummarino, M. Tortello, R. S. Gonnelli, V. A. Stepanov, C. Tarantini, P. Manfrinetti and E. Lehmann, *Phys. Rev. B* **74**, p. 174519 (2006).
43. R. Di Capua, H. U. Aebersold, C. Ferdeghini, V. Ferrando, P. Orgiani, M. Putti, M. Salluzzo, R. Vaglio and X. X. Xi, *Phys. Rev. B* **75**, p. 014515 (2007).
44. V. Ferrando, M. Affronte, D. Daghero, R. Di Capua, C. Tarantini and M. Putti, *Physica C* **456**, p. 144 (2007).
45. P. Brotto, I. Pallecchi, M. Putti, E. Galleani and D'Agliano, *Phys. Rev. B* **82**, p. 134512 (2010).
46. N. R. Werthamer, E. Helfand and P. C. Hohenberg, *Phys. Rev.* **147**, p. 295 (1966).
47. C. Tarantini, H. U. Aebersold, C. Bernini, V. Braccini, C. Ferdeghini, U. Gambardella, E. Lehmann, P. Manfrinetti, A. Palenzona, I. Pallecchi, M. Vignolo and M. Putti, *Physica C* **463–465**, p. 211 (2007).
48. M. Eisterer, M. Zehetmayer and H. W. Weber, *Phys. Rev. Lett.* **90**, p. 247002 (2003).
49. I. Pallecchi, C. Tarantini, H. U. Aebersold, V. Braccini, C. Fanciulli, C. Ferdeghini, F. Gatti, E. Lehmann, P. Manfrinetti, D. Marré, A. Palenzona, A. S. Siri, M. Vignolo and M. Putti, *Phys. Rev. B* **71**, p. 212507 (2005).
50. I. Pallecchi, V. Ferrando, C. Tarantini, M. Putti, C. Ferdeghini, Y. Zhu, P. M. Voyles and X. X. Xi, *Supercond. Sci. Technol.* **22**, p. 015023 (2009).
51. J. Wang, C. G. Zhuang, J. Li, Y. Z. Wang, Q. R. Feng and D. N. Zheng, *J. Appl. Phys.* **106**, p. 093915 (2009).
52. C. Krutzler, M. Zehetmayer, M. Eisterer, H. W. Weber, N. D. Zhigadlo and J. Karpinski, *Phys. Rev. B* **75**, p. 224510 (2007).
53. R. S. Gonnelli, D. Daghero, A. Calzolari, G. A. Ummarino, Valeria Dellarocca, V. A. Stepanov, S. M. Kazakov, N. Zhigadlo and J. Karpinski, *Phys. Rev. B* **71**, p. 060503(R) (2005).
54. H. W. Weber, Irradiation damage in superconductors, *Adv. Cryog. Eng.* **32**, pp. 853–864 (1986).
55. A. Nishimura, T. Takeuchi, S. Nishijima, K. Ochiai, G. Nishijima, K. Watanabe, T.Shikama, 14 MeV neutron irradiation effect on critical current and critical magnetic field of Nb_3Sn and Nb_3Al wires, *Advances in cryogenic engineering: Transactions of the International Cryogenic Materials Conference–ICMC*, vol. **56**, edited by U. Balachandran, American Institute of Physics, 2010.
56. T. Okada, M. Fukumoto, K. Katagiri, K. Saito, H. Kodaka and H. Yoshida, *J. Appl. Phys.* **63**, p. 4580 (1988).

Chapter 6a

MRI Magnets based on MgB_2

Leonardo Bertora

Paramed s.r.l., Magnet Technology
Corso F.M. Perrone 73r, Genova 16152
leonardo.bertora@paramed.it

Magnetic resonance imaging (MRI) is a well established technology in medical radiology, being the "gold standard" to diagnose many soft tissue pathologies and thus widely used in clinical research. For practical image quality, intense magnetic field strength is needed. The 4.2 K-MRI magnet production is by far the most relevant commercial application of superconductivity. Naturally, development of a new superconductor and MRI industry show reciprocal interest, for mutual beneficial growth. The main reasons for adopting MgB_2 in MRI scanners are linked to its higher T_c value, cryogenic simplification, little sensitivity to predicted helium outage in future and low raw material cost. MgB_2 at its present development stage is suitable for MRI niche markets, whereas many open issues have to be addressed for reaching a larger commercial arena. MgB_2 wires and tapes optimized for MRI applications have been commercially available for nearly ten years, and their properties have been largely improved in this period. Several MRI devoted magnets have been designed based on available MgB_2 wires, some projects having generated prototypes. So far, one of them has turned into a proper scanner available on the market.

1. Introduction

From the first pioneering papers[1–4] during the 1970s, MRI has become the criterion standard in many radiological specialties,[a] and is a continuous source of new discoveries about complex brain "mechanics" as through functional neuroimaging. It is also advancing in interventional modalities, where images are not only used for diagnostic purposes but also to improve healing rates. Every year, nearly 4,000 scanner units are sold within a very competitive market dominated by technological innovation, effectiveness, reliability and cost reduction. Fundamental, technical and clinical researchers are pursuing new medically relevant modalities, pushing the limit towards present cost-benefit boundaries as with the 11.7 T INUMAC–ISEULT project, with a 200 M€ range budget.[5]

In present day's commercial NbTi scanners operating at 4.2 K, mainly based on 1.5 T "mid-field" and 3 T "high-field" cylindrical magnets (~80%), all engineering and economic aspects have been optimized over 35 years, since the introduction of the first superconducting 0.18 T MRI whole body magnets during 1979–1980.[6,7] Some key issues related to magnets including cryogenics, electromagnetic design, magnet protection coil manufacturing, superconducting joints, superconducting switches and current leads have been highly industrialized in order to shrink system costs and, at the same time, to improve the reliability to guarantee typical scanner downtime of less than 1% a year for the end customer.

Modern industrial superconductivity has so far been driven by the MRI market. No other industrial application has consumed as many kilometers of superconducting wire. From Conectus data[8] the MRI is responsible for nearly 80% of the superconducting (SC) technology demand. Global MRI sales were valued at U.S.$ 3.9 billion in 2011, with a forecast for 2017 of $ 4.8 billion.[9] The magnet is the most expensive single part in a scanner and its performances define some of the most important image quality (IQ) factors; thus the basic

[a]This is what could be considered the clinical mainstream use, thanks also to the absence of ionizing radiation delivered to patient, typical of possibly rival methodologies as CT.

research and industrial development of a "new" superconductor (SC) technology, as the one MgB_2 entails, is paramount to the MRI industry.

Hereafter two main questions would be addressed: stating the hypothesis of its suitability, why should MgB_2 be introduced in an optimized industry such as MRI? Thus, does MgB_2 really fit MRI applications, and if yes, what are its limitations?

2. Reasons for MgB_2 Adoption in MRI

Within this industrial mature landscape, the introduction of a new superconducting material, such as MgB_2, could be justified if it will satisfy at least one of these two requirements:

- Creation of product value;
- Reduction of life cycle or product cost.

Alternatively or in addition, MgB_2 would spread in the MRI market if a disruptive event in the MRI scanner technology could drift value-cost of NbTi magnets cooled by liquid helium (LHe) bath. From this point of view, helium cryogenics "crisis" could be the most relevant factor.

2.1 *Commodities affecting MRI technology: MgB_2 as viable solution*

The relatively abundant availability of LHe in the last three decades has been a necessary factor for the proliferation of MRI superconducting scanners worldwide. In fact, with very few and limited exceptions (see Paragraph 2.2), all the commercialized superconducting scanners have been based on NbTi wire, cooled by a LHe bath.

With the introduction of re-condensing cryocoolers in MRI systems, a huge reduction of helium consumption has been possible, reaching zero boil-off scanners. Still, typically 1,000 to 2,000 liters of LHe are necessary to maintain superconductivity in a MRI scanner and an even larger quantity is needed to cool it from room temperature to 4.2 K. In case of any unlucky event that could lead to a quench, or any efficiency reduction of the cold head, a complete or partial LHe refill is necessary. Finally, a reduction of LHe market availability, i.e., a cost increase, could largely affect the MRI industry.

2.1.1 *Helium availability*

As commonly known, helium is the second most abundant chemical element in the universe, and is continuously generated by nuclear fusion in stars. To a much lesser extent, He has been generated since earth's formation by heavy nuclei alpha decays. Due to its chemical and physical properties, He is present only in a few ppm in the atmosphere; extracting it from air would be extremely costly. As with natural gas (NG), helium has been trapped by impermeable rock underground. Thus crude He is extracted as a byproduct of NG processing, up to 7% by volume.[10] The largest estimated He resources[11] are located in the U.S. (40%), Qatar (19%), Algeria (16%) and Russia (13%), where often it remains blended and finally lost, when NG is burned. While global demand is increasing (cryogenics, welding, etc.), extraction and spare capacity is limited by economical and physical considerations. Hence, it is reasonable to state that helium is a "finite" resource, whose cost is intrinsically destined to increase in the mid-long run and whose price dynamics can largely be affected by the NG market or upstream plant problems. Furthermore, the He market is not yet free, instead it is still heavily regulated. In fact, US government is the largest crude helium supplier worldwide, providing during 2012, 34% of global helium sold.[11] Since 1925 U.S. Federal Administration stored helium in the Federal Helium Reserve, near Amarillo, Texas. In order to reduce US federal debt, the "Helium Privatization Act" directed the U.S. Government to deploy the reserve within 10 years, starting from 2005. The price was fixed with the idea of recovering the invested money during the century, not by supply–demand equations. Eventually, the reserve (Cliffside Plant) will go offline, creating what is termed "helium cliff".

The helium market has been artificially dumped by many concomitant factors. The prospective of a helium crisis due to the upcoming Federal Reserve deployment could have been partially balanced by the regulated price in the short term. On the other hand, concomitant 2009 economic crises drastically reduced global demand of NG extraction[b] adding another relevant helium price increasing factor. Especially during 2013, scientific community[10] and industrial lobbies (MRI and Semiconductor manufacturer associations) pleaded US lawmakers to avoid the abrupt (October 2013) helium shortage due to complete pay off of helium debt as prescribed previously. On October 2, 2013 U.S. Congress approved "Stewardship Helium Act" that finally relieved helium market ensuring supply of crude helium by Cliffside plant till 2021. Moreover, to reduce helium price economic distortion, helium will be sold by auction and not at fixed price, at least partially.

At the end, comparing the price for grade A helium (Fig. 1) from 1996 to 2014, it is possible to note a factor 400% increase, with a

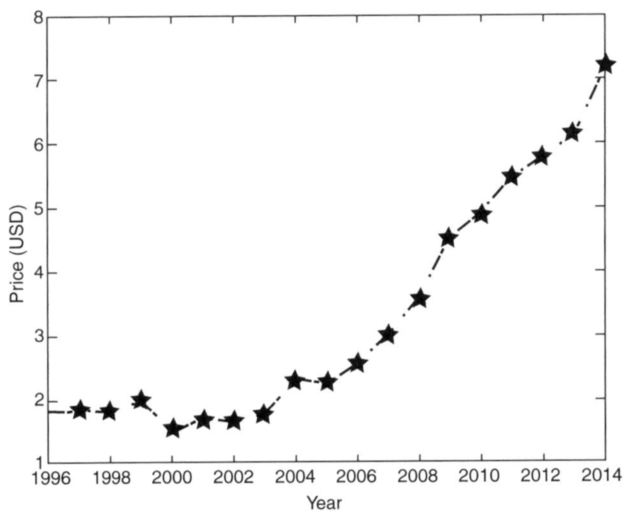

Fig. 1. Grade-A helium averaged estimated prices per cubic meter. Data collected from Ref. 12.

---

[b]Globally, NG consumption increased constantly after 2009, recovering 2008 data just immediately in 2010. In Europe and Asia, the NG consumption during 2013 was still less than during 2008. Algeria and Russia, third and fourth positions in helium resources, supply most NG (approximately 90%) to Europe and Asia region,[13] with reduced helium extraction investment appeal in the area, i.e., higher end user He prices.

22% yearly peak during 2009, and a 15% increase just in 2014. In the last two years, intermittent shortage have been experienced by many He users.[10] This already is a relevant obstacle in assuring short downtime to MRI scanners in case of quench, especially in developing or underserved areas.[14]

2.1.2 *Rare earth crisis case*

In this short discussion about the MRI market future dependence and trends versus LHe price and availability, it is interesting to look to what happened to the MRI economy during 2011 rare earth crisis. Nearly 80% of the MRI units are based on SC (NbTi) magnets, but every year, approximately 700–900 scanners using permanent magnets are still sold. Naturally, permanent magnets have a lower magnetic field strength than superconducting ones (typically 0.3–0.45 T versus 1.5–3 T), i.e., lower SNR. On the other hand, they are cheaper than SC scanners, concerning acquiring price, installation price and maintenance, so they maintain a relevant MRI niche market focused on small radiological centers and in underdeveloped areas. A typical magnet of this kind has a very simple structure compared with SC magnets. Most importantly an iron yoke, effective low losses pole plates and sintered NdFeB blocks are the only parts required for manufacturing this kind of MRI magnet. Because it needs tons of NdFeB, the magnet final price is substantially dependent on the Nd price, being by far the most expensive material in the magnet. Neodymium, as with several other rare earth material, is massively used in electrical motors, which are a key element of the new automotive trends. Between 2003 and 2007, the Nd price was nearly some tens (20–50) of dollars per kilogram. However, at the end of 2010 the Nd price passed 100 $/kg, and during August–September 2011 it went beyond 450 $/kg: a factor ten increase in less than four years. In other terms, just the NdFeB needed for a total body scanner could cost more than a whole superconductive magnet. In those difficult months many OEM magnet MRI manufacturers stopped their productions, the less competitive closed, and some started superconductive programs that are now ready to market.

In the meantime (end of 2012) Nd price stabilized below 100 \$/kg, but now it is clear that the rare earth based MRI magnet business has a high intrinsic risk, and manufacturers have to diversify with different technologies, such as superconductivity,[c] which improves magnet field strength at the same time.

The same interest for strategic risk reduction from raw material price fluctuation and shortage could be followed by superconducting MRI magnet manufacturers, gradually introducing "liquid helium free" magnets in their product ranges. In this case, MgB_2 could have a key role in the MRI world due to its higher T_c value with respect to low temperature superconductors (LTS) that entails, for conduction cooled solutions, reduced cooling power consumption, simpler cryogenics, higher stability and longer ride-through.

2.1.3 *SC raw material cost*

There is another common consideration favorable to MgB_2 the reduced intrinsic cost of magnesium and borate raw elements. This argument is easily provable considering HTS wires (not yet used in MRI market technology), where the costs of silver matrix in Bi2212 at 4 K/10 T (~100 \$/kAm) and of present YBCCO at 30 K/2 T and 77 K /0 T (~70 and 300 \$/kAm, respectively) and projected costs are higher than the present ones for MgB_2, where a comparison is possible.[6,15] The same argument is not so easy to prove considering present LTS prices: ~1–1.5 \$/kAm for 4 K/5 T NbTi, and ~3–4 \$/kAm for 4 K/10 T Nb_3Sn.[6] A more quantitative analysis, out of this work purposes, could start from the estimated 2013 price per ore kg: \$ 2.6 (Mg), \$ 0.57(Borate) versus \$ 68 (Nb) and \$ 13.6 (Ti) respectively.[11] Unfortunately, this is a far from practical use, because purity is relevant in the manufacturing process: high purity boron can cost more than 400 \$/kg. Stabilizer, shield and structural metals, such as nickel, are expensive material as well.

[c]Unfortunately, MgB_2 technology is not yet cheap enough to substitute neodymium in this favorable segment (low field), due to its present $J_c(B)$ properties.

Moreover, in same MgB_2 wire designs,[16] the same costly Nb can be used in a larger percentage (34%) than the MgB_2 compound itself (15%).

Finally, the only relevant considerations about conductor (global or raw) cost dynamics can be performed by manufactures and will be briefly reported in Sec. 4, where MgB_2 wire projected price between 1–3 \$/kAm at 4 K/4 T and 2–5 \$/kAm at 4 K/6 T are reported. This price level, together with predicted high performances, will allow a wider diffusion of MgB_2 magnets in MRI.[15]

By a more "green" prospective, raw material risks and fluctuations are often linked to environmental sustainability. Naturally, the extensive substitution of helium cryogenics would avoid the depletion of a not "quickly" renewable element such as He; less commonly pointed out, also the use of magnesium metal may be considered, to a limited degree, a renewable resource.[d]

2.2 *A MgB_2 mid-term introduction: "dry" systems overview*

In MRI magnet technology, the He bath has several advantages: it is a powerful and homogeneous cooler, and, at least temporarily, it is insensitive to external negative conditions, as power or cooling water outage, cryocoolers faults, etc. On the other hand, not considering what was discussed above about present and future cost or availability, LHe magnets need more complex cryogenics, special installation devices, as He gas evacuation line and space access for He refill tank. Finally, they have an intrinsic safety risk in case of severe malfunctioning.

For all these reasons, over the past few years, remarkable efforts have been made by several companies to offer cryogenless (or "dry," i.e., without liquid cryogens) magnets especially within nuclear-resonance technology. For example, just to name few, "cryogen-free"

---

[d]Magnesium metal is derived from seawater, and other minerals. To a limited degree, natural brines may be considered to be a renewable resource wherein any magnesium removed by humans may be renewed by nature in a short span of time.[11]

NMR systems up to 14.1 T by Cryogenic Limited, 3 T relaxometer magnets by HTS-110, preclinical MRI up to 7 T by MR Solutions; many other conduction cooled magnets are commercially available nowadays. Many of them are based on NbTi, others on Nb_3Sn and few on HTS. Cooling power (at the end a cost factor) versus whole-body MRI larger coil masses (acceptable cooling time), magnetic forces (thermal load through mechanical supports), radiating surfaces and thermal conductivity (enhanced and limited by large dimensions respectively), are the key issues to extend cryogen-free technology to whole body MRI magnets.

Nonetheless, still from 1994[17] with GE Signa SP 0.5 T scanner (often named double-donut due to its open split design), SC 10K "dry" MRI systems have been available on the market. GE SP Signa 0.5 T was based on Nb_3Sn wire and it was conduction cooled by the use of two cryocoolers. By a technology point of view, this scanner was ahead of its time: open split magnets, i.e., cylindrical magnets with an empty annulus (gap) in the center, are becoming more popular in recent years with the introduction of hybrid MRI techniques, such as PET/MRI, commercially available, or Linac/MRI, still under development. The scanner layout was designed for clinical interventional-MRI research segment, in a pioneering stage at that time. Totally 15 scanners were installed.[15]

Another effective example has been provided by Toshiba Opart. As with the GE double-donut, Opart was an open MRI scanner, but with a more standard four-pillar iron yoke structure, quite similar to "ninetieth" whole-body scanners based on permanent magnets, but with a significantly larger patient space. Opart had a central field strength of 0.35 T and was launched during 1996. In contrast to GE Signa SP 0.5 T, the scanner was devoted to regular clinical use, positioning between lower field (at that time) open permanent magnet and cylindrical SC mainstream. Toshiba sold roughly 500 Opart scanners in nearly 10 years of product life.[18] Again this conduction cooled MRI scanner was based on Nb_3Sn.[19]

Many other MRI prototypes or demonstrative cryogen-less magnets have been proposed, designed or built, but without reaching the market. Within the dedicated scanners (not suitable for scanning all the anatomical parts), a 0.5 T conduction cooled brain dedicated scanner was developed in 1995 by GE. The most relevant "total body" example was the first MRI magnet based on HTS (Bi-2233), a 0.2 T/16–20 K, C-shape, not-persistent magnet manufactured by Siemens/OMT during 2003. The still high conductor price forecast compared to market product positioning was the main reason for not pursuing the project.[20]

Nevertheless, continuous attempts to introduce Bi-2223 in MRI had been made. During 2013, a 3 T–20 K conduction cooled Bi-2223 head dedicated magnet has been tested at NIMS (Japan). The high-field MRI scanner was devoted to functional research.[21] MRI images have been acquired at 1.5 T to prove feasibility. After successfully reaching nominal field value (3 T), abnormal events described in Ref. 22 occurred, causing mechanical damages and unfortunately burning the HTS tape in one coil. In 2013, also YBCO reached the selected group of SC employed in a MRI scanner prototype. The conduction-cooled extremity dedicated, passively shielded magnet (manufactured by HTS-110, NZ) worked at 1.5 T–20 K in driven-mode. Drifts induced by screening currents and shimming instability have been managed in order to achieve images. Typical MRI requirements on field short-term stability, as those concerning gradient-magnet interactions, limited the reported results to SE-T1 image modalities, although implementation of standard solutions was envisaged.[23]

With respect to many "dry" systems summarized above, the MgB_2 based ones present several advantages indeed. As discussed previously, HTS are still too expensive for MRI mainstream and dedicated applications, while MgB_2 costs nearly between one tenth and one hundredth, and persistency is easier to achieve. In comparison to Nb_3Sn, some benefits can be stated: higher T_c, easier "react and wind" (R&W) technology, lower mid-term predicted wire cost. Finally, higher T_c is the MgB_2 plus versus MRI workhorse NbTi. This allows (1) higher operating temperature that can be translated in lower power requirements for optimized cryogenics, (2) larger enthalpy gap between operating condition and quenching

Table 1: Comparison between discovery of relevant MRI conductors and respective scanner market introduction dates. NbTi: first MRI was performed around 1973, so NbTi time should start 10 years later from its discovery, but naturally, between 1962 and 1980 relevant magnet technology was already developed. Nb_3Sn has been employed 16 years later, showing how raw conductor cost, brittle nature and W&R manufacturing process have an better impact. YBCO and BSCCO are still facing high conductor cost.

SC	Discovery date	First MRI scanner to market
NiTi	1962	1980
Nb_3Sn	1954	1996
MgB_2	2001	2009
YBCO	1987	Not yet introduced
BSCCO	1988	Not yet introduced

temperature, which, in case of persistent-mode operation, could significantly affect scanner downtime. Following the above consideration, MgB_2 introduction in MRI economy appears straightforward, and from a specific perspective it has been, as shown in Table 1.

Thanks to MgB_2's higher T_c values, other cryogenics solutions than LHe or direct conduction cooling are possible, both in dry configurations, and in liquid coolant baths. As reported hereinafter, non-traditional cryogenic modalities have been embraced in MRI magnet designs and Prototypes based on MgB_2. There are some cryogens useful to this purpose, and between them the most prominent is hydrogen (20.3 K Reasons for exploring and adopting MgB_2 in MRI could be summarized as following: MgB_2 technology is an economically viable and environmentally green solution to assist MRI industry to overcome predicted He cost increase or reduced availability). Its economy bodes a new fascinating era for wide-spreading of superconductivity,[15] on long-term scale, at least in medical applications. For clinical application, MgB_2 can become the first choice for dry large systems, due to its higher critical temperature than LTS, and more affordable price than HTS.

3. Issues in MgB_2 Introduction in MRI Systems

MgB_2 spreading in MRI is bounded to satisfactory fulfillment of several requirements that can be summarized in the following areas:

- Value
  - Magnet layout mainly impact clinical applications, product positioning and siting possibilities (stray field dependent also).
  - Field strength directly related to image SNR, i.e., IQ.
  - Field uniformity denote scanner field of view (FOV).
  - Field short-term temporal stability is necessary for image clearness, IQ.
- Cost
  - Manufacturing cost (e.g., SC, cryogenics, mechanics, labor hours, tooling, production failure rates, etc.).
  - Required maintenance and long-term magnetic field stability affect life and servicing costs.

Due to the limited space available, only a few of them would be analyzed in the following.

3.1 *Field intensity*

Magnetic resonance SNR (i.e., one of the most relevant factors defining IQ) increases with static magnetic field strength in a not so trivial way,[24] but limited to some value ranges, nearly proportionally. This physical fact is translated inside a high-field scanner into, for example, detailed and fast images.

Thus, the most diffuse interpretation of "creating product value" for an MRI magnet is field strength increase. Present MRI clinical market mainstream is focused on 1.5 T and 3 T multicoil active shielded total body scanners, where typically peak field on conductor can reach approximately. 5 T, depending on scanner central field, dimensions and required uniformity. Ultra-high field (>7 T) scanners (some tens worldwide) are restricted to research area. In this case, due to higher central field, solenoidal compensated design is adopted in order to reduce the B peak very near to central field value,[25] albeit increasing the amount of conductor required with respect to multicoil layout. Up to now, it has not been possible to use MgB_2 in such magnets due to poor $I_c(B)$ properties, being self-shielded whole-body 1.5 T magnet a target not reached yet. This is one of the most important factors limiting MgB_2 adoption in the present MRI industry, even if recent studies[26–28] show it will be possible to overcome these obstacles in the near future.

On the other hand, MgB_2 can be employed in clinically relevant whole body scanners, yet at low-field and in niche markets. Magnet value-added can be moved from pure IQ improvement (i.e., high central field), towards investigation of diseases not visible in the recumbent position, or towards interventional procedures, or towards the patient comfort (i.e., optimized magnet layout). Furthermore, present MgB_2 can be used in state-of-art high field scanners dedicated to anatomically restricted applications, where peak field could be limited on the coils.

3.2 *Long-term temporal stability*

Persistent magnet operation is strictly linked to scanner reliable performances, sale and maintenance costs. A small and homogeneous central field drift can slightly affect IQ for the most used sequences, but it can't be acceptable for more complex acquisition techniques. For example, in tissue MRI thermometry, a magnetic field variation of 0.01 ppm during examination-intervention time, some tens of minute, means an error of 1°C in temperature measurement.[29] Several techniques can avoid such a stringent limit, but this number gives the prospective of how much MRI technology is sensitive to magnet field stability.

Mainly, a magnetic field strength drift affects radio-frequency (RF) hardware quality response: both transmitting (TX) — with a high quality factor (Q) typically larger than 100 — and receiving RF antennas (coils) are tuned to magnet frequency. Being 42,58 MHz/T the frequency-magnetic field ratio for clinical hydrogen based MRI, an exponential law drift of 0.1 ppm/hr for a 1.5 T magnet means a frequency change of 56 KHz in a year. Supposing TX–Q values around 200, the frequency drift causes an efficiency loss of nearly 10%. Thus 0.1 ppm/hr is commonly considered as the acceptability decay limit that quantitatively defines what is called "*persistent*" *operation mode*. In terms of total magnet resistance, it means a limit of less than 10^{-9} Ohm, considering a typical MRI magnet inductance larger than some tens of Henry. A larger drift (for example, 1 ppm/hr) should be recovered by frequent scanner servicing (increasing maintenance costs) or by a more expensive RF design.

Alternatively the use of a high stability class magnet power supply (MPS) should be considered: again, scanner cost would increase. On the other hand, MPS allows end users to freely ramp up or down the magnet for whichever reason (reducing incident risks during regular cleaning, MRI room maintenance, marketing events as photographic sets, etc.) or to autonomously recover scanner functionality after a quench, instead of waiting for tool shipment and service engineer availability (again a relevant argument in subserved areas). In the author's experience, this feature has been appreciated by radiological center operators. Besides, by a system point of view, the adoption of an MPS, that controls precisely the current value thus obstructing induced voltages, can partially lighten specifications about spurious couplings between gradient coils (inserted in the magnet bore or pole plates and pulsing at audio frequency) and the SC magnet.[23] If an MPS solution is chosen, a balance between cost and value could limit its viability to low current magnets, i.e., low-mid field scanners. Other mechanisms have also been proposed to achieve long-term field stability or to improve MPS short-time scale stability, as reviewed during ISEULT design.[30] For example, flux pumping[31] has already been tested in HTS insert coils,[32] drift compensation equally tested for NMR,[33] and MPS filtering by temperature stabilized resistor[30] could be applied to MgB_2 MRI magnets as well.

In the end, the cheapest solution to be used in large scale still remains the employment of "*persistent*" *joints*, and as a matter of fact, all NbTi (and previously Nb_3Sn) commercially available MRI scanners employ this solution. Similarly, efforts have been made to achieve this goal with MgB_2 wire technology. Looking at the timing of disclosed methodologies, the first patent about the use of MgB_2 in "superconducting" joints[34] was issued just four months after the discovery of superconductivity in MgB_2,[35] years before any large scale application even faced a laboratory.

LTS joint techniques were primarily developed in the 1970s[36,37] for ductile NbTi. With a large number of proposed methodologies and experimentally proven, cold-pressed welding was one of the most industrially notable ones for its simplicity and reliability. Unfortunately, due to the brittle nature of MgB_2, it hasn't been possible to apply the same technique in joining MgB_2 wires. Nb_3Sn powder joint method[38] could prove more suitable for MgB_2 metallurgic properties, even if boron evaporation, oxidation and contamination are a few of the obstacles to be overcome. Essentially, SC compound powders are placed in dedicated shells or holders where superconducting filaments can be joined under determined pressure and thermal cycles.[39] For example, to list some of them, in Ref. 40 simple wire joints or coil joints[41] are manufactured adding "structural" non-superconducting binding powders to the superconducting powders; in Ref. 42, the use of a joining block with variable volume is claimed; in Ref. 43, a substance to lower the powder melting point is employed; in Ref. 44, B powder is infiltrated by magnesium; the beneficial help from HTS powders is advocated in Ref. 45; in Ref. 46, the sintered MgB_2 is pressed from a different direction with respect to the wire axis; in Ref. 47 the introduction of a titanium oxygen trap, or a vacuum pump, during the sintering process is claimed.

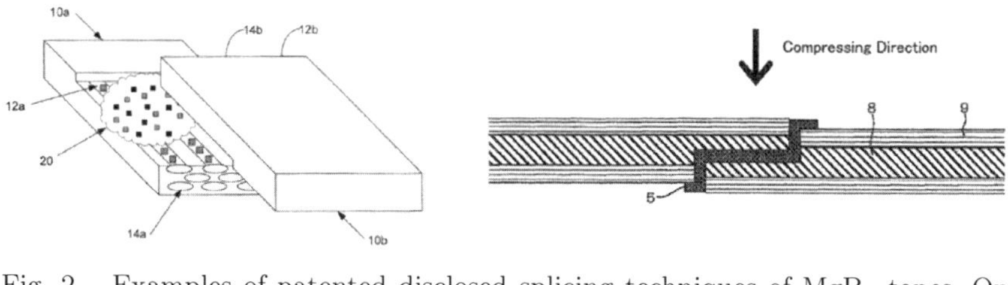

Fig. 2. Examples of patented disclosed splicing techniques of MgB_2 tapes. On the left, from Ref. 40 and from Ref. 48 on the right.

Considering the utility of persistent joints in MRI/NMR magnets, this patent list can be helpful in understanding expectation on MgB_2 in MRI industry.

About the effectiveness of present technology for MgB_2 joints, results and improvements have been reported by several authors. The possibility to join NbTi to MgB_2[49] could be useful in 4.2 K magnets, but also in higher temperature systems if dedicated cooling power is considered, a cost not so easily industrially justified. Superconducting joints extended to nearly the whole MgB_2 critical temperature range have been proved during 2006[50] up to 30 K in self-field, monocore wires (*ex situ*, R&W) and in 2009 up to 10 K but at much higher I_c[39] values, with multifilamentary wires (*in situ*, W&R). These were preliminary steps toward an industrial reliable joint process. Later Nardelli reported the possibility to join commercially available multifilamentary *ex situ* wires up to 300 A at 20 K self-field with a single joint resistance of less than 10^{-14} Ohm.[51]

Recently, Columbus Superconductors SpA. (CS) succeeded in manufacturing consecutively 30 single joints without any failure, always achieving a critical current above 70% of the short sample I_c at 20 K in self-field. This remarkable result clearly opens the use of superconducting MgB_2 joints to a pre-industrial phase, that CS is ready to apply to demonstrative magnets.[52]

Iwasa's group introduced an original solution adopting monofilamentary wires for MRI applications (see Paragraph 6.1), mainly because joints between monocore *in situ* wires, from Hypertech Research Inc. (HTR), have been optimized.[53] Persistence has been claimed at 98.5A at 11 K in a subassembly of a dedicated 0.5 T MRI magnet prototype.[54]

Within an MRI project,[14] GE developed joints up to 70A at 20 K in 0.23 T field strength perpendicular to wire direction employing *ex situ* multifilamentary wire.[55] This result, even if not sufficient, is quite important and promising. Joints can be naturally placed in very low magnetic field strength regions, when a steel yoke is present, i.e., low-mid field open scanners, otherwise they should withstand high magnet stray field. Typically, within a total-body 3 T self-shielded magnet cryostat, the lowest field strength region can be as low as 0.5 T. This value can be further reduced in magnet design optimization target functions, but increasing overall costs, or considering ferromagnetic screening structures albeit their negative effect on scanner homogeneity and center field stability. So joints with improved performance in-field are still under investigations.

As well known, even when low loss joints are industrially available, the MRI highly demanding persistent condition (<0.1 ppm/h) cannot be achieved in the presence of a residual voltage due to the wire poor n-performance. Because it is empirically possible to write the tape-dependent resistance as $R = R_c \cdot (I/I_c)^{(n-1)} < 10^{-9}$ Ohm, where $R_c = nV_c/I_c$, and an arbitrary reference limit of 10^{-9} Ohm is considered, a low n value would exponentially reduce the real usable current, equivalently it would strongly increase the final cost. Many papers have been published about this topic measuring several MgB_2 commercially available or research wires, most recently in Refs. 56, 57, to prove that persistence is an achievable target. In practical km-length conductors, any wire's manufacturing flaw, such as filament sausaging, impurities or cracks, can heavily deteriorate otherwise intrinsic good n values.[58]

As a matter of fact, even if MgB_2 MRI scanners working in the persistent mode are not yet proved, all the most recent research papers about MRI magnets based on MgB_2 envisage persistent mode operation. Technically it is more likely to see first the introduction of the persistent-mode in less demanding low-field scanners, but considering better site management and less production risks in driven-mode, versus hardware (MPS) cost reduction and higher field stability in persistent-mode, the balance between the two could play for introducing persistency in mid-high field scanner initially.

3.3 *MgB_2 and short-term temporal stability*

Magnetic field stability at the millisecond scale is essential in MRI acquisitions. Instability source can be a miscellaneous collection of electromagnetic interferences (AC unshielded lines at 50/60 Hz), magnet mechanical vibrations (for example induced by building natural frequencies) or more commonly interaction between static (SC magnet structure) and electromechanical excitations (in the acoustic frequency range) induced by gradient coils.

Cold heads introduce magnetic disturbances, as well. Field instability can come both from driven motor, magnetic material inside the cold finger, or mechanical vibrations. All these sources have to be carefully managed to avoid IQ reduction. As illustrative example, simulations of unrealistic enhanced artefacts ("ghosting") induced at typical cold head working frequency (~1 Hz) of a not resonating (a) or a resonating (b) mechanical structure are shown in Fig. 3. MRI field coherent fluctuation amplitude requirements are

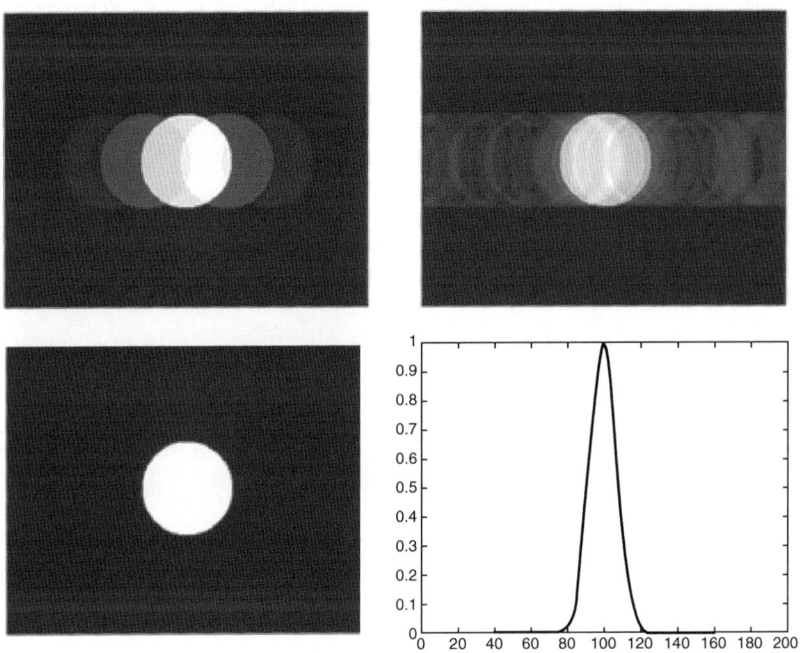

Fig. 3. Simulations of unrealistic enhanced image artefacts induced by short-term temporal magnetic field instability. Left bottom, ideal homogeneous phantom image; left top, simulated artefact induced by pure typical cold head revolution frequency (~1Hz); right bottom, hypothetical Gaussian magnet structural resonance (a.u.), centered at 100 Hz; right top, simulated artefacts based on "right bottom" mechanical response. Disturbance amplitudes and sequence parameters are arbitrarily chosen.

quite severe, with 0.05–0.01 ppm as typical demand, depending on magnet central field (being spin precession error phase a function of absolute error values) and disturbance frequency. It could be helpful to translate this limit in mechanical stability requirements in a simplified model. Taking into account an open split magnet, it is possible at first order,[e] to linearly link central field value to magnet gap. Considering a reference value of 0.5–1 m, field stability requirements can be expressed mechanically as fraction of microns.

Since MgB_2 use is more practical in dry MRI scanners, higher attention to coils conducted vibrations has to be paid with respect to recondensing cryocoolers. In fact, in the last case, no direct connection between cold finger and coils is necessary, whereas conduction cooling requires low thermal impedance between sources and loads, thus massive coupling. Accordingly, adequate compromises between the two requirements have to be pursued with dedicated design as in Ref. 59. The use of pulse-tube technology, instead of mechanical Gifford–McMahon cryocoolers, can reduce vibration spectra,[23,60,61] albeit increasing magnet cost.

Direct conduction cooled alternative approaches have been explored by the MIT group[62] and by GE GRC,[63] both on MgB_2 based MRI devoted magnets. In particular, the system studied by GE explicitly took into account artefacts induced by cold heads reestablishing re-condensing concept in a "dry" magnet (see Paragraph 6.2). Besides, the MRI research community is moving towards real-time feedback for spatio-temporal field stabilization, mainly devoted to cancel gradient coil–magnet interactions, or field fluctuations of physiological origin.[64] The same solution would neutralize any field imperfection induced in a non-vibrational optimized conduction cooled system, further reducing its cost per transmitted Watt. To fulfil stringent MRI requirements, MgB_2 technology should address not only intrinsic conductor properties, but also magnet engineering related issues, and last but not least possible scanner layouts, in order to best exploit MgB_2 potential and to meet market needs.

[e]Considering first odd vibrational mode.

4. MRI MgB_2 Conductors

The aim of this section is to briefly introduce present market available MgB_2 wires and tapes limited to those used in MRI products, prototypes or magnet designs. At present, Columbus Superconductor and Hypertech Research offer MgB_2 wire or tape of several km-lengths; for both of them, higher $J_e(B, T)$, better n values, high uniformity along conductor length, good stability design together with price attractiveness are the major goals. CS adopts the *ex situ* technique, focused on uniformity requirements and "react and wind" easier usability, while HTR embraces the *in situ* power-in-tube (PIT) method (first generation), and more recently Advanced Internal Mg Infiltration (AIMI) approach (second generation), to stress on high $J_e(B, T)$ performances.

In Fig. 4, cross-sections of typical CS magnetic resonance devoted tapes are shown. The 12-filament, 3.5 × 0.65 mm tape (LFCS-12) has a nickel matrix, with a copper core for stabilization, chemically insulated from MgB_2 by an iron barrier. It is optimized for low-field applications, thus favoring grain connectivity, with better performances around 1 T. The 19-filament, 3 × 0.5 mm, nickel matrix tape (HFCS) has a banded (copper, 0.2 mm thick) architecture for improved stability with higher operational currents. Powders are optimized for better performances in-field, with better pinning properties. A comparison between filling factors in both wire types is reported in Table 2.

At 1.2 T/20 K, the most recent I_c results for HFCS and LFCS tapes are 442 A and 329 A, respectively. From previous development

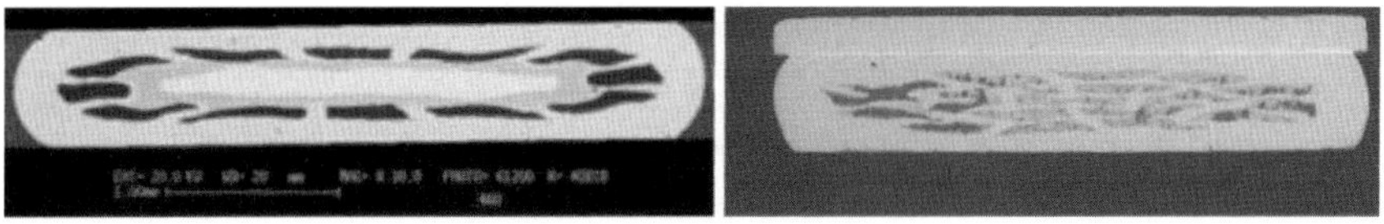

Fig. 4. Cross-section of CS MRI devoted tapes: "LFCS" 12-filament version on the left, and "HFCS" 19-filament copper banded on the right.[f] Courtesy of Columbus Superconductor S.p.A.[65]

[f]Conductor names created by the author.

Table 2: Typical layout of LFCS and HFCS tape families.[65]

Material	12-filament-LFCS (%)	19-filament-HFCS (%)
MgB_2	10	17
Nickel	55	55
Copper	15	28
Iron	10	—

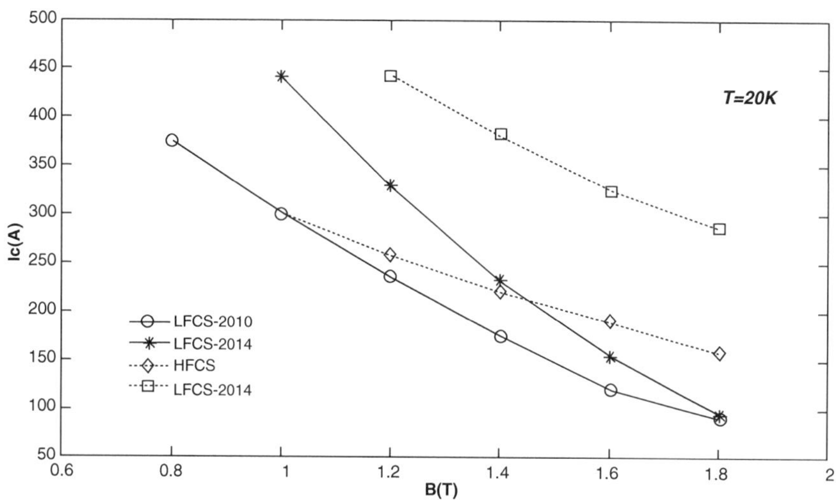

Fig. 5. I_c versus B at 20 K. Data for productions 2010 and 2014. R&D results are shown for samples of LFCS and HFCS tapes. Courtesy of Columbus Superconductors.[65]

stage measurements, shown in Fig. 5, the HFCS and LFCS crossing point was 300 A at 1 T/20 K. From LFCS short-sample production quality checks, at 20 K/1.8 T, n values larger than 30 are constantly found.[52]

The R&W technique is presently used in MRI industry with NbTi workhorse conductors. With respect to the W&R method, the main advantages are the use of (1) standard material in coils assembling, thus without costly parts withstanding high reacting temperatures, (2) lower investments and coil manufacturing production costs, without expensive large coil heat treatment, (3) lower defective risks due to insulation damages.[62] A tape with a layout similar to LFCS-12 is employed in Paramed's MROPEN scanner, whereas a tape of the HFCS family has been used by GE GRC for its limb-dedicated prototype as discussed in Paragraph 6.2

At the beginning of MgB_2 winding for MRI magnet applications, both with R&W and W&R methodologies, highest concerns were in handling the conductor without degrading its I_c performances,[62,66,67] mainly due to the brittle nature of MgB_2. Extensive research investigating stress–strain I_c degradation has been performed on LFCS tapes (14-filament version) since 2005 with a linear relation $I_c \sim 1 + 0.02 \cdot \varepsilon$, for ε in the range (−0.5;0.5), at 4.2 K/4 T.[68] A banded, 5 m length, HFCS tape has recently been tested for "easy bending" (regular tape bending, 0.17% strain) on a 30 cm OD disk, and "easy" plus additional 0.2% "hard bending" (turn-to-turn simulation, i.e., bending forces within straight tape plane). Under these conditions, n values around 40 have been obtained, thus proving minimal degradation effects induced by bending strain.

Furthermore, HFCS tapes have been harshly curved as depicted in Fig. 6. Two 90-degree-5 mm radius corners have been crooked, between points 5 and 4, in the "in-bent" (and actual coil winding) direction, and between points 3 and 2, in the "out-bent" direction. Notably, "in-bent" portion showed no linear V–I relation before 200 A at 24 K-self-field. Meanwhile, the same rough but "out-bending"

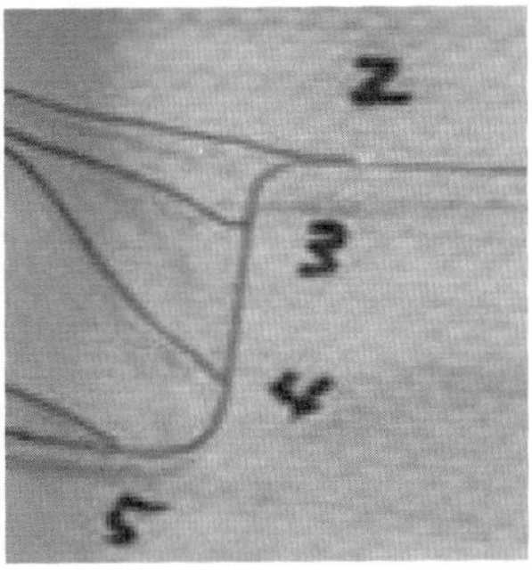

Fig. 6. A tape sample of the HFCS family harshly curved at GE GRC. The Cu strip faces lower semi-plane (labels 5, 4 and 3) and the nickel matrix faces upper semi-plane (label 2). Reproduced with permission from Ref. 55. © 2015 IEEE.[65]

tensile strain originated remnant voltage.[69] Respect to neutral axis, SC filament are asymmetrically disposed, due to banded tape layout.

As expected in this extreme condition, MgB_2 appears to better withstand compression than extension. HFCS strain tolerance has been further explored at 4.2 K/6 T, where a strain limit of 0.12% has been measured.[70]

As a matter of fact, no failure reported so far has been ascribed to improper handling or too high strain during the assembling or operating stage. Nevertheless, especially in a production workshop, a transition from ductile NbTi wires, on which rubber hammers can eventually be used for perfect turn placement, toward "very carefully"[67] handled MgB_2 wire or tape, is not straightforward.

Once the magnet is wound, concerns move toward proving performances and protecting coils from possible quench damages. Since the initial stages of magnet construction,[71] it was clear that MgB_2 conductor stability properties could have been largely enhanced with respect to those of 4.2 K/NbTi, thanks to metal enthalpy densities nearly 100 times greater than at 4.2 K,[54] possibly leading toward "quench free" magnets, i.e., lower scanner downtime. On the other hand, magnet protection from thermomechanical stresses is an issue, being the same enthalpy argument a slowing factor in normal zone propagation.[72,73] Without countermeasures, hot-spots could permanently damage the magnet, while too "quick" protection discharge (with high internal voltages) could be adverse as well, for example, causing damage to coil-to-ground insulation.

First predictions for LFCS-14 (14-filament version) tape indicated a minimum quench energy of the order of several hundreds of *mJ*, in the 15–25 K/0.7–2 T range.[74] Similar results (260–630 *mJ*) have been measured previously on monofilamentary Ni/Cu tape.[75] In Table 3 is reported an approximate comparison between different SC MQEs and normal zone propagation velocities (NPZV).

In a recent paper, MQE and propagation velocities for HFCS conductor have been modeled at 10 K/4.2 T, $I_{op}/I_c \sim 0.8$, considering both a single conductor straight length, and a solenoidal coil. In these

Table 3: Approximate comparison between protection parameters for MRI relevant superconductors. Based on Refs. 74 and 76 for Nb_3Sn NPZV and partially on Refs. 77–79 for NPZV LFCS tape.

SC (at temperature T)	Typical MQE range (mJ)	Typical NPZV range (m/s)
NbTi(4.2 K)	<10	20–40
Nb_3Sn(4.2 K)	<10	2.5–10 (12 T)
LFCS-14 (4:25 K-0:2 T-I_{op}/I_c =0.5)	100–630	3–120 10^{-2}
BSCCO 2223(0 T)	>4400	7–22 10^{-3}(10–40 K)

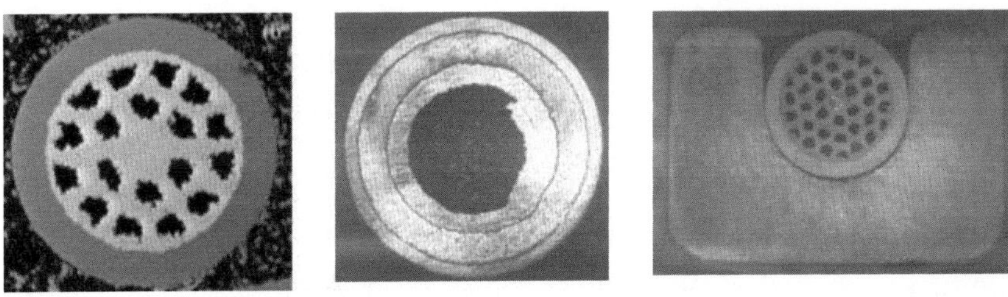

Fig. 7. Cross-sections of HyperTech Research Inc. wires used for MRI magnet design or prototypes. 18+1 wire (left), example of monofilament wire (center), example of multifilamentary WIC (wire in channel) (right). Courtesy of HyperTech Research Inc.[81]

conditions, due to the limited enthalpy difference $H(T_{sc}) - H(T_{op})$,[g] MQE of tens of *mJ* have been predicted for straight conductor, with propagation velocities larger than $20 \cdot 10^{-2}$ m/s. Remarkably, MQE coil simulation result is quite different, being more than 1 *J*. This is mainly caused by the long metastable state (2–3s) before longitudinal quench spreading, thus allowing transited conductor adjacent turns to contribute to stabilization process.[80]

Most recent data from HTR, for first (market available) and second (MRI useful lengths not yet available) generation J_e are reported in Fig. 8. Clearly the second generation outshines the first generation's performances: for external field lower than 4 T, second generation J_e at 20 K appears higher than first generation J_e at 4.2 K.

[g]At temperatures equal to T_{sc}, current is shared with the matrix and power dissipation increases linearly as $(T - T_{sc})$ up to T_c.

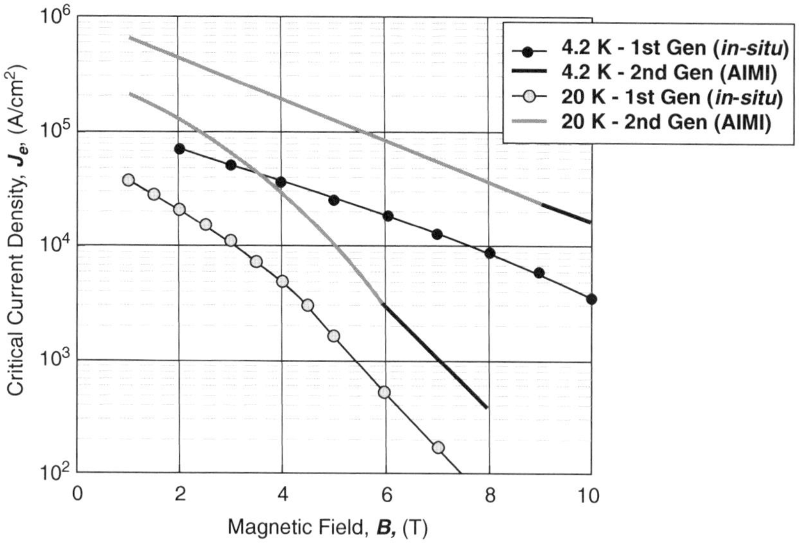

Fig. 8. MgB_2 performances of commercially available wires: *in situ* (first generation) and AIMI (second generation). Courtesy of Hypertech Research Inc.[81]

In the latest state-of-art MRI magnet review article, the target J_e values of MgB_2 wires for MRI applications have been pointed out: at 4.2 K/5 T, a J_e value of $\sim 3-4 \cdot 10^4\,\text{Acm}^{-2}$ would place MgB_2 in direct commercial competition with NbTi, as long as all other relevant issues (joints, n value, stability and winding technique) are competitive as well.[15]

At 20 K/2.5 T, J_e around $2.5 \cdot 10^4\,\text{Acm}^{-2}$ would be compatible with low-field scanners.[15] Both targets are met with second generation wires. First-generation short sample n values measured by HTR, are reported in Fig. 9.

Cross-sections of wires from HTR are reported in Fig. 7. Shown on the left is the 0.84 mm–monel sheath 18+1 multifilamentary wire (first generation used for MIT-DEMO), where each MgB_2 filament is Nb shielded from "+1" copper matrix. In the center, a monocore wire with its typical concentric architecture is shown, presenting, from outside, the monel sheath, a copper stabilization ring, a niobium barrier and the MgB_2 core. A 0.84 mm monofilamentary wire has been used in the MIT-0.5 T/240 project: the copper and the SC

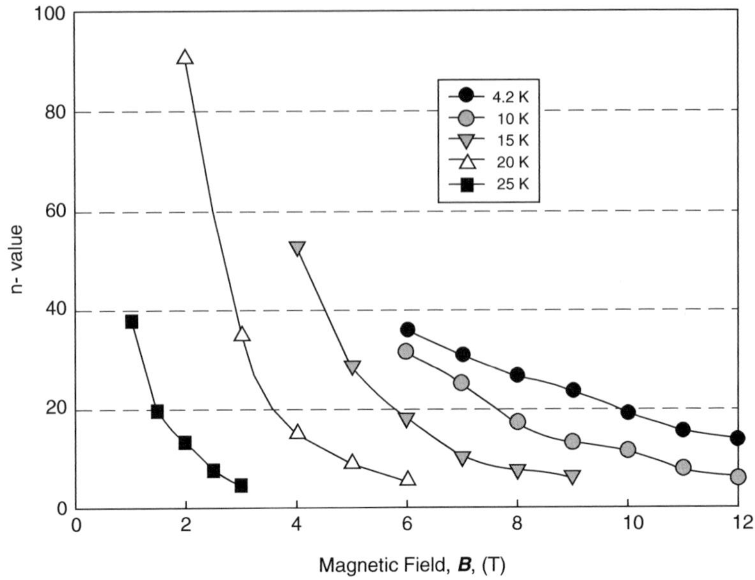

Fig. 9. Typical HTR first generation wire n values versus applied magnetic field. Courtesy of Hypertech Research Inc.[81]

occupy 35% and 12% of the total cross-section, respectively. AIMI (second generation) projected properties of a multifilamentary wire in channel (WIC), as the one depicted in Fig. 7 (right), have been accounted in 1,5 T Ohio magnet design.[28]

For *in situ* wires, Tomsic *et al.* reported strain limit values depending (positively) on the number (or reduction of size) of filaments present in the wire: a 0.4% tensile strain limit has been found for the 19-filament wire with a similar structure of the 18+1 conductor shown in Fig. 7.[82] For the same wire, stability properties have been measured, with MQE and NPZV around 5–20 mJ and $5\text{–}40 \cdot 10^{-2}$ m/s respectively, in approximately 4.2–25 K/1.6–5 T range (data from HTR1 sample in Ref. 79.

At 4.2 K/5 T, a typical reference value for MRI, the cost for NbTi is 1–1.5 \$/kA · m.[6] The MgB_2 HTR 1G/4 K wire is offered between 16 \$/kA · m at 4 T and 30 \$/kA · m at 6 T. Competitive projected prices in ~2020 for the AIMI (2G) wires are between 1.3 \$/kA · m at 4 T–2.5 \$/kA · m at 6 T. At 20 K, the present *in situ* (1G) wire cost is

10, 20 and 50 \$/kA · m at 1, 2 and 3 T, respectively, while projected (~2020) values are approximatively 0.7 \$/TkA · m between 1 and 3 T.[81]

5. Designed MRI Magnets

Thanks to the availability of the MgB_2 conductors discussed above, starting from Iwasa group's DEMO[62,83] and ASG team's MRO,[66,84,85] many MRI magnets have been proposed, some of those have become prototypes and one a product available on the market. In this section just the ones proposed but not yet or not completely proved will be briefly summarized.

Within a project funded by State of Ohio, an expert team at Ohio State University, Case Western Reserve University and HTR (Ohio) have advanced several MRI MgB_2 based magnets in the last two years. Actively shielded 1.5 and 3 T whole-body cylindrical magnets have been proposed during 2013.[26,27] A functional optimization technique, based on current l^2-minimization with Lagrangian multiplier constraint terms and subsequent discrete optimization[86] was applied to a large bore layout (patient gantry diameter larger than 700 mm), considering 2G AIMI HTR wire potential performances. The 3 T coil design exhibits a peak field of 5.15 T and a global length of 1.8 m (for a quantitative understanding of the link between the two quantities see Fig. 14 in Ref. 25), with a magnet homogeneity of 11 ppm on a 45 cm DSV. The magnet working temperature is 10 K.

At the ASC2014 conference, in Prof. Flükiger's plenary talk a second whole-body cylindrical 1.5 T magnet designed by OSU-CWRU-HTR team has been shown. Target values are quite in line with typical parameters commercially available in MRI market,[87] as summarized in Table 4.

The 8+2 coil design has the smallest coil ID of 1 m and forecasted uniformity of 1 ppm within 42 × 44 cm ($Z \times R$), fully consistent with present commercial available scanners. Notably, the peak magnetic field is 5.4 T. The magnet would be conduction cooled at a target temperature of 10 K.

Table 4: Comparison between typical 1.5 T MRI magnet, based on Ref. 87, an OSU-CWRU-HTR (Ohio) 1.5 T design, based on Ref. 81.

	Typical 1.5 T cylindrical MRI magnet	Ohio MgB_2 1.5 T magnet
Length (cm)	125–170	140
Outer Diameter (cm)	190–210	197
Stored Energy (MJ)	2–4	3.74
5-gauss line Z × R (m)	4 × 2.5	3.7 × 2.7
Amp-length (kA*km)	15–25	18

To achieve this remarkable result, which is not possible with the present off-the-shelf conductors, projected long length 2G AIMI wire performances have been considered (see Fig. 8), with a required J_{op} of 178 A/mm^2 (5.4 T/10 K). From reported stored energy, amp-length and inductance values (242 H),[28] it is possible to figure an approximate total need of 100 km of wire, naturally with uniform performances. Considering stability and stress-hoop requirements, a wire in channel conductor (WIC, as depicted in Fig. 7) has been chosen. Short length coils (100 m), wound with the chosen WIC, are under testing at OSU.[28] MgB_2 diffusion in MRI on the long run, has been predicted by many authors; the success of this project, the first 1.5 T total-body magnet based on MgB_2, would be a milestone in establishing MgB_2 as a concrete candidate for a wider range of MRI products than the presently proved ones.

Within the same group, an interesting MRI viable product analysis has been performed, comparing actively (or not) shielded, 1.5 T/0.5 T full-body or head/extremity dedicated scanners. 1.5 T actively shielded, head/extremity dedicated magnet has been chosen for offering higher value, while minimizing product cost. Besides, the most relevant impact was predicted within wealthiest citizens of developing areas, where higher liquid helium costs, and lower MRI scanner units per million people available, could favor local market acceptance.[88] Finally, HTR is developing an open split multicoil MgB_2 magnet for MRI imaged guided radiation treatment, see

Fig. 10. HTR active shield open-split magnet for radiotherapy purposes (left, courtesy of HyperTech Research Inc.), 3 T/10 K active shield whole-body magnet based on MgB_2 (on the right, reproduced with permission from Ref. 27 © 2013 ISMRM).

Fig. 10. As described in Ref. 89, it would be introduced in ViewRay Renaissance System 1000 MR, thus substituting present LHe bath cooled 0.35 T magnet based on NbTi. The operating temperature would be less than 10 K, dry cooled, a DSV of 40 cm is claimed with a peak field on the conductor of 3 T.[81]

6. Prototypes

6.1 *MIT Way: Demo*

The first attempt of making a "DEMO" coil for MRI magnet with *in situ* 1G wire was reported by Iwasa's group.[62,83] The magnet design was not focused on a real MRI scanner, but was driven to show feasibility of two new technical approaches: adopting the *in situ* 1 km-class MgB_2 wire, manufactured by HRT, and introducing solid nitrogen (SN) as thermal reservoir source. The inner bore claimed was 700 mm, i.e., compatible with a whole body system. On the other hand the total length of the assembled cylindrical coils was less than half of its internal diameter, highly limiting the possible field of view, compatibly with the demonstration nature of MIT MgB_2-MRI project in that stage.

The cold mass was directly cooled by the SN volume (50 liters, nearly five times the wire volume) using the same nitrogen to impregnate the wire in consideration of its superior thermal conductivity

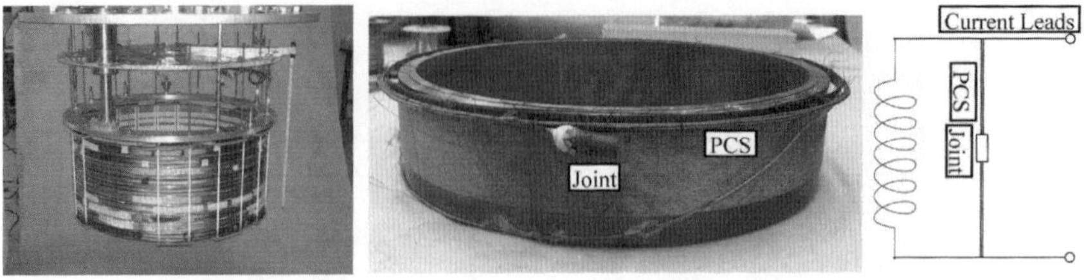

Fig. 11. MIT Demo prototype, on the left. Reproduced with permission from Ref. 62. © 2008 IEEE. On the right, conceptual scheme and module coil of the MIT 0.5 T/240 mm MRI dedicated magnet. Reproduced with permission from Ref. 54. © 2014 IEEE.

and similar stiffness compared to more common epoxy. The SN can was cooled through conduction by a double stage 10 K-class cryocooler. It is relevant to note the reported need of turning off the cold head to eliminate its vibrations in order to create a noise-free environment. Likely the stiffness of the cryocoolers–coil connection was too high to effectively damp vibrations even if a poor thermal contact was reported.[62] The presence of this kind of field fluctuations is not tolerable in an MRI scanner as discussed above, but the long ride-through allowed by SN thermal capacity could nicely satisfy the field stability requirements, once out of steady-state electromechanical interactions are managed and an optimized cryocooler duty-cycle is found.[h] For its superior specific heat in the temperature range of interest (10–20 K) solid Neon has been proposed for commercial systems, albeit at a higher cost.

The adopted wire was an "18 + 1" multifilamentary Nb/Cu/monel 1G wire provided by HTR (see Fig. 8), suitable for conduction cooled low-field MRI applications. The wire was insulated with s-glass sleeve, but facing the more technically challenging W&R modality, after heat treatment, the insulation broken in some parts exposing the coil to ground short-circuit. The magnet was composed by ten separated coils spliced trough SnPb joints, with a typical resistance value below 10^{-6} Ohm. The designed target central field

[h]Frequent On/Off compressor capsule cycles could damage cryocooler performances increasing cold head servicing.

was 0.61 T, corresponding to 1.2 T as peak field on the conductor at nominal current of 100 Amps. The estimated I_{op}/I_c ratio could have reached 0.64 at a working temperature of 14 K. However, the magnet showed resistive voltage starting at current value of 10 A (likely due to short length wire defects or a medium-long length inadequate n value), and quenched at 87.7 A. The quench was induced by some short circuits, which finally damaged some coils.[90]

Even if the "DEMO" system didn't reach target performances, it had the great merit of proposing and proving feasibility of an original design devoted to MgB_2 introduction in MRI.

6.1.1 *MIT way: phase II magnet (actually not a prototype)*

With the goal to proliferate the adoption of MgB_2, persistency has to be achieved. Starting from this consideration Iwasa's group developed joints for the unreacted "18 + 1" wire used in DEMO and proposed a 10 K/SN submerged/0.5 T whole body MRI magnet, named Phase II, based on the same wire. The magnet should have worked in persistent mode, and its design was based on eight coils, with wire lengths varying between 1.94 and 2.92 km. Coils would be resistively-series connected, but with SC switches on each separate coil (Coil–Module approach) to permit global persistency and easy substitution of eventually defective superconducting loops at the same time. Quench protection would have been based on "detect and heat", i.e., active detection quick enough to allow heating and transition ($\ll$1 sec) of a small magnet portion (2%) in order to maintain normal-state wire temperature within 300 K, thus preserving the conductor from too high thermal stresses.[90]

6.1.2 *MIT way: 0.5 T/240 mm magnet*

The same ideas explored with the Phase II design have been applied to a new 0.5 T/240 mm bore magnet,[54] optimizing several aspects.

(1) The multifilamentary "18 + 1" wire was abandoned in favor of a monocore wire, due to the higher reliability of SC joints between unreacted monofilament wires previously obtained

by Iwasa's group.[39] At the same time, this solution (monocore superconducting wire), normally not stable with LTS, has been proven to be flux-jump-free, thus suitable to be employed in a low-field MRI magnet.[53] The monel/Cu/Nb matrix conductor provided by HRT (1G), has a bare diameter of 0.84 mm and after s-fiber insulation reached 1 mm. For the Demo magnet, the same dimensions were 0.85/1.01 mm respectively.[53,62]

(2) The magnet size has been reduced from a whole body (coil ID 860 mm) to a dedicated magnet (coil ID 276 mm) with a final warm bore of 240 mm and a homogeneity designed value of less than 200 ppm on 120 mm DSV, before shimming. Just to give a comparison, the commercially available GE Optima MR430s 1.5 T orthopedic scanner (previously ONI MSK-Extreme) has a cylindrical patient gantry ID of 218 mm, without RF coils, as reported in Ref. 91; the 1.5/3 T limb dedicated GE project (see Sec. 6.2) has a magnet warm bore ID of 250 mm, for a final patient gantry of 200 mm diameter, while the 1.5 T-YBCO based orthopedic scanner prototype had a warm bore ID of 242 mm.[23] The MIT choice is quite in line with present commercial products and research efforts, thus suitable for practical clinical use.

The diameter reduction involved a coil length reduction of a factor ten, with the longest measuring 273 m. This choice is compatible with the development stage of MgB_2 wire and its detect-and-heat protection, being the inductance reduced from 25–0.74 H while the energy stored approximately passed from 120–4 kJ. Consequentially, the 10 m long superconducting switch, acting as dump resistor during quench discharge, would reach a lower peak temperature (70 K) with respect to Phase II design (300 K), with a discharge time of $\sim$3 s.[92] In a first attempt, two coils have been separately tested reaching persistency with a resistance of some 10^{-9} Ohm for each coil, but not at nominal current (70 and 50 A respectively, compared to target value of 102 A) due to two different problems on the two coils.[54] Recently, three coils (of the eight needed to assemble the MRI magnet) have been manufactured and tested, both singularly and noteworthy together, proving the concept (superconducting joints, switch, coil-module approach, active protection) illustrated in the Phase II design. The SC joint technology used guaranteed a splice resistance lower than $3 \cdot 10^{-12}$ Ohm, coherent with the MRI requirements discussed previously. The assembled coils maintained persistency at nominal current (100 A) up to 10 K, and figures showed a slightly lower value (95 A) in the designed temperature operational range of 10–15 K. The field decay measured at nominal current and $T > 10$ K was due to a resistive behavior of one coil module, with a resistance $R \sim 2.8 \cdot 10^{-7}$ Ohm. Solid nitrogen will be added in next phase of the project.

Some aspects have to be proved in future, for example: due to low inductance value, MRI requirements for persistency could need a coil-module resistance of one order of magnitude lower (better n value of long conductor lengths or lower I_{op}/I_c), the coil-module approach could reveal some difficulties when applied to an MRI scanner, such as different decay rates in different modules, or gradient EM and thermal interactions with separate loops. As for all MRI conduction cooled magnets, cryocooler vibrations should be managed; even if the LN thermal reservoir will contribute positively, it has to be experimentally optimized and proved in a clinical setup.

All the efforts and original techniques introduced at MIT Francis Bitter Laboratory, from cryogenics to protection, are an example of the not straightforward way which has to be opened to spread MgB_2 use in MRI magnets. At the same time, the project tries to address the healthcare demand in underdeveloped or underserved areas.

6.2 *The GE way: 1.5 T–3 T/10"*

The application of MgB_2 in MRI appears to be linked to the introduction of new cryogenic design in magnet scanners. In fact, at least at present, MgB_2 cannot compete with the cheaper and well-grounded LHe NbTi technology. Therefore, the higher T_c of MgB_2 has to be exploited either in terms of higher thermal stability when used at 4.2 K or directly at higher working temperatures. In both cases, however, MgB_2 adding value to the system: simpler and cheaper cryogenics, installation, recovering, etc.

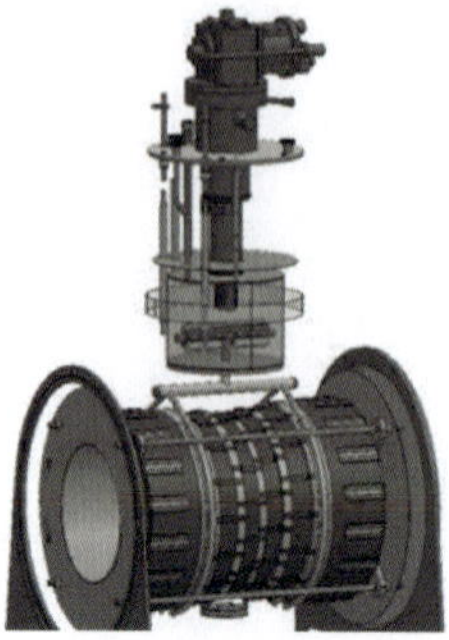

Fig. 12. GE GRC 1.5–3 T MRI dedicated magnet concept (left), reproduced with permission from Ref. 93. (© 2014 ELSEVIER). Assembling of some coils on the right, reproduced with permission from Ref. 55. © 2015 IEEE.

It is interesting that both temperature modes have been merged in the GE 1.5 T–3 T/10" warm bore MRI limb-dedicated magnet.[55] The cryogenic design has been widely reported by Stautner *et al.* in Ref. 63. It is based on a double-circuit thermosiphon heat-pipe design, where sealed primary cooling tubes surround the cold mass, while the secondary directly faces the re-condensing cryocooler 4.2 K cold finger (see Fig. 12). The heat exchanger, between the two, benefits of a finned surface, and it is located on the top of the magnet, the secondary functioning as thermal reservoir at the same time. The system has been engineered to work at 4.2 K (3 T central field) or at 20 K (1.5 T central field). In the first case, both circuits are filled with He (only 2 liters of LHe for the secondary), whereas the primary is filled with hydrogen and the secondary with He when operating at 20 K. The heat exchanger would be thermally controlled by appropriate resistors to avoid hydrogen solidification. Reasonable ride through capability for (1) cryogenic self-recovering of the magnet in case of quench, (2) short power outage and (3) cold head maintenance is expected. A limited volume (four liters as proposed, i.e., small enough to be hidden within scanner structure) external reservoir would temporarily store evaporated gas from the primary circuit that would successively reclaim it when functioning is restored.

The proposed design takes into account a mechanical decoupling between the cryocooler structure and the cold mass, thus reducing vibration artefacts as discussed previously. R&W technique was chosen from the beginning of the project, being considered more practical for industrial use.[67] The first conceptual design was based on a 3 T/28 cm warm bore (not actively shielded) cylindrical magnet, with relatively low hoop stresses (around 20 MPa). A 600 m long 24-filament MgB_2 wire, 0.72 × 1.53 mm, monel-sheathed with a Ni and Cu matrix by HTR, has been adopted for the first test coil — of 300 mm ID, wet wound with epoxy. The solenoid was conduction cooled by flexible Cu braids, at that stage. Quench conditions were quite near to short sample values, but when the coil was held at the target current of 120 A at 20 K, a 15 mV remnant voltage was measured, corresponding to a n value of six, apparently ascribed to wire performance.

In a second attempt a 990 m long lower J_e tape from CS, with the same HFCS layout depicted in Fig. 14, has been employed to test a second and a third solenoid of 300 mm ID, wound and cooled with the same technique as the first coil.[69] Cu laminated wire design has been chosen to keep wire stability against magnet quench.[55] Here too, a low n value type remnant voltage of 6 mV was found holding the coil number 2 at 120 A/20 K, while similar values for coil number 3 were measured. By testing medium length samples (10 m) lowest I_c portions have been located. After the detection, the wire was visually investigated and large abnormal particles, likely responsible for disrupting connectivity, have been identified.[69] After monitoring the boron particle size, the tape performances increased both in terms of I_c and of n values. CS supplied the improved wire for magnet manufacturing, with over 3 km single length wire delivered. The not actively shielded magnet (Fig. 12) designed by GE GRC consisted of six coils of three different dimensions, with the two smallest ones being 595 m long (S-coil), for a global length of 8,080 m.

At last, the first S-coil has been realized, again cooled through direct conduction by a sandwich of two Al plates, and it was ramped up to 120 A/21 K showing no remnant voltage on the millivolt scale. The coil quenched at 308 A and 14 K under a peak self-field of nearly 1.9 T, quite comparably to short sample values, thus confirming that no degradation has been added during the winding process. Fast

Table 5: Double-coolant 1.5–3 T GE GRC MRI prototype main parameters as described in Refs. 55, 67, 69, 93, 94.

Primary cooling circuit filled with He (or H)			
Operating Temperature (K)	4.2(20)	Homogeneity on 15 cm DSV (peak to peak, ppm)	5
Central field (T)	3(1.5)	Coil ID — all coils (mm)	300
Peak field (T)	3.85(1.92)	Warm bore Diameter (mm)	250
Operating Current (A)	250(125)	Patient Gantry diameter (mm)	200
Stored Energy (KJ)	247(62)	Coil OD-max (mm)	412
Inductance (H)	7.9	Magnet length (mm)	558

ramping rates have been used, 1 A/s (corresponding to a peak field rate of 6 mT/s), without untimely quenches.[94] In Table 5, some magnet properties have been summarized.

Recently, test results for the other five coils have been reported.[55] As for S coil, the Al plate sandwich conduction cooled technique has been adopted during single coil assessment. However, all the coils showed a larger remnant voltage per unit length. The worst coil of these last wire batches exhibited a smaller voltage for unit length, 3.5 mV/1000 m compared to 6 mV/990 m previously reported for Coil number 2, whereas the best one was within the 0.1 mV/1000 m range.

The large coil (OD 412 mm) responsible for the higher remnant voltage did not show the poorest performances in the innermost layers, where higher peak field appears, but inside the coil. In conclusion, tape non-uniform performances have been measured in the five new coils. In view of operating the 1.5–3 T magnet in the persistent mode, superconducting joints have been developed reaching 150 A at 20 K in self-field. As outlined in Sec. 3.2 a large degradation of the joint performances in-field has been reported.

The assembled magnet would be tested soon, yet with simple direct conduction cooling at first. At 20 K, even the best coil manifested a remnant voltage that, once applied to entire magnet length, would result in a global resistance of some 10^{-6} Ohm, i.e., not allowing 1 ppm/hour normally required decay rate.

In the light of recent CS results for its HFCS tape at 20 K/1.4 T (see Fig. 5), a critical current of nearly 375 A is claimed, whereas approximately 225 A is found in same conditions with an older wire used for the S-Coil.[93] By analogy, extrapolating the 1.9 T/20 K critical current values, it is possible to figure a transition from a magnet working point I_{op}/I_c larger than 0.7 to something less than 0.5. This wire improvement could mitigate the problems faced.

The GE GRC project cryogenics is groundbreaking with respect to the present MRI technology, pursuing 4.2 K-limited cryogen volumes for next generation scanners. Thanks to the superior ride-through with respect to LTS materials, MgB_2 appears as the optimal choice for exploiting this technology. Besides, a migration from He cryogenics to cheaper and greener H 20 K cryogenics would be linked to the availability of new superconductors: again, MgB_2 would be first choice, but to achieve this result, better wire performance is the *sine qua non*. It is clear that the introduction of hydrogen as a coolant in a medical environment initially faces objective safety concerns and emotive subjects as the "Hindenburg syndrome." On the other hand, the H wide-spreading in automotive industry could largely help, also in terms of industrial parts and engineering.

The success of GE prototype, with a central field of 3 T, would set the highest value ever reached for an MgB_2-based MRI magnet. Moreover, the magnet would exceed present commercial available scanners in the limb dedicated segment (GE Optima MR430s, 1.5 T NbTi and Esaote O-Scan, 0.31 T NdFeB) laying down in the higher field race mainstream.

7. A Product: Paramed MROPEN

At the 2006 Radiological Society of North America (RSNA) conference in Chicago,,[95] Paramed s.r.l. (a small-cap Italian company presently owned by ASG Superconductors) presented as a work-in-progress its MROPEN system, derived from the ASG MRO prototype magnet[84] based on MgB_2. The scanner was born combining Paramed founder's, Roberto Duttos' will to move from extremity imaging towards a larger arena, ASG management's bet to push

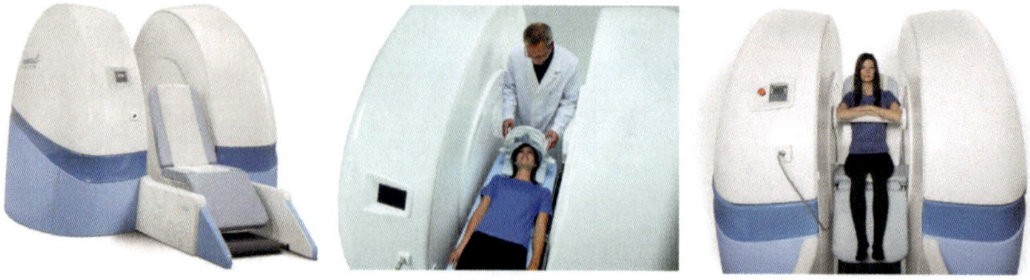

Fig. 13. First MgB_2 based scanner reaching the market: Paramed MROPEN in different examination modalities.

MgB_2-cryogenless frontier towards MRI, and finally Columbus' interest to step into larger scale production of *ex situ* MgB_2 wire.

The goal with MROPEN was an open scanner with a new U-shape design (see Fig. 13), with free access both from rear and front side, establishing a new standard for open-sky MRI systems.[i] Transversal field[j] allows higher efficiency RF architecture (solenoid), not possible with classical cylindrical magnets. Moreover, because the patient coronal plane is not perpendicular to magnetic field, the U-shape permits to use effective superficial coils, a configuration being impossible with C-shaped systems.

Looking to CS conductor properties along the years (see Fig. 14 and Table 6) during magnet design definition (2003–2005), it was possible to outline a low-field open magnet with iron yoke, producing a higher central field and a wider gap with respect to typical permanent magnets (0.35 T/41 cm), suitable to comfortably accommodate 99-percentile US-man shoulders (52.3 mm).[96] MROPEN central field is 0.5 T with a magnet gap of 60 cm, 2 cm (for each part) used for housing exceeding gradient, transmission boards and covers.[k] Two separate coils, defining the patient gantry space, form the

[i]Its usability was designed to be in the middle between SC GE 0.5 T SP Signa, plus free open space above the FOV and resistive 0.6 T Fonar StandUp MRI, plus larger gantry gap and interventional access to patient form scanner rear.

[j]With respect to patient axis direction.

[k]Using the same geometric constrains, it is still possible to consider a permanent magnet (with some more tons), at the cost, just for sintered NdFeB as "raw" blocks, that could vary between nearly 100 K$ to more than 800 K$, considering commodity market fluctuations between 2009 to now, as discussed previously.

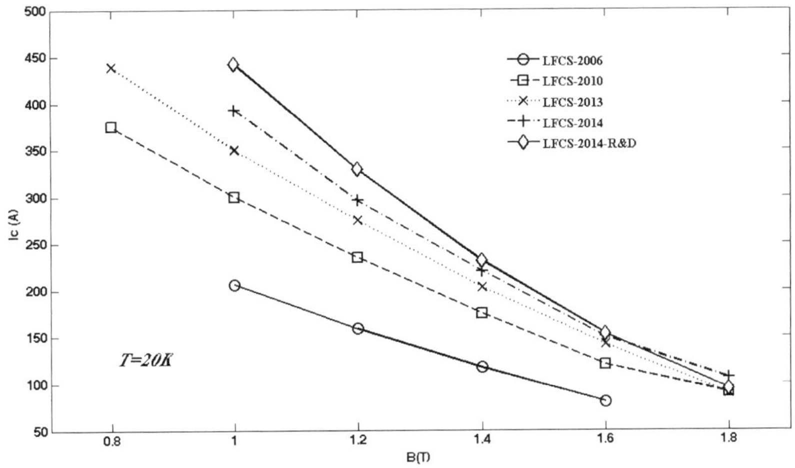

Fig. 14. Typical conductor performances at 20 K for a tape belonging to the LFCS family, along the years. Data set courtesy of Columbus Superconductors.[65]

Table 6: MgB_2 *ex situ* tape families used in the MROPEN MRI magnet production along the years.

	14-filament-LFCS (used before 2010)	12-filament-LFCS (used after 2010)
MgB_2 Filling Factor (%)	10	10
Copper/MgB_2 ratio	1.7	1.5
Twist pitch (cm)	None	30

superconducting magnet, which is conduction cooled by two-stage "10 K class" GM-cryocoolers, one for each coil. The scanner working temperature is in the 11–20 K range. Taking into account the average properties of the wires, the magnet field peak on MgB_2 filament contours reaches 1.3 T on the 14-filament untwisted tape used for magnets produced before 2010.[51,85]

Protection is safeguarded by a digital multichannel customized unit capable of processing events from 0.005–20 s. It manages dynamic magnet-gradient coil interactions as well. In case of quench or emergency, the magnet energy is discharged on external dump resistors. The magnet is driven by an ultra-stable power supply unit at a current of approximately 90 A. This value can be fine adjusted on site, taking into consideration the magnetic flux increase due to extra-passive shielding (steel plates) hidden within MRI suite walls. The same shielding is useful to screen off low frequency magnetic interferences (i.e., elevators, moving cars in neighboring streets, etc.), even in case of active shielded magnet settings. None of the installed MgB_2 coils had any damage, even if some scanners are frequently ramped up and down by end users for their convenience, thus increasing SC coil stress.

Often, radiological centers are housed in building basements, so that the system (weight: 28 Ton), can be installed without many restrictions.[1] In case of cryogenic system faults, MROPEN is perfectly functional up to nearly one hour, giving a reasonable ride-through time considering the absence of any proper thermal reservoir. Remarkably, the scanner is ready to scan no longer than one hour after having recovered the maximum allowed coil temperature. Considering to swap NbTi with MgB_2, fixed magnet design, it would be necessary to add roughly two other cryocoolers (see Fig. 21 in Ref. 15). Thus the costs reduction due to conductor and power supply (assuming persistent mode) would be nearly compensated by the cryocooler costs, while servicing cost would clearly increase and ride-through would be wasted. Swapping standard W&R Nb_3Sn with MgB_2, fixed magnet design, coils would be cured after winding, increasing the production costs for both processing and raw materials, balancing MgB_2 extra conductor length and power supply costs; ride-through would be largely reduced too.

As in the first prototypes,[84,85] no training has been observed on series production. Magnet homogeneity has been largely improved with respect to the one reported in Ref. 85, with all Paramed installed systems within 2 ppm in 300 mm DSV (FWHM). MROPEN 0.5 T has been the first commercial MRI scanner based on MgB_2 available on the EU market in the last quarter of 2009 and in North America (FDA approval) six months later. As of 2014, 20 scanners have been shipped to customers in Italy, UK, North America and recently also Middle East.

[1]MROPEN has been installed in historical buildings also.

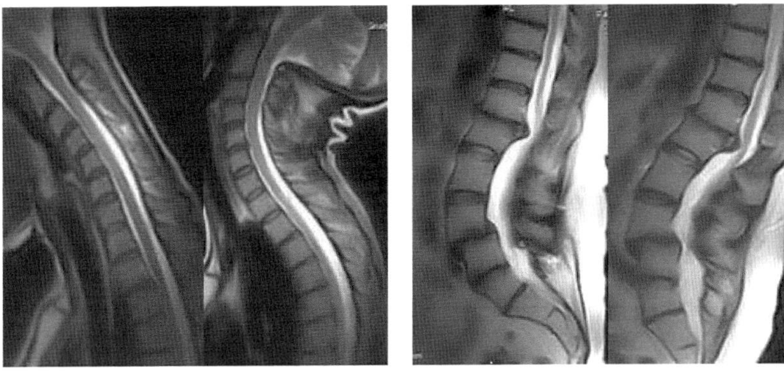

Fig. 15. Images acquired on the MROPEN MRI system in non-conventional patient positions, from left to right: upright cervical spine in flexion and extension, comparison between supine and upright lumbar spine acquisitions on the same patient.

Table 7: MROPEN scanner parameters.

Issue	Value or Method	Issue	Value or Method
Central Field (T)	0.5	Patient Clearance (cm)	56
Peak field (T)	0.9	FOV — Imageable DSV (cm)	36
Magnet Gap (cm)	60	Field Homogeneity on 30 cm DSV (FWHM ppm)	>2 ppm
Magnetic Energy (MJ)	0.25	MgB_2 Coils Steady State Temperature range (K)	11–15
Amp-Length (kA × km)	1.6	MgB_2 Coils max. allowed working temperature (K)	20
Protection	Active with external dump resistor	Operating mode	Driven-mode
Operating mode	Driven-mode	Cooling	Conduction cooled: two double-stages GM cryocoolers

Due to its unique design, the scanner is suitable, other than for regular diagnostic use where it faces competition from higher field (IQ) magnets, for multiposition and upright MRI. Besides, the U-shape layout helps with claustrophobic patients, who have no difficulties to undergo MRI procedures otherwise. Extended interventional applications are forecasted within the near future such as thermoablation, vertebroplasty, cervical and lumbar nucleolysis, pulsed RF treatment, needle biopsy of the musculoskeletal system and ozone therapy.[97] Furthermore, MR procedures combined with spine surgery and neurosurgery are expected shortly. All these applications clearly go beyond the present use of mainstream MRI scanners, such as diagnostic supine examinations, revealing the necessity of suitable niche markets for scanners outside cylindrical 1.5–3 T design.

Towards MgB_2 proliferation in the MRI market, the commercial introduction of the first scanner based on MgB_2 is a milestone already achieved, even if it is limited to a niche.

7.1.1 *MROPEN magnet next steps*

In the very competitive MRI market, magnet cost, normally the most expensive single part within a MRI scanner, has to be shrunk to a minimum to allow product profitability. Due to the large improvement in CS wire performances since 2006, the I_{op}/I_c ratio largely dropped from the initial range to present MROPEN production values, leaving room for series product cost reduction. On the other hand, any change in the magnet design implies engineering and tool investments, which also affect profitability, especially in a niche market. Completed MROPEN strategic marketing positioning, the new sale perspective induced a step forward towards magnet cost reduction goal, decreasing the number of double pancakes needed in a MROPEN magnet. In December 2014, the first prototype of such a magnet was successfully tested.[98]

Once the operation current has been increased, the introduction of superconducting joint becomes a more pressing issue even if, as previously pointed out, the driven mode has proven its value, allowing both solutions.

7.1.2 *Special applications*

An interesting application of conduction cooled MgB_2 magnets has been conceived and built by Prof. Fallone's group[100] at Cross Cancer Institute (CCI), Alberta (Canada), which combines radiotherapy with MRI. A linear accelerator (Linac) has been coupled with a modified MROPEN (forming Phase II assembling), onto a single gantry that rotates around the patient (see Linac-MRI in Fig. 16). This allows contiguous MR imaging during radiation delivery to permit real-time MR guidance of the radiation beam to the tumor even if the latter moves due to breathing. In this way radiation closely conforms to the shape and position of the tumor thus eliminating the irradiation of healthy tissues. The system places the direction of the MR's magnetic field to be parallel to the Linac beam to control the dose distribution of scattered secondary electrons.[101] The use of simple direct conduction cooled technique instead of a liquid He bath greatly simplifies the design of a rotating magnet, thus rendering MgB_2 the ideal candidate for the MRI.

In the CCI LINAC–MR design, there is a large hole in the middle of one of the magnet poles for the Linac beam to have unobstructed access to the patient, thus lowering the magnetic field homogeneity (inducing visible banding artefacts in Fig. 16) with respect to the standard MROPEN. A larger gap and a new bi-planar MRI MgB_2

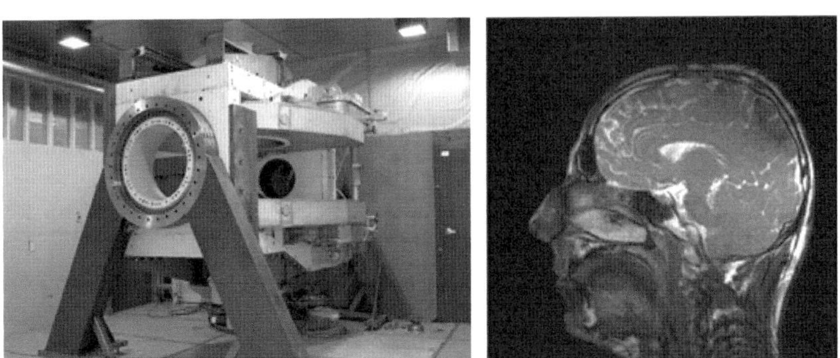

Fig. 16. Linac-MR II Phase II rotating scanner (left) and first human image acquired in July 2014 (right). Reproduced from Ref. 99.

magnet with improved homogeneity is being constructed (by ASG and Paramed) within the Phase III CCI project.[100]

8. Conclusions

MRI magnet technology, based on LHe–NbTi coils is highly stabilized, achieving remarkable results in terms of performances and economic competitiveness required for massive production. The swap of established LHe–NbTi technology with one based on MgB_2 is not straightforward, mainly for its present $I_c(B)$ performances allowing low-field whole body scanners and mid-high field dedicated systems. On the other hand, clinical needs outside the present MRI mainstream are not easily satisfied by a standard production, leaving room for magnets tailored to specific applications based on MgB_2. This material has been proven to fit MRI's highly demanding requirements in a commercial product (MROPEN), which shows its specific value for claustrophobic patients and multiposition imaging, with high forecasted impact in interventional MRI. The constant improvement rate in conductor properties, shown in recent years, will allow MgB_2 to reach its full potential and to enter in direct competition with NbTi.

Up to now, MgB_2 has proven to be the most promising candidate to support the MRI industry in case of unfavorable economic drivers for established LHe cryogenics. Studying, understanding and solving open technical issues is both a social need and an industrial opportunity. MRI's multibillion dollar market can attract the necessary investment capitals to achieve the desired performance criteria for proliferation of a new technology such as MgB_2. At the same time, social and public health considerations can boost this process by public capital.

Acknowledgment

The author is in deep debt with many colleagues and friends for useful discussions and contributions. In particular, Darren Bruce, Alessandro Carrozzi, Giorgio Ricca, Andrea Serra and Stefano Tassano. A special gratitude to all those who helped by sharing their work and knowledge, especially from ASG Superconductors, Columbus Superconductors, Cross Cancer Institute, GE GRC and Hypertech Research.

References

1. R. Damadian, Tumor detection by nuclear magnetic resonance, *Science* **171**, pp. 1151–1153 (1971).
2. P. C. Lauterbur, Image formation by induced local interactions: examples employing nuclear magnetic resonance, *Nature.* **242**, pp. 190–191 (1973).
3. A. N. Garroway, P. K. Grannell and P. Mansfield, Image formation in NMR by a selective irradiative process. *J. Physics C: Solid State Physics.* **7**, p. L457 (1974).
4. W. A. Edelstein, J. Hutchison, G. Johnson and T. Redpath, Spin warp NMR imaging and applications to human whole-body imaging, *Physics in Medicine and Biology* **25**, p. 751 (1980).
5. INUMAC. Retrieved from http://portal.uni-freiburg.de/inumac/copy_of_inumac-?set_language=de [27/01/2015].
6. M. N. Wilson, 100 years of superconductivity and 50 years of superconducting magnets. *Applied Superconductivity IEEE T. Appl. Supercon.* **22**, p. 3800212 (2012).
7. M. Lakrimi, A. M. Thomas, G. Hutton, M. Kruip, R. Slade, P. Davis *et al.*, The principles and evolution of magnetic resonance imaging, *J. Physics, Conference Series* **286**, pp. 12–16 (2011).
8. CONECTUS. Retrieved from http://www.conectus.org/market.html [27/01/2015].
9. Transparency-Market-Research. http://www.transparencymarketresearch.com/pressrelease/magnetic-resonance-imaging-market.htm [27/01/2015].
10. W. J. Nuttall, R. H. Clarke and B. A. Glowacki, Resources: stop squandering helium, *Nature* **485**, pp. 573–575 (2012).
11. USGS. Mineral Commodity Summaries, In: Survey USG, Ed. 2014.
12. USGS. Mineral Commodity Summaries. In: Survey USG, Ed. from 1996 to 2014.
13. VV. AA. British Petroleum Statistical Review of World Energy. 2014.
14. TAGGS. http://taggs.hhs.gov/AwardDetail.cfm?s_Award_Num=R01EB009359&n_Prog_Office_Code=57 [27/01/2015].
15. Y. Lvovsky, E. W. Stautner and T. Zhang, Novel technologies and configurations of superconducting magnets for MRI, *Superconductor Science and Technology* **26**, p. 093001 (2013).

16. M. Tomsic, M. Rindfleisch, J. Yue, K. McFadden, D. Doll, J. Phillips *et al.*, Development of magnesium diboride (MgB2) wires and magnets using *in situ* strand fabrication method, *Physica C: Superconductivity* **456**, pp. 203–208 (2007).
17. E. T. Laskaris, R. Ackermann, B. Dorri, D. Gross, K. Herd and C. Minas, A cryogen-free open superconducting magnet for interventional MRI applications. *IEEE T. Appl. Supercon.* **5**, pp. 163–168 (1995).
18. J. Zimmer, Personal communication. 2014.
19. M. F. Reiser, W. H. Semmler and H. Hricak, *Magnetic Resonance Tomography* (2008).
20. G. Gilgrass, K. Marken and T. Holesinger, http://www.htspeerreview.com/2004/pdfs/presentations/d_oxford_hts_mri_spi_review_04.pdf [27/01/2015].
21. H. Kitaguchi, O. Ozaki, T. Miyazaki, N. Ayai, KI. Sato, S. Urayama *et al.*, Development of a Bi-2223 HTS magnet for 3 T MRI system for human brains. *IEEE T. Appl. Supercon.* **20**, pp. 710–713 (2010).
22. Y. Terao, O. Ozaki, S. Kawashima, K. Saito, T. Hase, H. Kitaguchi *et al.* Analysis of an abnormal event in a 3-T MRI magnet wound with Bi-2223 tape conductors. *IEEE T. Appl. Supercon.* **24**, pp. 1–5 (2014).
23. R. A. Slade, B. J. Parkinson and R. M. Walsh, Test Results for a 1.5 T MRI system utilizing a Cryogen-free YBCO Magnet, *IEEE T. Appl. Supercon.* **24**, pp. 1–5 (2014).
24. T. W. Redpath, Signal-to-noise ratio in MRI, *The British Journal of Radiology* **71**, pp. 704–707 (1998).
25. Y. Lvovsky and P. Jarvis, Superconducting systems for MRI-present solutions and new trends. *IEEE T. Appl. Supercon.* **15**, pp. 1317–1325, (2005).
26. M. Martens, T. Baig, M. Cara, R. Brown, D. Doll and M. Tomsic, An actively shielded 1.5 T MgB2 MRI magnet design, APS March meeting (2013).
27. M. Martens, T. Baig, M. Cara, R. Brown, D, Doll and M. Tomsic, An actively shielded 3 T MgB2 MRI magnet design, 21stISMRM annual meeting and exhibition (2013), p. 2769.
28. R. Flukiger, ed. *Advances in MgB2 conductors*, Applied Superconductivity Conference. Charlotte, NC, USA (2014).
29. V. Rieke, K. Butts Pauly, MR thermometry, *J. Magn. Reson. Im.* **27**, pp. 376–390 (2008).
30. A. Sinanna, S. Bermond, A. Donati, P. Gros, C. Hugon, J. F. Jacquinot *et al.*, Field stabilization of an MRI magnet operating in driven mode, *IEEE T. Appl. Supercon.* **19**, pp. 2301–2304 (2009).

31. Y. Iwasa, Microampere flux pumps for superconducting NMR magnets Part 1: basic concept and microtesla flux measurement, *Cryogenics* **41**, pp. 385–391 (2001).
32. S. Jeong, H. Lee and Y. Iwasa, Superconducting flux pump for high-temperature superconductor insert coils of NMR magnets, *AIP Conference Proceedings.* **614**, pp. 305–312 (2002).
33. A. Otsuka, T. Kiyoshi, S. Matsumoto, K. Kominato and M. Takeda, Drift Compensation of 600 MHz NMR Magnet, *IEEE T. Appl. Supercon.* **17**, pp. 1442–1445 (2007).
34. H. Morita, K. Tanaka, Y. Suzuki and M. Okada, inventors; Hitachi Ltd., assignee. Superconductor connection structure US6921865–2005.
35. J. Nagamatsu, N. Nakagawa, T. Muranaka, Y. Zenitani and J. Akimitsu, Superconductivity at 39 K in magnesium diboride, *Nature* **410**, pp. 63–64 (2001).
36. M. J. Leupold and Y. Iwasa, Superconducting joint between multifilamentary wires 1. Joint-making and joint results, *Cryogenics* **16**, pp. 215–216 (1976).
37. Y. Iwasa, Superconducting joint between multifilamentary wires 2. Joint evaluation technique, *Cryogenics* **16**, pp. 217–219 (1976).
38. P. McIntyre, W. Yu, L. Gan and C. R. Meitzler, Study of Nb3Sn superconducting joints for very high magnetic field NMR spectrometers. *IEEE T. Appl. Supercon.* **5**, pp. 238–241 (1995).
39. Y. Weijun, J. Bascunan, H. Seungyong and Y. Iwasa, A superconducting joint technique for MgB2 round wires, *IEEE T. Appl. Supercon.* **19**, pp. 2261–2264 (2009).
40. D. Nardelli, inventor; ASG Superconductors s.p.a, assignee. Granular superconducting junction-WO2009127956–2008.
41. D. Nardelli, R. Marabotto and A. Laurenti, inventors; ASG Superconductors s.p.a., assignee. Superconducting coil having a granular superconducting junction-WO2009127958–2008.
42. D. Doll and M. Tomsic, inventors; Hyper Tech Research Inc., assignee. Low loss joint for superconducting wire-WO2010088254–2009.
43. Mh. Lakrimi, inventor; Siemens Plc, assignee. Methods for forming joints between magnesium diboride conductors-WO2013107537–2012.
44. F. Tenbrink, A. Aubele, V. Gluecklich, B. Sailer and K. Schlenga, inventors; Bruker Eas Gmbh, assignee. Method for superconducting connection between MgB2 superconducting wires via a MgB2 matrix made from a boron powder compressed element infiltrated with Mg-US8709961–2010.
45. F. Tenbrink, A. Aubele, V. Gluecklich, B. Sailer and K. Schlenga, inventors; Bruker Eas Gmbh, assignee. Superconducting connection

between MgB2 superconducting wires via a compressed element made from HTS powder-US8420558–2010.
46. M. Kodama and T. Wakuda, inventors; Hitachi Ltd, assignee. Superconducting joint and method for manufacturing same-US8385994–2013.
47. M. P. Oomen, J. J. Rabbers, W. Goldacker, S. Schlachter and A. Drechsler inventors; Siemens Aktiengesellschaft, assignee. Procédé de fabrication d'une structure de liaison entre deux supraconducteurs et structure permettant de relier deux supraconducteurs — WO20110392229–2011.
48. Y. Ichiki, T. Wakuda and K. Tanaka, inventors; Hitachi Ltd., assignee. Joint of superconducting wires and method for joining superconducting wires — US8644897–2012.
49. M. Takahashi, K. Tanaka and M. Okada, H. Kitaguchi and H. Kumakura Relaxation of trapped high magnetic field in 100 m-long class MgB2 solenoid coil in persistent current mode operation, *IEEE T. Appl. Supercon.* **16**, pp. 1431–1434 (2006).
50. R. Penco and G. Grasso, Recent development of MgB2-based large scale applications, *IEEE T. Appl. Supercon.* **17**, pp. 2291–2294 (2007).
51. D. Nardelli, S. Angius, A. Capelluto, D. Damiani, R. Marabotto, M. Modica *et al.*, persistent mode MgB2 short windings, *IEEE T. Appl. Supercon.* **20**, pp. 1998–2001 (2010).
52. D. Nardelli, ASG Superconductors Spa, Genova, Italy: private communication (2014).
53. L. Jiayin, J. Voccio, K. Youngjae, H. Seungyong, J. Bascunan, D. K. Park *et al.*, Monofilament MgB2 Wire for a Whole-Body MRI magnet: superconducting joints and test coils, *IEEE T. Appl. Supercon.* **23**, p. 6200304 (2013).
54. L. Jiayin, J. Voccio, H. Seungyong, K. Youngjae, S. Jungbin, J. Bascunan *et al.*, Development of a 0.5-0T/240-mm MgB2 MRI Magnet: assembly design and module coils, *IEEE T. Appl. Supercon.* **24**, pp. 1–5 (2014).
55. S. Mine, M. Xu, Y. Bai, S. Buresh, W. Stautner, C. Immer *et al.*, Development of a 3 T-250 mm bore MgB2 magnet system, *IEEE T. Appl. Supercon.* **25**, pp. 1–4 (2015).
56. L. Guangze, J. B. Zwayer, C. J. Kovacs, M. A. Susner, M. D. Sumption, M. A. Rindfleisch *et al.*, Transport critical current densities and n-values of multifilamentary MgB2 wires at various temperatures and magnetic fields, *IEEE T. Appl. Supercon.* **24**, pp. 1–5 (2014).
57. D. Patel, M. S. A. Hossain, A. Motaman, S. Barua, M. Shahabuddin and J. H. Kim, Rational design of MgB2 conductors toward practical applications, *Cryogenics* **63**, pp. 160–165 (2014).

58. V. Braccini, D. Nardelli, R. Penco and G. Grasso, Development of *ex situ* processed MgB2 wires and their applications to magnets, *Physica C: Superconductivity* **456**, pp. 209–217 (2007).
59. D. Damiani, A. Laurenti, R. Marabotto and M. P. Segre, inventors; ASG Superconductors spa, assignee. Coil with superconductive windings cooled without cryogenic fluids - US80227980–2009.
60. T. Tomaru, T. Suzuki, T. Haruyama, T. Shintomi, N. Sato, A. Yamamoto *et al.*, Development of a cryocooler vibration-reduction system for a cryogenic interferometric gravitational wave detector, *Classical and Quantum Gravity* **21**, p. S1005 (2004).
61. J. M. van Oort, E. T. Laskaris, P. S. Thompson, B. Dorri and K. G. Herd, A Cryogen-Free 0.5 TESLA MRI magnet for head imaging, In: Kittel P, Ed., Advances in Cryogenic Engineering. 43: Springer US; 1998. pp. 139–147.
62. Y. Weijun, J. Bascunan, K. Woo-Seok, H. Seungyong, L. Haigun and Y. Iwasa, A Solid Nitrogen cooled MgB2 demonstration coil for MRI applications, *IEEE T. Appl. Supercon.* **18**, pp. 912–915 (2008).
63. W. Stautner, M. Xu, S. Mine and K. Amm, Hydrogen cooling options for MgB2-based superconducting systems, *AIP Conference Proceedings*, **1573**, pp. 82–90 (2014).
64. Y. Duerst, B. J. Wilm, B. E. Dietrich, S. J. Vannesjo, C. Barmet, T. Schmid *et al.*, Real-time feedback for spatiotemporal field stabilization in MR systems, *Magnetic Resonance in Medicine* **73**, pp. 884–893 (2015).
65. M. Tropeano, Columbus Superconductors, Genova, Italy: private communication (2014).
66. R. Marabotto, L. Bertora, M. Modica, D, Nardelli, D. Damiani, G. Grasso *et al.*, eds., Development of a new MgB2 superconducting open MRI magnet, *Proc Intl Soc Mag Reson Med*, 2006.
67. S. Mine, H. Song, M. Xu, J. Marte, S. Buresh, W. Stautner *et al.*, Test coil for the development of a compact 3 T MgB2 magnet, *IEEE T. Appl. Supercon.* **22**, p. 4400604 (2012).
68. H. Kitaguchi and H. Kumakura, Superconducting and mechanical performance and the strain effects of a multifilamentary MgB 2 /Ni tape, *Supercond. Sci. Tech.* **18**, p. S284 (2005).
69. S. Mine, X. Minfeng, S. Buresh, W. Stautner, C. Immer, E. T. Laskaris *et al.*, Second Test coil for the development of a compact 3 T MgB2 magnet, *IEEE T. Appl. Supercon.* **23**, p. 4601404 (2013).
70. P. Kovac, L. Kopera, T. Melisek, G. Sarmiento, S. Sanz, S. Brisigotti *et al.*, Tensile and bending strain tolerance of *ex situ* MgB2/Ni/Cu tape superconductor, *IEEE Trans. Appl. Supercond.* **99**, p. 1 (2014).

71. R. Musenich, P. Fabbricatore, S. Farinon, C. Ferdeghini, G. Grasso, M. Greco *et al.*, Behavior of MgB2 react & wind coils above 10 K, *IEEE T. Appl. Supercon.* **15**, pp. 1452–1456 (2005).
72. Y. Iwasa, Case studies in superconducting magnets: design and operational issues, Springer, Berlin, 2nd edition (2009).
73. M. N. Wilson, Superconducting Magnets, Clarendon Press, Oxford OX2 6DP, UK (1987).
74. A. Stenvall, A. Korpela, R. Mikkonen and G. Grasso, Stability considerations of multifilamentary MgB2 tape, *Superconductor Science and Technology.* **19**, p. 184 (2006).
75. M. Fu, X. Xu, Z. Jiao, H. Kumakura, K. Togano, L. Ding *et al.*, Minimum quench energy and normal zone propagation velocity in MgB2 superconducting tape, *Physica C: Superconductivity* **402**, pp. 234–238 (2004).
76. H. van Weeren, N. C. van den Eijnden, W. A. J. Wessel, P. Lezza, S. I. Schlachter, W. Goldacker *et al.*, Adiabatic normal zone development in MgB2 Superconductors, *IEEE T. Appl. Supercon.* **15**, pp. 1667–1670 (2005).
77. V. Cavaliere, A. Matrone, G. Masullo, R. Quarantiello, A. Saggese, S. Pace *et al.*, Normal Zone Propagation in a MgB2 Conduction Cooled Test Magnet, *IEEE T. Appl. Supercon.* **18**, pp. 924–927 (2008).
78. U. Gambardella, A. Saggese, P. Sessa, A. Guarino, S. Pace, G. Masullo *et al.*, Stability Measurements in Multifilamentary MgB2 Tapes, *IEEE T. Appl. Supercon.* **17**, pp. 2937–2940 (2007).
79. Hv. Weeren, Magnesium diboride superconductors for magnet applications. Enschede2007.
80. F. P. Juster, C. Berriaud, A. Bonelli, R. Pasquet, H. Przybilski, T. Schild *et al.*, The preliminary study of the quench protection of an MgB2, *AIP Conference Proceedings* **1573**, pp. 746–753 (2014).
81. M. Tomsic, Hyper Tech Research, Inc., Columbus, Ohio, USA: private communication (2014).
82. M. Tomsic, M. Rindfleisch, J. Yue, K. McFadden, J. Phillips, M. D. Sumption *et al.*, Overview of MgB2 Superconductor Applications, *International Journal of Applied Ceramic Technology* **4**, pp. 250–259 (2007).
83. J. Bascunan, H. Lee, E. S. Bobrov, H. Seungyong, Y. Iwasa, M. Tomsic *et al.*, A 0.6 T/650 mm RT bore solid nitrogen cooled MgB2 demonstration coil for MRI: a status report, *IEEE T. Appl. Supercon.* **16**, pp. 1427–1430 (2006).
84. M. Modica, S. Angius, L. Bertora, D. Damiani, M. Marabotto, D. Nardelli *et al.*, Design, construction and tests of MgB2 Coils for

the development of a cryogen free magnet, *IEEE T. Appl. Supercon.* **17**, p. 2196–2199 (2007).
85. M. Razeti, S. Angius, L. Bertora, D. Damiani, R. Marabotto, M. Modica *et al.*, Construction and operation of cryogen free MgB2 magnets for open MRI systems, *IEEE T. Appl. Supercon.* **18**, pp. 882–886 (2008).
86. Y.-C. Cheng, T. Eagan, R. Brown, S. Shvartsman and M. Thompson, Design of actively shielded main magnets: an improved functional method, *Magn. Reason. Mater. Phy.* **16**, pp. 57–67 (2003).
87. T. C. Cosmus and M. Parizh, Advances in Whole-Body MRI Magnets, *IEEE T. Appl. Supercon.* **21**, pp. 2104–2109 (2011).
88. D. C. Kara, Production of a viable product in magnetic resonance imaging using MgB2: Case Western Reserve University; 2014.
89. NCI, MRI Image Guided Radiation Therapy (IGRT) system with low cost, highly stable mag 2011 [27/01/2015].
90. Y. Weijun, J. Bascunan, H. Seungyong and Y. Iwasa, MgB2 coils for MRI applications, *IEEE T. Appl. Supercon.* **20**, pp. 756–759 (2010).
91. J. Tkach, N. Hillman, A. Jobe, W. Loew, R. Pratt, B. Daniels *et al.*, An MRI system for imaging neonates in the NICU: initial feasibility study, *Pediatric Radiology.* **42**, pp. 1347–56 (2012).
92. J. Ling, J. P. Voccio, S. Hahn, Y. Kim, J. Song, J. Bascunan *et al.*, Construction and persistent-mode operation of MgB2 coils in the range 10–15 K for a 0.5-T/240-mm cold bore mri magnet, *IEEE T. Appl. Supercon.* **25**, pp. 1–5 (2015).
93. S. Mine, M. Xu, S. Buresh, W. Stautner, C. Immer, E. T. Laskaris, *et al.*, Development of a 3 T-10' bore MgB2 magnet at GE, *Physics Procedia*, **58**, pp. 224–227 (2014).
94. S. Mine, X. Minfeng, S. Buresh, W. Stautner, C. Immer, E. T. Laskaris *et al.*, Development of a compact 3 T MgB2 magnet, *IEEE T. Appl. Supercon.* **24**, pp. 1–4 (2014).
95. B. Casey, Italian firm turns MRI on its head with new magnet design 2007 [27/01/2015]. Retrieved from http://www.auntminnie.com/index.aspx?sec=ser&sub=def&pag=dis&ItemID=74210.
96. A. R. Tilley and H. D. Associates, The Measure of Man and Woman: Human Factors in Design, Revised Edition, New York (2002).
97. Auntminnie.com. Paramed to debut new MROpen applications in 2015 [27/01/2015]. Retrieved from http://www.auntminnie.com/index.aspx?sec=sup&sub=mri&pag=dis&ItemID=109689.
98. R. Marabotto, Columbus Superconductors Spa,16133 Genova, Italy: private communication (2014).
99. LINAC-MR, http://www.mp.med.ualberta.ca/linac-mr/2015 [29/01/2015].

100. B. G. Fallone, The Rotating Biplanar Linac–Magnetic Resonance Imaging System, *Seminars in Radiation Oncology* **24**, pp. 200–202 (2014).
101. J. Yun, E. Yip, K. Wachowicz, S. Rathee, M. Mackenzie, D. Robinson *et al.*, Evaluation of a lung tumor autocontouring algorithm for intrafractional tumor tracking using low-field MRI: a phantom study, *Medical Physics.* **39**, pp. 1481–1494 (2012).

Chapter 6b

Bulk MgB_2 Permanent Magnets

Akiyasu Yamamoto[*,†,‡] and Kohji Kishio[*]

[*] *Department of Applied Chemistry, The University of Tokyo*
7-3-1 Hongo, Bunkyo, Tokyo 113-8656, Japan

[†] *PRESTO, Japan Science and Technology Agency*
4-1-8 Honcho Kawaguchi, Saitama 332-0012, Japan

[‡] *yamamoto@appchem.t.u-tokyo.ac.jp*

The compound MgB_2 has attractive features for magnet applications, such as high T_c, small anisotropy, long coherence length and transparent grain boundaries. Especially, the absence of weak-links at grain boundaries enables easy and reproducible production of MgB_2 materials with high J_c in randomly oriented polycrystalline forms. In this chapter, we will introduce bulk MgB_2 in permanent magnets. After briefly introducing the general features of bulk superconducting magnets, various fabrication techniques for bulk MgB_2 materials as well as their trapped field properties are discussed. Finally, the trapped field properties of MgB_2 bulks are compared with those of other bulk permanent magnets and prospects for applications of MgB_2 bulks in compact cryo-permanent magnets of the 5 Tesla class are discussed.

1. Introduction

1.1 *Superconducting bulk magnets*

Trapped field in superconducting bulk magnet is a direct manifestation and beautiful example of quantum phenomena on the macroscopic scale. By magnetizing below its transition temperature (T_c), the entire bulk material acts as a compact Tesla class magnet due to the presence of remotely induced macroscopic circulating

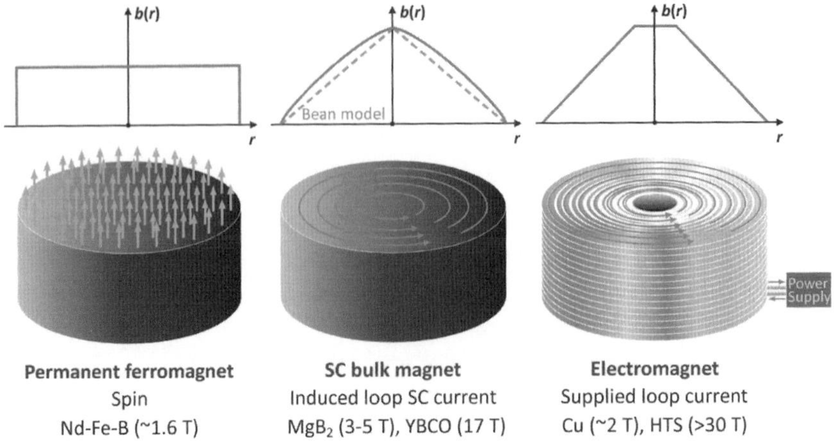

Fig. 1. Schematic images of the radial flux density distribution $b(r)$ and origin of field for permanent ferromagnets, superconducting bulk magnets and solenoid electromagnets.

supercurrents (Fig. 1). Once a magnetized bulk is kept cold, it will behave like a permanent magnet, since the induced persistent current does not decay. Another interpretation about the origin of trapped fields consists in quantum vortices which are compacted and pinned in vortex pinning centers. Generally, the maximum trapped field B_{T} of a cylindrical superconducting bulk can be expressed using the Bean model[1]

$$B_{\mathrm{T}} = \mu_0 J_{\mathrm{c}} \times r \times g, \tag{1}$$

where μ_0 is the vacuum permeability, J_{c} is the circulating bulk current density, r is the radius of the bulk and g is geometrical factor which is 1 for a cylinder of infinite length and $\sim$0.88 for $t = r$ (t: thickness). In analogy to a solenoid electromagnet, the trapped field of a bulk magnet is roughly proportional to J_{c} and to the size (current cross-section). Owing to a high current density and a size dependent nature, one could expect very high fields (>10 T) in bulk superconducting magnets, substantially higher than those of conventional spin based ferromagnets, where the field strength is limited by the intrinsic remanent magnetization (B_r), typically inferior to 2 T. These unique features, *i.e.*, compactness, lightweight property and very high field, make superconducting bulk magnets interesting for novel permanent magnet applications.

For comparison: single-grain bulk magnets based on rare-earth barium copper oxide (REBCO) have been studied as a model material for bulk superconductors. High trapped fields of 2–3 T at 77 K and over 17 T at lower temperatures ~20 K have been achieved by controlling vortex pinning and mechanical strength.[2–6] Various potential applications of bulk magnets have emerged,[7–9] such as NMR, motors and particle accelerators.

1.2 *MgB_2 bulk magnets*

Due to its high T_c (39 K)[10] and recent progress in high performance cryocoolers, MgB_2 is a strong contender for future liquid helium-free applications in the intermediate temperature range between 5 and 30 K. In practice, the distinct advantage of MgB_2 over HTS materials resides in its metallic superconductivity, *i.e.*, simple and symmetric pairing mechanism, long coherence length, weak link-free supercurrent flow across grain boundaries (even in untextured polycrystalline forms[11]), leading to a high carrier density. The nanoscale grain boundaries lead to strong pinning of quantum vortices,[12,13] resulting in uniform, stable and high-field trapping in a bulk magnet.[14]

2. Fabrication Techniques

Polycrystalline MgB_2 bulk materials can be synthesized either by *in situ* (thermally reacting Mg and B)[14–20] or by *ex situ* (sintering pre-reacted MgB_2)[21–24] processes under ambient or high pressure (Table 1). Dense MgB_2 bulk materials with packing factors exceeding 90% can be obtained by employing high pressure or Mg infiltration processes.[18–24] High pressure processed samples are known to yield higher trapped fields.[20–24] On the other hand, the size of those samples is so far limited to values of a few centimeters, while pressureless processes can produce considerably larger bulk superconductors.

Table 1: Overview on synthesis techniques, bulk dimensions and trapped fields for MgB_2 bulk superconducting magnets.

Reference	Synthesis technique	Bulk dimension	Trapped field
Yamamoto *et al.*[14]	*in situ*, pressureless	30 mmϕ × 20 mm	4.02 T (10 K)
Sugino *et al.*[15]	*in situ*, pressureless	30 mmϕ × 10 mm	3.72 T (5 K)
Naito *et al.*[16]	*in situ*, pressureless	30 mmϕ × 9 mm	1.50 T (16 K)
Muralidhar *et al.*[17]	*in situ*, pressureless	20 mmϕ × 10 mm	1.50 T (20 K)
Perini *et al.*[18]	reactive Mg liquid infiltration	55 mmϕ × 15 mm	1.27 T (15 K)
Bhagurkar *et al.*[19]	infiltration and growth	32 mmϕ × 6 mm	2.19 T (5 K)
Fujishiro *et al.*[20]	*in situ*, hot-isostatic-pressing	26 mmϕ × 6.5 mm	2.5 T (13 K)
Durrell *et al.*[21]	*ex situ*, hot-pressing	25 mmϕ × 11 mm	3.14 T (17 K)
Noudem *et al.*[22]	*ex situ*, spark plasma sintering	20 mmϕ × 5 mm	2.5 T (10 K)
Aldica *et al.*[23]	*ex situ*, spark plasma sintering	20 mmϕ × 3.5 mm	3.3 T (12 K)
Fuchs *et al.*[24]	mechanical alloy, hot-pressing	20 mmϕ × 8 mm	5.4 T (12 K)

3. Trapped Field Properties

3.1 *Magnetic hysteresis loop*

For practical applications of bulk superconductors as permanent magnets, the magnetic response and stability, including the change in the applied external field during the initial magnetization process and the residual magnetization value as well as its corresponding time dependent creep after field removal, must be determined. Figure 2(a) shows cyclic hysteresis measurements performed on a zero-field-cooled MgB_2 bulk pair at 10, 20 and 30 K. A pair of disk-shaped MgB_2 bulk materials of 30 mm in diameter and 10 mm in thickness was vertically stacked across a spacer containing a Hall sensor to measure the magnetic flux density. In Fig. 2, H_p, B_r and H_c represent the most important parameters describing the hysteresis loops. H_p, called the *penetration field*, is the external field occurring when the penetrating magnetic field front reaches the center of the disk. B_r corresponds to the remnant local magnetization

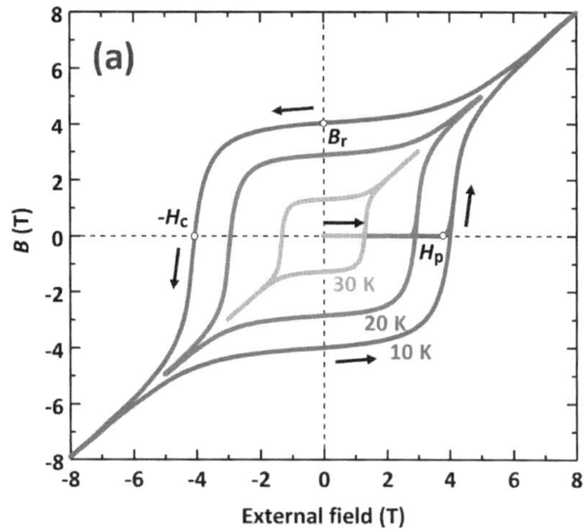

Fig. 2. (a) Magnetic hysteresis loops of a pair of bulk MgB_2 magnets. Arrows indicate the counter-clockwise direction of the loop. Reproduced with permission from Ref. 14. © 2014, AIP Publishing LLC.

at zero field of the descending step and is commonly referred to as the trapped *remnant field*. Similarly, H_c is the *coercive field*. Although the microscopic origins of magnetism are entirely different, the results showed that MgB_2 superconducting bulk magnets are macroscopically equivalent to conventional ferromagnets.

3.2 *Trapped field strength*

Figure 3 shows the temperature dependence of the magnetic field trapped in the pair of bulk MgB_2 magnets. After field cooling magnetization at 10 K, the trapped field was measured as a function of the temperature. A maximum trapped field of 4.02 T at 11 K was recorded, which is 3 times larger than the remnant magnetization (B_r) of neodymium–iron–boron (NIB). As the temperature increased, the field decreased due to the decreasing critical current density of the bulk magnet and vanished at 38 K, the T_c value of MgB_2. The trapped field values at 15, 20, 25 and 30 K were 3.6, 2.9, 2.1 and 1.3 T, respectively. The difference in trapped fields between surface and center is due to geometrical details, the ratio $B_{\text{center}}/B_{\text{surface}} \sim 1.35$ being considered as reasonable when

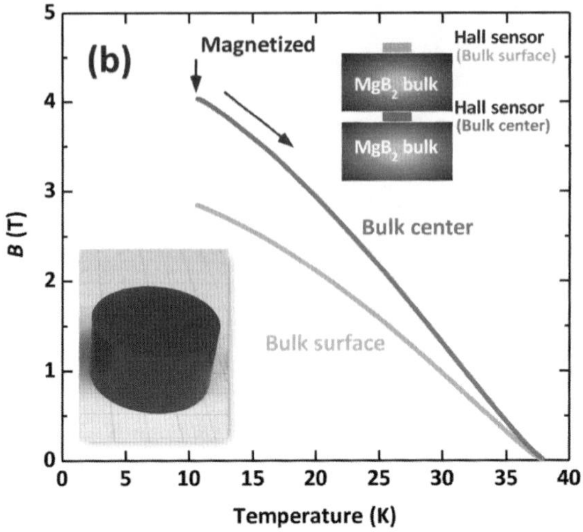

Fig. 3. Temperature dependence of the trapped magnetic field measured at the center and the surface of the MgB_2 bulk pair. The upper right inset shows the schematic cross-sectional configuration of the bulk pair, including two MgB_2 bulk disks and two Hall sensors. The lower left inset shows the appearance. Reproduced with permission from Ref. 14. © 2014, AIP Publishing LLC.

compared with ~1.6, the value obtained from the Biot–Savart law (assumes a constant current density in the bulk material. These results suggest that MgB_2 bulk magnets are promising for use in cryomagnet applications, such as NMR and MRI, at medium temperatures (between 5 and 30 K).

3.3 *Spatial uniformity of trapped field*

Figure 3 shows the local trapped field distribution of a magnetized MgB_2 disk at 20 K. The spatial distribution of trapped field displays a conical shape (Fig. 4(a)), just as expected for an ideal superconducting magnetized disk in which the circulation of supercurrent $J = 1/\mu_0 \times \partial B/\partial r$ is circumferentially uniform. The local field as a function of the position was consistent in different radial directions (Fig. 4(b)), suggesting that the magnitude of the circulating supercurrent is homogeneous. As a result, contour lines for the equivalent local field (Fig. 4(c)) show a true circular shape, which is similar to that of an NIB disk. The superior field uniformity is

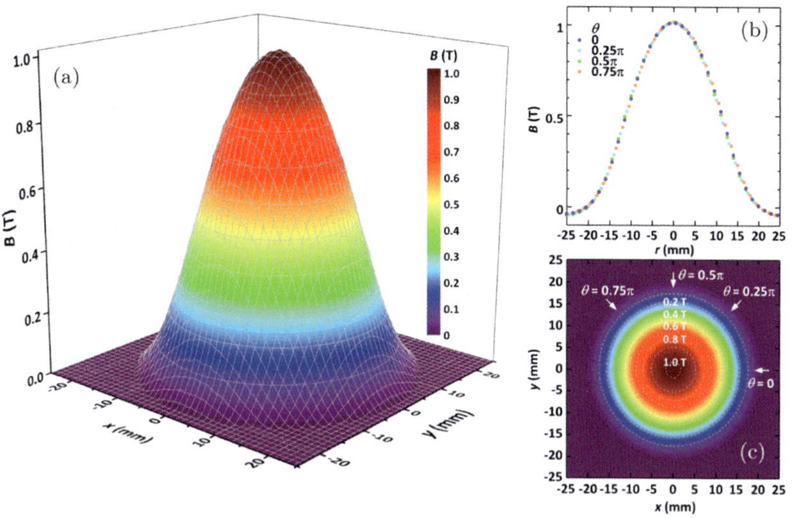

Fig. 4. Trapped magnetic field distributions of a MgB_2 bulk disk magnet at 20 K measured 3 mm above the bulk surface. (a) A three-dimensional (3D) representation of the measured local trapped field distribution, (b) Local magnetic field as a function of the relative position in different radial directions, including $\theta = 0, 0.25, 0.5$ and $0.75\ \pi$, as shown in (c), showing the contour lines of equivalent local magnetic fields. Reproduced with permission from Ref. 14. © 2014, AIP Publishing LLC.

attributed to the metallic superconductivity of MgB_2, i.e., the weak-link-free uniform supercurrent flow and strong pinning occurring at the nanoscale, uniformly distributed grain boundaries.

3.4 *Vortex creep and temporal uniformity of trapped field*

The trapped magnetic field in a bulk MgB_2 magnet pair with different magnetic–vortex–stability conditions is shown in Fig. 5. The time dependence of the trapped field in the bulk pair is shown in Fig. 6. After magnetization at 20 K, a continuous decay in the trapped field was observed. The trapped field after magnetization was 2.87 T and decreased to 2.82 T after 3 days, which corresponds to a loss of 1.7%. After one hour, the trapped field decreased linearly over time in a logarithmic scale and was well fitted with the

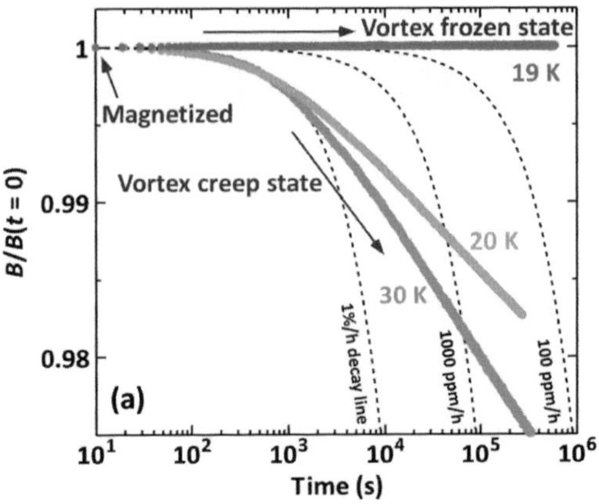

Fig. 5. Trapped magnetic field in a bulk MgB_2 magnet pair with different magnetic–vortex–stability conditions. The field is normalized by the remnant trapped field $B(t = 0)$ immediately after magnetization at 20 K and at 30 K and isothermally maintained for 3 days (vortex creep state). The magnetized bulk magnet pair was magnetized at 20 K and maintained for 7 days at 19 K, which is 1 K less than the magnetizing temperature (vortex frozen state). Reproduced with permission from Ref. 14. © 2014, AIP Publishing LLC.

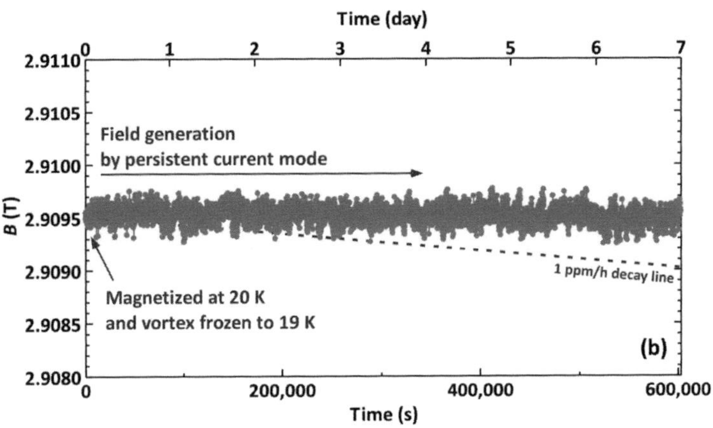

Fig. 6. Trapped magnetic field in a bulk MgB_2 magnet pair: stable trapped field over 1 week generated by circulating supercurrent in persistent current mode. Reproduced with permission from Ref. 14. © 2014, AIP Publishing LLC.

equation $B/B(t=0) = 1.018 - 2.84 \times 10^{-3} \ln(t)$ for $t > 10^4$, where t is dimensionless time standardized by second. When the sample was magnetized and held at a higher temperature the decay rate of the trapped field increased and, at 30 K, was fit by $B/B(t=0) = 1.027 - 4.06 \times 10^{-3} \ln(t)$. These results suggest that the creep is due to the motion of trapped quantum vortices.[25] Thermally activated vortices of pinning potential U are de-pinned by the Lorenz force $F_L = \Phi_0 \times J$ at a rate of $R = \omega_0 \exp(-U/k_B T)$, where Φ_0 is the vortex quantum, ω_0 is the vibration frequency of the vortex and $k_B T$ is the thermal energy.

The thermal activation of vortices can be suppressed by operating at a temperature lower than that used to establish the critical state.[26] To further suppress the thermal activation of vortices, the bulk pair was maintained at 19 K, which is 1 K below the magnetizing temperature. By slightly lowering the temperature, the stability of trapped field was significantly improved, showing a decay of less than 1×10^{-4} T in 1 week (Fig. 6). The fitted decay rate $B_T(t)/B_T(0) = 1.00 - 4.79 \times 10^{-6} \ln(t)$ confirmed the extremely slow magnetic creep rate of <0.1 ppm/h. This creep rate is identical to that specified for MRI scanners. Thus, MgB_2 bulk magnets are quasi-permanent magnets with ultimate stability at operating temperatures.

4. Conclusions

The trapped field as a function of temperature, position, time and external field clearly shows that bulk polycrystalline MgB_2 magnets represent novel strong Tesla-class of quasi-permanent magnets that can produce magnetic fields exceeding 5 T with excellent field uniformity, very high magnet stability and a response to cyclic fields similar to that of spin-based magnets. The mechanism by which the high, uniform and stable field in bulk MgB_2 develops is in contrast with the one observed in HTS materials and is attributed to grain boundaries acting as homogeneous vortex pinning sites rather than as current-blocking defects. Given irreversibility fields exceeding 10 T at 20 K,[27] we expect that even higher fields can be trapped by increasing the diameter of bulk MgB_2 magnets and their current density.[15] This can be achieved either by tuning the electronic- and microstructure or by hybridizing with other HTS materials. On the other hand, iron and/or rare-earth elements, which are considered as essential for spin-based permanent magnets, are not critical for superconducting bulk magnets because their high magnetic fields originate from macroscopic electromagnetic induction due to a persistently circulating supercurrent. Therefore, bulk liquid helium-free superconducting quasi-permanent magnets using MgB_2 are very interesting for novel permanent cryomagnet applications.

Acknowledgement

We thank A. Ishihara, T. Akasaka, M. Tomita of RTRI and K. Iwase, J. Shimoyama of UT for experimental assistance and discussion. This work was supported by a Grant-in-Aid for Scientific Research from Japan Society for the Promotion of Science under Grant Nos. 22860019 and 23246110, and by Japan Science and Technology Agency, PRESTO.

References

1. C. P. Bean, Magnetization of high-field superconductors, *Rev. of Mod. Phys.* **36**, p. 31 (1964).
2. D. A. Cardwell and N. H. Babu, Processing and properties of single grain (RE)-Ba-Cu-O bulk superconductors, *Physica C* **445**, p. 1 (2006).
3. N. H. Babu, Y. H. Shi, K. Iida and D. A. Cardwell, A practical route for the fabrication of large single-crystal (RE)-Ba-Cu-O superconductors, *Nat. Mater.* **4**, p. 476 (2005).
4. J. H. Durrell, A. R. Dennis, J. Jaroszynski, M. D. Ainslie, K. G. B. Palmer, Y. H. Shi *et al.*, A trapped field of 17.6T in melt-processed, bulk Gd-Ba-Cu-O reinforced with shrink-fit steel, *Supercond. Sci. Tech.* **27**, p. 082001 (2014).
5. G. Fuchs, P. Schatzle, G. Krabbes, S. Gruss, P. Verges, K. H. Muller *et al.*, Trapped magnetic fields larger than 14 T in bulk YBa2Cu3O7-x, *Appl. Phys. Lett.* **76**, p. 2107 (2000).
6. M. Tomita and M. Murakami, High-temperature superconductor bulk magnets that can trap magnetic fields of over 17 tesla at 29 K, *Nature* **421**, p. 517 (2003).

7. A. M. Campbell and D. A. Cardwell, Bulk high temperature superconductors for magnet applications, *Cryogenics* **37**, p. 567 (1997).
8. Y. Iwasa, S. Y. Hahn, M. Tomita, H. Lee and J. Bascunan, A "persistent-mode" magnet comprised of YBCO annuli, *IEEE T. Appl. Supercond.* **15**, p. 2352 (2005).
9. T. Oka, Processing and applications of bulk HTSC, *Physica C* **463**, p. 7 (2007).
10. J. Nagamatsu, N. Nakagawa, T. Muranaka, Y. Zenitani and J. Akimitsu, Superconductivity at 39 K in magnesium diboride, *Nature* **410**, p. 63 (2001).
11. D. C. Larbalestier, L. D. Cooley, M. O. Rikel, A. A. Polyanskii, J. Jiang, S. Patnaik *et al.*, Strongly linked current flow in polycrystalline forms of the superconductor MgB2, *Nature* **410**, p. 186 (2001).
12. R. Flükiger, H. L. Suo, N. Musolino, C. Beneduce, P. Toulemonde and P. Lezza, Superconducting properties of MgB2 tapes and wires, *Physica C* **385**, p. 286 (2003).
13. T. Matsushita, M. Kiuchi, A. Yamamoto, J.-I. Shimoyama and K. Kishio, Essential factors for the critical current density in superconducting MgB2: connectivity and flux pinning by grain boundaries, *Supercond. Sci. Tech.* **21** p. 015008 (2008).
14. A. Yamamoto, A. Ishihara, M. Tomita and K. Kishio, Permanent magnet with MgB2 bulk superconductor, *Appl. Phys. Lett.* **105**, p. 032601 (2014).
15. S. Sugino, A. Yamamoto, J.-I. Shimoyama and K. Kishio, Enhanced trapped field in MgB2 bulk magnets by tuning grain boundary pinning through milling, *Supercond. Sci. Technol.* **28**, p. 055016 (2015).
16. T. Naito, T. Sasaki and H. Fujishiro, Trapped magnetic field and vortex pinning properties of MgB2 superconducting bulk fabricated by a capsule method, *Supercond. Sci. Technol.* **25**, p. 095012 (2012).
17. M. Muralidhar, A. Ishihara, K. Suzuki, Y. Fukumoto, Y. Yamamoto and M. Tomita, Optimization of the fabrication process for high trapped field MgB2 bulks, *Physica C* **494**, p. 85 (2013).
18. E. Perini, G. Giunchi, L. Saglietti, A. F. Albisetti, A. Matrone and V. Cavaliere, Magnetic Field Trapping in MgB2 Bulks and Inserts, *IEEE T. Appl. Supercond.* **21**, p. 2690 (2011).
19. A. G. Bhagurkar, A. Yamamoto, N. H. Babu, J. H. Durrell, A. R. Dennis and D. A. Cardwell, Synthesis of dense bulk MgB2 by an infiltration and growth process, *Supercond. Sci. Technol.* **28**, p. 015012 (2015).
20. H. Fujishiro, T. Naito and T. Yoshida, Numerical simulation of the trapped field in MgB2 bulk disks magnetized by field cooling, *Supercond. Sci. Technol.* **27**, p. 065019 (2014).

21. J. H. Durrell, C. E. J. Dancer, A. Dennis, Y. Shi, Z. Xu, A. M. Campbell *et al.*, A trapped field of >3 T in bulk MgB2 fabricated by uniaxial hot pressing, *Supercond. Sci. Technol.* **25**, p. 112002 (2012).
22. J. G. Noudem, M. Aburras, P. Bernstein, X. Chaud, M. Muralidhar and M. Murakami, Development in processing of MgB2 cryo-magnet superconductors, *J. Appl. Phys.* **116**, p. 163916 (2014).
23. G. Aldica, M. Burdusel and P. Badica, Trapped magnetic field in a (NdFeB)-(MgB2) pair-type bulk magnet, *Physica C* **505**, p. 18 (2014).
24. G. Fuchs, W. Haessler, K. Nenkov, J. Scheiter, O. Perner, A. Handstein *et al.*, High trapped fields in bulk MgB2 prepared by hot-pressing of ball-milled precursor powder, *Supercond. Sci. Technol.* **26**, p. 122002 (2013).
25. P. W. Anderson and Y. B. Kim, Hard superconductivity — theory of motion of Abrikosov flux lines, *Reviews of Mod. Phys.* **36**, p. 39 (1964).
26. M. R. Beasley, R. Labusch and W. W. Webb, Flux creep in type-2 superconductors, *Phys. Rev.* **181**, p. 682 (1969).
27. V. Braccini, A. Gurevich, J. E. Giencke, M. C. Jewell, C. B. Eom, D. C. Larbalestier *et al.* High-field superconductivity in alloyed MgB2 thin films, *Phys. Rev. B* **71**, p. 012504 (2005).

Chapter 6c

Applications for very Fine MgB_2 Wires

Sonja I. Schlachter* and Wilfried Goldacker

Karlsruhe Institute of Technology, Institute for Technical Physics
Hermann-von Helmholtz-Pl. 1, 76344 Eggenstein-Leopoldshafen, Germany
**sonja.schlachter@kit.edu*

This chapter gives an overview on various applications realized with very fine MgB_2 wires produced with the powder-in-tube (PIT) technique. The high critical temperature, the possibility to tailor the sheath composite according to the requirements of an application and the light weight makes MgB_2 an ideal candidate for different applications. After a short introduction some considerations about the production and properties of very fine PIT MgB_2 wires are presented showing that the influence of inhomogeneities can become large when their size is in the order of the small filament diameter.

In the main part details about already realized and possible application of very fine MgB_2 wires will be presented. In the first application these wires performed as current leads for adiabatic demagnetization refrigerator (ADR) magnets, and as valves on satellites or as LH_2 level sensors. For advanced ADRs used in space-flight, fine MgB_2 wires will be an ideal candidate once the current carrying capability is further improved. Finally, the possibility to use fine MgB_2 wires in high-current low ac-loss cables will be discussed, as the thin MgB_2 strands can easily be assembled using standard cabling techniques.

1. Introduction

The first MgB_2 wires, produced shortly after the discovery of superconductivity in this compound[1] were produced by exposing $100\,\mu$m thin boron (B) fibers with a tungsten core to Mg vapor at temperatures of 950°C.[1] Even though the wires were inhomogeneous and very brittle after the reaction, they carried macroscopic transport currents of approximately 100 A/cm^2 in a magnetic field of 8 T at $T = 5$ K. Due to their brittleness these short wires were not suitable for any application, however, they demonstrated well the potential of MgB_2 material and gave hope that quick advances could render MgB_2 an ideal candidate for applications. Only a few weeks after the first brittle MgB_2 fibers had been produced, the first labs succeeded with the production of MgB_2 powder-in-tube (PIT) wires and tapes (see also Chapter 4).[2–7] A variety of sheath materials and sheath composites was used by the different labs and it soon became clear that advanced tailoring would lead to conductors fulfilling the requirements of various applications. For example, Cu could be used as electrical stabilizer, whereas a stainless steel component in the sheath composite could mechanically reinforce the conductors. The quick advances in wire and tape production made MgB_2 conductors an interesting candidate for new applications. A group at NASA Goddard Space Flight Center was looking for thin MgB_2 wires that could be used as current leads for magnets and valves for the XRS instrument on the satellite ASTRO-E2 (renamed "Suzaku" after launch). Mechanically reinforced MgB_2 PIT wires with a diameter of only 310 μm developed at the former Forschungszentrum Karlsruhe (now Karlsruhe Institute of Technology, KIT) for this purpose were successfully tested and space qualified. With the launch of "Suzaku" this became the first technical application of MgB_2 wires. The availability of these thin mechanically stable MgB_2 wires quickly allowed the adaptation to another application requiring MgB_2 wires with low current carrying capability: thin MgB_2 wires, which were identified as ideal candidates for the use as LH_2 or LNe level sensors.[8–10] In order to realize those applications, MgB_2 wires with diameter down to 50 μm have been produced meanwhile.[11] Thin MgB_2 wires with diameters below 100 μm can still carry currents of several A at 4.2 K, self-field, or can be used with reduced currents at elevated temperatures or at higher magnetic fields. Thin MgB_2 wires can also be used for small coils by the react and wind process. Indeed, the critical bending diameters below which the critical current decreases due to crack formation is markedly smaller than for thicker wires. Bending experiments on stainless steel reinforced wires with a diameter of 400 μm did not show any hints for irreversible degradation for bending radii down to 10 mm.[12]

With decreasing wire and filament diameter the bending properties become even better. Using a 0.5 mm thick wire with a core thickness of 0.32 mm,[13] a coil with a diameter as small as 10 mm, built by "wind & react," carried a critical current of about 100 A. Thin MgB_2 wires could also be used as strands for the fabrication of low ac loss cables. Due to the fact that they can be produced with round cross-sections and small diameters, usual cabling processes similar to those used for NbTi and Nb_3Sn conductors can be applied.

It is much easier to tailor MgB_2 wires than cuprate high-temperature superconductors (HTS) for fulfilling the requirements of special applications. In addition, their considerably higher high T_c value with respect to that low temperature superconductors (LTS), e.g., NbTi and Nb_3Sn makes them attractive for particular applications. Some examples of already existing fine MgB_2 wires and possible applications will be reviewed after considerations regarding the production of fine MgB_2 wires or fine MgB_2 filaments.

2. Considerations for Production of Very Fine MgB_2 Wires

It has been observed that the superconducting properties of mono and multifilamentary MgB_2 wires and tapes can strongly vary with the diameter and thickness of the filaments. The choice of precursor powders and sheath material, as well as the whole deformation process including the final heat treatment are strongly interlinked, so that the change of only one parameter may lead to considerably different properties of the final wires. The conductors which are produced in order to study the influence of precursor powder quality, grain size or doping on the superconducting properties are often monocore wires with diameters in the range of 1 mm. Their filament diameters are often in the range of hundreds of microns, so that reaction layers at the filament-to-sheath interface or impurity phases with the size of several microns play only a minor role. In many cases, these preliminary studies are performed on short wires, in order to avoid starting the fabrication with very thick metal tubes, requiring a large number of deformation steps. Short prototype wires have another advantage: sausaging effects occurring during strong deformation from larger initial diameters can often be avoided.

When long or very thin monocore or restacked multifilamentary MgB_2 wires and tapes are produced, the deformation ratios (starting diameter/final diameter) are considerably higher. Additional deformation steps lead to better crushing, mixing and better compaction of the powder, thus improving homogeneity and grain connectivity. This can lead to an increase in critical current density.[14–16] A disadvantage of additional deformation is the occurrence of sausaging effects when the flow of powder is hindered above a certain degree of compaction. Especially conductors with very fine precursor powders, e.g., milled or mechanically alloyed powders, often suffer from sausaging effects. The resulting variations in filament diameter along the conductor lead to reduced current carrying capability and sometimes also to the formation of cracks in the sheath material.

The sheath material itself is also strongly influenced by the deformation process. The increased number of deformation steps necessary to achieve thin wires or tapes causes hardening of the sheath composite.[17] This can cause cracks in the sheath material if no intermediate heat treatments are applied to release the stresses. On the other hand, work hardening of the sheath material during cold deformation can enhance the resistance of the wire against tensile stresses. However, this strongly depends on the sheath material and the temperature of the final heat treatment applied to the wires. If the temperature of heat treatment is low enough to preserve the stresses due to work hardening, the mechanical properties can be strongly improved, as has been observed for SS-sheathed MgB_2 wires, for example.[11,12,18]

With decreasing filament size the influence of impurity phases or reaction layers at the filament-to-sheath interface increases tremendously. An impurity phase with a much smaller size than the filament diameter reduces the effective cross-section for current transport only slightly. However, if the size of such an impurity phase is comparable to the filament diameter in very thin or multifilamentary wires, the current flow can be partly or completely blocked.[15] During the deformation process a large and hard B grain could easily cause a crack if it gets close enough to the sheath material in a very fine filament, whereas in a large filament it could be surrounded by smaller or softer material, thus preventing a damage of the sheath material during deformation.

Similarly, thin reaction layers at the filament-to-sheath interface of thick filaments might have only a slight influence on the transport properties of the conductor. The consequence of a reaction of the precursor powder with the sheath material, is that either the latter diffuses into the filament, forming impurity phases there, or that the diffusion from the filament to the sheath causes a composition change in the filament. In thick filaments only a small section of the filament close to the sheath material might be affected, leading only to a small reduction of the critical current. In small filaments, however, the whole filament might be influenced, resulting in a strong reduction of the critical current.[12,15]

In summary there are many effects influencing the superconducting and mechanical properties of thin MgB_2 wires. In some cases the high deformation rates that are applied to those wires may help to improve the properties (better compaction, better milling and mixing of precursor powders, higher strain tolerance due to work hardening of sheath composite). However, it also happens that the properties become worse, due to a stronger influence of impurity phases or to reaction layers in thin filaments. This explains why the trends for the dependence of critical current on filament diameter presented in the literature are sometimes quite controversial.[15,17,19]

3. Applications for Very Fine MgB_2 Wires

The first technical applications of MgB_2 conductors were realized with thin wires below one meter length. In the early days, most research labs concentrated their efforts on the improvement of superconducting properties rather than on long wire lengths. Since

MgB_2 conductors were at that time not commercially available, it was not yet possible to produce large components, e.g., MRI magnets, motors, generators or power cables. Nevertheless there were some applications where MgB_2 turned out to be the best choice due to its relatively high T_c of 39 K and to the possibility to tailor the conductor fitting the special requirements of an application, even though only small quantities could be produced by single research labs. In the following sections some applications requiring the use of thin MgB_2 wires will be reviewed.

3.1 *Current leads*

A common challenge for all superconducting high current applications consists in feeding the current from a high temperature level (mostly room temperature) to the superconducting component at cryogenic temperature. A current lead should have a high electrical conductance to minimize Ohmic losses but at the same time a low thermal conductance in order to keep the heat losses due to heat transfer along the current lead at a small level. As most metals with high electrical conductivity also show a high thermal conductivity,[20] superconductors are the ideal choice for this application. They can transfer high currents on very small cross-sections without ohmic losses. Due to the small cross-section the heat losses are also small, especially if the used sheath or substrate materials have a low thermal conductance, too. The temperature levels between which the current shall be transferred determine the choice of the superconducting materials. When coming from room temperature, the first section of the current lead is generally made of copper. The copper sections lead the current to a temperature level well below the current sharing temperature of a superconducting material. From there a superconducting section can lead the current to the component at cryogenic temperature. For the superconducting section being at 40 K or above only HTS materials like Bi–Sr–Ca–Cu–O (BSCCO) or *RE*-Ba–Cu–O (*RE*BCO, with *RE* = *rare earth elements like Y, Nd, Sm, Gd,*...) can be used. For applications where the warm end of the superconducting section is at temperatures well below 39 K, MgB_2 can also be an adequate choice. MgB_2 conductors can be produced with a variety of sheath materials with low thermal conductance, thus minimizing the heat transfer from the warm to the cold end.

3.1.1 *Current leads for XRS-2 on the satellite ASTRO–E2/Suzaku*

One example for a current lead application is the XRS–2 instrument that was built for the Japanese satellite ASTRO–E2 (renamed "Suzaku" after the launch).[11,21,22] The high resolution X-ray spectrometer was designed to measure X-rays arising from black holes, centers of galaxies or to study the death and birth of stars in the energy range of 0.3–12 keV with a resolution of about 10 meV. In order to achieve such a high resolution, the detector array consisting of 32 microcalorimeters needed to be cooled to a temperature of 60 mK. To reach this low temperature a cryogenic system with solid Ne, superfluid helium and an adiabatic demagnetization refrigerator (ADR) were used. The anticipated lifetime of 2.5–3 years for the XRS instrument could only be reached by fulfilling the requirement of a minimized heat load. Therefore a special current lead assembly had to be designed, both to power the ADR magnet and for ground operation of fill/vent valves. The required current carrying capability between temperature levels of 17 K (solid Ne) and 1.3 K (superfluid helium) was 2 A for the magnet leads and up to 1 A for the current leads to the valve motors. For these low currents at relatively low temperatures standard BSCCO or YBCO current lead conductors were completely overdimensioned.

At the time the XRS–2 instrument was designed and built, the standard width of BSCCO tapes was approximately 4 mm with a thickness of a few hundred microns. In self-field at 77 K, the current carrying capability of these conductors was in the range of 100 A. Standard REBCO coated conductor tapes had widths in the range of a few millimeters and thicknesses around 50–100 μm with a current carrying capability around 100 A at 77 K, self-field, for a 4 mm wide tape. The large current carrying capability would have been appropriate for the application giving a giant safety margin, however, due to the large cross-section the heat load would have exceeded the tolerable values by far. Cutting coated conductor tapes to reduce the cross-section would have been possible, but at that time the homogeneity of the available tapes was much poorer than today so that reductions in width could have led to disproportionately high reduction of the current carrying capability, thus justifying the choice of MgB_2.

A predecessor of XRS–2, the XRS–1 spectrometer on the satellite ASTRO–E1, that never reached the orbit due to a failure of the rocket during launch, used a current lead assembly with thin YBCO filaments that were bonded lengthwise to two thin-walled fiberglass tubes. At one end those YBCO filaments were heat sunk to 4 K and soldered to Cu-clad NbTi wires which served as electric connection to 1.3 K. In order to minimize the heat transfer from 4 K to 1.3 K a short section of Cu was etched away.[21,23,24] The disadvantage of this current lead design was that the Cu-clad NbTi wires looses superconductivity at around 8 K. Even though a failure has never been observed during the ground tests of XRS–1, thermo-acoustic oscillations in the vent line that sometimes increased the temperature of the 4 K heat sink could have led to a failure of the assembly due to the low transition temperature of the NbTi wires at the low-temperature end. Another disadvantage of this assembly was the brittleness of the YBCO filaments which could have been damaged by the strong vibrations occurring during the launch of the satellite.

For the XRS–2 instrument a new solution that would minimize the risk of failure was desired.[25] Shortly after the discovery of superconductivity in MgB_2 in 2001 it was clear that wires based on this material could be a candidate for current leads. Several labs have tried to produce thin wires being suitable as current leads for XRS–2. Advances in conductor development enabled the fabrication of thin MgB_2 wires with optimized properties right on time to allow the installation in the XRS-2 instrument before ground tests and launch. The wires that fulfilled the criteria of current carrying capability of 1–2 A in the temperature range of 1.3–17 K, low thermal conductance, high mechanical strength in order to withstand the vibrations during the launch of the satellite and

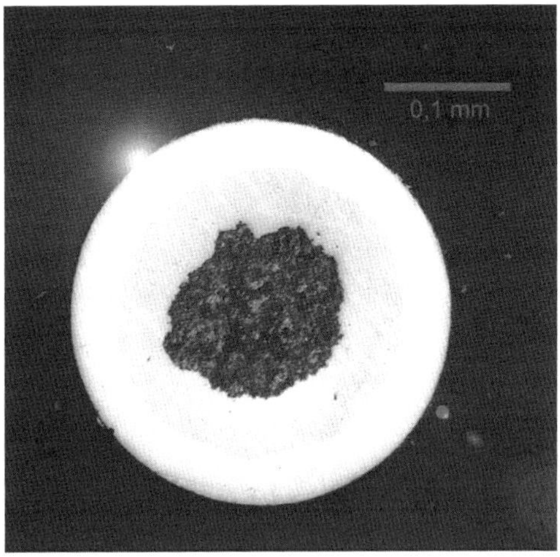

Fig. 1. Cross-section of a thin MgB_2/Fe/SS wire (310 μm diameter) used as current lead for valve motors and ADR magnet of the XRS–2 instrument on the satellite Suzaku. (Reprinted from Cryogenics, Ref. 46, pp. 201–207 (2006), S.I. Schlachter *et al.*, "Properties of MgB_2 superconductors with regard to space applications," © (2006), with permission from Elsevier).

stability during thermal cycling were developed at KIT. Figure 1 shows the cross-section of such a 310 μm thin stainless steel reinforced MgB_2 wire. Besides the steel reinforcement, the sheath composite contains a component with pure iron as compromise between good electrical conductivity to serve as electrical shunt (in case of loss of superconductivity in the filament) and bad thermal conductivity to keep the heat flux along the wire low. A further condition was the chemical compatibility to the MgB_2 core to prevent formation of thick reaction layers and secondary phases.

The field and temperature dependence of the transport current carrying capability of short wire samples was measured up to maximum currents of 2 A with a 9 T Physical Property Measurement System (PPMS) from Quantum Design. Additional measurements were performed at $T = 4.2$ K in a bath cryostat with a maximum current of 300 A shows $J_c(T)$ at different magnetic fields.

The required current of up to 2 A at 17 K to power the ADR magnet of the XRS–2 instrument corresponds to $J_c \sim 13.1\,\mathrm{kA/cm^2}$. As can be seen, in self-field this value can be achieved even for

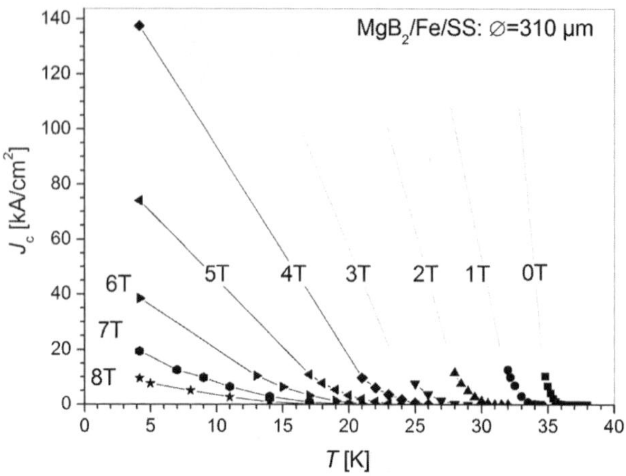

Fig. 2. J_c versus T of XRS–2 MgB_2 current lead wires at magnetic fields from 0 to 8 T. Measurements above 4.2 K were performed in a Quantum Design PPMS with a maximum current of 2 A corresponding to J_c of ~13.1 kA/cm^2. All measurements at 4.2 K were performed in a bath cryostat allowing application of higher currents up to 300 A. Dotted lines are guides to the eye. (Reprinted from Schlachter *et al.*[12] with permission from Elsevier).

temperatures exceeding 30 K, giving a large temperature margin for the current lead assembly. Besides the short sample test, all 0.35 m long current lead wires were tested in LHe up to currents of 12 A to see if there was a damage in the wire.

The mechanical performance of the 310 μm thick MgB_2/Fe/SS wires was excellent: (a) shows the stress-strain dependence of one of the 310 μm thin MgB_2/Fe/SS wires and (b) shows the strain dependence of the normalized critical current $I_c(\varepsilon)/I_c(\varepsilon = 0)$. With increasing tensile strain ε the critical current first increases due to the release of precompression of the filament caused by the stronger thermal contraction of the sheath composite compared to the MgB_2 filament. At $\varepsilon = 0.7$ the critical current starts to decrease, reaching the initial value $I_c(\varepsilon)/I_c(\varepsilon = 0) = 1$ at tensile strain of 0.8%. The tolerable stress for which $I_c(\varepsilon)/I_c(\varepsilon = 0) \geq 1$ is more than 800 MPa. For MgB_2 wires with single component Fe sheath, the tolerable strains above which the critical current starts to decrease are only in

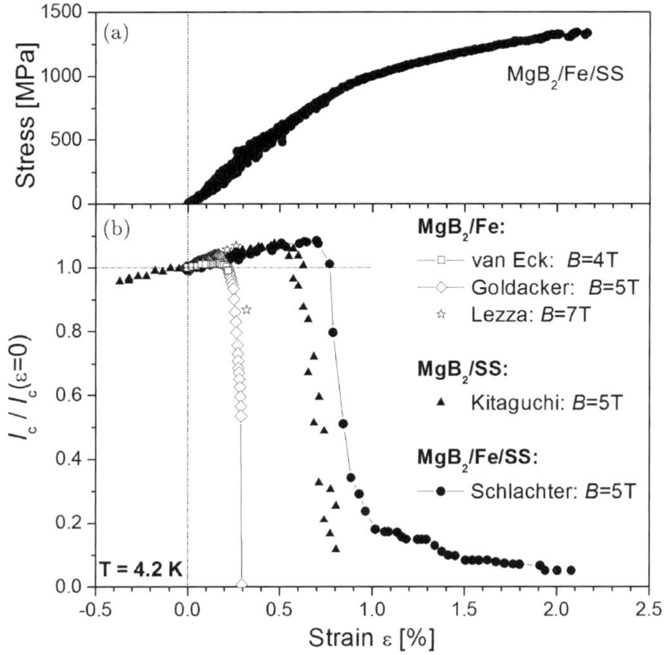

Fig. 3. (a) Stress/strain dependence for a thin MgB_2/Fe/SS wire ($\varnothing = 310\,\mu$m) and (b) normalized critical current $I_c(\varepsilon)/I_c(\varepsilon = 0)$ of this MgB_2/Fe/SS wire and also for MgB_2/Fe and MgB_2/SS versus applied strain (Reprinted Schlachter *et al.*,[12] with permission from Elsevier).

the range of 0.2–0.4%.[26–28] A conductor with single component SS sheath showed a tolerable strain around 0.6%.[29]

The high temperature margin for the anticipated transport currents and the excellent mechanical performance of the 310 μm MgB_2/Fe/SS wires raised hope that those wires would not be influenced by slight temperature changes due to thermoacoustic oscillations in the He vent line and that they could withstand the strong vibrations during the launch of the satellite. This was a clear advantage compared to the old current lead assembly design with brittle YBCO filaments. Similarly, to the old YBCO CL assembly the new MgB_2/Fe/SS CL assembly consisted of 12 single wires. Due to the high mechanical stability there was no need to bond them to a fiber glass tube. Instead, they were assembled in a cylindrical arrangement with G-10 washers and a central Kevlar suspension and additionally fixed to two fiberglass cylinders with epoxy. The two fiberglass cylinders also served as heat sinks. The heat sink closer to the warm end was connected to the 11 K heat exchanger of the He vent line, while the other heat sink closer to the middle of the CL assembly was connected to the 4 K heat exchanger of the He vent line. At the cold end the MgB_2/Fe/SS wires were directly soldered to NbTi wires. The solder points were anchored to the He tank so that the cold end temperature of the current lead assembly was held close to 1.3 K. During the test phase, the CL assembly was tested with nominal currents up to temperatures of 30 K without any problems.

The only drawback of this new type of CL assembly was the higher heat load of 435 μW to the cold heat exchanger and 34 μW to the 1.3 K stage. The heat load of the old YBCO CL assembly to the cold heat exchanger was only 175 μW, the heat load along the NbTi wires to the 1.3 K stage even being negligible.[22]

Nonetheless, the design team decided to include the new MgB_2 CL assembly into the XRS–2 instrument in order to guard against heating problems discovered by the operation team during ground tests of XRS–1. As there was no experience with long-term stability of the new MgB_2 superconductors, and as this material had never been used in any applications before, the design team decided to use both types of current lead assemblies, the new and the old one in parallel, for redundancy. With this redundancy it was clear that the XRS–2 instrument could operate as long as at least one of the CL assemblies would work — even though it was clear that the higher heat load of both assemblies would reduce the lifetime of the XRS–2 instrument.

The ASTRO–E2 satellite was launched on July 7, 2005 and then renamed "Suzaku." During the first few weeks after launch, the XRS–2 instrument performed perfectly. On July 27, 2005 the first cycle of the ADR was performed. A temperature of 60 mK was reached — a world record in space.[21] The cryogen consumption seemed to be low enough to ensure a lifetime exceeding three years.

However, two weeks after opening the dewar guard vacuum vent the LHe coolant was completely vaporized — an inadequate venting of He and Ne gases out of the spacecraft into space has been identified as the reason for the malfunction afterwards.[22]

3.1.2 *Current leads for 50 mK cooling solution with an ADR precooled by a sorption cooler (KIT/CEA)*

The International X-ray Observatory/Advanced Telescope for High Energy Astrophysics (IXO/ATHENA) and Space Infra-Red Telescope for Cosmology and Astrophysics telescope (SPICA) missions will contain detectors that need to be cooled to temperatures as low as 50 mK. In earlier missions temperatures down to 1.5 K have been provided by liquid coolants. In order to extend mission lifetimes to more than five years the liquid coolants shall be replaced by mechanical coolers. In the framework of a European Space Agency Technological Research Program CEA/SBT in France has developed a cooling solution for 50 mK with an ADR (adiabatic demagnetization refrigerator) precooled by adsorption cooler.[30–33] In principle, also an ADR or a chain of ADRs could be used to precool the ADR that is supposed to provide 50 mK. However, at higher temperatures ADR magnets need to produce higher fields and are therefore bigger in size and mass. In the approach of CEA/SBT the higher temperature ADR stage, i.e., the stage with the highest mass is replaced by a ^{3}He sorption cooler that is substantially lighter than the ADR. The sorption cooler is used to pre-cool the 50 Mk ADR stage to a temperature around 300 mK. The pre-cooled ADR can be kept small. In an engineering model focused on the IXO mission[30,32] the NbTi ADR magnet is designed to produce a magnetic field of 0.8 T with a current of 5 A. Two heat sinking stages at 2.5 K and 15 K are available. The current leads of the ADR magnet have been built with thin MgB_2 wires — similar to the XRS–2 instrument. The cold and warm ends are connected to the 2.5 K and 15 K stage, respectively.

The MgB_2 wires were produced by the same group at KIT as the current leads that had been applied for the XRS–2 instrument on the satellite Suzaku. In order to keep the heat load to the cold

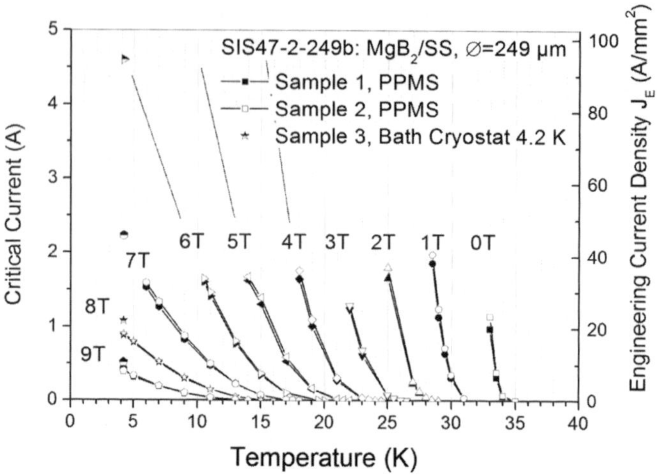

Fig. 4. J_c and J_{ceng} of 2 short MgB_2/SS wire samples with a diameter of 249 μm at various temperatures and applied magnetic fields. The measurements were performed in a 14 T PPMS system with a maximum current of 2 A. Data of a third sample measured to higher currents at 4.2 K in a bath cryostat are shown for comparison.

stage as low as possible, KIT fabricated MgB_2 wires with single component stainless steel sheath without the Fe component and different diameters. While the XRS–2 CL wires had an MgB_2 area fraction of only 20%, the area fraction in the MgB_2/SS CL wires for the ADR of the IXO engineering model could be increased to 34.5%, leading to higher engineering critical current density. Due to the larger superconductor fraction the diameter for the wires that were supposed to carry 5 A at 15 K could be reduced down to 249 μm. Figure 4 shows the critical current of two short 249 μm thin MgB_2/SS wire samples at different temperatures and magnetic fields. The measurements could only be performed up to a maximum current of 2 A due to the limitations of the PPMS system. Nevertheless an extrapolation of the curves suggests that in self-field it should be possible to transfer a current of 5 A through the wires up to temperatures close to 30 K.

Before delivery of the 2 m and 2.6 m long, 249 μm thin wire pieces to CEA/SBT transport current measurements up to a current of 20 A were performed in LHe at KIT. As the wires are not well stabilized due to the lack of a highly conductive sheath component, no higher currents were applied in order not to destroy the wires in case of a quench. During the measurements both wire samples did not show any voltage increase up to the maximum current of 20 A. In order to see if thermal cycling or bending influences the current carrying capability up to this current, the wires were warmed up to room temperature and then measured again in LHe. For the bending tests short pieces of the wire were bent in and against the 'natural bending direction' (the 2 and 2.6 m long 'green' wires were heat treated on a cylinder for MgB_2 phase formation resulting in a curvature after the heat treatment) followed by $V(I)$ measurements up to $I = 20$ A. No sign of degradation could be observed in all of these tests.

The thermal conductivity and an estimation of parasitic heat loads were carried out at KIT and CEA/SBT with MgB_2/SS wires of 720 mm diameter. Measurements performed on thinner MgB_2/SS wires ($\emptyset = 270\,\mu$m) with the thermal conductivity option of a 9 T PPMS system at KIT could only give an upper limit due to limited resolution. The measurements of CEA/SBT and KIT coincide quite well as can be seen in Fig. 5.

CEA/SBT also measured the parasitic heat load to 4.2 K for a 50 mm long sample from the same batch and for a comparable steel wire as a function of the hot end temperature. From this measurement it was calculated that the parasitic heat load would be about 5 μW for two 140 mm long and 249 μm thin MgB_2/SS CL wires.[31] Another source for additional heat load at the cold end of the current leads is the interface between the MgB_2/SS wire and the NbTi wire at the solder joint. Resistivity measurements on 10 mm and 40 mm long joints of NbTi to MgB_2/SS wires with a diameter of 270 μm showed a resistance of 13 and 5 $\mu\Omega$ and parasitic heat loads of about 325 and 125 μW, respectively. In the engineering model cooler for IXO the anticipated joint length was about 100 mm, so that the parasitic heat load from the current leads to the 2.5 K stage would be in the range of 0.1 mW. This is much less than the available cooling power of 10 mW at this temperature. Indeed tests on the assembled engineering model cooler showed that the thermal performance was as predicted.[32]

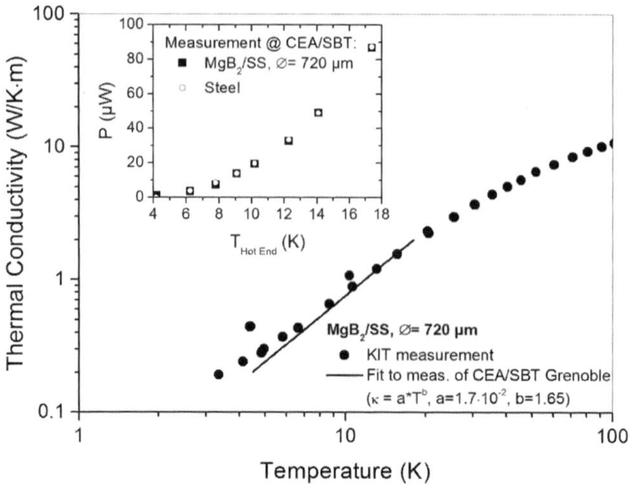

Fig. 5. Thermal conductivity of a MgB_2/SS wire with a diameter of 720 μm measured at KIT[34] and fit to measurement of CEA/SBT.[35] The inset with data extracted from Ref. 31 shows a measurement of parasitic heat load to 4.2 K through a 50 mm long MgB_2/SS wire of the same diameter and from the same batch (black squares) as a function of the hot-end temperature. Comparison with an equivalent stainless steel wire is shown (white circles).

For the SAFARI instrument which will be flown on the SPICA mission CEA/SBT is developing a 50 mK engineering model cooler which is similar to the IXO engineering model.[33] It also consists of a 300 mK sorption cooler combined with a 50 mK ADR stage. The ADR magnet is wound with 102 μm thin NbTi wire and mounted on the 4.9 K cooler structure. The magnet leads are connected to thicker NbTi wires which are joined to 2 Cu interfaces mounted on the cooler external structure. The Cu interfaces serve as heat sinks and connections to MgB_2 current lead wires. The warm end of the CL wires is thermally shunted to the 20 K heat sink. The NbTi magnet produces a magnetic field of 1.1 T with a current of 2 A. This current needs to be carried by the MgB_2 current lead wires which will be of the same type as the 249 μm thin wires used for the IXO engineering model. In principle the MgB_2 CL wires are able to carry the required current up to approximately 30 K. However, for redundancy the two MgB_2 CL wires will be doubled (2×2 wires).

3.1.3 *Current leads for electronic devices*

In principle, very fine MgB_2 wires are also interesting as current leads for superconducting electronics where often only currents in the mA range are flowing. There has been some interest in the past, however results have not been published so far.

3.2 *Level sensors for liquid hydrogen (LH_2) and liquid neon (LNe)*

Hydrogen (H) is one of the future candidates for storage and transport of energy. As the mass and energy density of H at atmospheric pressure and temperature is very low, it is mainly stored and transported in a pressurized or liquefied state. Due to the higher energy density the liquid state is very attractive for many applications, e.g., for the automotive area or as propellants for rockets. For these applications it is essential to be able to determine the remaining content in a vessel with a high precision. For this purpose capacitive level sensors are commercially available. They often consist of co- or tri-axial metal tubes which are vertically placed in the liquid. However, they are often not very accurate due to the small capacity changes caused by the small difference between the dielectric constant of the liquid and vapor phase. Signal changes in the pF range must be handled with a poor signal-to-noise ratio. Furthermore they often suffer from capillary forces which change the height of the liquid-vapor interface in the sensor. Therefore, the resolution and reliability of these capacitive sensors is often very limited.

An alternative solution for measuring the level of cryogens is the use of superconductors with a T_c above the boiling point. A wire/tape consisting of the superconductor material is vertically placed in the liquid. If the T_c value is close enough to the boiling temperature of the liquid the wire/tape is superconducting in the liquid phase ($R_{sc} = 0$) and normal conducting ($R_{nc} > 0$) in the vapor phase. By measuring the resistance of the wire, the height of the liquid can easily be determined. Such level sensors are widely used for liquid He using NbTi with a T_c of $\sim$9 K as a superconducting

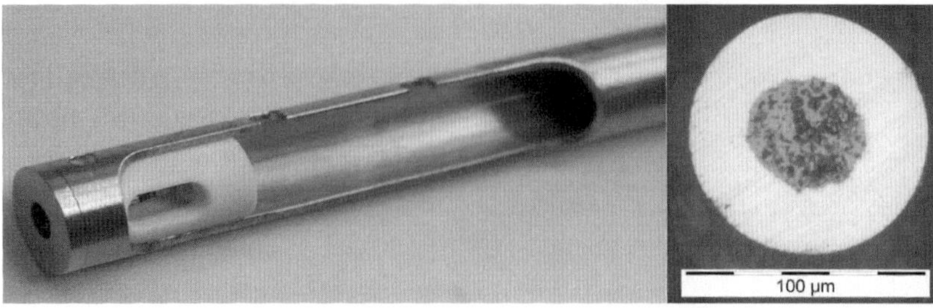

Fig. 6. LH_2 level gauge embodiment with the MgB_2 filament suspended centrally in a SS shelter tube (left, courtesy of Ch. Haberstroh, TU Dresden). MgB_2/SS wire with a diameter of 100 μm (right).

material. With the discovery of superconductivity at 39 K in MgB_2 a superconducting material with a T_c in an appropriate range to build resistive level sensors for LH_2 was available. For saturation pressures between 0.1 and 0.7 MPa the LH_2 boiling temperature varies between 20 and 29 K. In a collaboration of TU Dresden and KIT the first resistive type LH_2 level probes using thin MgB_2 wires with stainless steel sheath and diameters between 50 and 200 μm were built.[8–10]

Figure 6 (left) shows a LH_2 level sensor with a design that is geared to the needs of application in dewar vessels, e.g., for the automotive industry. Here the MgB_2 wire is centrally fixed to a stainless steel tube. The probe contains an initial heater element that raises the temperature of the wire in the vapor phase above T_c. In order to prevent damage of the MgB_2 wire in case of thermal stresses, the wire is fixed to a mechanical spring. The probe design is made for slight damping of level fluctuations as they could easily occur in driving vehicles, for example.

The MgB_2 wires used to build the first LH_2 level probes consisted in a single component stainless steel tube with filament diameters between 50 and 200 μm (see Fig. 6, right). Stainless steel was chosen as sheath material for its high mechanical stability and the relatively high resistance, which allows precise measurements. The specific resistance for the 100 μm thin MgB_2/SS composite wire shown in Figure 6 is $\rho = 59\,\mu\Omega \cdot \mathrm{cm}$.[36] The resistance per length ratio is $R/L = 0.75\,\Omega/\mathrm{cm}$ at 37 K with a variation of only 5% between

37 K and 100 K. For comparison, a 270 μm thin MgB_2/Fe/SS wire showed a specific resistance of 4.7 $\mu\Omega \cdot$ cm and a variation of more than 50% between 40 and 100 K.[37] It can be assumed that this large variation is due to the relatively high conductivity of the Fe component. Another LH_2 level probe wire with SiC-doping, CuNi (7:3) sheath and a diameter of 320 μm showed a resistance per unit length of ~0.05 Ω/cm,[38] i.e., a specific resistivity of ~40 $\mu\Omega \cdot$ cm. A low normal state resistance that goes along with high thermal conductance should be avoided for LH_2 level sensors, because this leads to smaller temperature gradients between liquid and vapor phase and therefore to less precise measurements. Another advantage of the stainless steel sheath of the wire shown in Fig. 1.

Figure 6 is the slight contamination of the MgB_2 filament with Ni, leading to a reduction of T_c to about 34 K with a transition width of ~1.7 K.[39] This brings T_c closer to the boiling point of LH_2 and reduces the need for heating of the wire in the vapor phase. Other methods to reduce T_c of wires for LH_2 level sensors include the addition of dopants like SiC[38,40] or Al.[41] Unfortunately, the addition of dopants can easily lead to a broadening of the transition from normal to superconductivity, resulting in less precise measurements if the temperature of a larger section of the wire in the vapor phase is in the range of the transition. In order to drive the sensor wire normal in the vapor phase the wire can either be self-heated by the measurement current[42] or an external heater can be applied at the warm end of the wire[8] or on the whole wire length.[38,40] In some cases even the solder contact resistance to sheath of the MgB_2 wire can be sufficient for heating. By winding a thin insulated manganin wire as heater around the whole length of 320 μm thin and 200 mm or 800 mm long MgB_2/NiCu sensor wires, a short response time could be achieved by Takeda *et al.*[38] allowing even the detection of level changes during filling and removal of LH_2 from the storage tank. The heating on the whole length of the wire has the advantage that a high precision is obtained, even at low liquid levels when a long portion of the sensing wire is in the vapor phase.

Tests on resistive type LH_2 level sensors made of thin MgB_2/SS wires have shown good results for LH_2 saturation pressures in the range of 0.1–0.5 MPa, i.e., boiling temperatures in the range of 20–27 K. The tested sensors with working lengths of 200–500 mm showed a high linearity and were able to detect level variations distinctly below ±1 mm. The time resolution was better than 1 Hz so that even level variations due to the shaking of the dewar were detected. The power deposition was in the range of 0.2 W for a full tank and 2.2 W for an empty tank.[8,9] A first set of probes for road tests in automotive dewar tanks could be delivered by TU Dresden to the industry in June 2006.[9]

Kajikawa *et al.*[43,44] proposed a new type of LH_2 level sensor with two wires having a similar normal state resistivity, but one being a superconductor and the other one a non-superconductor in the liquid phase. The non-superconductor can either consist only of the sheath material or being from the same wire as the superconductor but either without the phase forming heat treatment or with a second heat treatment that decreases T_c below the boiling point of the liquid. Both wires are immersed in the liquid and the current is flowing through both wires. By subtracting the voltage drops along both wires from each other a signal is obtained that is affected only by the normal state resistivity of the non-superconducting wire in the liquid phase. By using this technique, the obtained signal is not affected by variable conditions of the vapor. In 2012, Kajikawa *et al.*[45] successfully tested a pump system composed of a MgB_2 motor and MgB_2/CuNi level sensors to transfer LH_2. Two level sensors with MgB_2 wire diameters of 155 μm and 92.5 μm which were mounted in a metal dewar and a glass vessel, respectively, were used to detect the liquid levels, these signals being used for controlling the LH_2 transfer achieved by the MgB_2 motor.

The feasibility of MgB_2 wires for the use as LH_2 level sensors has been successfully demonstrated by different groups. In principle such sensors can also be used for measuring the liquid level of Ne, which has a boiling point of 27 K at atmospheric pressure.

3.3 *ADR magnets for space applications*

Many current and future space missions require cooling temperatures below 1 K for detectors. To reach this temperature range adiabatic demagnetization refrigerators can be used in combination with liquid coolants or mechanical coolers. As described in section 3.1., temperatures around 50–60 mK have already been reached by a single stage ADR coupled to a LHe bath at 1.3 K (XRS–2 instrument on the satellite "Suzaku") or by a single stage ADR connected to a sorption cooler (engineering model cooler for IXO).

ADR coolers magnetically cool a paramagnetic salt pill inside the bore of the magnet. The salt pill is magnetized and connected to a heat sink to which the heat caused by the magnetization process is dumped. After opening the heat switch between salt pill and heat sink the salt pill is slowly demagnetized resulting in adiabatic cooling of the salt pill and a device thermally connected to the salt pill. With single stage ADRs continuous cooling is not possible, due to the need to be thermally disconnected periodically from the device to be cooled when the salt pill is magnetized and heat is dumped to the heat sink.

The use of multistage ADRs[46,47] allows continuous cooling when the first (continuous) stage which is thermally connected to the device to be cooled is itself cooled by a second stage operating at slightly lower temperature (Fig. 7). The second stage transfers heat either to a third stage or to a cascade of stages operating at higher temperatures. The various stages can use different paramagnetic materials which magnetize to different temperatures. The last ADR stage operating at the highest temperature can be cooled by the cold

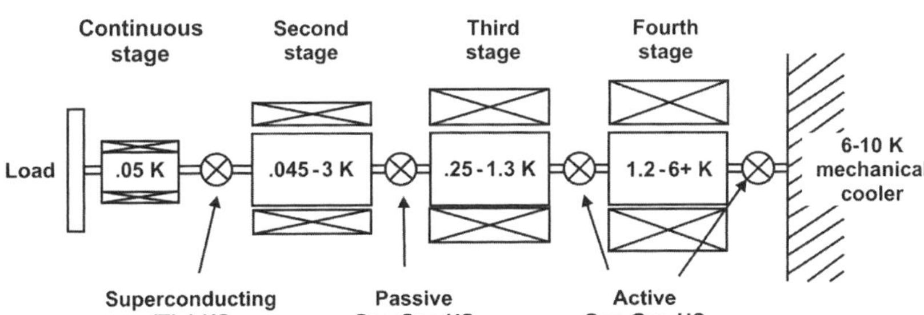

Fig. 7. Schematic of a four-stage continuous ADR (CADR) operating between 50 mK and 6–10 K (With kind permission from Springer Science and Business Media: Fig. 1 in Ref. 47).

stage of a mechanical cooler. As the efficiency of mechanical coolers decreases strongly with decreasing temperature, it is desirable to use ADR magnets operating at temperatures above the T_c of NbTi for the warm stage. In principle Nb_3Sn, MgB_2 and HTS materials are candidates for high temperature space-flight ADRs. As the mass and energy budget for most missions is strictly limited, space-flight ADR magnets need to be light-weight and are generally run with low currents. To achieve the required field of a few Tesla with low currents, high winding numbers and high engineering current densities are required. Commercial HTS materials like *RE*BCO coated conductors or Bi2223 conductors are only available with relatively large cross-sections rendering the magnets too heavy. Tuttle *et al.*[48,49] have designed a light-weight low-current 10 K magnet for space-flight ADRs. They used 180 μm thin Nb_3Sn wires with Cu stabilization and Vanadium reinforcement to fabricate a 3 T magnet operating at a current of 8 A.

Hopkins *et al.*[50] pointed out that MgB_2 has the potential to be applied in space-flight ADR magnets. The big advantages of this material are the low mass density of 2.6 g/cm^3 and the high critical temperature of 39 K. In their paper, Hopkins *et al.*[50] discuss the requirements for an MgB_2 magnet designed to meet the specifications for the European Space Agency's XEUS (X-ray Evolving Universe Spectroscopy) project. The requirements of a central field exceeding 3 T at an operating temperature of 15–20 K and an operating current of less than 15 A places high demands on the conductor to be used. For a wire with a radius of 0.1 mm and an MgB_2 core radius of 75% of the wire radius the critical current density would have to be of the order of 800 A/mm^2 to meet the requirements. Hopkins *et al.*[50] suggested a monocore conductor design with a reinforced Cu or Cu-alloy sheath, the core consisting of an inner *in situ* MgB_2 filament and an outer *ex situ* MgB_2 shell. This conductor design combines the high current density of *in situ* MgB_2 conductors and the low reactivity between the *ex situ* material and the Cu sheath. However, the authors say that the production of such wires with the required small diameter and high fill factor is very challenging.

In the framework of a NASA SBIR phase I project[51,52] Hyper Tech, one of the commercial producers of MgB_2 wires, aimed to develop thin MgB_2 wires which could be used to build a magnet for an advanced multistage ADR in a later SBIR phase II project. The design parameters of the advanced MgB_2 ADR magnet required currents in the range of 5–20 A at a temperature of 15 K in a magnetic field of 3 T, corresponding to engineering current densities in the range of 50–100 kA/cm^2. For full scale ADR magnets wire lengths up to 60 km were envisioned.

In the framework of the project, Hyper Tech investigated different wire drawing techniques and produced thin wires with *in-situ* Mg + B precursor powders and various strand architecture designs. They demonstrated that the superconductor performance of MgB_2 wires can get close to the required current densities J_c and J_e at 15 K, 3 T. Multifilamentary strands with Nb barrier, Cu matrix and monel sheath with diameter of 0.2 mm and small filament diameters of 24–30 μm were able to carry a current of 7 A at 3 T, 4.2 K, corresponding to an engineering current density of 20 kA/cm^2. MgB_2 wires with diameters down to 0.07 mm were successfully fabricated. A small ADR demonstration coil was fabricated using 10 m of 0.22 mm thin undoped MgB_2/Nb/monel wire (Fig. 8).

The coil had an inner bore of 3.2 cm and the engineering critical current density was 56 kA/cm^2 at 3 T, 4.2 K. Even though the achieved engineering current densities were slightly below the requirements, the NASA grant helped to expand Hyper Tech's fine wire manufacturing capabilities. For other applications, Hyper Tech has been able to fabricate wires with lengths exceeding 1,000 m in sizes smaller than 0.3 mm.

An engineering critical current density of 55 kA/cm^2 at 3 T, 4.2 K has also been reported for a 164 μm thin monofilamentary MgB_2 wire with Nb/Cu/SS composite sheath.[11] The MgB_2 fraction of the wire is only 14.2%. By enhancing the superconductor fraction of such a wire, sufficiently high engineering current densities for the use in MgB_2 ADR magnets could be achieved.

The fabrication of low-current MgB_2 ADR coils with fields of 3 T at temperatures around 15 K remains challenging. The engineering

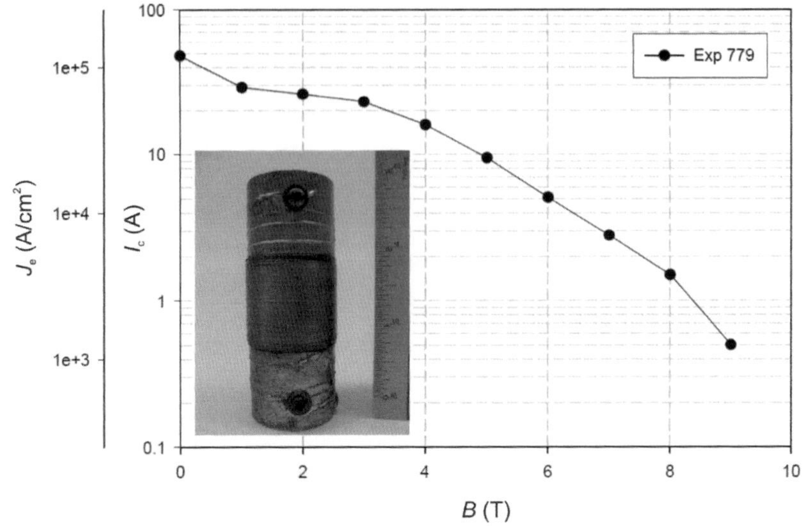

Fig. 8. Performance of an MgB_2 ADR demonstration coil measured at 4.2 K (courtesy of Hyper Tech.[52]).

current densities of thin wires have to be increased and wire insulation has to be applied. However, the techniques for MgB_2 conductor fabrication have been improved constantly in the last years (see Chapters 4e and 4f) and if techniques like internal Mg diffusion[53,54] can be applied to thin wires, MgB_2 magnets for space flight ADRs could be feasible.

3.4 *Thin wires for cabling*

Many superconducting devices require higher currents than can be carried by single wires. This leads to the request for cables assembling single wires or tapes. To reduce AC losses a transposition of the strands is desirable. While cabling is very challenging for BSCCO and *RE*BCO conductors due to the tape geometry, thin MgB_2 wires can easily be assembled with standard cabling techniques. MgB_2 cables were developed with different geometries: twisted and braided cables,[55–61] Roebel[60] and Rutherford style[62] geometries as collected in Fig. 9. The most attractive version is a braided cable which gives opportunity for very complex structures (see Chapter 6d), the use of

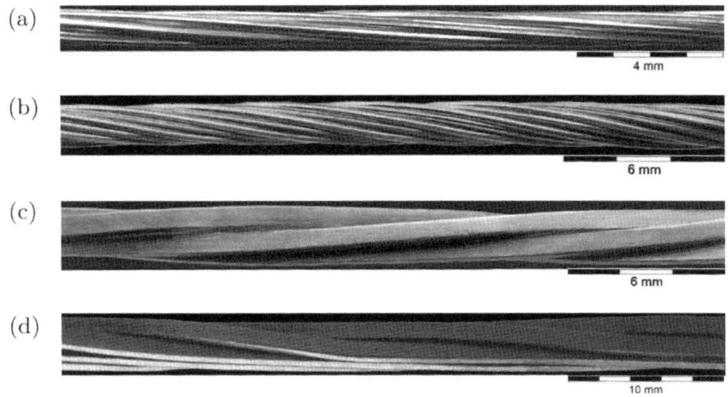

Fig. 9. Longitudinal view of (a) 7-strand braided MgB_2/Ti/Cu cable, (b) 19-strand braided MgB_2/Ti/Cu cable, (c) 4-strand continually transposed cable (CTC, Roebel type) and (d) 6-strand CTC (Roebel type) (Reprinted from; *Cryogenics* **49**, I. Hušek, P. Kováč*, T. Melišek, L. Kopera, "Transport current densities of MgB_2 wire, cable and continually transposed conductor," pages 366–370, © (2009), with permission from Elsevier).

mono or multicore wires and the mixture of strands with stabilizing Cu and reinforcing SS wires to address all possible application requests.

AC losses have mainly three contributions, hysteresis losses being suppressed by applying filaments, coupling losses being lowered by high interfilament or interstrand resistivity and eddy current losses in matrix components. To reduce AC losses in transposed cables as much as possible the use of multifilamentary MgB_2 wires is preferred. However the achievable diameter of the filaments in multifilamentary wires is limited due to inherent problems of the PIT wire technology to keep an accurate regular geometry of the composite after deformation. Depending on the individual wire architecture, a strong degradation of the transport currents may be observed when reducing the size to very small filament diameters. This problem can be solved in a better way using cables made of thin wires with less filaments. In braided cables this leads to a much less complex composite structure. Use of thin wires allows a more complex cable structure, a better distribution of stabilizing Cu strands and the option to add SS strands for mechanical reinforcement.

Technical cables profit from moderate coupling between the strands which allows some current sharing for a more stable overall performance. Cu strands are used as external stabilizers, but are effective only with a sufficiently large contact area, to the superconducting strand the latter can be improved and adjusted by compaction treatments, i.e., by careful swaging of the whole cable. This was successfully shown by Hossain *et al.*[61]

For cabling purposes, *in situ* MgB_2 wires are the best candidates to realize a transposition of strands, allowing hard bending before reaction treatment without damage. Unreacted wires can be bent to very small bending radii. Such "*in situ*" cables can easily be reacted in long lengths, the reaction treatment being performed at moderate temperatures (630–700°C) and short times (10–20 min), thus allowing a reel-to-reel continuous process. In contrast to *in situ* wires, the use of *ex situ* wires is limited by their bending behavior when being assembled to cables. Since the tolerable bending strain is a function of wire diameter, very thin *ex situ* wires are better suitable for cables, favoring short transposition lengths.

Balancing the final different properties of a cable to an optimum is still a challenge and requires more systematic work to control the different parameters. In particular the application specific requirements determine the detailed cable design and preparation approach.

4. Conclusions

Very fine MgB_2 wires have been successfully applied in current leads or LH_2 level sensors. The relatively high critical temperature, combined with a light weight and high mechanical strength as well as the possibility to produce MgB_2 conductors in various geometries and with various sheath composites makes this material an ideal candidate for various applications, for example in aerospace. For some applications of very fine wires, for example in advanced ADRs or cables, further optimization of superconductor properties or conductor fabrication and cabling process is still progressing.

References

1. P. C. Canfield, D. K. Finnemore, S. L. Bud'ko, J. E. Ostenson, G. Lapertot, C. E. Cunningham and C. Petrovic, Superconductivity in Dense MgB_2 wires, *Phys. Rev. Lett.* **86**, pp. 2423–2426 (2001).
2. S. Jin, H. Mavoori, C. Bower and R. B. van Dover, High critical currents in iron-clad superconducting MgB_2 wires, *Nature* **411**, pp. 563–565 (2001).
3. B. A. Glowacki, M. Majoros, M. Vickers, J. E. Evetts, Y. Shi and I. McDougall, Superconductivity of powder-in-tube MgB_2 wires, *Supercond. Sci. Technol.* **14**, pp. 193–199 (2001).
4. W. Goldacker, S. I. Schlachter, S. Zimmer and H. Reiner, High transport currents in mechanically reinforced MgB_2 wires, *Supercond. Sci. Technol.* **14**, pp. 787–793 (2001).
5. S. Soltanian, X. L. Wang, I. Kusevic, E. Babic, A. H. Li, M. J. Qin, J. Horvat, H. K. Liu, E. W. Collings, E. Lee, M. D. Sumption and S. X. Dou, High-transport critical current density above 30 K in pure Fe-clad MgB_2 tape, *Physica C* **361**, pp. 84–90 (2001).
6. G. Grasso, A. Malagoli, C. Ferdeghini, S. Roncallo, V. Braccini, A. S. Siri and M. R. Cimberle, Large transport critical currents in unsintered MgB_2 superconducting tapes, *Appl. Phys. Lett.* **79**, pp. 230–232 (2001).
7. H. L. Suo, C. Beneduce, M. Dhallé, N. Musolino, J.-Y. Genoud and R. Flükiger, Large transport critical currents in dense Fe- and Ni-clad MgB_2 superconducting tapes, *Appl. Phys. Lett.* **79**, pp. 3116–3118 (2001).
8. Ch. Haberstroh and G. Zick, A Superconductive MgB_2 Level Sensor for Liquid Hydrogen, *AIP Conf. Proc.* **823**, pp. 679–684 (2006).
9. Ch. Haberstroh, G. Dehn and D. Kirsten, Liquid Hydrogen Level Sensors based on MgB_2, *21st ICEC/ICMC Conf. Proc.* pp. 357–360 (2007).
10. Ch. Haberstroh, Superconductive level indicator for liquid hydrogen and liquid neon, and measuring method for liquid level measurement, Patent US2008-014844 A1 (2008).
11. S. I. Schlachter, W. Goldacker, A. Frank, B. Ringsdorf and H. Orschulko, Properties of MgB2 superconductors with regard to space applications, *Cryogenics* **46**, pp. 201–207 (2006).
12. W. Goldacker, S. I. Schlachter, B. Obst, B. Liu, J. Reiner and S. Zimmer, Development and performance of thin steel reinforced MgB_2 wires and low-temperature in situ processing for further improvements, *Supercond. Sci. Technol.* **17**, pp. 363–S368 (2004).
13. S. Soltanian, J. Horvat, X. L. Wang, M. Tomsic and S. X. Dou, Transport critical current of solenoidal MgB_2/Cu coils fabricated using

a wind-reaction *in situ* technique, *Supercond. Sci. Technol.* **16**, pp. L4–L6 (2003).

14. S. I. Schlachter, A. Frank, B. Ringsdorf, H. Orschulko, B. Obst, B. Liu and W. Goldacker, Suitability of sheath materials for MgB_2 powder-in-tube superconductors, *Physica C* **445–448**, pp. 777–783 (2006).
15. S. I. Schlachter, A. Frank, W. Goldacker, B. Liu, H. Orschulko, B. Ringsdorf and A. Will, Filament microstructure and superconducting properties of MgB_2 wires with thin filaments, *IEEE T. Appl. Supercond.* **17**, pp. 2842–2845 (2007).
16. A. Malagoli, V. Braccini, C. Bernini, G. Romano, M. Vignolo, M. Putti, C. Ferdeghini, Study of the MgB_2 grain size role in *ex situ* multifilamentary wires with thin filaments, *Supercond. Sci. Technol.* **23**, p. 025032 (2010).
17. Y. Yamada, M. Kanazawa, T. Ohno, K. Tachikawa, K. Kajikawa, T. Inoue, I. Aoki, A. Matsumoto and Kumakura, H. PIT processed MgB_2 thin wires sheathed with stainless steel, *IEEE T. Appl. Supercond.* **22**(3), p. 6200304 (2012).
18. P. Kováč L. Kopera, T. Melišek, M. Rindfleisch, W. Haessler and I. Hušek, Behaviour of filamentary MgB_2 wires subjected to tensile stress at 4.2 K; *Supercond. Sci. Technol.* **26**, p. 105028 (2013).
19. P. Kováč, I. Hušek, T. M. Melišek and L. Kopera, Current densities of thin filament MgB_2/Ti/GlidCop® wire, *Supercond. Sci. Technol.* **24**, p. 105006 (2011).
20. R. Franz and G. Wiedemann, Ueber die Wärme-Leitungsfähigkeit der Metalle., *Ann. Phys.* **165**, pp. 497–531 (1853).
21. P. J. Shirron, M. J. DiPirro, J. Panek, R. Kelley, K. Mitsuda, R. Fujimoto, M. Hirabayashi and D. McCammon, The Astro-E2/XRS-2 helium insert system, *Nuclear Instr. and Methods in Physics Research A* **559**, pp. 666–668 (2006).
22. R. L. Kelley, K. Mitsuda, C. A. Allen, P. Arsenovic, M. D. Audley, T. G. Bialas, K. R. Boyce, R. F. Boyle, S. R. Breon, G. V. Brown, J. Cottam, M. J. DiPirro, R. Fujimoto, T. Furusho, K. C. Gendreau, G. G. Gochar, O. Gonzalez, M. Hirabayashi, S. S. Holt, H. Inoue, M. Ishida, J. Ishisaki, C. S. Jones, R. Keski-Kuha, C. A. Kilbourne, D. McCammon, U. Morita, S. H. Moseley, B. Mott, K. Narasaki, Y. Ogawara, T. Ohashi, N. Ota, J. S. Panek, F. S. Porter, A. Serlemitsos, P. J. Shirron, G. A. Sneiderman, A. E. Szymkowiak, Y. Takei, J. L. Tveekrem, S. M. Volz, M. Yamamoto and N. Y. Yamasaki, The suzaku high resolution X-Ray spectrometer, *Publ. Astron. Soc. Japan* **59**, pp. 77–112 (2007).
23. S. Breon, P. Shirron, R. Boyle, E. Canavan, M. DiPirro, A. Serlemitsos, J. Tuttle and P. Whitehouse, The XRS low temperature cryogenic

system Ground performance tests results, *Cryogenics* **39**, pp. 677–690 (1999).
24. J. G. Tuttle, T. P. Hait, R. F. Boyle, H. J. Muller, J. D. Hodge and S. R. Breon, A high T_c superconducting current lead assembly for the XDS helium cryostat, *Adv. Cryog. Eng.* **43**, pp. 965–972 (1998).
25. J. S. Panek, J. G. Tuttle, V. Marrero, S. Mustafi, R. Edmonds, A. Gray and S. Riall, Astro-E2 Mg diboride high current leads, *AIP Conf. Proc.* **710**, p. 952 (2004).
26. P. Lezza, V. Abächerli, N. Clayton, C. Senatore, D. Uglietti, H. L. Suo and R. Flükiger, Transport properties and exponential n-values of Fe/MgB_2 tapes with various MgB_2 particle sizes, *Physica C* **401**(1–4), pp. 305–309 (2004).
27. W. Goldacker, S. I. Schlachter, H. Reiner, S. Zimmer, B. Obst, H. Kiesel and A. Nyilas, Mechanically reinforced MgB_2 wires and tapes with high transport currents, *Studies of High Temperature Superconductors* **44** (2003); NY Nova Science Publishers; pp. 169–185, Eds. A. Narlikar.
28. H. J. N. Van Eck, D. C. van der Laan, M. Dhalle, B. ten Haken and H. H. J. ten Kate, Critical current versus strain research at the University of Twente, *Supercond. Sci. Technol.* **16**(9), pp. 1026–1030 (2003).
29. H. Kitaguchi, H. Kumakura and K. Togano, Strain effect in MgB_2/stainless steel superconducting tape, *Physica C* **363**, pp. 198–201 (2001).
30. J. M. Duval, N. Luchier, L. Duband and T. Tirolien, Progress in the Development of the IXO 50 mK Sorption-ADR stage, Presented at the *16th International Cryocooler Conf.*, held May 17–20, 2008 in Atlanta, Georgia; (2008).
31. N. Luchier, J. M. Duval, L. Duband, P. Camus, G. Donnier-Valentin and M. Linde, 50 mK cooling solution with an ADR precooled by a sorption cooler, *Cryogenics* **50**, pp. 591–596 (2010).
32. N. Luchier, J. M. Duval, L. Duband and T. Tirolien, Performances of the 50 mK ADR/sorption cooler, *Cryogenics* **52**, pp. 152–157 (2012).
33. L. Duband, J. M. Duval and N. Luchier, SAFARI engineering model 50 mK cooler, *Cryogenics* **64**, pp. 213–219 (2014).
34. M. Schwarz, KIT, private communication (2009).
35. J. M. Duval, CEA/SBT, private communication (2009).
36. S. I. Schlachter, KIT, unpublished data (2009).
37. K. Kajikawa, T. Inoue, A. Jokinen, Y. Yamada and I. Aoki, Experimental Study on Superconducting Level Sensors for Cryogen Composed of MgB_2 Wires with Stainless Steel Sheath Material, *Extended Abstract of Internat. Supercond. Electronics Conference 2009*, Fukuoka, Japan, 16–19 June 2009, TD-P43 (2009).

38. M. Takeda, Y. Matsuno, I. Kodama, H. Kumakura and C. Kazama, Application of MgB_2 wire to liquid hydrogen level sensor — external-heating-type MgB_2 level sensor; *IEEE T. Appl. Supercond.* **19**, pp. 764–767 (2009).
39. S. I. Schlachter and B. Obst. KIT, unpublished data (2008).
40. M. Takeda, Y. Matsuno, I. Kodama and H. Kumakura, Characteristics of MgB_2 sensor for detecting level of liquid hydrogen, *AIP Conf. Proc.* **985**, p. 933 (2008).
41. H. Yamada, N. Koshizuka, K. Mochizuki and S. Shimura, MgB_2-based wiring material for liquid hydrogen fluid level sensors, liquid hydrogen fluid level sensor, and fluid level meter for LH2, Patent No. WO2014076981 A1, 22. Mai 2014 (2014).
42. K. Kajikawa, K. Tomachi, N. Maema, M. Matsuo, S. Sato, K. Funaki, H. Kumakura, K. Tanaka, M. Okada, K. Nakamichi, Y. Kihara, T. Kamiya and I. Aoki, Fundamental investigation of a superconducting level sensor for liquid hydrogen with MgB_2 wire, *Journal of Physics: Conference Series* **97**, p. 012140 (6pp) (2008).
43. K. Kajikawa, T. Inoue, K. Watanabe, M. Kanazawa, Y. Yamada, H. Kobayashi, H. Taguchi and I. Aoki, Proposal of a new type of superconducting level sensor for liquid hydrogen with MgB_2 wires, *Physics Procedia* **36**, pp. 1396–1401 (2012).
44. K. Kajikawa, T. Inoue, K. Watanabe, Y. Yamada and I. Aoki, Numerical optimization of monofilamentary MgB_2 wires with different metal sheath materials for liquid hydrogen level sensor, *AIP Conf. Proc.* **1573**, p. 905 (2014).
45. K. Kajikawa, H. Kuga, I. T. Takuro, K. Watanabe, Y. Uchida, T. Nakamura, H. Kobayashi, M. Hongo, T. Kojima, H. Taguchi, Y. Naruo, T. Wakuda and K. Tanaka, Development of a liquid hydrogen transfer pump system with MgB_2 wires, *Cryogenics* **52**, pp. 615–619 (2012).
46. P. Shirron, E. Canavan, M. DiPirro, M. Jackson, T. King and J. Tuttle, Progress in the development of a continuous adiabatic demagnetization refrigerator; *Cryocoolers* **12**, pp. 661–668 (2003).
47. P. Shirron, E. Canavan, M. DiPirro, J. Francis, M. Jackson, J. Tuttle, T. King and M. Grabowski, Development of a cryogen-free continuous ADR for the constellation-X mission, *Cryogenics* **44**, pp. 581–588 (2004).
48. J. Tuttle, S. Pourrahimi, P. Shirron, E. Canavan, M. DiPirro and S. Riall, A 10 K magnet for space-flight ADRs, *Cryogenics* **44**, pp. 383–388 (2004).
49. J. Tuttle, S. Pourrahimi, E. Canavan, M. DiPirro and P. Shirron, A lightweight low-current 10 K magnet for space-flight ADRs, *Cryogenics* **46**, pp. 196–200 (2006).

50. S. C. Hopkins, A. Morawski, B. A. Glowacki, I. D. Hepburn, C. Brockley-Blatt, S. Milward, S. Harrison, M. H. Hancock, N. Bay and A. J. Zaleski, Design Considerations for MgB_2 ADR Superconducting Coils, *Proc. of ICMC '06* **2**, 157 (2007), *CryoPrague* 2006, incorporating I*CMC '06: International Cryogenic Materials Conf.* 2006, Prague, Czech Republic, 17–21 July 2006.
51. HyperTech, Magnesium Diboride Superconducting Coils for Adiabatic Demagnetization Refrigerators (ADR's); contract NNG05CA56C, fiscal year 2005.
52. M. Rindfleisch, Hyper Tech, private communication (2014).
53. J. M. Hur, K. Togano, A. Matsumoto, H. Kumakura, H. Wada and K. Kimura, Fabrication of high-performance MgB_2 wires by an internal Mg diffusion process, *Supercond. Sci. Technol.* **21**, p. 032001 (2008).
54. G. Z. Li, M. D. Sumption, M. A. Susner, Y. Yang, K. M. Reddy, M. A. Rindfleisch, M. J. Tomsic, C. J. Thong and E. W. Collings, The critical current density of advanced internal-Mg-diffusion-processed MgB_2 wires, *Supercond. Sci. Technol.* **25**, p. 115023 (2012).
55. R. Musenich, M. Greco, M. Razeti and G. Tavilla, Electrical characterization of a multi-strand MgB_2 cable; *Supercond. Sci. Technol.* **20**, pp. 235–238 (2007).
56. S. I. Schlachter, U. Braun, A. Drechsler, W. Goldacker, T. Holúbek, A. Kling and C. Schmidt, Cabling of thin MgB_2 strands for high current conductors with reduced ac losses, *AIP Conf. Proc.* **1219**, pp. 302–309 (2010).
57. P. Kováč, I. Hušek and T. Melišek, MgB_2 cable made from two-axially rolled wires; *Supercond. Sci. Technol.* **21**, p. 125003 (2008).
58. T. Holúbek, S. I. Schlachter and W. Goldacker, Fabrication and transport properties of superconducting MgB_2 cables, *Supercond. Sci. Technol.* **22**, p. 055011 (2009).
59. P. Kováč, I. Hušek, A. Rosová, T. Melišek, L. Kopera, Fine-filamentary *in situ* MgB_2 wires, *Supercond. Sci. Technol.* **23**, p. 105006 (2010).
60. I. Hušek, P. Kováč, T. Melišek and L. Kopera, Transport current densities of MgB_2 wire, cable and continually transposed conductor, *Cryogenics* **49**, pp. 366–370 (2009).
61. M. S. A. Hossain, A. Motaman, S. Barua, D. Patel, M. Mustapic, J. H. Kim, M. Maeda, M. Rindfleisch, M. Tomsic, O. Cicek, T. Melišek, L. Kopera, A. Kario, B. Ringsdorf, B. Runtsch, A. Jung, S. X. Dou, W. Goldacker and P. Kováč, The roles of CHPD: superior critical current density and n-value obtained in binary *in situ* MgB_2 cables, *Supercond. Sci. Technol.* **27**, p. 095016 (2014).
62. L. Kopera, P. Kováč, I. Hušek, T. Melišek, Rutherford cable made of single-core MgB_2 wires, *Supercond. Sci. Technol.* **26**, p. 125007 (2013).

Chapter 6d

MgB_2 Transmission Lines for the Large Hadron Collider

Amalia Ballarino*, Bernardo Bordini and Sebastiano Giannelli

CERN, TE-MSC
Route de Meyrin 385
1223 Meyrin, Switzerland
**amalia.ballarino@cern.ch*

This paper summarizes the work performed at CERN on development of high-current MgB_2 electrical transfer lines conceived for feeding the High-Luminosity LHC magnet circuits. The activity includes development of: (a) round MgB_2 wires produced in industry, with mechanical properties enabling cabling after reaction, (b) high-current MgB_2 cables, (c) the cryogenic–electrical system optimized for the powering of the magnets. The work is performed for accelerator technology, but it finds use in any other application requiring power transmission.

1. Introduction

Since the early 1960's, when the first application of the newly discovered type II superconductors appeared, the potential of use of superconducting materials for electrical transmission was identified. Already in 1967, Garwin and Matisoo envisioned a 100 GW DC Nb_3Sn power transmission line operated at 4 K via a dedicated He refrigeration system.[1] The discovery of High Temperature Superconductors (HTS) relaunched the development of potentially more economic superconducting lines relying on cooling by liquid nitrogen and pilot installations in electrical networks took place in the U.S., Europe, Japan, China and Russia. The earlier development of 1G BSCCO 2223 conductors funneled the initial effort on this material.

Continuous progress on 2G conductor opened the way to *RE*-Ba–Cu–O (*RE*BCO, *RE* = rare earth elements)-based transmission lines, which then entered the market of industrial demonstrators, taking advantage of the existing widespread know-how.

In accelerator technology, electrical transmission till now relied on Nb–Ti bus bars operated in liquid He, connected at one end to the magnets and at the other end to the current leads, which provide the link to the room temperature electrical equipment of the associated circuit. In the Large Hadron Collider (LHC), there are more than 3,000 current leads[2] connected to an equivalent number of Nb–Ti bus bars. The leads are located in cryostats in line with the accelerator magnets, the bus bars run from the current leads to the magnet cold masses, and the power converters are in underground areas near the accelerator ring. All this equipment is exposed to significant levels of radiation.

For the High Luminosity upgrade of the LHC machine,[3] it is proposed to use superconducting transmission lines (Superconducting Links)[4] replacing the Nb–Ti bus bar. These lines bring more flexibility into the electrical circuits, enabling the location of power converters and current leads in radiation free areas. Away from the accelerator ring, electronics is protected from radiation effects and access of personnel for maintenance, repair and routine tests on electrical and cryogenic equipment is easier and safe.[4]

The Superconducting Links for the High-Luminosity LHC will feed tens of magnets operated in quasi-DC mode at currents of up to 18 kA. The cables will be cooled by He gas and, to relax constraints on cryogenic operation, they will be made with HTS materials able to guarantee a temperature margin with respect to nominal conditions. The LHC cryogenic system offers the possibility of cooling the Superconducting Links with He gas at temperatures below 20 K. MgB_2 is therefore a candidate for this application, and its potentially low cost renders it attractive and competitive with respect to Bi–Sr–Ca–Cu (BSCCO) and REBCO conductors.

Development of MgB_2 wires, MgB_2 cables and MgB_2 electrical transmission lines was launched at European Organization for Nuclear Research (CERN)[4] to demonstrate the feasibility of adopting

MgB_2 Superconducting Links for the LHC machine and for high-current MgB_2 electrical transfer in general.

2. Superconducting Links for the LHC Accelerator

The specificity of the accelerator circuits calls for development of MgB_2 cables rated at currents much higher than those used today in superconducting power transmission lines (<4 kA). Cables rated at up to 20 kA are required, and tens of cables operated at different currents — ranging from 0.1 kA–20 kA — are contained in the same compact cryogenic envelope having a total current capacity of over |150| kA at about 25 K.[4] The development launched by CERN focuses on: development of MgB_2 round wires with high critical current and with mechanically properties enabling cabling of reacted conductor; development of 20 kA MgB_2 cables; design of a test station suitable for qualification of 2×20 m long cables in a transmission line configuration; design of a system incorporating the Superconducting Links for the powering of the LHC High-Luminosity magnets.

2.1 *MgB_2 wire specification and development*

When the Superconducting Link project started, MgB_2 round wire with the required characteristics was not commercially available. In 2008, CERN and Columbus Superconductors[5] entered into a collaborative R&D activity aimed at the development of powder-in-tube (PIT) MgB_2 round wire with the electrical and mechanical properties needed for the project. Key specification targets were a minimum bending radius of 100 mm and a minimum critical current of 320 A at 25 K and 0.5 T.

Different MgB_2 wires were produced at Columbus Superconductors SpA. In parallel, an intense characterization program was carried out at CERN with the purpose of qualifying the conductor and providing feedback on the measured performance and development requirements. The first attempts to make *ex situ* MgB_2 wires led to the production of a conductor with a quasi-square cross-section and a width of initially 1.6 mm, and later 1.1 mm.[4] The composition of this wire was the same as that of the tape commercialized by Columbus SpA i.e., 12 quasi-rectangular superconducting filaments (130–170 μm width) were embedded in a Ni matrix, and the stabilization was provided by a central Cu core surrounded by a Fe barrier. In-depth characterization showed that homogeneity of critical current and mechanical performance of this wire had to be improved. It was therefore decided to introduce a Nb barrier around each superconducting filament to avoid formation, during wire reaction, of brittle MgB_2–Ni intermetallic layers.

By early 2012, the state of the art MgB_2 conductor for the Superconducting Link project was a 1.17 mm diameter wire composed of thirty MgB_2 filaments embedded in a Ni matrix; the Ni was surrounded by a Monel annulus; each MgB_2 filament was separated from the Ni via a Nb barrier; a central Cu core was acting as stabilizer. The Figs. 1(a) and 1(b) show the layout of this wire, with the SEM images taken from wire cross-sections extracted from a R&D billet. This wire reached engineering critical current densities of 550 A/mm^2 at 4.3 K and 1 T, but homogeneity still had to be improved: a typical J_e variation of up to 30% was measured at CERN on samples extracted from the same wire unit length. SEM and EDX analysis identified local failures of the Nb barriers that favored generation of cracks in the MgB_2 filaments and formation of the brittle MgB_2–Ni intermetallic layers (Fig. 1(b)). These failures were identified as the cause of the inhomogeneity of the electro-mechanical performance of the wire.

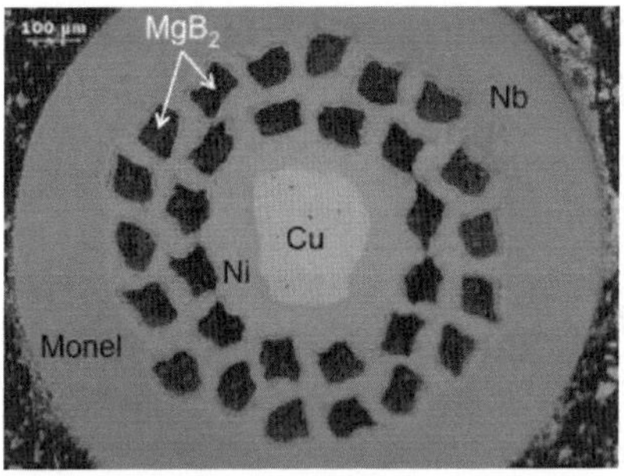

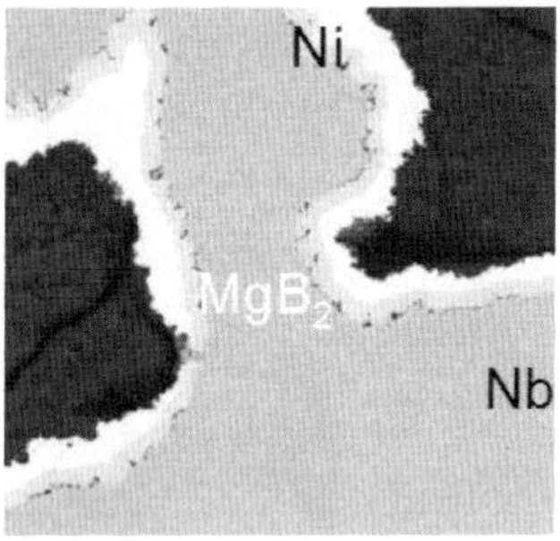

Fig. 1(a). SEM of MgB_2 wire developed by Columbus in collaboration with CERN. Left: the overall cross-section of the wire, Right: MgB_2 filaments with Nb barrier.

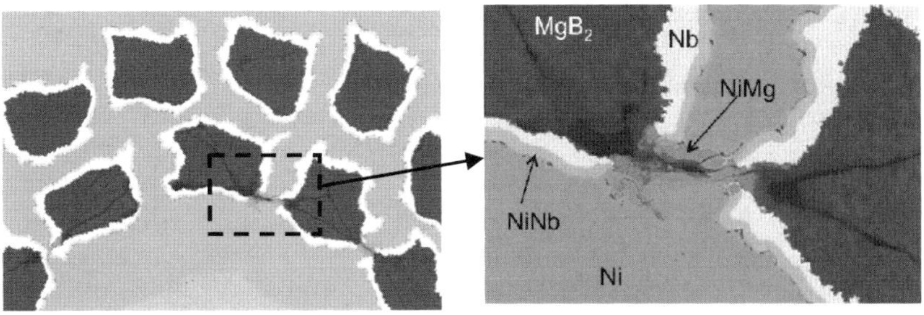

Fig. 1(b). Cracks in the MgB_2 filaments of an earlier generation of MgB_2 wire. The cracks were associated to a failure of the Nb barrier.

Table 1: Composition of Reference 1 MgB_2 wire developed by Columbus for the Superconducting Link project.

Material	Volume fraction
MgB_2	10.4%
Monel 400	44.6%
Ni	26.8%
Nb	13%
Cu	5.2%

To overcome the problem described above, CERN and Columbus decided to double the thickness of the Nb barrier, and this modification was implemented in a 1 mm diameter wire. The composition of this new wire, containing MgB_2 filaments twisted with a twist pitch of 300 mm, is summarized in Table 1. The wire, hereafter referred to as Reference 1, was delivered to CERN in September 2012 in a unit length of 1.5 km, and it was used for the production of the first high current cables. More than a hundred short samples extracted from the 1.5 km piece length were characterized at CERN, and a good homogeneity of the critical current was found (RMS 4%). The critical current measurements performed at CERN on a 2 m long sample are reported in Fig. 2. Bending tests also showed that the wire could sustain at room temperature a bending radius of 100 mm without degradation of its critical current. A bending radius of 75 mm generated a critical current degradation of 25%. CERN received an

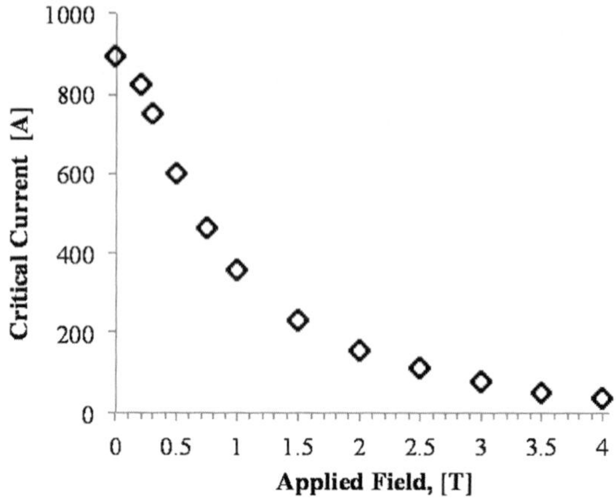

Fig. 2. Critical current measurement at 4.2 K of a 2 m long sample of a 1 mm diameter Reference 1 MgB_2 wire. The measurements were performed in the FRESCA test station at CERN.

additional 24 km of Reference 1 wire, which also incorporated Cu plating (about 20 μm thick) and tinning (about 5 μm thick) of the Cu surface. More than a 100 critical current measurements were made on short samples and the results confirmed the soundness of the wire design.

The next step of the wire development aimed at increasing the number of filaments (from 30 to 37), reducing the strand diameter to 0.85 mm and reducing the fraction of the resistive Monel in the matrix. In June 2013, Columbus delivered to CERN several piece lengths of this 0.85 mm diameter Reference 2 wire. Measurements at CERN confirmed a significant increase in electrical performance (J_e equal to 750 A/mm^2 at 4.3 K and 1 T). The wire also incorporated a shorter twist pitch (100 mm) of the superconducting filaments.

The last step of the wire development focused on improvement of the electrical performance. Higher quality B powder was introduced in the wire manufacturing process. This change enabled achievement of J_e values as high as 1,150 A/mm^2 at 4.3 K and 1 T. In order to meet the mechanical requirements, the Monel thickness had to be increased with respect to the Reference 2 wire, and the wire diameter was brought back from 0.85 mm to 1 mm (see Fig. 3). The critical current

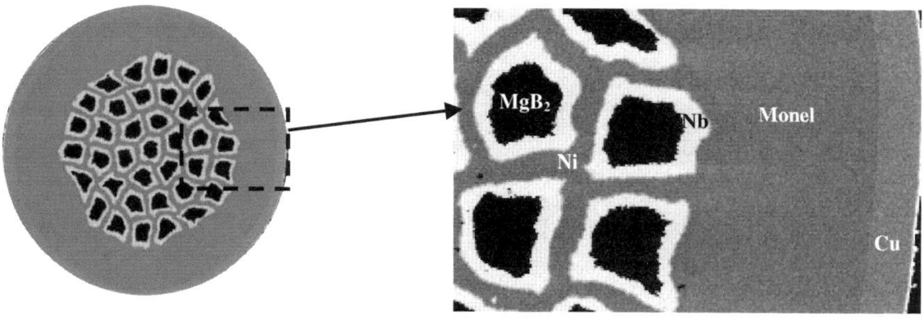

Fig. 3. State of the art Reference 3 MgB_2 wire developed for the Superconducting Link project.

of the wire did not change, but the J_e went down to 830 A/mm^2 (at 4.3 K, 1 T). This Reference 3 wire is today the baseline for the Superconducting Link project.

2.2 MgB_2 cables development and characterization

As mentioned in Sec. 2, in the Superconducting Link tens of cables feeding different electrical circuits are housed in the same cryostat. Each cable is wrapped with polyimide tape for electrical insulation. Different cable configurations are foreseen for different current ratings. In all cable topologies, MgB_2 strands are co-wound and twisted with additional Cu stabilizer. The number of MgB_2 strands per cable is set by the current capability required at the highest temperature (25 K), which includes the temperature margin (5 K) with respect to nominal operating conditions. The baseline cable topologies are illustrated in Fig. 4. The highest-rated cables, designed to carry 20 kA at 20 K, are composed of six parallel subunits, each made from 18 MgB_2 strands wound around a central Cu core. This subcable topology is planned to be used also as the inner conductor in 3 kA coaxial cable configurations, as shown in Fig. 4. For the lowest-rating cables, two or three MgB_2 wires will be co-wound with the required number of Cu strands.

In all cable configurations illustrated in Fig. 4, the MgB_2 strands describe coaxial helices around the centerline of the (sub) cable. This configuration allows nominally homogeneous strain conditions along

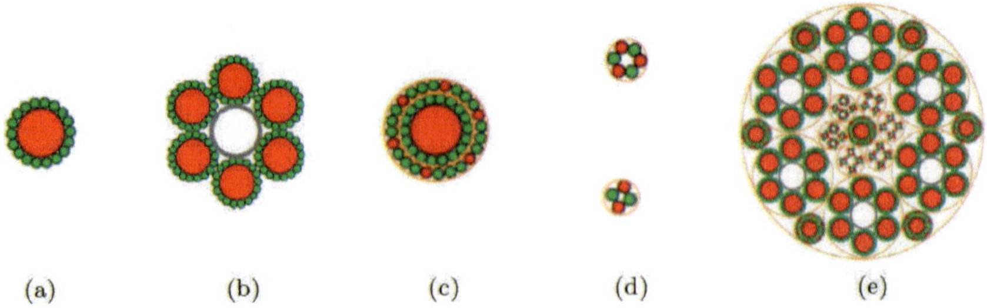

Fig. 4. Cable assemblies for the Superconducting Link [8]. From left: (a) Sub-unit of a 20 kA cable, $\Phi \sim 6.5$ mm, (b) 20 kA cable, $\Phi \sim 19.5$ mm, (c) concentric 2×3 kA cable, $\Phi \sim 8.5$ mm, (d) 0.4 kA cable (top) and 0.12 kA cable (bottom), $\Phi < 3$ mm, (e) 165 kA cable assembly (6×20 kA, $7 \times 2 \times 3$ kA, 4×0.4 kA, 18×0.12 kA), $\Phi \sim 70$ mm. The cables are made of Cu stabilizer (dark) and MgB_2 wire (clear).

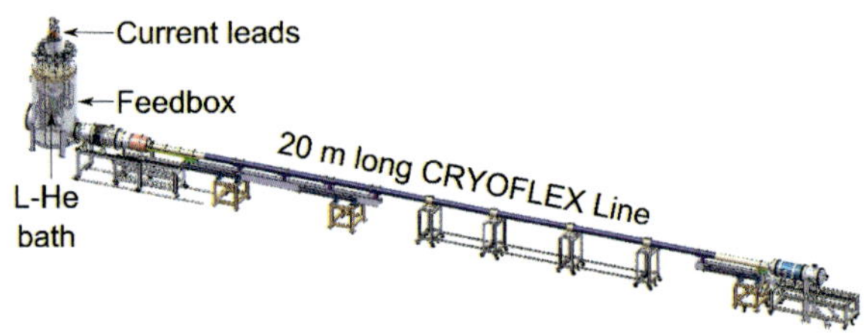

Fig. 5. Superconducting link test station at CERN.

the cable, and it enables use of MgB_2 wires with the mechanical characteristics reported in Sec. 2.1. The twist pitch of the MgB_2 strands in each cable is 300 mm. The cables are operated in forced-flow gas cooling conditions. The maximum operating temperature is 25 K, and the peak field on the conductor when the cables are operated at nominal current is 1 T. The envelope diameter of the grouped cables (Fig. 4(e)) in the Superconducting Link is about 70 mm.

In order to test cable prototypes, a test station was designed and built at CERN. A schematic drawing of this test station is reported in Fig. 5. A 20 m long CRYOFLEX® cryostat procured from Nexans was connected to a cryogenic feed-box, which contains a pair of current leads dipping in liquid He and a mixer generating the He gas used for the cooling of the MgB_2 cables. The installation

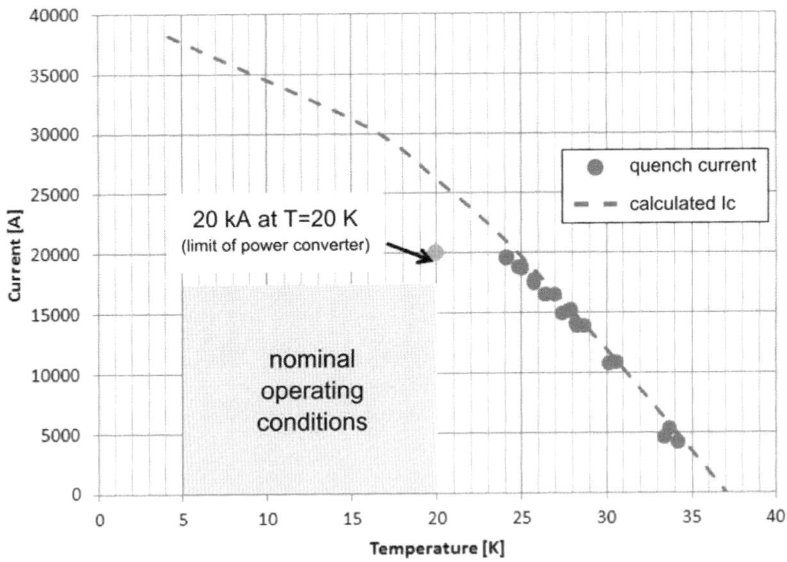

Fig. 6. Comparison between the calculated critical current and the measured quench current of the MgB_2 cable in Fig. 4(b) at different temperatures.

enables operating temperature of the cables, which are housed in the CRYOFLEX® cryostat, in the range of 5 K–70 K, with He gas mass flow rates of up to 10 g/s, and a maximum DC current of 20 kA. The required He inlet temperature and the mass flow rate are obtained by mixing in the cryogenic feedbox ambient temperature He gas with He gas at about 5 K. For any inlet temperature of the gas, a temperature gradient of less than 1 K can be maintained along the line, enabling precise characterization of the cables. An insulating plate separates the liquid He from the He gas environment, and electrical feedthroughs made from MgB_2 wires (for temperatures below 30 K) or BSCCO 2223 tapes (for temperatures above 30 K) provide the electrical continuity across the plate.

In February 2014, a 40 m long MgB_2 cable (Fig. 4(b)), made at CERN from Reference 1 wire, was successfully tested in the Superconducting Link test station at currents of up to 20 kA.[6] The cable consisted of two 20 m long units, connected in series to form the go and return line of the electrical circuit. The MgB_2 cable reached successfully the target value of 20 kA — the maximum current that could be delivered by the power converter — at 20 K, with a peak field on the conductor of 0.9 T. The current carrying capability was tested by quenching the cable several times at different temperatures and comparing the measured quench currents with those calculated with a FEM model. The measured and calculated values are reported in Fig. 6. A quench current of 20 kA was reached at an operating temperature of 24 K.[7]

This test showed for the first time the feasibility of reaching currents as high as 20 kA in MgB_2 cables operated at above 20 K. It also showed the potential of MgB_2 round wire for use in cables optimized for electrical power transmission.

3. Conclusions

Significant progress was made on development of high current MgB_2 round wire and cables for the Superconducting Link project at CERN.

A record current of 20 kA at 24 K was achieved in a prototype electrical transfer line. In the meantime, a large (80 km) procurement of MgB_2 wire, delivered in long (1 km) unit lengths, was launched by CERN, and the development activity continues with the objective of finalizing the system design. The integration of the Superconducting Links in the LHC machine is presently foreseen in 2023–2024, when the High Luminosity upgrade will take place.

The work on the Superconducting Link is performed for accelerator technology, but it finds use in any other application requiring power transmission. The 20 kA prototype tested at CERN was also a demonstrator for an innovative transmission line proposed for long-distance transport of green power.[8] This specific activity, which was performed in the framework of a collaboration agreement between CERN and the Institute for Advanced Sustainability Studies in Potsdam, is presently continuing with a FP7 European initiative (BESTPATHS) to which CERN is contributing. With this EU funded project, laboratories and industry joined forces with transmission systems operators for studying the potential of MgB_2 high voltage power transmission lines. In the framework of this activity, a

340 kV MgB_2 demonstrator is planned to be developed and tested in industry within the next three years.

References

1. R. L. Garwin and J. Matisoo, Superconducting lines for the transmission of large amounts of electrical power over great distances, *Proc. IEEE* **55**(4), pp. 538–548 (1967).
2. A. Ballarino, Large capacity current leads, *Physica C* **468**, pp. 2143–2148 (2008).
3. L. Rossi, *LHC Upgrade Plans: Options and Strategy Proc. IPAC11* http://www.jacow.org/ (2011).
4. A. Ballarino, Development of superconducting links for the LHC machine, *Supercond. Sci. Technol.* **27**, p. 044024 (2014).
5. http://www.columbussuperconductors.com/.
6. http://cds.cern.ch/journal/CERNBulletin/2014/17/News%20Articles/1693853.
7. S. Giannelli, A. Ballarino, B. Bordini, J. Hurte and A. Jacquemod, *First measurements of MgB_2 cables operated in helium gas up to 35 K*, CERN Internal Note 2015-03, EDMS N.1476839 (2015).
8. http://www.iass-potsdam.de/en/research-clusters/earth-energy-and-environment-e3/news/long-distance-transport-green-power-first.

Chapter 6e

Hydrogen Cooled MgB_2 Cables

Vitaly Vysotsky

Russian Scientific R&D Cable Institute
111024, Moscow, Russia
vysotsky@ieee.org

The first experiments with prototype MgB_2 power cables for hybrid energy transfer lines cooled by liquid hydrogen (LH_2) are described. LH_2 can be used as a coolant in MgB_2 cables, the temperature margin being sufficient for thermally stable operations. In addition, the cable can be used as a transfer line for H as a fuel. Various cryogens are compared from the point of view of their use in long MgB_2 power cables. Simultaneous use of LH_2 as a cryogen and as a source of chemical energy could increase both total energy transfer volume and power transfer density.

1. Introduction

One of the most attractive applications of superconductivity is high power energy transfer with no or minimal loss. Various superconducting power cables based on High Temperature Superconductors (HTS) have been already developed and tested with liquid nitrogen as a coolant.[1,2] There are strong arguments for envisaging a replacement of the HTS by MgB_2, due to the considerably lower costs and the availability of round MgB_2 multifilamentary wires in kilometer lengths. The margin between the operation temperature of the coolant and the properties of the SC should be sufficient to ensure thermal stability during the operation. In this chapter, various cryogens are compared from the point of view of their possible use in long power cables based on MgB_2. The use of LH_2 appears as the best choice of MgB_2 cables, particularly since it could be used as a fuel, too. The simultaneous use of LH_2 as a cryogen and as a source of chemical energy could enhance the total transferred energy and the power transfer density. In the following, the first experiments with LH_2 cooled MgB_2 power cables as prototypes of hybrid energy transfer lines are described.

2. Possible Cryogens for MgB_2 Cables

A coolant for long superconducting power cables should be operational at sufficiently low temperature with respect to the transition temperature of the used SC to ensure a stable operation. In addition it should be able to remove a sufficient amount of heat to ensure thermal stability of the cable. It should finally have the adequate dielectric properties or breakdown voltage to ensure stable operation of a power cable at high voltages. Various cryogenic liquids suitable for cooling of MgB_2 are listed in Table 1.[3] Another option is to use gaseous He cooled below 30 K and this will be discussed as well.

All mentioned cryogens have boiling temperatures well below 39 K, the superconducting transition temperature of MgB_2. The most important cryogenic parameters to ensure the stable work of a superconducting cable are the heat of vaporization and the maximum boiling nucleate heat transfer. The maximum boiling nucleate heat

Table 1: Properties of cryogenic liquids with boiling temperatures below 30 K.[3]

Cryogen	Boiling T at atmospheric pressure, K	Volume heat of vaporization J/cm^3	Mass heat of vaporization, J/g	Density at boiling T, g/cm^3	Nucleate boil. heat transfer W/cm^2
Liquid helium, LHe2	4.2	2.6	20.9	0.125	~1
Liquid hydrogen, LH2	20.4	31.3	447	0.070	~10
Liquid neon, LNe	27.1	104	86	1.210	~15

transfers for both LH_2 and LNe are one order of magnitude higher than those for LHe_2. LH_2 has the highest vaporization heat per unit of mass, while its value per unit volume is below the value of LNe, due to the low density of LH_2. It has to be noted that LN_2 has a specific mass heat vaporization of 199 g/cm^3, i.e., twice less than that for LH_2. LHe_2 has the advantage of low temperature, but its cooling is expensive, thus eliminating the advantage of the high critical temperature of MgB_2.

Although LNe appears in Table 1 as an attractive cryogen, it presents some clear disadvantages. The first one is the lowest LNe temperature in the liquid state (before its solidification): with 24.5 K, it is too high to reach the best operation parameters for MgB_2. In addition neon is rare and also considerably more expensive than H_2. A possible solution of cryogenic problems consists in using gaseous He at temperatures below ~20–25 K. However, the cooling properties of gaseous He are limited, its thermal conductivity of 25.3 mW/m·K at 20 K[3] being two orders of magnitude lower than any liquid cryogen. In addition, gaseous He has rather bad dielectric properties in comparison with other liquid cryogens. *Gaseous* He_2 and H have so-called arcing voltages V_{mn} of 156 V and 253 V respectively: it follows that even *gaseous* H has better dielectric properties than *gaseous* He.[3] *Gaseous* He at temperature ~20 K has a breakdown voltage of ~300 V/mm, in contrast to *liquid* LH_2 which at 20 K has the voltage of ~100 kV/mm.[4] High dielectric properties are very important for high power high voltage superconducting cables made of MgB_2. Resuming all the above mentioned properties, LH_2 appears as the best candidate for the cooling of high current MgB_2 cables for energy transfer.

Hydrogen has other advantages: (a) it has the highest energy-to-weight ratio of all fuels (1 kg H corresponds to energy ~124 MJ/kg, i.e., the same as 2.1 kg gasoline), (b) it is renewable and (c) it is the most abundant element on the earth. Thanks to its good cryogenic properties LH_2 can easily be transferred over rather long distances.[5] Once the H transfer line is installed, it appears as quite natural to use it in parallel for the transmission of electricity, too. This led to the idea of hybrid energy line discussed below.

3. Hybrid Energy Transfer

LH_2 appears as an ideal candidate for the delivery of chemical energy, an example being the transfer through pipes. At present, the possibility[5] to deliver LH_2 from the Latrobe Valley to the Australian coast through a 80 km pipeline is studied in Japan. This liquid would then be transferred by special ships to Japan, similarly to liquid natural gas. It is a convincing argument that in a LH_2 line already cooled at ~20 K a MgB_2 cable could be cooled without any supplementary costs. This idea was in the wind after the discovery of high temperature superconductivity. The concept of a simultaneous delivery of chemical and electrical power employing just MgB_2 wire cooled by LH_2 through a single "cable corridor" was first mentioned by P.M. Grant as early as 2001–2002.[6,7] Later on, there were more theoretical and simulation works discussing possible hybrid energy transfer lines using LH_2 both as a fuel and cryogen, and a superconducting power cable to deliver extra electrical energy.[8–11] However, no experimental works have been performed so far to proof this concept. The first experimental works in this direction started in Russia in 2011.

4. Russian Program on Hybrid Energy Transfer Line — First Stage

At the first stage, the Russian program for development of Hybrid Energy Transfer Lines was supported by the Russian Academy of Science.[12] The plan was to develop a 10 m long prototype of a hybrid energy transfer line for the transport of LH_2 as well as electricity through a superconducting cable. The major goal of this work was to get first experimental data about hybrid energy transport systems with LH_2 and superconductivity.

4.1 *SC cable and cryogenic line*

The very first task was to choose the proper SC. The most common superconductors available on the market and used in applications are listed in Table 2.[13] After excluding common LTS superconductors, the choice was limited to superconductors that can work at LH_2 at

Table 2: Properties of most common superconductors.

Type — (Wire technology)	Basic material, T_c	Cryogen and typical temperature	Approximate prices US$ per 1 kA·m (as of mid–2014)
LTS — (Metallurgy)	NbTi (10 K)	LHe at 4.2 K and below	Up to 3–5$ at 4.2 K
LTS — (Metallurgy)	Nb_3Sn (18 K)	He up to 8–10 K and below	Up to 15$ at 4.2 K
HTS 1 Generation (Powder in tube, metallurgy)	Ceramic $Bi_2Sr_2Ca_{n-1}Cu_nO_{2n+4}$ (Bi-2223, Bi-2212) (~90–110 K)	LN_2 up to 77 K (with other cryogens)	120–150$ at 77 K About 40–50$ at 20 K
HTS 2 Generation (Coated conductors, electronics)	Ceramic $ReBa_2Cu_3O_{7-d}$ (~90 K)	LN_2 up to 77 K (with other cryogens)	300–500$ at 77 K 80–150$ at 20 K
MgB_2 (Powder in tube, metallurgy)	MgB_2 — compound (~39 K)	LH_2 and below (with other cryogens)	3–8$ at 20 K

~20–25 K, e.g., either HTS or MgB_2. HTS of both generations (1G and 2G) are available in the market, but for economic reasons (see Table 2) the preference was given to MgB_2. This material exhibits a high stability at 20 K[13] and was found to be a good choice for a LH_2 cooled system.

As described in Chapter 4 of the present book, there are three main methods to fabricate long lengths of MgB_2 wire with sufficiently high critical current densities for cables: the "*in situ*" and the "*ex situ*" processing and the Internal Mg Diffusion (IMD or AIMI). For the present cable, the "*ex situ*" MgB_2 wires of Columbus Superconductors SpA, Genoa (Italy) were chosen, the only available ones on the market in 2011.

In 2005–2009, we developed the technology for HTS power cables based on flat HTS tapes.[14] The same geometry was maintained for the present project, which was achieved on the basis of flat "*ex situ*"

MgB_2 tapes, using the same cabling and insulation technology as for HTS power cables. The selected flat tape had a cross-section of $3.64 \times 0.65\,\text{mm}^2$. The MgB_2 cross-section occupied 12%, and the Cu cross-section 15% of the total area. The minimum bending diameter with no degradation of critical current density specified by Columbus Superconductors was 110 mm. The expected critical current per tape at self-field was 520–540 A at 20 K. The prototype design of a superconducting MgB_2 cable based on flat tapes consists of three elements: a former, the current carrying layers and the insulation, as shown in Fig. 1(a). The former is a central element performing the supporting and protection functions. The superconducting current carrying path consists of two serially connected layers; each of them consists of five MgB_2 tapes helically wound on the former. The number of tapes was selected to ensure a maximum current not exceeding 3 kA, the limit of the available DC power supplies.

The insulation consisted of 20 layers of a polyimide tape with a thickness of 50 μm. The total insulation thickness of ~1 mm allowed, in principle, to operate the cable at voltages exceeding 20 kV. The cable cross-section with all sizes is shown in Fig. 1(a), a picture of the MgB_2 cable model in Fig. 1(b). The total cable length was about

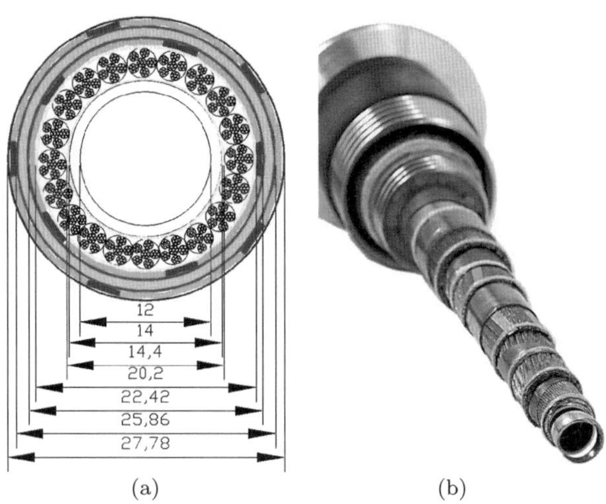

Fig. 1. MgB_2 cable design: (a) Sketch of cross-section with sizes in mm; (b) photo of the cable model. From Ref. 13, reproduced with permission of IEEE.

Fig. 2. Cable production illustrations: (a) pay off wires from a cabling machine, (b) insulation of the inner layer, (c) outer layer of the cable before final stainless steel tape covering. From Ref. 13, reproduced with permission of IEEE.

10 m. At one end both current layers were connected by jumpers to provide the returning current. Thus, the total length of the current carrying element considering the two layer assembly was ~20 m. The expected critical current of the cable was ~2.5–3 kA at a temperature of ~20 K.

The cable has been manufactured with the standard cabling equipment in JSC "VNIIKP," with the same technology as previously developed for the HTS power cable production,[14] and some details of the cable manufacturing process are shown in Fig. 2. After production the cable has been delivered to the Moscow Aviation Institute (Technical University) to be installed in the cryostat developed there. The design of the first prototype cryostat is described in detail in Refs. 12 and 18. This cryostat had a total length of 12 m and was of a coaxial "pipe-in-pipe" type ordinary vacuum super insulation (VSI). The diameters of the cryostat outer shell and of the inner cryostat shell were 80 and 40 mm, respectively. The power cable with outer diameter 28 mm was installed inside the cryostat.

Fig. 3. General view of the hybrid energy transfer line installed at the test facility (1) current leads; (2) cryostat; (3) inlet part; (4) mounting frame. From Ref. 13, reproduced with permission of IEEE.

During the experiments the LH_2 was flowing to the cryostat and in the outlet it was draining to the environment. Draining has been performed with permanent burning gas in a flare via a jet nozzle. The jet nozzle keeps the H flow discharge constant. Current leads without high voltage insulation were connected by ordinary cables to the power supplies. The general view of the cryostat installed on the loading frame is shown in Fig. 3.

4.2 *Test results*

The experiments with the prototype of the hybrid power transmission line cooled by LH_2 forced flow were carried out in November 2011. They were performed at the special facility intended for testing O_2–H liquid propellant rocket engines at the LH_2 production plant of the KB "Khimavtomatika" (Voronezh City). The detailed results of the

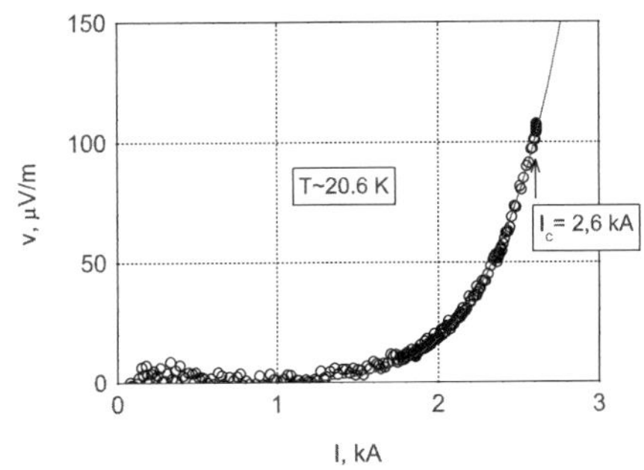

(a)

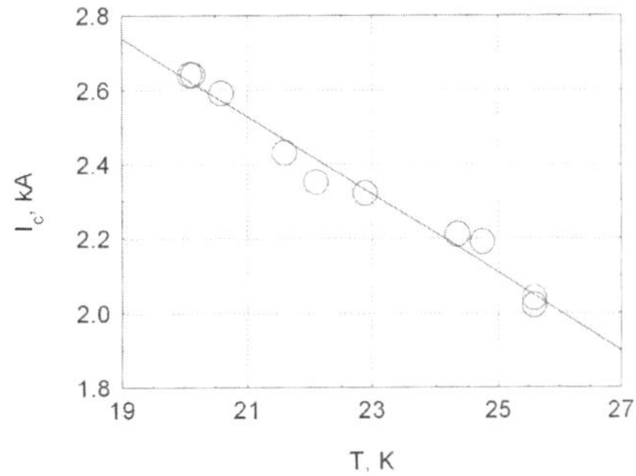

(b)

Fig. 4. Tested parameters of MgB_2 cable 10: (a) example of V–I characteristic at T = 20.6 K (critical current, $I_c \sim 2600$ A); (b) plot of I_c versus temperature T of LH_2 (averaged over the cable length).

cryogenic tests are presented in Ref. 18. Figure 4(a) shows the typical V–I characteristics measured at $T = 20.6$ K. The value of the critical current reached ~2640 A. Figure 4(b) presents the corresponding temperature dependence of the critical current. The temperature change along the cable varied from 0.2–0.9 K, depending on the mass flow rate of LH_2. The maximum LH_2 flow in our experiments was up to 250 g/s, which corresponds to the chemical energy transfer line of ~31 MW. The superconducting cable at 2.5 kA and 20 kV would thus be able to deliver an extra power of 50 MW, the total being thus ~80 MW with only 5 MgB_2 tapes.

As shown in Fig. 1, it would be easy to add at least 10 more tapes to the current carrying layers: that would increase the transport current by a factor of 3, the corresponding electrical power to 150 MW and the total power to ~180 MW. The cross-section of our energy line being limited to 50 cm^2, the prototype of the tested hybrid energy transfer line has the potential to deliver a power exceeding 100–150 MW with a power flow density exceeding $p \sim 3 \cdot 10^6$ W/cm^2.

With the result of the tests performed in Refs. 12, 13 and 18 the concept of a *hybrid energy transmission system* with high energy flow *was* experimentally proven. This was the major result of the first stage of the Russian hybrid energy transfer line project.

5. Russian Program on Hybrid Energy Transfer Line — Second Stage

Further research and developments have been performed at the second stage of the Program partially supported by JSC "Rosseti" — "Russian Grids." At the second stage of the program our tasks were:

— to develop a longer MgB_2 cable (~30 m) with high voltage insulation;
— to develop a flexible H cryogenic line of at least 30 m length;
— to develop and to compare different thermal insulation methods for a hydrogen cryogenic line;
— to perform current test of the superconducting cable with higher currents;
— to perform high voltage test of the current leads and of the cable.

To complete these tasks a new 30 m long MgB_2 superconducting cable with high voltage insulation was developed. The superconducting cable was inserted into the newly developed flexible 30 m long LH_2 line, which had three sections with different types of thermal insulation in each section: simple vacuum superinsulation, vacuum superinsulation with LN_2 precooling and *active evaporating cryostat (AEC)* thermal insulation (this term is explained subsequently). High voltage current leads with rating current up to 4 kA and voltage 20 kV were developed.

The superconducting cable, the cryogenic line and the current leads were successfully tested in October 2013. Details about the cryogenic line and current leads and their cryogenic test results are reported in Refs. 19 and 20, the superconducting cable test results being described in Ref. 21.

5.1 *The cable and cryogenic line design*

The cryogenic transfer line consisted of three serially connected 10 m sections. The first section with a total length of 10 m is an insulated cryostat "pipe-in-pipe" with vacuum super-insulation (VSI). This section just represents the cryostat tested at the first stage of the program. The second section is a flexible cryostat made of three corrugated tubes with reinforcement. The auxiliary channel between the main H flow and the main VSI shield is cooled by LH_2 which has a lower temperature due to the reduced pressure, being pumped out by the ejector pump. We call this *active evaporative cryostat system* (AEC). The third section is also a flexible cryostat made of four corrugated tubes with reinforcement. It has an active thermal insulation with LN_2 shield as insulation.

The current in the superconducting power cable was supplied by current leads, with cryostats having VSI installed at the input and the output of the transfer line. All elements are mounted on a 11 m long load frame made of welded steel profiles, providing a rigid attachment of all elements of the facility. In addition, this frame allowed the handling and transportation by ordinary trucks. The second hybrid energy transfer line prototype installed at the test facility is shown in Fig. 5. More details about the cryogenic system are given in Refs. 19 and 20.

The superconducting power cable design was similar to that one described in Refs. 12, 13 and 18. It is a coaxial design made of six

Fig. 5. The photo of the second prototype of hybrid energy transfer line installed at the test facility.

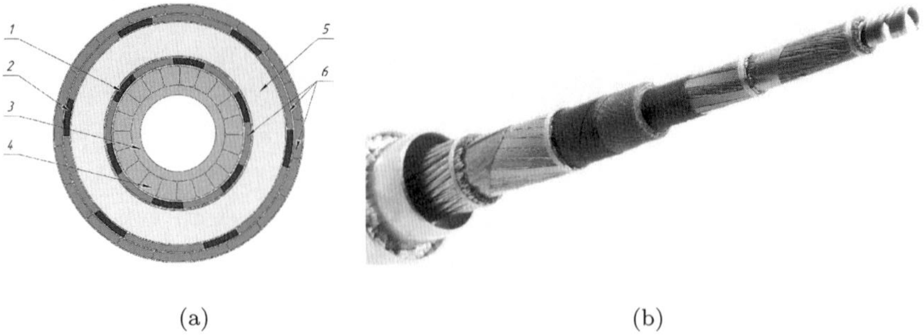

Fig. 6. (a) Cross-section of the cable, (b) Photo of its model: 1 — inner superconducting layer; 2 — outer superconducting layer; 3 — central spiral; 4 — bunch of Cu wires; 5 — high voltage insulation; 6 — separators made of flattened Cu bunches. From Ref. 21, reproduced with permission of IEEE.

flat MgB_2 tapes. A general view of the cable is shown in Fig. 6. In addition to the larger quantity of MgB_2 tapes, a major difference between this cable and the one tested in the first experiment resides in the high voltage insulation between the layers. This insulation consisted of crepe cable paper of 3.7 mm thickness. The total cable diameter was 28 mm, the length 30 m.

The second hybrid energy transfer line prototype was been tested at the same facility of KB "Khimavtomatika" in Voronezh city in October 2014. During all tests the main LH_2 flow temperature was held below the saturation temperature, i.e., in subcooled conditions under pressure. In the nominal mode the pressure in the channel was kept between 0.25 and 0.5 MPa, the pressure drop along the transfer line being always $\leq$0.05 MPa. The LH_2 flow varied between 70 and 450 g/s.

A major goal of the cryogenic tests was to compare three different methods of thermal insulation. The results are described in detail in Refs. 19 and 20. A very high efficiency of the AEC system could be demonstrated when compared to a simple vacuum superinsulation or a LN_2 shield. It was found that as little as $\sim$1% of the LH_2 flowing in the auxiliary channel is already able to completely eliminate the thermal inflow to the main channel.[16,17] At the 10 m section with AEC shield, the temperature of the main H flow channel was reduced

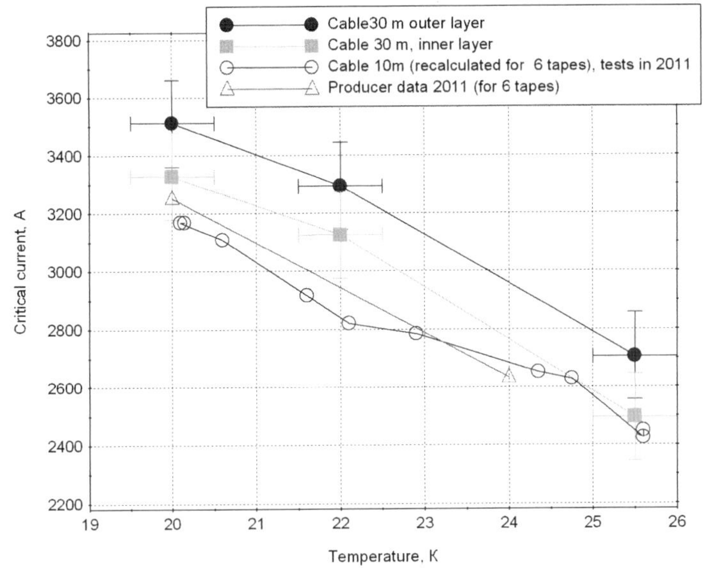

Fig. 7. Current test of the second prototype MgB_2 cable. Data from experiments in 2011 are shown also for comparison. From Ref. 21, reproduced with permission of IEEE.

by ~1–2 K.[16,17] This means that the AEC is very promising for long cryogenic transfer lines.

The results of current tests of the 30 m MgB_2 cable are shown in Fig. 7 as a function of critical current of each layer versus temperature. Due to the longer cryostat, some uncertainties still subsist about the temperature distribution along the cable (error bars in Fig. 7). One can see that at 20 K the critical current of both layers well exceeds 3 kA, reaching 3.3 kA in the inner layer and 3.5 kA in the outer layer. It follows that the cable can be safely operated up to ~3 kA. The data from the experiments in 2011 were recalculated for six tapes and are shown in Fig. 7 as well. It is seen that with a better quality of the MgB_2 basic wires in 2013 the current carried by the cable could be markedly enhanced.

A very important new test was the high voltage test of both cable and current leads. The cryostat body was grounded and the high voltage power source was connected to the inner layer of a cable via

(a) (b)

Fig. 8. High voltage test of the second prototype of hybrid energy transfer line: (a) general arrangement, (b) view from high voltage source display. From Ref. 21, reproduced with permission of IEEE.

the corresponding current lead (Fig. 8). The voltage was increased up to DC 50 kV in 10 kV steps with stops of ~1/4 h. at each level. The leakage current remained at values below 10 μA — the sensitivity limit of our device. This reflects very good insulation properties of the paper insulation impregnated by LH_2. This test allowed setting the limit for operation voltage to 25 kV. Electric tests confirmed the validity of the designed parameters of both the 30 m MgB_2 cable and the current leads. We firmly believe that the design of cable and current leads can safely be extended to longer hybrid energy transfer lines.

The 50 kV high voltage tests of LH_2 cooled current leads and superconducting cable are the first ones ever performed. They confirmed the feasibility of transfer of large hybrid energy flow of LH_2 and electricity. In the tested transfer line, the transferred chemical power could reach 50–60 MW and the electrical power 50–75 MW, thus 100–135 MW in total. After increasing the number of superconducting tapes at the same medium voltage, such a system could transfer ~0.3–0.4 GW with a power flow density of $\sim 3 \cdot 10^6$ W/m^2, i.e., the same as in oil or gas transfer lines.

6. Conclusions

In the 21st century, the transfer of energy could reach a power of dozens of GW over distances of thousands of kilometers. Hybrid energy transfer lines would allow higher power flow densities: the power would be delivered in its chemical form by LH_2 while electricity would be supplied by a superconducting power cable operating at LH_2 temperatures. Being already available in the market, the low cost MgB_2 superconductors (tapes or wires) could significantly reduce the cost of a system. Recently, practical MgB_2 SC production and several experiments performed demonstrated the feasibility of this idea. In Russia, the team of researchers performed successful tests of two prototypes of hybrid energy transfer lines: 10 m in 2011 and 30 m in 2013. High voltage tests up to 50 kV at LH_2 environment performed on a second cable prototype confirmed the good dielectric properties of H. The advanced AEC system used in one section of the second prototype allowed a drastic reduction of the heat transfer from room temperature to the LH_2 channel, with very little increase of H flow. The test runs described above confirmed the feasibility of high power and, long length hybrid energy transfer systems; their practical feasibility is expected in the near future.

Acknowledgments

This work is the review of big collective efforts and great teamwork and the author has to acknowledge the great contribution to this work of his co-authors: E. V. Blagov, V. V. Kostyuk, A. A. Nosov, S. S. Fetisov, S. Yu. Zanegin, I. V. Antyukhov, V. P. Firsov, G. G. Svalov, B. I. Katorgin and V. S. Rachuk.

References

1. P. M. Grant, Superconducting lines for the transmission of large amounts of electrical power over great distances: Garwin–Matisoo Revisited forty years later, *IEEE T. Appl. Supercond.* **17**, pp. 1641–1647 (2007).

2. S. S. Fetisov, Superconducting cables for electrical energetics — state of works in the world and in Russia, *Kabeli i provoda* **346**(3), pp. 22–29 (2014) (in Russian).
3. Y. Iwasa, *Case studies in superconducting magnets*, Springer, 2nd Edn, New York (2009).
4. M. S. Naidu, V. Kamaraju, *High voltage engineering*, Tata McGraw-Hill Education, p. 474 (2009).
5. S. Kamiya, M. Nishimura, E. Harada *et al. Study on Introduction of CO_2-Free Energy to Japan with Liquid Hydrogen*; Paper OO-2_Shoji_Kamiya-TUE-MO-Plenary_R2, presented at ICEC-25-ICMC2014, the Netherlands, 2014.
6. P. M. Grant, Will MgB2 work, *The Industrial Physicist* pp. 22–23 (2001).
7. P. M. Grant, Energy for the city of the future, *The Industrial Physicist*, pp. 22–25 (2002).
8. P. M. Grant and S. Eckroad, Functional requirements of a Hydrogen-Electric SuperGrid: Two Scenarios — SuperSuburb and SuperTie, 2006, EPRI Report 1013204.
9. C. Rubbia, The future of large power electric transmission, available at: http://rubbia.web.cern.ch/rubbia/SCWorkshop1_May2011.ppt
10. S. Yamada, Y. Hishinuma, T. Uede *et al. J. Phys.: Conf. Ser.* **234** p. 032064 (2010).
11. T. Nakayama, T. Yagai, M. Tsuda, T. Hamajima, Micro power grid system with SMES and superconducting cable modules cooled by liquid hydrogen, *IEEE T. Appl. Supercond.* **19**, pp. 2062–2065 (2009).
12. V. V. Kostyuk, E. V. Blagov, V. S. Vysotsky *et al.*, *Tech. Phys. Lett.* **38**, pp. 279–282 (2012).
13. V. S. Vysotsky, A. A. Nosov, S. S. Fetisov *et al.*, Hybrid energy transfer line with liquid hydrogen and superconducting MgB_2 cable — first experimental proof of concept, *IEEE T. Appl. Supercond.* **23**, p. 5400906 (2013).
14. V. E.Sytnikov, V. S.Vysotsky, S. S.Fetisov *et al.*, Development of HTS power cable on the base of HTS technology, *Kabeli i provoda* **321**, pp. 3–10 Moscow, Nawka (2010) (in Russian).
15. V. Kostyuk, I. V. Antyukhov, E. V. Blagov *et al.*, First in the world prototype of the hydrogen — superconducting energy transport system, *Proc. ICEC 24-ICMC 2012*, Fukuoka, Japan, pp. 247–250 (2012).
16. V. V. Kostyuk, E. V. Blagov, I. V. Antyukhov *et al.*, Cryogenic design and test results of 30-m flexible hybrid energy transfer line with liquid hydrogen and superconducting MgB_2 cable, *Cryogenics* **66**, pp. 34–42 (2015).

17. V. S. Vysotsky, I. V. Antyukhov, V. P. Firsov *et al.*, Cryogenic tests of 30 m flexible hybrid energy transfer line with liquid hydrogen and superconducting MgB_2 cable, Paper #Wed-Mo-O8-2, presented at *ICEC 25–ICMC 2014, the Netherlands* (2014).
18. V. Vysotsky, E. Blagov, V. Kostyuk *et al.*, New 30 m Flexible hybrid energy transfer line with liquid hydrogen and superconducting MgB_2 Cable — development and test results, *IEEE T. Appl. Supercond.* **25**, p. 201 (2015).

Chapter 6f

Wind Generator Projects based on MgB_2 Superconductors

Asger Bech Abrahamsen

Technical University of Denmark, Department of Wind Energy
Frederiksborgvej 399, DK 4000 Roskilde, Denmark
asab@dtu.dk

The challenges of upscaling wind turbines in terms of the demand on the drive train are outlined and the possibility to use MgB_2 for superconducting generator at 10 MW is discussed by reviewing a number of concepts proposed in the literature. Generator concepts with either the field windings or both field and armature windings are found. The concepts using only superconducting field windings are generally using more superconducting wire than the fully superconducting concepts, but the latter require more complicated cooling systems. A roadmap on how to obtain a market fraction of 5 GW superconducting offshore turbines by 2035 is proposed and it seems that the production volume of MgB_2 would be sufficiently large to fulfill this scenario. Further improvement of the critical current density at moderate magnetic fields as well as a reduction of AC losses in the fully superconducting machines will be needed to make the MgB_2 wind generators economically feasible.

1. Introduction

Wind energy from onshore wind turbines in Europe in 2014 have been reported to provide a lower cost of energy (CoE) than new installations of conventional sources (gas and coal fired power plants) if the subsidies are taken into account[1] (see Fig. 1). This is, however, only true for the onshore turbines; the cost of energy from offshore wind is about two times more expensive and will have to

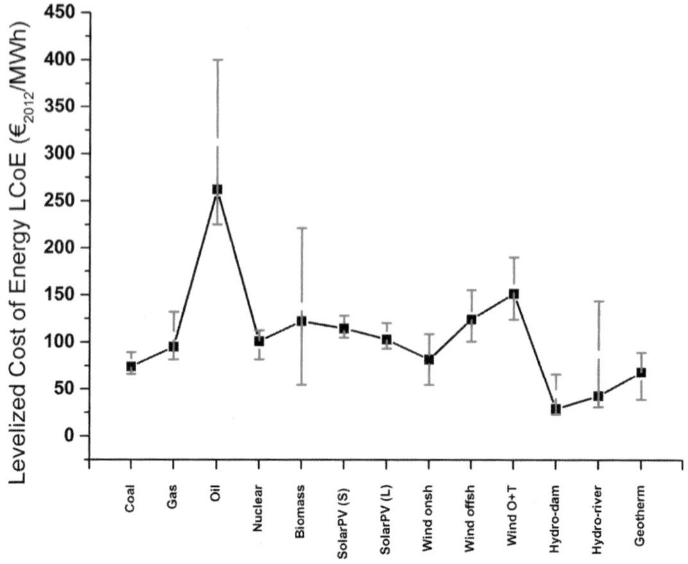

Fig. 1. Levelized cost of energy (LCoE) of different energy sources when realized in EU.[1] The CoE is determined as the sum of the Capital Expenditure (CAPEX) and the Operational Expenditure (OPEX) divided by the energy produced in the life time of the equipment. The LCoE is including all additional cost of financing and in the above case also subsidies. It should be noted that the LCoE of onshore wind power became comparable with the coal production in 2014, but offshore wind power being about twice as expensive will need further cost reductions to compete with the conventional sources. Reproduced from Alberici *et al.*[1]

be considerably reduced. Thus wind power is believed to play an important role for future electricity production in most countries. In this chapter, the technology development of wind power will be analyzed in terms of lowering the cost of energy. The question will be raised whether there is a chance of further lowering the energy costs in superconducting wind turbine generators, in particular in connection with MgB_2 superconductors.

The advantage of MgB_2 is an operation temperature between that of Low Temperature Superconductors (LTS) used for Magnetic Resonant Imaging (MRI) in hospitals and that of High Temperature Superconductors (HTS) having the potential to provide power cables cooled with LN_2. One of the challenges of superconducting wind generators is the relatively large magnetic field on the superconductors, which will decrease the performance of the magnets, thereby increasing the cost. Since the superconductors have to be kept cold, reliability and cooling cost will also have a large impact on the energy cost.

The present chapter is organized as an introduction of wind energy and a discussion of the trends of drive train developments. Several concepts of superconducting wind turbine generators will be introduced, and the question will be discussed whether only field winding or both field and armature winding should be made superconducting. Finally a roadmap for the development of superconducting wind generators in the future is discussed, before the conclusion.

2. Wind Power Development

The development of modern wind power was initiated by the oil crisis of the 1970s and resulted in commercial turbines with power ratings below 300 kW in the 1980s. A wind turbine is a device that converts the kinetic energy of the air into mechanical energy of a shaft and finally to electricity using an electrical machine. This can be done using many different kinds of geometries and with the turbine shaft being horizontal or vertical.[2] Out of the many concepts investigated in the 1970s–1980s it was the three-bladed horizontal axis turbine concept that turned out to be commercially most viable, because the aerodynamic efficiency of converting the kinetic energy of air into mechanical energy is superior and with lesser usage of materials. In the following decades the three-bladed Horizontal Axis Wind Turbine (HAWT) concept was scaled up to the MW power range, by producing longer blades. This upscaling is however not trivial, which can be illustrated by the so called "Square-Cube" challenge of wind power. The challenge can be illustrated by estimating the cost of energy from a theoretical three-bladed HAWT with a rotor diameter of D (inset in Fig. 2). The mechanical power provided to the shaft of the turbine is the product of the energy density E_{wind} of the air flowing with a wind velocity v, the flow volume V_{flow} passing

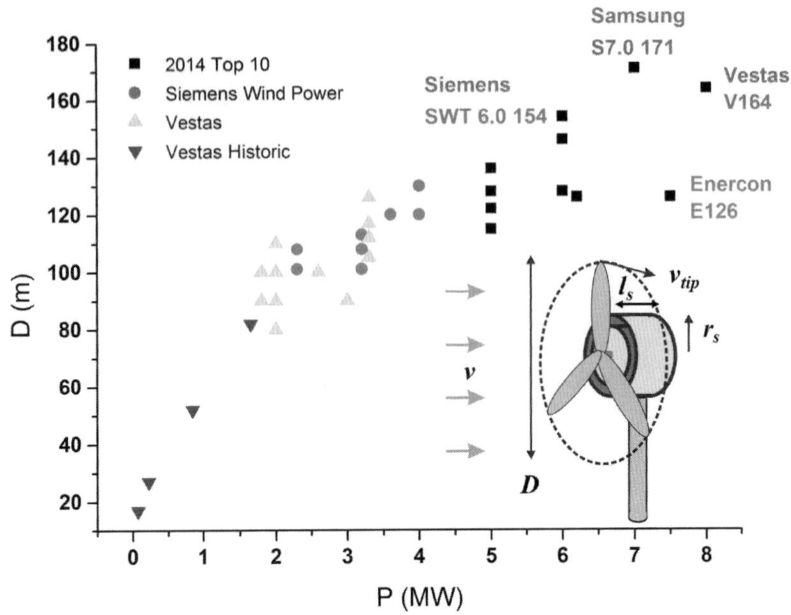

Fig. 2. Rotor diameter of historic[3] as well as the top 10 largest wind turbines[4] in 2014. It should be noticed that a large rotor can be used for a high power turbine under strong wind conditions or for medium power turbine for low- and medium wind speed conditions. The latter is giving more days of full power production and more installation sites. Inset: illustration of turbine with rotor size D, a generator air gap radius of r_s and length of l_s and exposed to a wind velocity v. The blade tip velocity is denoted v_{tip}.

the rotor area of the turbine and the aerodynamic power coefficient of the turbine blades C_P, which has a maximum between 20–50% depending on blade design. It can be written as

$$P_{\text{shaft}} = E_{\text{wind}} V_{\text{flow}} C_p = \frac{1}{2}\rho v^2 \cdot \pi \left(\frac{D}{2}\right)^2 v \cdot C_p = \frac{\pi}{8}\rho C_p v^3 D^2, \qquad (1)$$

where ρ is the mass density of air and the power coefficient can be adjusted between zero and $C_{P,\max}$ by turning the angle of the turbine blades. The turbine will produce power as function of the incoming wind speed according to Eq. (1) up to the rated power of the turbine, after which the controller will keep the power constant by turning the blades (called pitching). At high wind speeds, exceeding 25 m/s, the controller will stop the turbine to protect it from damage.

Thus the produced power of a turbine plotted as function of the wind speed is called the power curve.

From Eq. (1), it is seen that the power of a turbine scales with the area of the rotor or with the square of the blade length. The cost of the turbine roughly scales with the weight and this in turn scales with the volume of the component. In case the blade is just scaled up linearly in all dimensions then the width w and the thickness t of the blade will scale with the blade length $D/2$. The blade volume and thereby cost will therefore scale as $m \sim cost \sim D^3$ or with the cube of the blade length. Similar arguments can be formulated for the turbine tower.

From the above argument, one can estimate the CoE by simplifying that the energy produced is proportional to the maximum power of the turbine and that the capital (CAPEX) and operational (OPEX) expenditures are proportional to the mass and the energy produced.

$$CoE = \frac{CAPEX + OPEX}{E} \propto \frac{D^3 + D^2}{D^2} \propto D. \tag{2}$$

This shows that upscaling a turbine would linearly increase the electricity cost with the size of the turbine: that upscaling would never pay off, as a consequence of the "Square-Cube" challenge mentioned above.

Historically the size of the three-blade HAWT has increased by almost a factor of 10 since the 1980's, the improvements of the blade technology resulting in longer and more flexible blades with a weight scaling exponent smaller than three (see Fig. 2). It should, however, be noted that the long blades with a length above ~60 m are targeted towards the offshore market, while shorter lengths are generally used for the onshore market, to limit the interaction with neighbors.

2.1 *Installed capacity and future demand*

The growth of the rotor blade technology has mainly been driven by a constant expansion of the installed power capacity of wind turbines, illustrated in Fig. 3. It can be seen that wind power was mainly

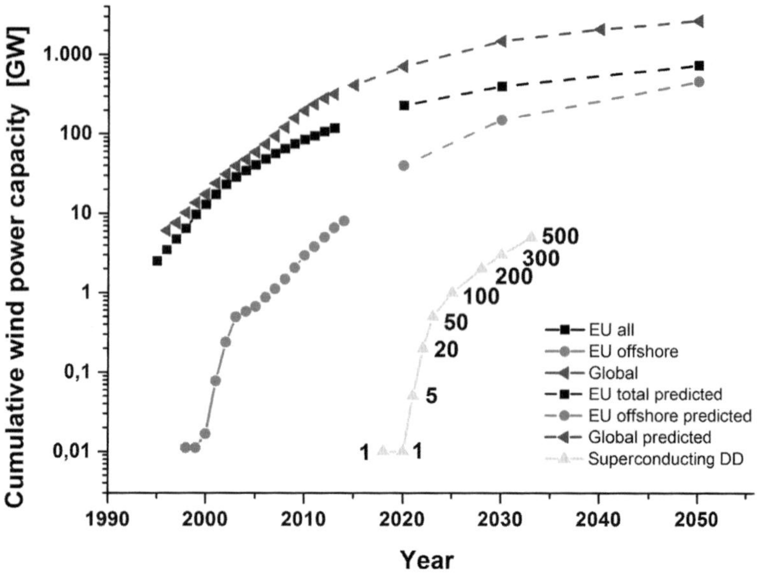

Fig. 3. Installed total wind power capacity of Europe and offshore capacity for Europe. Scenarios for the wind power development towards 2050 have been added[5,6] together with a scenario[7] for introducing 5 GW of 10 MW superconducting wind turbines by 2035.

developed for the onshore market until about the year 2000, where considerable offshore activity was initiated in Europe. The figure also contains some predictions of the future developments[5,6] of the wind sector towards 2050, where initiatives to limit the CO_2 emissions from fossil fueled energy sources are expected to become significant. It is generally believed that the onshore market of Europe will double in 2030 compared to 2015, but that the offshore market is expected to grow by a factor of 10, because the space for more onshore wind power is limited due to relatively large population densities in Europe.[5] The offshore market is expected to provide the majority of wind power in 2050.[6]

2.2 *Drive train technologies: from gearbox to direct drive*

Along with the development of longer blades there has also been a huge development of the turbine drive trains, which are converting the rotation of the turbine shaft into electricity for the electricity grid. Historically the drive train consisted of a standard gearbox and an induction generator (IG) connected directly to the grid. This is however causing the turbine to rotate at constant speed, since the generator rotation speed must match the frequency of the grid. In order to allow some speed variation the induction generator was replaced by a double feed induction generator (DFIG), where the magnetization currents of the rotor could be controlled by power electronics. This concept is still widely used, but as a first challenge, the armature windings are still directly connected to the grid, which is giving problems in case of faults on the grid. A second challenge of the DFIG concept is represented by failing gearboxes and considerable repair expenses. This has initiated the development of hybrid drive trains, where the gearbox has as few gear stages as possible, which is compensated by larger and slower rotating generators. Such a configuration will need a full power converter between the generator and the grid, because the frequency of the generator will not match the grid. The concept of completely removing the gearbox is called a *direct drive train* and has been realized by several turbine manufactures up to 6 MW using a multipole permanent magnet generator. The advantage of the direct drive concept is the smaller number of moving parts, but the generator tends to become rather large and has to be considered as a completely mechanically integrated part of the nacelle holding the drive train and the blades.

The last important point to understand is that some additional constrains of the turbine will be imposed on the drive train in terms of rotation speed and torque. In addition to Eq. (1), the tip speed ratio λ between the blade tip velocity v_{tip} and the wind velocity v must be maintained constant at an optimal value in order to harvest the wind energy as efficiently as possible due to the aerodynamic properties of the blade

$$\lambda = \frac{V_{tip}}{v} = \frac{R_{blade}\omega}{v} \Rightarrow \omega = \frac{\lambda}{R_{blade}} v, \tag{3}$$

where R_{blade} is the length of the blade and ω is the angular rotation speed. Thus the rotation speed will increase linearly with the wind speed up to the rated wind speed. The torque of the shaft T is finally found from the rotation speed and the power by

$$T = \frac{P_{shaft}}{\omega} = \frac{\pi\rho}{16}\frac{C_p}{\lambda}v^2D^3, \tag{4}$$

with similar variables as Eq. (1).

Summarizing the demand for the drive trains it can be seen from Eq. (1) that the power scales with the rotor size, $P \sim D^2$, while the rotation speed scales as $\omega \sim D^{-1}$ from Eq. (3), which results in a scaling of the drive train torque as $T \sim D^3$. The fast growth of the torque of the drive train, is a major challenge for the generator design of multi-MW turbines and the superconducting technology is believed to be able to provide these high torque/slow speed direct drive generators, but the question arises about cost and reliability.

3. Superconducting wind Turbine Generators

What is the impact of introducing superconductors into wind turbine generators?

Generally one can state that for a radial flux generator, the torque T provided by the machine will scale with the radius of the air gap, the area of the air gap and the shear stress F_d provided by the active materials of the field and the armature rings (active materials are defined as the materials taking part of the magnetic flux circuit: these can be permanent magnets, electrical wires in coils or magnetic steel laminates).

$$T = r_sF = 2\pi r_s^2 l_s F_d, \tag{5}$$

where r_s is the radius of the air gap, l_s is the active length of the generator. The shear stress F_d is related to the electromagnetic properties

$$F_d = \frac{1}{2}B_gA_s cos\gamma, \tag{6}$$

where B_g is the peak flux density [Tesla] provided by the field generating ring (permanent magnets or coils), A_s is the current loading in the unit [A/m] of the armature windings and γ is the angle between the field and current loading maxima.[8] The peak field B_g is often limited below about 1 T due to the saturation of steel laminates and the intrinsic properties of permanent magnet materials. Similarly the current load of the armature is limited by the available generator cooling and also by the losses imposed by Joule heating. Thus the shear stress in conventional machines is submitted to inherent limitations. Superconducting wires can provide large critical current densities J_C with no resistive heating and in some cases also in quite high magnetic flux densities B_a. This can be used

- to pin magnetic flux in bulk samples, thus acting as permanent magnets with a magnetic flux density at the surface largely exceeding the performance of a permanent magnet, or
- to support a transport current in a wire holding the SC.

In both cases superconductors have the potential to increase the magnetic flux density in the airgap, B_g, and thereby the shear stress as formulated by Eq. (6), whereby more compact machines can be realized following Eq. (5) for a similar torque. One could also consider increasing the current load of the armature A_s using superconducting windings, further enhancing the shear stress. The major question is, however, whether it will be economically feasible to utilize these high shear stress force densities. This question is still open.

A major challenge when using superconductors is that they must be maintained at a considerably lower operation temperature than their critical temperature T_C, below which the material changes from being in a resistance state and to the superconducting state with a vanishing resistance (see Fig. 4). The bulk superconductors will have to get magnetized by a magnetic flux density of several Tesla, which will require the use of superconducting coils. Most generator designs are therefore directly based on coils of superconducting wires to create the magnetization inside the generators.

A review of superconducting wind turbine generators has been given by Qu *et al.*[9] showing the possible combinations of superconducting field and also armature windings either being fixed or rotating in the nacelle frame.

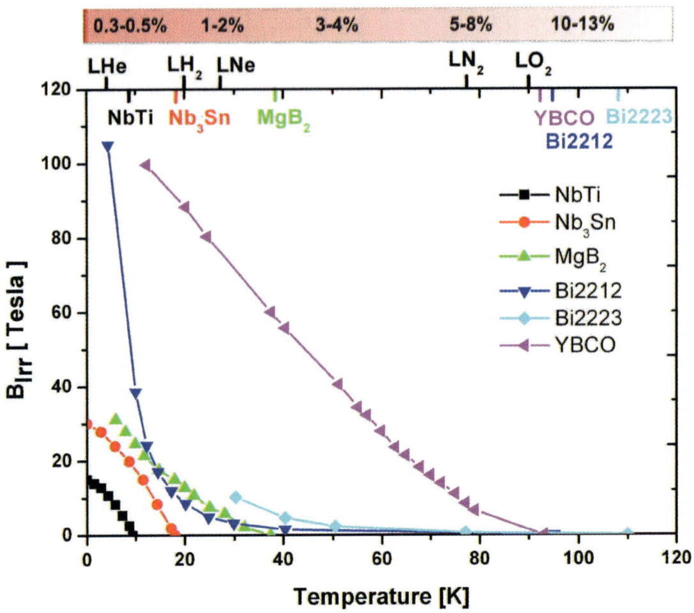

Fig. 4. Irreversibility field B_{irr} versus T for various superconductors.[14] The markers at the top of the figure show their critical temperature T_C and the boiling points of several cryogenic liquids. Finally, the bar at the top is indicating typical efficiencies of cryogenic cooling equipment as function of the cold operation temperature. Reproduced with permission from Jensen *et al. J. Renewable Sustainable Energy* **5**, p. 023137 (2013). © 2013, AIP Publishing LLC.

The first topologies of the superconducting wind turbine generators where based on rotating inner windings and a conventional armature winding operated at ambient temperature. The Sea Titan 10 MW generator from American Superconductors based on coated high temperature superconducting field windings is a good example for this concept.[10] A similar design was investigated by Converteam for a 1.7 MW hydropower plant, but the topology would also be applicable as a wind turbine generator providing high torque and low speed.[11]

A variation of the topology was introduced by GE Global Research for a 10 MW NbTi-based wind turbine generator, where the inner superconducting field windings are fixed and conventional armature windings are rotated outside.[12] The advantage of this topology is that the cryogenic system used for cooling the superconducting field winding can be simplified, but a major disadvantage is that the totality of the 10 MW power has to be transferred in rotating coupling from the rotating frame into the hub. Such a rotating coupling will need maintenance and might reduce the reliability of the wind turbine.

Finally one can consider using both a superconducting field and armature winding in order to utilize the high current density also in the armature, but the additional AC losses in the superconducting armature must be kept sufficiently low not to reduce the overall efficiency of the generator system. The first proposal of a fully superconducting wind turbine generator was given by American Magnetic Lab[13] and was based on the double helix winding method applied to MgB_2 wires.

In the following it will be discussed why MgB_2 wires might be an interesting choice for superconducting wind generators and several examples of generator concepts will be outlined.

3.1 *Choices of superconductors: challenges and advantages of MgB_2*

There are several types of superconducting wires for the application in superconducting windings of a wind turbine generator. Figure 4 illustrates the irreversibility field B_{irr} of most of the superconducting wires and tapes available on the market.[14] At the irreversibility field the pinning of flux lines is vanishing and the critical current goes to zero. Superconductors for generators must thus be used at fields well below B_{irr} and thereby also at temperatures well below T_C.

Figure 4 shows that the critical temperature T_C of MgB_2 is situated between that of HTS superconductors (e.g., Bi–Sr–Ca–Cu–O and R–Ba–Cu–O with $T_C \sim 93-110$ K) and that of LTS superconductors (e.g., the workhorse NbTi with $T_C \sim 10$ K and Nb_3Sn with a $T_C \sim 18$ K). NbTi is the most widely used wire for MRI scanners and also accelerator magnets, and it is often kept cold using LHe boiling at 4.2 K. He is considered a limited resource and is also quite expensive, whereby the availability might be questioned. The HTS can be cooled by the abundant and less expensive LN_2, but the B_{irr} of the Bi-HTS shows that it has to be cooled to about $\frac{1}{4}$ of the T_C before it can be operated in a reasonable magnetic field of a few Tesla. The HTS compounds R–Ba–Cu–O seem very promising for LN_2 cooling, but it often turns out that they have to be cooled to $\sim\frac{1}{4}$ of T_C, too, for providing sufficiently high current densities for generator applications. Thus MgB_2 provides an alternative to HTS materials in the temperatures 15–25 K, where LTS materials are no longer operational. From Fig. 4, it can also be concluded from the typical efficiencies of cryogenic cooling equipments that the system efficiency can be improved by a factor of 3–4 by increasing the operation temperature from 4.2 to 20–25 K. Finally it is important to mention the unit prices (€/m) of the wires shown in Fig. 4, which in 2012 were 30, 20, 4, 1 and 0.4 €/m for R–Ba–Cu–O, Bi-2223, MgB_2, Nb_3Sn and Nb–Ti, respectively. These prices reflect the raw material cost as well as the manufacturing cost of the different wires. They are expected to decrease as the production volume of a specific wire type is increased to mass production, but it should be noted that the HTS tapes are approximately 20–50 times more expensive than the low temperature superconductors, due to complex technical procedures and to a small production volume. The MgB_2 wires are produced from low cost material and the processing is based on the PIT technique, where powder is filled in tubes and drawn into thin wires of several km length. Thus the MgB_2 wire is thereby 4–10 times as expensive as the low temperature wires. One will however also have to consider the performance of the wires and the unit €/(kA·m) is often introduced. This comparison obviously makes sense only if the cost is quoted for identical magnetic fields and operation temperatures, which have to be optimized individually for the different kind of wires. It can therefore be said that MgB_2 is a cheap alternative to HTS wires in the temperature interval 15–25 K if the limited magnetic field range of a few Tesla can be tolerated, which is often the case for certain superconducting generators.

3.2 *Superconducting field windings*

The first topology of the superconducting wind turbine generators was based on superconducting field windings and a more conventional armature operated at ambient temperature. Generator concepts based on MgB_2 field winding coils are investigated in the FP7 projects called INNWIND.EU and SUPRApower. A very different way of utilizing a field winding in a claw-pole generator is also discussed.

3.2.1 *INNWIND.EU superconducting generator nacelle layout*

The INNWIND.EU project is investigating how to design the future P = 10–20 MW offshore wind turbines to be installed above 50 meter water depth in the North Sea off the coast of European member states.[15] The project is organized into a work package providing the conceptual designs and weather/wave conditions, a work package on the turbine rotor and blades, a work package on drive train converting the mechanical energy of the rotor into electricity and finally a work package on the tower and offshore foundations. The goal of the project is to make several demonstrations of technologies and to determine the Levelized Cost of Energy (LCoE) provided by this large offshore wind turbine. Superconducting direct drive (SCDD) generators are considered as part of the work package on drive trains and they are compared to a pseudo magnetic direct drive (PDD) combining a magnetic gearbox and a ring generator, as well as to a conventional permanent magnet direct drive (PMDD) generator.

Figure 5 shows the INNWIND.EU 10 MW MgB_2 nacelle which is based on the King-pin concept of DNV–GL.[16] A major concern when designing a nacelle for a P = 10–20 MW turbine is whether the main bearings holding the hub where the turbine blades are sitting will last for the 25 year lifetime of the turbine. For turbines larger than $P \sim 10$ MW it is believed that two main bearings are needed and in the King-Pin design they are placed on each side of a rotating tube holding the three-blades of the turbine. The idea with the design is to transfer the mechanical loads of the blades as directly as possible into the tower. The question of where to put the generator

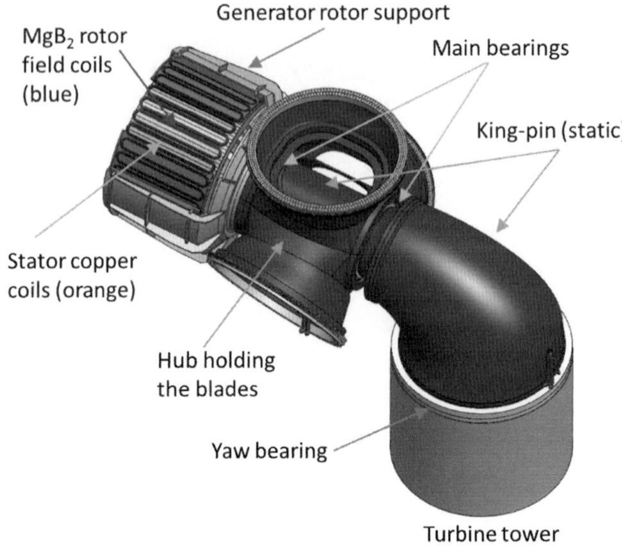

Fig. 5. INNWIND.EU 10 MW front mounted MgB_2 generator nacelle concept[16] based on a King-pin configuration holding two main bearings on each side of the hub, where the blades are mounted. Superconducting race track coils (dark) are shown to be placed outside the armature (clear) and will be rotated by the hub.

then arises and there are basically two possibilities: (a) behind the rotor or (b) in front of the rotor. Option (a) is what is done with most permanent magnet direct drive wind turbine generators and has the advantage that the generator can be large without creating a shadow for the rotor blades. Option (b) allows replacing the generator without first un-mounting the turbine rotor, but the size of the generator should be small enough not to create a shadow for the rotor blades. This option (b) has been selected in the INNWIND.EU project for further investigation, as superconducting generator being more compact than conventional generators. A generator based on option (a) will be a direct cost competition with the PMDD, which will be quite hard to beat due to the current technological maturity of the PMDD.[17]

The INNWIND MgB_2 generator is based on a 19-filament $3.0\,mm \times 0.5\,mm$ MgB_2 *ex situ* tape stabilized by a 0.2 mm thick Cu strip from Columbus Superconductors SpA.[18] This tape has been chosen being a standard product having the potential to be upscaled for mass production. Thus there has been no attempt to custom design the tape for the generator.

A concept generator with the intention to investigate a relatively high magnetic flux density of 1.5 Tesla in the air gap has been introduced and is based on stacks of double race track coil making the field coil of a multipole generator.[18] The field coils are supported by non-magnetic materials and the armature teeth are also considered as non-magnetic. The armature is enclosed in a steel laminate back-iron. The double race track coils are insulated with Kapton tape and impregnated using Stycast. The dimensions of the concept generator are adjusted in such a way that the generator diameter $D_{\text{gen}} \sim 6\,\text{m}$ is slightly larger than the hub of the nacelle. The length of the generator $L_{\text{gen}} \sim 3$–$4\,\text{m}$ is chosen in order to limit the aspect ratio of the generator to $L_{\text{gen}}/D_{\text{gen}} \sim \frac{1}{2}$, whereby the shadowing of the rotor blades is kept limited. The 10 MW turbine will result in a torque of $T \sim 10\,\text{MNm}$ on the shaft, since the turbine blades will rotate at 10 rpm at rated power. From Eq. (4) one can then calculate the shear stress needed to provide this torque, $F_d = T/(2\pi(D_{\text{gen}}/2)^2 L_{\text{gen}}) \sim 60\,\text{kN/m}^2$. This can then be related to the generator technology using Eq. (5). If $B_g \sim 1.5$ Tesla is assumed, the current load of the armature should be $A_s \sim 2F_d/B_g \sim 80\,\text{kA/m}$.

One can now propose a generator design with a certain number of pole pairs and amounts of superconducting and armature windings as well as steel laminates to shape the magnetic field and also structural materials to support the windings. By assuming some general cost of the materials as shown in Table 1, one can estimate the cost of the active materials and try to minimize the cost of the system. An initial proposal of the generator resulted in race track coils with an opening

Table 1: Cost of active materials for wind generators.[17]

Material	Cost including manufacturing
MgB_2 conductor	4 €/m (expected to decrease towards 1 €/m)
Cu	15 €/kg
Steel laminates	3 €/kg

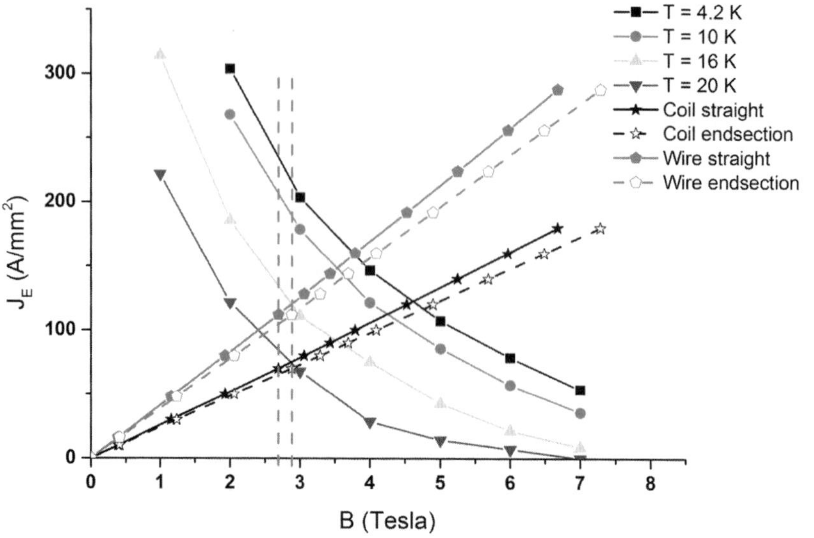

Fig. 6. Engineering critical current density of the 3.0 mm × 0.5 mm MgB_2 tape with an added 0.2 mm thick Cu strip (Columbus SC) as function of magnetic flux density and operation temperature. The load line of the INNWIND.EU field coil is indicating that an operation temperature between 10 and 15 K is needed in order to have a current density of 113 A/mm^2 in the wire at B ∼ 2.8 Tesla.[16,18] Originally published under a CC BY licence by IOP Publishing.[31]

of 0.3 m, which is three times the critical bending diameter of the MgB_2 tape, a length of the straight section of 3.1 m and a coil cross-section of 80 mm × 80 mm. The magnetic field strength on the most loaded part of the end section of the race track coil reaches about 2.8 T with a J_c in the MgB_2 wire of 113 A/mm^2. A challenge for the MgB_2 tape is that the operation temperature must remain between 10 and 15 K. In particular, the operation at $T = 20$ K operation will not be possible without changing the design for a lower magnetic field as shown on Fig. 6 or without using MgB_2 wires with higher J_C values.

An estimate of the cost of capacity[17–19] can be obtained from the cost of proposed usage of active material[17,19] (SC, Cu and laminates) by dividing with the power rating of the turbine. Assuming a MgB_2 tape cost of 4 €/m (see Table 1), one finds 226 €/kW; since this cost is expected to be reduced by a factor of 4 for a mass production, the cost of capacity could be reduced to 84 €/kW. The target for the cost of capacity for the generator has an upper limit, which can be found from the target cost of offshore wind power being about 1.5 M€/MW and the expected fraction needed by the drive train, which is approximately 20%. This results in an upper target of 300 €/kW. Thus the MgB_2 generator solution might have a chance to fulfill this condition, even after adding the cost for the cooling system and the power electronics.

The declared goal of the INNWIND project is to demonstrate that the high-field race track coil described above can be constructed and tested in a downscaled version, with a straight section of only 0.5 m length and a coil opening of 0.3 m, identical to the coil of the generator concept model. The coil is mounted in a large circular cryostat previously used for demonstrating a MgB_2 coil for an Al billet heater.[18] The coil demonstration will be used as input for the final generator concept design for power ranges P = 10–20 MW.

3.2.2 *SUPRApower generator concept*

The SUPRApower project is investigating a 10 MW, 8.1 rpm and 11.8 MNm MgB_2 direct drive generator, which is intended to be mounted behind the turbine rotor blades as illustrated in Fig. 7. This allows for a much larger diameter of the generator $D_{\text{gen}} = 11.9$ m and a reduced length of $L_{\text{gen}} = 0.52$ m.[20] Using Eq. (5) the shear stress can be estimated to be $F_d \sim 102\,\text{kN/m}^2$. This is obtained from race track coils based on the same MgB_2 tape as used in the INNWIND.EU project (19-filament MgB_2 in Ni matrix with dimension 3.0 mm × 0.5 mm and a 3.0 mm × 0.2 mm Cu strip as supplied by Columbus Superconductors). The race track coils are placed over 60 steel laminate poles operated at ambient temperature, whereby the cryostats holding the MgB_2 coils must have a hole in the center. An air cored Cu armature is placed outside the rotor coils. The peak air gap flux density is estimated to be $B_g \sim$ 1.5 Tesla.[20] Using Eq. (6) the current loading of the armature is estimated to $A_S = 2 \cdot F_d/B_g \sim 136\,\text{kA/m}$. This generator is similar

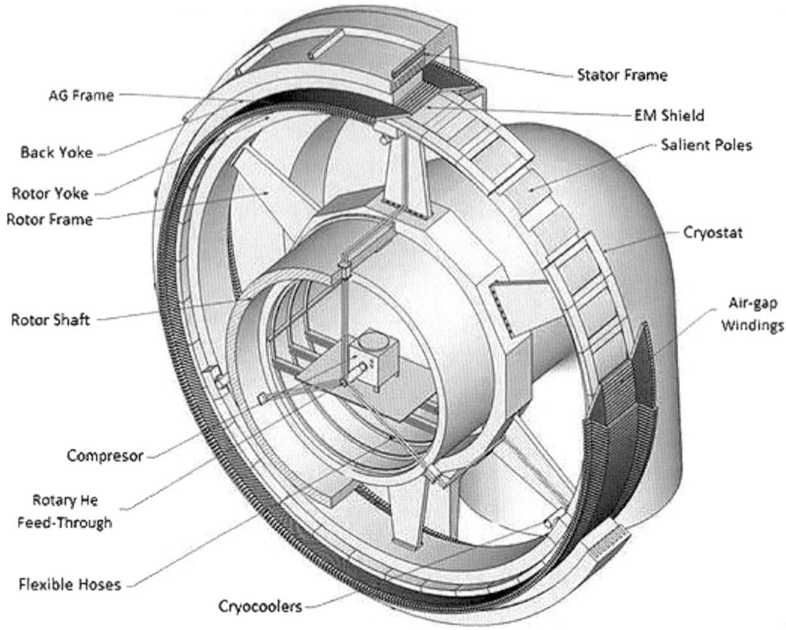

Fig. 7. SUPRApower 10 MW MgB_2 wind turbine generator concept with superconducting coils placed around salient Fe poles being at ambient temperature. A He gas coupling is providing pressurized He lines to cryocooler cold heads mounted on the superconducting coils.[20] Originally published under a CC BY licence by IOP Publishing.[31]

to the INNWIND.EU generator, but more steel laminates have been included in order to reduce the amount of SC tape needed, because the magnetic flux density at the SC is in the order of 1 T. This concept is thus mainly using the large current density of the MgB_2 and not the ability to operate in high magnetic fields. The cost of capacity is estimated to be 280 €/kW when including the cryogenic system.[20] The generator mass including the structural mass is expected to be less than 180 MT and the use of MgB_2 tape in a single rotor coil is 3.2 km resulting in a total of 192 km of tape in the generator.[21]

The SUPRApower project is demonstrating a 500 kW and 4-pole version of the generator, where the MgB_2 coils have sizes equal to those of the 10 MW generator. One cryocooler is cooling the four coils to an operation temperature of $T = 20\,\mathrm{K}$ and the 10 MW generator will have 15 cryocoolers installed in the rotating frame and be supplied with He gas from compressors in the stationary frame as illustrated in Fig. 7.

3.2.3 *Double-sided claw pole concept*

A double sided claw pole superconducting 10 MW, 10 rpm and 9.6 MNm wind turbine generator has been proposed by Keyzan and Muller[22] as illustrated in Fig. 8. This concept is based on a stationary superconducting DC coil placed inside a 3D claw

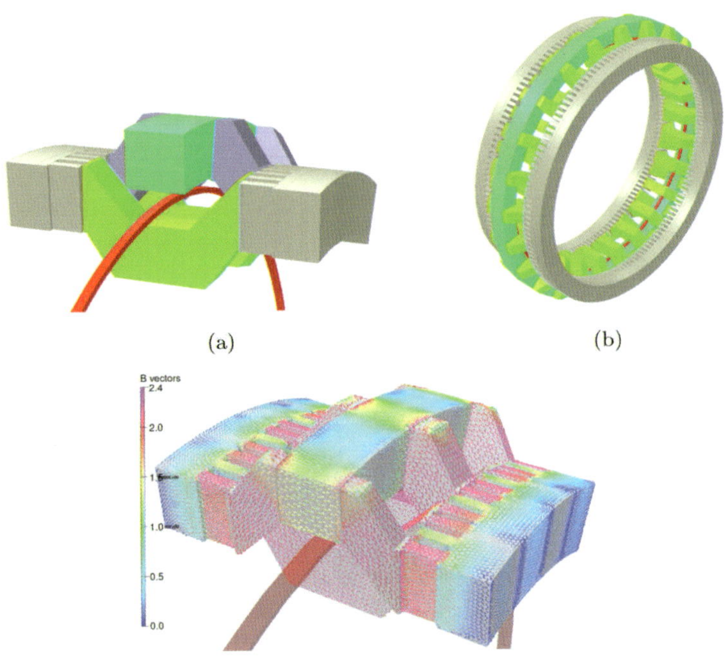

Fig. 8. 10 MW double claw pole wind turbine generator concept.[22] (a) A stationary superconducting winding (red) is magnetizing the two sets of moving claw poles (green and blue), which are creating an alternating magnetic flux in the two armature sections at the sides (gray), (b) Full generator ring, (c) Magnetic flux density in claw pole structure. © IOP Publishing. Reproduced with permission. All rights reserved.

pole moving structure making the magnetic flux change in two independent stationary armature windings on each side of the claw pole structure. A high saturation magnetic flux density steel called VacuFlux50 from Vacuum Schmelze is proposed as laminate in order to obtain a saturation of 2.35 Tesla. The diameter of the armature is $D_{\text{armature}} = 5.9\,\text{m}$ and $L_{\text{gen}} = 1.4\,\text{m}$, which is giving the equivalent shear stress of $F_d = 130\,\text{kN/m}^2$. The mass calculated to be 184 tons and the efficiency is 94.5%. If the SC winding is based on 3.7 mm $\times$ 0.7 mm *ex situ* MgB_2 tape and is operated at T = 30 K, a tape length of 15 km is needed for the entire generator. This is considerably smaller than what has been shown for the previous machines and will thus provide a large cost saving.

The advantage besides the reduced SC cost also includes a simplified cryostat and cooling system, which is proposed to hold four cryocoolers only. A calculation of the total cost of the generator is however not available and the cost of the VacuFlux50 material is not specified. A considerably higher cost than Fe–Si laminates is expected, since VacuFlux50 consists of 49 wt.% Cobalt, 2 wt.% Vanadium, the rest being Fe.[23]

3.3 *Superconducting armature windings*

The possibility to use a superconducting armature winding in a wind generator is motivated by the large current density of the superconducting wires compared to Cu, but the challenge consists in reducing the AC losses dissipated in the superconductors due to AC currents and fields of the armature. Examples of fully superconducting MgB_2 generators will be presented in this chapter.

3.3.1 *Advanced Magnet Lab concept*

The wind turbine generator concept of Advanced Magnet Lab[13] is based on the double helix coil geometry, where a radial magnetic flux density modulation is created by two tilted helical windings on a tube shaped support structure as illustrated on Fig. 9. A synchronous generator is obtained by inserting two such helical coils into one another and rotating one of them.[24] Both the field and armature

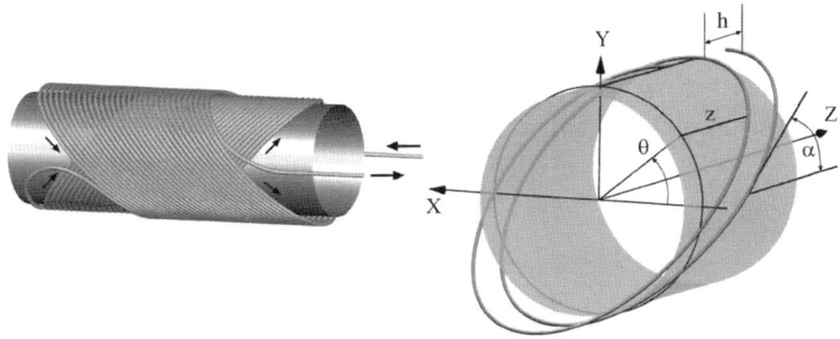

Fig. 9. Illustration of the double helix coil geometry used by American Magnetic Lab to create the field and armature windings of electrical machines.[24] © 2008 IEEE. Reprinted, with permission, from *IEEE T. Appl. Supercond.*

windings are proposed to be made from superconducting wires and MgB_2 is considered the best candidate due to the relatively low price compared to the high temperature superconductors. Additionally one can make a large number of MgB_2 filaments in the metal matrix of the wire, whereby the AC losses can possibly be reduced to an acceptable level dictated by the target efficiency of the generator system. This will involve a consideration of the cryogenic penalty of the cooling system as illustrated in Fig. 4.

The 10 MW AML wind turbine generator is mounted behind the turbine rotor and the main bearings.[25] A rotation speed of 10 rpm and result in a torque at rated power of $T = 9.6$ MNm. The dimensions are $D_{\text{gen}} = 3.5$ m and $L_{\text{gen}} = 7.6$ m, whereby the shear stress is estimated to be $F_d = 65.3\,\text{kN/m}^2$. A magnetic flux density at the MgB_2 wires is expected to be of the order of 2 Tesla and the operation temperature will be between 15 and 20 K. The estimated mass of the generator is 140 tons, which is lower than the SUPRApower generator. There are no available estimations about the cost of capacity of the double helix generator.

One of the major challenges when realizing the double helix machine consists in controlling the AC losses that will be dissipated in the armature windings. Thus the requirement for the MgB_2 wire does not only comprise a low cost wire with a high critical current density at 2 T and 15–20 K, but also small MgB_2 filaments. In addition, a rather complicated cryostat will be needed to hold the two helix coils. Finally it might also be a problem to make the mechanical support of the field and armature tubes stiff enough to maintain the physical air-gap between the two rings without collisions during extreme events, e.g., an emergency shutdown of the turbine.

3.3.2 *Swarn Kalsi's fully superconducting generator concept*

A 10 MW 24 pole fully superconducting generator concept has recently been introduced by Swarn Kalsi and is based on round AIMI MgB_2 wires from Hypertech.[26] A quite detailed design of the cryostat and cryogenic system has been developed; Fig. 10 illustrates how the superconducting coils of the field and armature rings are supported, both mechanically and thermally. Two types of Ø 0.83 mm wire with either 18- or 36-filaments are wound into strands of three or seven wires for the field and armature windings, respectively. The three-wire strand is wound into race track coils to form the field windings, whereas the seven-wire armature strand is wound into a special geometry, shown in Fig. 11. The power level of the generator is 10 MW with a rotation speed of 10 rpm resulting in a torque of 9.6 MNm. The dimensions of the air gap are $D_{\text{gen}} = 4.74\,\text{m}$ and $L_{\text{gen}} = 0.69\,\text{m}$, resulting in a shear stress of $F_d = 392\,\text{kN/m}^2$, which is considerably higher than for the superconducting machine discussed above. This is obtained by aiming for a magnetic flux density in the field and armature windings approaching 3.5 Tesla and 1.7 Tesla respectively, but secondly the current loading of the armature is $A_S = 2 \cdot F_d/B_g \sim 462\,\text{kA/m}$, reflecting that the current density of the superconducting armature winding is a factor of 5–10 times higher than that of the Cu armature windings. The amount of wire used for the field and armature windings is 114 km and 32 km, giving a total of 146 km.

The cooling of the field coils is suggested to be provided by eight cryocoolers removing the heat by conduction cooling, whereas the armature windings contain SS tubes holding a circulating He gas, which is cooled by 20 cryocoolers as shown on Fig. 10. A detailed evaluation of the heat load due to conduction through

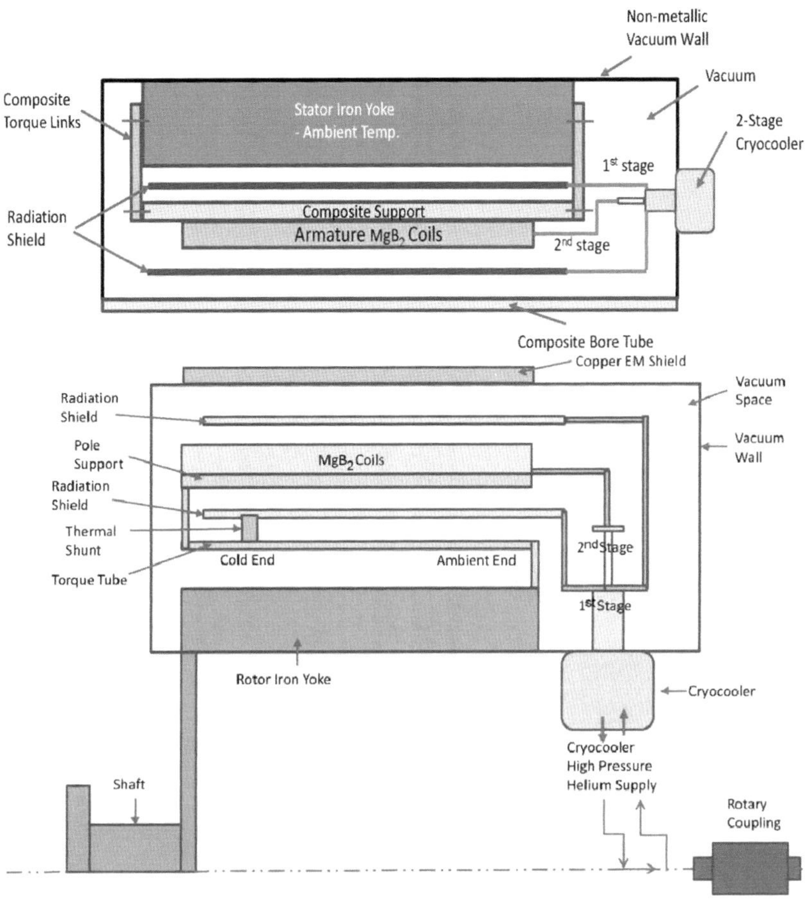

Fig. 10. Topology of the fully superconducting MgB_2 10 MW generator as proposed by S. Kalsi.[26] © 2014 IEEE. Reprinted, with permission, from *IEEE T. Appl. Supercond.*

the support, the radiation and the current leads is indicating that the efficiency of the generator can approach a full load efficiency of 98%.

The cost of the fully superconducting MgB_2 wind turbine generator is determined from the list of components and was estimated to U.S. $2.2 million. If the power electronics for interfacing to the grid is also included one has to add another million U.S. $. The average dollar to Euro ratio $/€ in 2014 was approximately 0.75 making

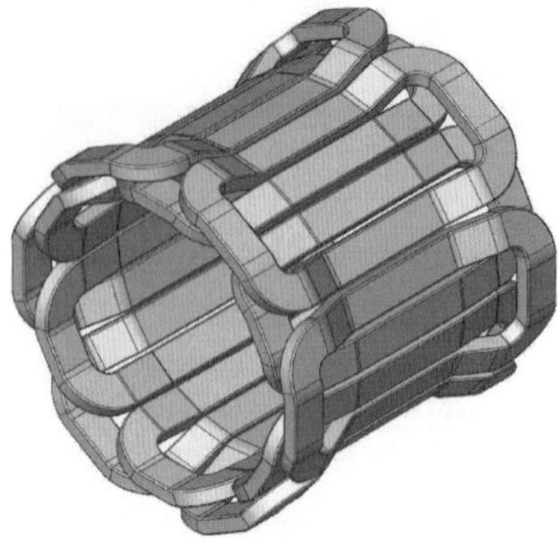

Fig. 11. Topology of the superconducting armature winding of the concept by Kalsi.[26] © 2008 IEEE. Reprinted, with permission, from *IEEE T. Appl. Supercond.*

the equivalent cost 1.65 M€ (2.4 M€) in 2014. From these numbers the cost of capacity becomes 165 €/kW (240 €/kW). In addition, the total weight of the generator was determined to 53 metric tons only, which is considerably lower than most other superconducting direct drive generators.

There are a number of potential barriers on the way towards the fully superconducting generator of Kalsi. First of all the critical currents of the wires as shown in Fig. 12 are not the current performance of large scale commercial *in situ* wires from HyperTech, but the expected performance of AIMI wires by 2015 as well as a reduction of the cost of the wire. These properties have been realized, but the production of long, homogeneous wires is still under development. The heat load analysis includes some assumption on the expected AC losses in the superconducting armature windings, yielding 1 W per coil. However, this has to be confirmed experimentally and has to be linked to the filament diameter, other studies predicting considerably higher AC losses in fully superconducting generators.[27] Finally, the cryogenic system of the turbine contains both cold heads and a closed loop of circulating He gas driven by a cryofan. The reliability of this entire system will have to be brought to a level, where maintenance is only done once to every second year of the 25 year lifetime of the turbine.

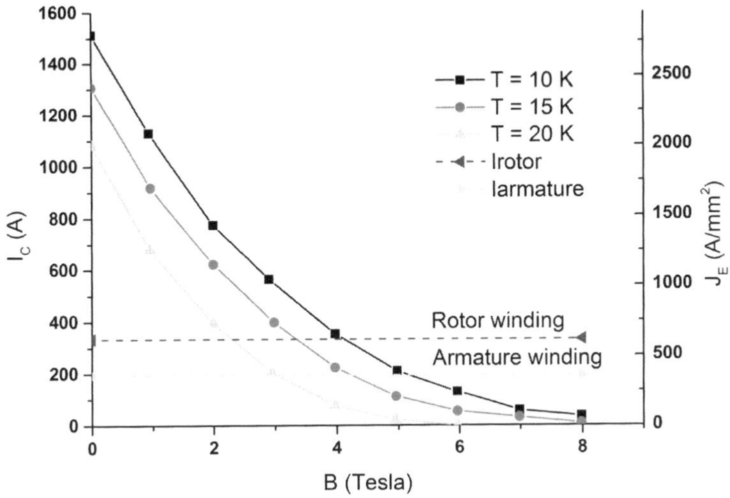

Fig. 12. Conservative expected critical current I_C and engineering critical current density J_E of HyperTech 0.83 mm MgB_2 wire used in the generator design corresponding to the development targets of Hypertech by 2015. Horizontal lines indicate the target operation current of the rotor and armature wires respectively. Reproduced from Ref. 26.

Additionally one should be able to repair all these components without long downtime (but this is valid for all types of wind generators).

4. Road Map for Superconducting Generators

When will superconducting wind generators be introduced into the drive train technology mix of wind turbines, presently based on gearboxes and of direct drive solutions? It is clear that this will be the case at the moment when the LCoE coming from these machines will be considerably lower than the currently used solutions. The related question about the year at which this will happen is more difficult to answer. With the fast increasing number of scientific and industrial proposals on superconducting wind turbine generators the consensus on a sound approach is being established. But the largest challenge will probably be to convince a major wind turbine company to take the risk of introducing such a non-proven technology at the largest power levels P >10 MW of the turbine fleet. In that context it should be remembered that the development cost of a new turbine is in the order of a few hundred times the cost of one turbine, which for a 10 MW turbine will be about 10–15 M€. The de-risking of the superconducting wind generator technology will therefore be a major task for a combination of the community of superconductivity, cryogenics, electrical machines and the wind sector.

A second challenge consists in establishing the supply chain of the vital components such as superconducting wires, cryostats and cooling machines. Figure 3 includes a scenario of introducing superconducting wind turbines into the offshore market. Assuming that the first 10 MW turbine will be constructed in 2020, it will probably have to be tested for several years to reveal any problems. After the successful tests, a considerable number of turbines has to be produced in order to cover the development cost of the turbine in a not too long time frame.

In order to make any impact on the offshore turbine market share a large number of turbines should be produced before 2035. The proposed dimensions, the shear stress, the weight, the amount of MgB_2 wire needed and the expected cost of capacity are listed in Table 2. By the amount of expected wire needed for one turbine one can determine the total amount of wire needed to reach a total of 5 GW installed offshore power capacity by 2035. The claw pole

Table 2: Comparison of 10 MW MgB_2 wind turbine generator concepts: rotation speed ω, torque T, generator diameter D, generator length L, shear stress F_d, mass M, generator MgB_2 wire length L_W, number of cryocoolers in the generator $\#_C$, Cost of Capacity (CoC) and total MgB_2 wire length L_{5GW} for 5 GW offshore wind turbines. The concepts: INNWIND, SUPRA, claw pole, AML and Kalsi are reviewed in 3.2.1–3 and 3.3.1–2.

Concept	ω rpm	T MNm	D m	L m	F_d kN/m^2	M tons	L_W km	$\#_C$	CoC €/kW	L_{5GW} Km
INNWIND	9.7	10	6.0	3	60	52[a]	474	—	226[b]	$2.4\cdot10^5$
SUPRA	8.1	11.8	11.9	0.5	102	180	192	15	280	$9.6\cdot10^4$
claw pole	10	9.6	5.9	1.4	130	184	15	4	—	$7.5\cdot10^3$
AML	10	9.6	3.5	7.6	65	140	—	—	—	—
Kalsi	10	9.6	4.7	0.7	392	53	146	28	240	$7.3\cdot10^4$

[a]Only active mass, [b]Only generator whereby power electronics is missing.

generator[22] will only need about 7,500 km of MgB_2 wire for the first 500 machines and that should be possible with the current production volumes of Columbus SC and HyperTech (about 3,000 to 10,000 km/year.[28,29] Thus the claw pole concept looks quite promising, but an estimate of the cost still remains to be seen, since the use of Co–Fe steel laminates might be rather expensive. The concepts of Kalsi and SUPRApower will need about 10 times as much wire, but this still seems to be within the production capacity of Columbus Superconductors and Hypertech. The INNWIND.EU concept will however demand 237,000 km of wire, thus exceeding the current production capabilities. If the INNWIND.EU concept could reduce the wire needs below the production capacity, this would result in 10,000 km/year × 15 years / 500 turbines = 300 km/turbine, thus removing the bottle neck of the wire supply. The optimization of the INNWIND.EU concept is ongoing, a further reduction of the wire needs by adding more Fe to the machine is envisageable.[30]

Finally what kind of performance improvements of the MgB_2 wires are needed to make the MgB_2 wind generator feasible? From the above review it is clear that both the INNWIND and the Kalsi concept are assuming both a performance improvement with I_C values in the magnetic field range around 2–4 Tesla and 10–20 K of a about a factor of 2–3 and a cost reduction also in the range of a factor 2–4. The low AC losses assumed in the fully superconducting machines will need further experimental evidence. The SUPRApower concept will be able to work with current wire specifications, but the new clawPole concept promises a considerable lower wire quantity and thereby an economical advantage. Thus the constant thrive to improve the generator topologies and the wire performance seems to provide a synergy to make the MgB_2 wind generators still interesting for the wind sector.

5. Conclusion

The challenges of upscaling wind turbines have been outlined in terms of the demand on the drive train consisting of either gearboxes and/or generators. Several 10 MW wind turbine generator concepts have been reviewed and it seems that the fully superconducting generators with both field and armature windings will have very interesting properties, but the cryogenic systems will be more complicated and will need a considerable effort to de-risk before being introduced into the wind turbine fleet. The concepts with only the field windings being superconducting might be easier to construct, but the total wire length is rather high, which will need a serious wire cost reduction and/or improved current carrying properties. Finally a new claw pole concept is indicating a generator topology with a considerably smaller length of superconducting wire and holds the potential of a quite small total cost of the wind turbine generator. Thus the constant thrive to improve the generator topologies and the wire performances seems to provide a synergy to make the MgB_2 wind generators still interesting for the wind sector.

Acknowledgments

I would like to acknowledge contributions from co-workers of the INNWIND.EU project. The research leading to some of these results has received funding from the European Community's Seventh Framework Programme FP7-ENERGY-2012-1-2STAGE under Grant No. 308974 (INNWIND.EU).

References

1. S. Alberici *et. al.*, *Subsidies and costs of EU energy*. Report ordered by the European Commision, Ecofys (2014). Retrieved from http://ec.europa.eu/energy/en/content/final-report-ecofys (visited Jan 2015).
2. F. Rasmussen *et. al.*, *Chapter 8: Emerging Wind Energy Technologies*, DTU International Energy report 2014: Wind Energy — drivers and barriers for higher shares in the global power generation mix, ISBN 978-87-550-3969-8 (2014).
3. Vestas A/S, Technical Specifications of turbines. Retrieved from www.Vestas.com (visited Jan 2015). Siemens Wind Power, Technical specifications of turbines. Retrieved from http://www.energy.siemens.com/hq/en/renewable-energy/wind-power/ (Visited Jan 2015).
4. Eize de Vries, *The top 10 biggest turbines in the world*, Wind Power Monthly (2014).

5. J. Moccia and A. Arapogianni, European Wind Energy association (EWEA) PurePower report 2011. Retrieved from http://www.ewea.org/publications/reports/pure-power-2011/.
6. K. Rave, S. Teske and S. Sawyer, Global Wind Energy Counsil (GWEC): GLOBAL Wind Energy Outlook 2010 report 2011. Retrieved from http://gwec.net/wp-content/uploads/2012/06/GWEO-2010-final.pdf.
7. A. B. Abrahamsen and B. B. Jensen, Superconducting direct drive wind turbine generators: advantages and challenges, Wind Energy Conversion Systems (Green Energy and Technology Series) (Ed.) S Muyeen (2012) (London: Springer), pp. 53–80 Ch. 3.
8. H. Polinder, J. A. Ferreira, B. B. Jensen, A. B. Abrahamsen, K. Atallah and R. A. McMahon, Trends in Wind Turbine Generator Systems, *IEEE J. Emerging and Selected Topics in Power Electronics*, **1**(3), pp. 174–185 (2013).
9. R. Qu, Y. Liu and J. Wang, Review of superconducting generator topologies for direct-drive wind turbines, *IEEE T. Appl. Supercond.* **23**(3), p. 5201108 (2013).
10. G. Snitchler, B. Gamble, C. King and P. Winn, 10 MW Class Superconductor Wind Turbine Generators, *IEEE Trans. Appl. Supercond.* **21**(3), p. 1089 (2011).
11. R. Fair, C. Lewis, J. Eugene and M. Ingles, Development of an HTS Hydroelectric Power Generator for the Hirschaid Power Station, *J. Phys. Conf. Series* **234**, p. 032008 (2010).
12. R. Fair and R. Duckworth, Superconductivity for Large Scale Wind Turbines, DOE final project report, DE-EE0005143 (2011). Retrieved from http://www.osti.gov/scitech/servlets/purl/1052970/.
13. Advanced Magnet Lab. Retrieved from http://www.magnetlab.com/ (Visited March 2015).
14. B. B. Jensen, N. Mijatovic and A. B. Abrahamsen, Development of superconducting wind turbine generators, *J. Renewable Sustainable Energy* **5**, p. 023137 (2013).
15. INNWIND.EU project web site. Retrieved from http://www.innwind.eu/ (visited March 2015).
16. A. B. Abrahamsen, N. Magnusson, D. Liu, E. Stehouwer, B. Hendriks and H. Polinder, Design study of a 10 MW MgB_2 superconductor direct drive wind turbine generator, Proc. of EWEA 2014, 2014, European Wind Energy Association (EWEA), Presented at: European Wind Energy Conference & Exhibition 2014, Barcelona.
17. D. Liu *et al.*, First assessment of performance indicators of Superconducting direct drive and Pseudo magnetic direct drive generators, INNWIND deliverable report D3.42 (2013).

18. A. B. Abrahamsen *et al.*, Design of an MgB_2 race track coil for a wind generator pole demonstration, *J. Phys.* Conference Series **507**, p. 032001 (2014).
19. A. B. Abrahamsen, B. B. Jensen, E. Seiler, N. Mijatovic, V. M. Rodriguez-Zermeno, N. H. Andersen and J. Østergård, Feasibility study of 5 MW superconducting wind turbine generator, *Physica C* **471**, pp. 1464–1469 (2011).
20. S. Sanz *et al.*, Superconducting light generator for large offshore wind turbines, *J. Phys. Conf. Ser.* **507**, web page: p. 032040 (2014) and SUPRApower.
21. G. Sarmiento and S. Sanz, Report on Design and manufacturing of an MgB_2 coil, Deliverable report D2.3 of the SUPRA power project (2014).
22. O. Keysan and M. Mueller, A modular and cost-effective superconducting generator design for offshore wind turbines, *Supercond. Sci. Technol.* **28**, p. 034004 (10pp) (2015).
23. Vacuum Schmelze web page. Retrieved from http://www.vacuum schmelze.com/ (visited March 2015).
24. S. Ishmael *et al.*, Flux Pump Excited Double-Helix Rotor for Use in Synchronous Machines, *IEEE T. Appl. Supercond.* **18**(2), pp. 693–696 (2008).
25. Advanced Magnetic Lab webpage on fully superconducting generator. Retrieved from http://amlsuperconductivity.com/applications/doe-approved-wind-turbine-solution/ (visited 2015).
26. S. S. Kalsi, Superconducting Wind Turbine Generator Employing MgB2 Windings Both on Rotor and Stator, *IEEE T. Appl. Supercond.* **24**(1), p. 5201907 (2014).
27. D. Kostopoulos *et al.*, Feasibility Study of a 10 MW MgB_2 Fully Superconducting Generator for Offshore Wind Turbines, Proceedings of European Wind Energy Association 2013 offshore conference, Frankfurt, Germany, 19–21 November 2013. Retrieved from http://repository.tudelft.nl/view/ir/uuid%3A42184015-8cac-4be2-8084-a2217 44d0452/
28. Columbus Superconductors web-page. Retrieved from http://www.columbussuperconductors.com/ (visited 2015).
29. HyperTech. Retrieved from http://www.hypertechresearch.com/ (visited 2015).
30. L. Dong, H. Polinder, A. B. Abrahamsen and J. A. Ferreira, Comparison of 10 MW superconducting generator topologies for direct-drive wind turbines, submitted to IEEE International Electric Machines & Drives Conference (IEMDC) 2015.
31. CC BY licence by IOP Publishing. Retrieved from http://creative commons.org/licenses/by/3.0/.